T0314029

Flight Simulation Software

Aerospace Series

Visit www.wiley.com to view more titles in the Aerospace Series.

Flight Simulation Software

Design, Development and Testing

David Allerton
Emeritus Professor, University of Sheffield
Visiting Professor, Cranfield University
Visiting Professor, Queen Mary University of London
England, UK

Registered Offices
John Wiley & Sons, Inc., 111 River Street, Hoboken, NJ 07030, USA
John Wiley & Sons Ltd, The Atrium, Southern Gate, Chichester, West Sussex, PO19 8SQ, UK

Editorial Office
111 River Street, Hoboken, NJ 07030, USA

For details of our global editorial offices, customer services, and more information about Wiley products visit us at www.wiley.com.

Wiley also publishes its books in a variety of electronic formats and by print-on-demand. Some content that appears in standard print versions of this book may not be available in other formats.

Library of Congress Cataloging-in-Publication Data
Names: Allerton, David, author.
Title: Flight simulation software : design, development and testing / David Allerton.
Description: Hoboken : John Wiley & Sons, 2023. | Series: Aerospace series |
    Includes bibliographical references and index.
Identifiers: LCCN 2022034801 (print) | LCCN 2022034802 (ebook) | ISBN 9781119737674 (hardback) |
    ISBN 9781119737698 (pdf) | ISBN 9781119737650 (epub) | ISBN 9781119738251 (ebook)
Subjects: LCSH: Flight simulators--Computer programs. |
    Airplanes--Piloting--Computer simulation.
Classification: LCC TL712.5 .A398 2023 (print) | LCC TL712.5 (ebook) |
    DDC 629.132/52078--dc23/eng/20220909
LC record available at https://lccn.loc.gov/2022034801
LC ebook record available at https://lccn.loc.gov/2022034802

Cover Image: © Frasca Flight Training Device
Cover Design: Wiley

Set in 9.5/12.5pt STIXTwoText by Integra Software Services Pvt. Ltd, Pondicherry, India
Printed and bound by CPI Group (UK) Ltd, Croydon, CR0 4YY

C9781119737674_291122

For my grandchildren Clémentine, Eléonore, Constance, Eugénie, Bonnie and Emmeline.

If, one day, one of my grandchildren were to thumb through this book and smile, the effort will have been time well spent.

Contents

Preface

The use of simulation has become an integral activity in many branches of engineering, particularly in aerospace, where flight simulation is recognised as an essential part of training and checking for airline and military pilots. Less well-known is the development of engineering flight simulators, mostly by aerospace companies, who have invested in simulation to support the design, development and testing of aircraft and systems, prior to manufacturing, reducing the development costs and increasing the design options.

Nowadays, the components of a flight simulator are mostly commercial off-the-shelf items. The situation is very different for flight simulation software. Although it may be possible to connect simulator hardware to flight simulation games, the software may be proprietary, restricting access to the code and obscuring simplifications in the aircraft dynamics. Alternatively, simplified models of aircraft dynamics can be developed using modelling packages to visualise outputs. However, complex models produced by such methods are likely to run far too slow for interaction with a human pilot. In order to achieve real-time performance, a third option is to develop the software based on open software, where the simulator modules are compiled to executable code. While this approach is advocated in this textbook, it is not without its challenges. Producing software for a flight simulator, the developer is confronted with the breadth of engineering, covering aerodynamics, propulsion, mechanics, avionics, control engineering and electronics and the depth of software covering mathematics, software engineering, distributed computing, real-time systems and computer graphics.

The challenge for the simulator developer, and the focus of this book, is to understand the techniques and methods used to design, develop and test software for an engineering flight simulator. The author's knowledge and experience in developing software for flight simulation form the topics of the chapters of the book. All the software covered in the book is used in operational engineering flight simulators and is available from the publishers for free download at http://www.wiley.com/go/flightsimulationsoftware. This software has run on Linux and Windows platforms for the PC and currently runs in real-time on Raspberry Pi computers.

This approach to software development is very much in line with the open systems community, where source code is made available to users who are encouraged to modify or improve the software, the associated libraries and support packages. The majority of the software in this book is written in the C programming language, increasing its portability for Windows and Linux platforms and extending its use to the many compatible libraries.

The book is aimed at a wide cross-section of users:

- Pilots, regulators and maintainers of flight simulators needing to understand the operation of software modules used in simulation;
- Computer scientists with a background in software engineering intending to implement aircraft dynamics and systems in a simulator;

- Engineers with a background in aeronautics developing software for aircraft dynamics, displays and visualisation;
- Systems engineers needing to test hardware designs in a synthetic environment prior to installation in an aircraft;
- Developers asked to produce simulation modules, needing to appreciate the techniques used in simulation, including the interconnecting of computers, modelling of dynamics and the generation of real-time graphics for displays and visualisation;
- University teachers, providing hands-on access to enable students to undertake experiments in aerodynamics, mechanics, propulsion, structures, avionics, flight dynamics and human factors, where the simulator software can contribute to student projects, enabling software modules to be developed offline and directly integrated with the software of a flight simulator;
- Students with a desire to develop simulation software, in order to better understand the principles underlying flight simulation.

A word of caution: the breadth of engineering covered in this book does not allow the background to all these topics to be explained in detail and it is assumed that the reader has an understanding of the material introduced in each chapter. As far as possible, links are provided to essential reading. But, in order to restrict the size of the book to a reasonable number of pages, many of the equations and algorithms used in the book are stated rather than derived. After a career teaching in universities, I am convinced that engineering is best learnt by practice and application to real problems in engineering and this paradigm extends to the software for engineering flight simulators.[1]

Chapter 1 covers the building blocks of simulators used in engineering design, particularly the concepts of real-time simulation, distributed computing and parallel processing. Chapter 2 provides a 'software toolbox' needed by most simulator developers, including data-fitting methods, the implementation of transfer functions and essential numerical methods. Chapter 3 covers the equations of motion, including axes systems, derivation of aerodynamic forces and moments and introduces the modelling of aircraft dynamics, engines and landing gear. Chapter 4 covers the design and testing of aircraft flight control systems including PID controller design and includes representative Airbus control laws. The principles and simulation of aircraft navigation systems including inertial navigation and satellite navigation systems are described in Chapter 5. The emphasis in Chapter 6 is on computer graphics, mostly illustrated with OpenGL examples, using modern graphics processing units (GPUs). Software is developed to emulate EFIS displays and traditional aircraft instruments. Image Generation (IG) systems are presented in Chapter 7, which describes techniques used with visual databases and provides interfaces to use the renderers in OpenSceneGraph and X-Plane. Chapter 8 covers the use of synthetic sound generation, using the OpenAL API to generate the range of sounds required in flight simulation. The instructor operating station (IOS) is introduced in Chapter 9 and includes the design of user interfaces, map displays and flight data recording. Chapter 10 concludes with the validation of flight simulation software covering aircraft performance and dynamics and the modelling of linearised systems using Octave and outlines the procedures used in the qualification of airline flight simulators.

I am indebted to many colleagues who have helped with simulator development over the years and have contributed to the book. Ed Zaluska's support and generosity in allowing me to use material owned by Aerosoft Ltd is much appreciated. Malcolm Blackwood provided guidance on

---

1 'Programming is a skill best acquired by practice and example rather than from books.' Alan Turing.

International Regulations. I also owe a huge debt to Graham Spence; his ability to find obscure bugs in my software and his readiness to help with the simulator software has contributed to much of the software covered in this book. Finally, I would like to thank my wife Clare, for her continuous and unstinting encouragement of all my activities and for her instinctive kindness, warmth and generosity.

Aerospace Series Preface

The Aerospace Series aims to be a practical, topical and relevant series of books that reflects the multidisciplinary nature of the field of aerospace, not only in design and engineering but also in many related supporting activities. The topics are intended to be wide-ranging, aimed at a broad audience that includes engineering professionals and operators, engineers in academia, and allied professions such as commercial and legal executives.

This book – *Flight Simulation Software: Design, Development and Testing* – examines the techniques and methods used to design, develop and test software for an engineering flight simulator. The content is a reflection of the author's working life in flight simulator development and covers the modelling of the dynamics of aircraft and aircraft systems including navigation systems, emulation of aircraft displays and visualisation of detailed 3D imagery. The book explains the underlying algorithms and methods used in flight simulation and their implementation in real-time software, with numerous examples taken from operational flight simulators. The enthusiasm of the author for his subject shines through and should be an inspiration to all practitioners in this field.

In addition to being a primer for designers of flight simulators, this book is essential reading for engineers designing aircraft systems, particularly developing software modules used in simulation, for flight crews needing a deeper understanding of flight simulation, for regulators testing and qualifying flight training devices for use in pilot training and for students in engineering and computer science needing to understand and develop algorithms used in simulation. With the increased use of simulation in engineering design, the ability to simulate dynamics, visualise outputs and validate designs will become essential requirements for the current and future generations of systems designers. The lessons to be learned are contained in this book.

Peter Belobaba, Jonathan Cooper and Allan Seabridge

Glossary

AC	advisory circular
ACK	acknowledge
ADC	analogue-to-digital conversion
ADF	automatic direction finding
ADS	automatic dependence surveillance
AFDX	Avionics Full-Duplex Switched
AGARD	Advisory Group for Aeronautical Research and Development
AI	attitude indicator
AIAA	American Institute of Aeronautics and Astronautics
AM	amplitude modulation
API	application programming interface
APU	auxiliary power unit
ARINC	Aeronautical Radio, Incorporated
ASCII	American Standard Code for Information Interchange
ASI	airspeed indicator
ASL	above sea level
ATC	air traffic control
ATIS	automatic terminal information service
ATM	air traffic management
BADA	Base of Aircraft Data
BCD	binary-coded decimal
BSP	binary-spaced partition
CAA	Civil Aviation Authority
CAD	computer-aided design
CAS	calibrated airspeed
CDI	course deviation indicator
CDU	control display unit
CFD	computational fluid dynamics
C/G	centre of gravity
CPU	central processing unit
CRC	cyclic redundancy check

CRS	course
CRT	cathode ray tube
CSMA/CD	carrier sense multiple access – collision detection
DCM	direction cosine matrix
DMA	direct memory access
DME	distance measuring equipment
DOF	degrees of freedom
DOP	dilution of precision
DTED	Digital Terrain Elevation Data
DTEM	Digital Terrain Elevation Map
ECEF	earth-centred earth-fixed
EFIS	electronic flight instrument system
EGT	exhaust gas temperature
EICAS	engine indicating and crew alerting system
EPR	engine pressure ratio
ESDU	Engineering Sciences Data Unit
FAA	Federal Aviation Administration
FADEC	full authority digital engine control
FAR	fuel–air ratio
FCS	flight control system
FCU	Flight Control Unit
FD	flight director
FFT	fast Fourier transform
FL	flight level
FMS	flight management system
FPA	flight path angle
FSTD	flight simulation training device
GA	general aviation
GDOP	geometric dilution of precision
GLSL	GL Shading Language
GLFW	Graphics Library Framework
GNSS	Global Navigation Satellite System
GPIO	general purpose input and output
GPS	Global Positioning System
GPU	graphics processing unit
GPWS	ground proximity warning system
GUI	graphical user interface
HDG	heading
HDMI	high-definition multimedia interface
HP	horsepower
HSI	horizontal situation indicator
HUD	head-up display
IAS	indicated airspeed
IC	integrated circuit
ICAO	International Civil Aviation Organisation
IDE	integrated development environment
IEEE	Institute of Electrical and Electronics Engineers

IFR	instrument flight rules
IG	image generation
ILS	instrument landing system
INS	inertial navigation system
IOS	instructor operating station
IP	internet protocol
IQGT	International Qualification Test Guide
ISA	International Standard Atmosphere
LCD	liquid crystal display
LED	light-emitting diode
LOD	level-of-detail
LVDT	linear variable differential transformer
MCDU	multi-purpose display unit
MCP	mode control panel
MP	manifold pressure
MUX	multiplexer
NACK	negative acknowledge
NACA	National Advisory Committee for Aeronautics
NASA	National Aeronautics and Space Administration
NDB	non-directional beacon
NED	north-east-down
NFD	navigation flight display
NOAA	National Oceanic and Atmospheric Administration
NTSB	National Transportation Safety Board
OAT	outside air temperature
OBS	omni-bearing selector
OSG	OpenSceneGraph
OSI	open systems interconnection
PC	program counter
PCM	pulse-coded modulation
PDOP	position dilution of precision
PFD	primary flight display
PID	proportional-integral-derivative (control)
QDM	magnetic heading
QDR	magnetic bearing
QFE	pressure relative to field elevation
QNH	pressure relative to nautical height (mean sea level)
QTG	qualification test guide
RAE	Royal Aircraft Establishment
RAF	Royal Air Force
RAeS	Royal Aeronautical Society
RBI	radio bearing indicator
RMI	radio magnetic indicator
RPi	Raspberry Pi
RPM	revolutions per minute
SDK	software development kit
SHP	static horsepower

SID	standard instrument departure
SP	stack pointer
SSH	secure socket shell
STAR	standard terminal arrival route
TACAN	tactical air navigation (system)
TAS	true airspeed
TCAS	Traffic Collision Avoidance System
TCP	transmission control protocol
UDP	user datagram protocol
USAF	United States Air Force
USB	universal serial bus
VAO	vertex array object
VBO	vertex buffer object
VFR	visual flight rules
VHF	very high frequency
VOR	vhf omni range
VSI	vertical speed indicator
WAV	waveform audio file
XML	Extensible Mark-up Language
ZFT	zero flight time

About the Author

David Allerton obtained a BSc in Computer Systems Engineering from Rugby College of Engineering Technology in 1972 and a Postgraduate Certificate in Education (PGCE) in Physical Education from Loughborough College of Education in 1973. He obtained his PhD from the University of Cambridge in 1977 for research on hardware simulation of differential equations and then worked in the UK aerospace industry, developing software for embedded systems. He was appointed as Lecturer in the Department of Electronics at the University of Southampton in 1981 and was promoted to a Senior Lectureship in 1987. He moved to the College of Aeronautics at Cranfield University as Professor of Avionics in 1991, establishing the Department of Avionics. In 2002, he was appointed to the Chair in Computer Systems Engineering at the University of Sheffield. He has been an Emeritus Professor of the University of Sheffield since 2009 and is currently a Visiting Professor at Cranfield University and Queen Mary University of London.

Professor Allerton has developed flight simulators at the universities of Southampton, Cranfield and Sheffield and is a past member and Chairman of the Royal Aeronautical Society's Flight Simulation Group. He has served on the UK Foresight Panel for Defence and on the Aerospace and National Advisory Committee for Avionics and the Committee for Synthetic Environments. In 1998, he was awarded a major grant by the Higher Education Funding Council for England (HEFCE) to establish a research centre in flight simulation at Cranfield University, where he was Director of the Annual Short Course in Flight Simulation from 1992 until 2001. He is a Chartered Engineer, a Fellow of the Institution of Engineering Technology and a former Fellow of the Royal Aeronautical Society. He is author of the textbook *Principles of Flight Simulation*, which is published in the Wiley Aerospace Series and the AIAA Education series.

His interests include gardening, walking in the Peak District, playing veteran tennis for Yorkshire, spending time in the Dordogne and writing software for the Raspberry Pi computer. He was previously a part owner of a Beagle Pup light aircraft and holds a private pilot licence with an IMC rating and a night rating.

About the Companion Website

This book is accompanied by a companion website:

www.wiley.com/go/flightsimulationsoftware

The website includes:

- Open-source software
- Data files
- Graphics

1

Design of an Engineering Flight Simulator

I have my hopes, and very distinct ones, too, of one day getting cerebral phenomena such that I can put them into mathematical equations.

Ada Lovelace

1.1   The Evolution of Flight Simulation

The Link Trainer is generally regarded as the forerunner of flight simulation. Ed Link had worked in his father's factory in Binghamton, where they manufactured air-driven pianos and church organs. Having gained his pilot's licence in the late 1920s, Link applied his knowledge of pneumatics to the construction of a flight trainer (Link, 1930), using compressed air to tilt and swivel the cockpit and to drive pressure gauges to replicate aircraft instruments. His invention was remarkable in several ways:

- It was the first time pilots could undertake instrument training in a synthetic device rather than an aeroplane.
- The flight trainer was based on pneumatics.
- The aircraft motion was based on an empirical model rather than a mathematical model.

Six Link Trainers were purchased by the US Army Flying Corps in the early 1930s, following several fatalities attributed to a lack of skill in instrument flying, establishing the benefits of a synthetic training device. The case for simulation was further reinforced during the Second World War with many allied pilots trained in instrument flying on the 'Blue Box', as the Link Trainer was affectionately known.

The limitation of the Link Trainer was that its model of aerodynamics and flight dynamics was based on a simple approximation to aircraft performance. It was the development of the operational amplifier in the 1940s, using thermionic valves, that enabled the differential equations in aircraft dynamics to be modelled. Analogue computers, constructed from operational amplifiers connected via patch boards, enabled complex sets of differential equations to be solved in many branches of engineering (Korn and Korn, 1965), although these computers required daily calibration and considerable care was needed to scale the equations to operate within the voltage range of the equipment.

It was not until the 1970s that the speed of digital computers was sufficient to solve the differential equations in flight simulation 50 or 60 times per second. This iteration rate, often known as the frame

Flight Simulation Software: Design, Development and Testing, First Edition. David Allerton.
© 2023 John Wiley & Sons Ltd. Published 2023 by John Wiley & Sons Ltd.
Companion Website: www.wiley.com/go/flightsimulationsoftware

rate, underpins all flight simulation and consequently, flight simulation demands high-performance computers. During the 1970s and 1980s, this performance was met by the minicomputers of the period.

Since that time, the performance has been increased as a result of developments in computer architecture and microelectronics (Moore, 1965). Firstly, processors the size of a postage stamp are capable of executing hundreds of millions of instructions per second, while at the same time the cost of processors has reduced dramatically. Secondly, with the availability of computer networks, the overall processing speed required in simulation can be achieved by connecting computers as a distributed architecture using a local network. Thirdly, the graphics needed for aircraft displays and image generation is now available from off-the-shelf graphics cards with multiple cores.

It is primarily these advances in processing speed that have enabled airlines and military organisations to provide very realistic pilot training (Allerton, 2000), albeit with simulators costing over $10 million, but where the hourly training costs are often less than one-tenth of the cost of airborne training. These advances in computer technology and flight simulation have been remarkable (Allen, 1993). Flight simulators are used by all major airlines, and regulations have been approved for worldwide training using flight simulators. Similarly, the training of military pilots in flight simulators has increased while reducing both the cost of training and impact on the environment.

Concomitant with these developments in flight training, simulation has taken on a pivotal role in engineering design and development. In industry, control systems are designed with the aid of analysis tools to determine the stability and response of complex systems. In electronics, circuit simulation tools enable circuits to be evaluated prior to the relatively expensive process of manufacturing integrated circuits. In mechanical engineering and aerospace, computer-aided design packages enable designs to be captured on computer screens, visualising and animating designs to facilitate the rapid development of concepts.

These developments in engineering have coincided with changes in the aerospace industry. Aircraft manufacturers have moved away from the traditional departments of aerodynamics, structures, propulsion and avionics towards much more integrated teams with a systems approach to design. Aircraft are viewed as platforms of sensors and computers to enable an aircraft and its flight crew to complete a mission in terms of efficiency and reliability. For many aerospace companies, synthetic environments, including flight simulation, are nowadays a major component in the design of aircraft, covering proof-of-concept and feasibility studies and enabling comparative studies to be undertaken (Allerton, 1996). In some organisations, full mission analysis is undertaken in synthetic environments and the flight simulator is just one of a set of synthetic tools to develop and analyse complex scenarios, which would be impractical with live aircraft.

For manufacturers of civil aircraft, an *iron bird rig* (Jacazio and Balossini, 2005) is used in the development and testing of aircraft systems and actuators. The aircraft systems are set out in a large building with actuators mounted in test rigs. Although the flight deck is a synthetic component, actual aircraft equipment is connected to the simulator, including cables, connectors, pipes, power supplies, avionics equipment, databuses and actuators. During simulated flight, the actuators respond as they would in the aircraft and each actuator can be monitored or loaded or failed in order to test the response of the aircraft systems. The iron bird rig is the final stage before flight testing and is largely used to prove that the software models developed in a laboratory meet the aircraft requirements for performance, stability and reliability.

In this modern role of simulation in the design, development and testing of aircraft and aircraft systems, the simulator is often referred to as an engineering flight simulator rather than a flight simulator training device (FSTD). Its use is not to train flight crews but to provide a tool to improve the design of systems and to validate these designs thoroughly prior to manufacturing. Of course, in tests involving a pilot, the engineering flight simulator has many of the characteristics of an FSTD.

In summary, as aircraft systems have increased in size and complexity, the dependence on simulation as an essential tool in developing and testing prototype systems has also increased. The alternative method, of designing an aircraft and flight testing a prototype, with the possibility of faults only becoming evident once the aircraft is in service (Cohen, 1955), is no longer seen as a viable option. Simulation is very much the focal point of modern system design and, consequently, the quality and accuracy of the simulation software will have a major impact on the success of the design. In further sections, the software used in simulation will be explored in more detail, but for now it is fair to assume that simulation is here to stay (Allerton, 2010) and all system designers need to appreciate both the capabilities and pitfalls of using flight simulation in aircraft design and development.

1.2   Structure of a Flight Simulator

The structure of the majority of flight simulators is shown in Figure 1.1. Although this is a hardware diagram of modules and interconnections, it can also be viewed as a model of the software modules and interfaces. The main modules are shown as rectangles and the databases are shown as ellipses. Note the direction of the arrows, implying that some modules generate data for the equations of motion, whereas others use data produced by the equations of motion. The most significant point of this diagram is that the equations of motion module is the focal point of the simulator and is connected to all the other modules.

Although the use of flight simulators in civil and military training differs considerably from flight simulators used in engineering research, the core software is common to most flight simulators. The major variation is the number of visual channels and methods of projection. Invariably, the motion platform is omitted in an engineering simulator and is usually replaced with a G-cueing seat in military simulators (White, 1989).

The modules are shown as individual systems, but many comprise sub-systems specific to the module function. Note also that the term *database* is not used in the common usage used in computing; rather, they are databases containing data that is specific to a module and, in this sense, they are application-specific files loaded when the simulation starts.

From a modelling perspective, the equations of motion module is the core of the simulator. It contains the state of the simulation, updates the aircraft dynamics at the frame rate of the simulator and acquires the inputs to compute the forces and moments applied to the vehicle. In turn, the forces and moments are used to compute the accelerations, velocities and positions of the vehicle. Often these equations are referred to as six-degree-of-freedom (or 6-DOF) equations because they compute the linear state of the vehicle in three axes and the rotary state of the vehicle in three axes. The equations of motion module is also responsible for transforming forces, accelerations, velocities and position between axes, to provide the simulator state data in an appropriate form for the various modules. Generally, the equations of motion are applicable to any form of aircraft, where the aircraft-specific information is retained in the aerodynamic model, the gear model and the engine model.

The aerodynamic model contains information to compute the aerodynamic forces and moments of the airframe, which is unique to a particular aircraft. For many aircraft, the database is provided by the manufacturer and contains the aerodynamic data for the aircraft for the complete flight envelope in all configurations. In addition to the aerodynamic data, it includes test data used to validate the simulation. The quality of the flight model is dependent on the quality of the flight data, and during qualification of a simulator, the performance and handling of the simulated aircraft will be compared with the validation data provided for the simulator.

The gear model is a mechanical model of the undercarriage assembly of the aircraft, in particular the springs, oleos (dampers), brakes and tyres. During the take-off and landing roll and taxiing, the

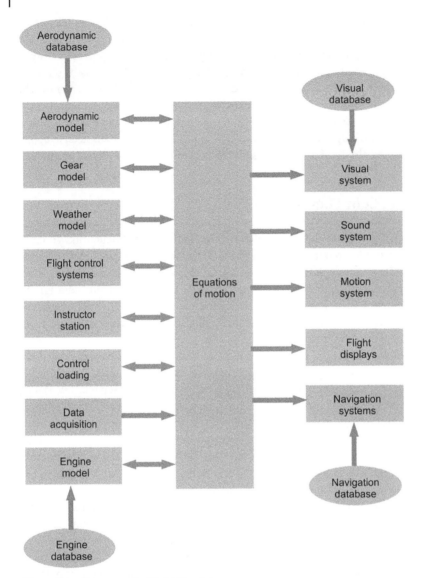

Figure 1.1   Structure of a Flight Simulator.

interaction between the rudder pedals, toe brakes and tiller results in forces and moments in the undercarriage assemblies, which are transformed to the aircraft frame. The models will include tyre scrubbing and scuffing, brake fading and overheat and possibly tyre burst. As the take-off and landing are critical phases of flight, the gear model is an important component of simulation and extensive tests of take-offs and landings are undertaken during qualification of a flight simulator.

As temperature, pressure and density of air in the troposphere vary with altitude, these parameters must be modelled correctly. Air temperature affects engine performance, aerodynamic performance varies with air density and air pressure is used in the air data computer computations and flight instruments. In addition, a weather model provides winds, from either meteorological data or a generic model, turbulence, microburst data, precipitation (used in modelling weather radar) and wind shear.

A modern civil transport aircraft has a flight management system (FMS) and flight control systems (FCS) to provide automatic modes to aid navigation. In a flight simulator, these systems are simulated to enable flight crews to operate the equipment in exactly the same way as the aircraft. The FMS commands the FCS, which drives the primary aircraft flight controls to manage airspeed, altitude, rate of descent and heading. The response of the FCS must match the aircraft systems very closely and, in some cases, an avionics database is also provided for the FMS and FCS.

A flight simulator for a civil transport aircraft used for pilot training will be managed by an instructor who sets flight conditions, introduces failures, monitors pilot performance and provides debriefing for flight crews. In a civil simulator, the instructor is positioned towards the back of the flight deck at the instructor operating station (IOS) and is provided with a screen to monitor the aircraft track, enter settings and introduce environmental conditions, typically via menu selection. The role of the simulator instructor is very important – they monitor the flight crew operations closely, to intervene as part of instruction but not to interfere with operations in an unnatural way. The design of the IOS and its capability can influence the quality of instruction considerably. Although there is no official standard for instructor station design, most modern instructor stations are very similar in terms of their capabilities and user interface.

In an aircraft, the aerodynamic loads on control surfaces vary with airspeed and acceleration and this dynamic loading of the primary controls is achieved in flight simulators by means of hydraulic or electrical actuation. In addition, the throttles and trim wheel may also be driven by an actuator, typically a servo-motor. The response of the control loading system, which is a dedicated system on most flight simulators, is much faster than the main frame rate. A major concern with active controls is the safety of the flight crew being physically close to active controls and safety interlocks are provided in both hardware and software to ensure that the movement of the controls cannot cause harm to a pilot as a result of inadvertent activation.

For the engine model, the manufacturer will provide detailed data of engine performance and engine dynamics. A modern turbofan engine includes a full authority digital engine control (FADEC) system and details of its operation are also provided in the data package. In addition to the provision of data to compute thrust, RPM, fuel flow and engine temperatures, the starting and shutdown procedures (both on the ground and in flight) must be modelled correctly, in order to introduce engine-start problems and to ensure flight crews have a full understanding of engine operation to cope with partial and full engine failures. The engine model is a combination of thermodynamics (gas flows) and the dynamics of turbomachinery rotating at very high speeds. Much of this data is proprietary and licensed to the airline operating the simulator.

A flight simulator may have several hundred wires connecting the levers, knobs, selectors, switches and controls. These connections are a combination of analogue and digital inputs which must be acquired every frame. Often dedicated input/output (I/O) hardware is provided to capture input signals at high data rates and with sufficient resolution. The actual sampling and conversion may take several microseconds per signal and consequently, special-purpose hardware is used for data acquisition with high bandwidth transfers to the memory of the simulator computers, in order to minimise delays associated with data acquisition.

In a modern flight simulator, three or four independent image generators (IGs) render the external scene viewed from the flight deck or cockpit. The actual scenery is stored in a visual database and loaded by each IG at run-time. The scene is typically formed from millions of coloured and textured triangles and the IG performs 3D graphics operations to display these images at 50 frames per second (fps) (or higher) with resolutions approaching 2000 × 2000 pixels per channel. The images are projected, merged and blended to form a continuous wrap-around display of scenery which includes dynamic entities. The conditions displayed by the IGs should match the weather conditions, varying from thick fog to perfect visibility, and produce lighting according to the (simulated) time of day. A considerable amount of development goes into the production of detailed databases for airports and terrain, and real-time 3D graphics algorithms ensure highly detailed images are rendered at the visual system frame rate.

In addition to providing visual cues, sounds that are audible in flight need to be replicated in a flight simulator. Nowadays, the accepted method of sound generation is to carefully record sounds on the flight deck and then play back these sounds using a sound card which generates a sound signal from data stored in a buffer in a recognised format, for example, a *wav* file. Fortunately, most of the requirements for sound generation are met by the capabilities of modern sound cards. Typically, a database of sounds is accessed to generate sounds appropriate to the flight conditions.

Many of the accelerations on the human body occurring in flight are very different from everyday movement and motion cues in a simulator are matched to visual cues to emphasise the sense of motion in a simulator. This is achieved in simulation by attaching the flight deck to six hydraulic (or, more recently, electrical) linear jacks, which move independently in order to replicate the three linear movements and the three angular rotations. The main problem is that the length of movement of the jacks is of the order of 2 m and there is an inevitable difference between motion perceived in an aircraft and motion in the simulator, which cannot be sustained. Nevertheless, for pilot training, a motion platform is a requirement for the qualification of full flight simulators. For engineering flight simulators, the motion platform is usually omitted on the basis that it contributes little to the design and testing of aircraft systems.

Modern military and transport aircraft have flat screen displays, known as electronic flight instrument systems (EFIS) rather than the mechanical instruments used in smaller aircraft. In flight simulation, it is possible to use 2D computer graphics to emulate EFIS displays. The difficulty with emulation is to ensure that the frame rate is maintained at all times. An option is to use actual aircraft displays and generate the appropriate inputs but, invariably, this is a far more expensive solution.

Finally, navigation is an essential part of aircraft operations and, in simulation, the aircraft may fly several thousand miles using en-route navigation and guidance from radio transmitters, satellites and inertial sensors. The operation of the navigation equipment and alignment of worldwide radio aids with aircraft routing must be accurate to the tolerances of airborne navigation. The navigation database contains all the beacon locations, frequencies and ranges, runway layouts and information that is loaded into an FMS, including airways and departure and arrival information (SIDs and STARs). In addition, failure modes must also be simulated correctly.

This brief description of a typical flight simulator outlines the modules and their functions. There is no unique organisation of a modern flight simulator. Nevertheless, these modules are common to most flight simulators and the software used in these modules is designed to meet the requirements of each module. Once the overall structure is determined, consideration can be given to the design of the software structures, in particular to establish the inputs required by software modules and their connection to the simulator computers.

1.3 Real-time Flight Simulation

1.3.1 The Concept of Real-time Computing

Most programs and applications running on computers are captured as source code written in a high-level language, compiled to machine instructions, linked with relevant libraries and then loaded into memory and executed. For most users, the emphasis is on correctness rather than performance. For example, in computational fluid dynamics, the computation of flow fields may take several hours and the speed of execution is mainly influenced by the efficiency of the compiler and the speed of the processor (or cores). In other applications, particularly where there is human interaction with the software, performance is a major consideration.

In the 1980s, several vendors produced flight simulator games for home computers. Many experienced pilots found these programs to be much harder to fly than airline simulators or the actual aircraft. The controls were flimsy and the PC screens were small but the main criticism was the speed of the PC graphics hardware to update the displays, particularly the external view. The games developers traded off visual fidelity (realism of the displays) for the frame rate to the extent that many games updated at only 3–5 fps, causing pilots to over-control their inputs. From an engineering perspective, this was a well-known problem in sampled-data theory.

There are many examples of systems that must meet tight timing constraints, which are referred to as *real-time* systems. A system is a real-time system if the inputs are captured and responded to within a defined time. For example, a computer in a bottling factory may be required to detect the position of a bottle on a conveyor belt, select a cap and actuate a mechanism to secure the cap to the bottle within one-tenth of a second. This repetitive process must be guaranteed to apply a cap within 100 ms, at all times and under all conditions, and the software will be validated to ensure that the real-time requirement is fully met.

The situation just outlined is also common in flight simulation. The positions of the flight controls and the various levers, knobs and switches are acquired, the equations of motion are computed for the airframe and the engines, and the results are output as computer graphics on displays and possibly to actuators to position the motion platform. A typical frame rate in real-time simulation is 50 Hz, giving a frame time of 20 ms. By completing the computations within this frame time, the perceived motion is smooth with no apparent discontinuities, the simulation of the dynamics is computed to an acceptable accuracy and the sampling rate is matched to the time constants of the flight dynamics.

There is one significant difference from the industrial example – the pilot interacts with the software. In a flight simulator, the pilot acquires information from the simulator displays and responds to move the inceptors. The delay resulting from acquiring visual information, cognitive processing and actuating muscles in the hands, arms and legs is of the order of 0.2 s (McRuer, 1995). At 50 Hz, a pilot will probably not detect any discontinuities or irregularities in the perceived motion of a transport aircraft but, with the simulation of agile fighter aircraft, the time constants of the dynamics may be much smaller, necessitating a faster frame rate. Similarly, hydraulic actuation of a motion platform may necessitate an update rate in excess of 1000 Hz to ensure smooth (and indiscernible) movement of the platform.

In real-time software, the time to respond to an input event must be guaranteed at all times (Burns and Wellings, 2001). However, synchronisation and timing of code in high-level languages are mostly provided by operating system functions. This distinction is important because responsibility for the real-time response is delegated to the operating system rather than user software, ensuring that:

- the operating system is responsible for the interface with hardware devices, providing optimal performance of data transfers;
- transfers and errors are managed in a clear and consistent manner;
- the operating system decides the order and priority of accesses to external devices.

The major constraint in a real-time system is that there is a finite number of instructions that can be executed during one frame. Consequently, in real-time systems, emphasis is focused on the speed of executing software, and advances in processor architecture have benefited flight simulation considerably:

- Processor cycle times have reduced from several microseconds to tens of nanoseconds over the last 20 years or so.
- Modern processors contain multiple cores operating in parallel. In applications where parallel streams can be identified and allocated to specific cores, considerable gains in processing speed are achievable.
- Instruction caches allow pending instructions to be fetched while the processor is executing the current instruction; fetching an instruction from the cache is much faster than fetching instructions from memory (Wilkinson, 1996). However, when functions are called or there is a jump in the execution path of the code, the instruction caching restarts.
- Similarly, data caches are used to pre-fetch data from memory so that frequently accessed data is available in the cache rather than via memory. When the region of accessed data changes, the data caching restarts.
- Direct memory access (DMA) allows data to be transferred between memory and external devices while the processor is executing instructions. The processor is responsible for initiating the transfer and responding to the completion of the transfer but the processor executes instructions in parallel with DMA transfers.
- Modern compilers generate code that is optimised for specific instruction sets and register sets of a processor, significantly increasing processing speeds as variables are mostly accessed via machine registers rather than memory.

However, these advances do not ensure the real-time requirement to complete operations within a frame; they enable more computing to be completed during the frame. The performance of the operating system in a real-time system is as important as the performance of the user code.

1.3.2 Operating Systems

A main role of an operating system is to facilitate data transfers between applications and external devices. The user interface is simplified because the user software needs no detailed knowledge of the hardware, the software interface is well-defined and much of the software interface is common to all devices. Typically, transfers comprise five stages:

1) Establishing linkage to the hardware.
2) Setting the device to perform specific transfers, for example, the direction of the transfer and the number of bytes of data to transfer.
3) Initiating a transfer.
4) Completing the transfer, particularly to establish if the transfer succeeded.
5) Closing the linkage to the hardware.

Stages 2–4 enable transfers to be repeated while the device is, in effect, connected to the user software. Between stages 3 and 4, while a transfer is under way, the processor executes other instructions rather than simply waiting for the completion of the transfer. The occurrence of stage 4 is normally an interrupt, where the processor stops its current process in order to respond to the interrupt, checking the status of the transfer and possibly initiating another transfer.

This sequence of events is common to all operating systems. For example, in a desktop computer, file transfers to the hard drive are performed by establishing a channel between an area of user memory and a region of the hard drive, defining the size of the data to be transferred and the direction of the transfer and activating the transfer. On completion of the transfer, the operating system checks that no errors occurred and can signal to the user process that the transfer is completed and the user code can proceed.

Depending on the operating system, it may be managing hundreds of processes. However, only one process can be active at any time. Many processes may be dormant, waiting for some event to occur or being unable to proceed until an I/O transfer is completed. Within the operating system, the *scheduler* is responsible for switching between processes and the characteristics of the scheduler define the behaviour or response of the operating system. The scheduler is able to start and stop processes, saving critical information, so that a process can subsequently resume. Of course, this saving and restoring of the process state is an overhead. Within a frame, each rescheduling of processes may take several hundred instructions which, in terms of a user process, is lost time.

The other responsibility of the scheduler is to select the most appropriate process to run and there are many variations of operating systems. For example, in a time-shared system, the scheduler keeps track of the amount of time spent by each process and tries to give each process a fair share of the overall processing. In most operating systems, I/O processes are given a higher priority than user processes to maximise the throughput of data transfers. However, in real-time applications, it is likely that some inputs are more important than others and should be responded to more quickly. A strictly real-time system allows the user to assign the response to inputs in terms of priority, where the interrupts from specific devices are responded to, not in chronological order, but in order of predefined priority.

There is an alternative methodology to achieving a real-time response which is used in most flight simulation software. In this case, a fixed frame rate is defined and all inputs, all processing and all outputs are completed within this frame time, for every frame. For example, the inceptor inputs for the elevator, ailerons and rudder are acquired, the flight dynamics are computed as a result of the new inputs and the displays and visual system are updated. Similarly, the engine controls are sampled, the engine dynamics are computed and the outputs are used to update the engine displays. It is essential that all the software must be completed within the defined frame time; it is not acceptable for any software to exceed this limit under any conditions. Any other processes which consume processor time must be minimised to avoid overrunning the frame time limit.

1.3.3 Latency

In flight simulation, the pilot control inputs are sampled, the system dynamics are updated and the outputs are visualised. If the delays between a pilot input and the updating of the displays and the visual system exceed a threshold, the response of the simulator can differ appreciably from the response of the aircraft. For most flight simulators, the maximum frame time is 20 ms, giving a minimum frame rate of 50 fps. The challenge for the simulator developer is to organise the software to ensure that the simulation updates at 50 fps, under all conditions.

The inputs may be movements of a mouse and the outputs may be to a graphics display. Alternatively, inputs can be sampled from analogue or digital devices where the output drives an actuator. In both cases, the input, computation and output must occur within a single frame. From the user's perspective, there is a finite delay between entering an input and the result being applied to a screen or an actuator, which is termed *latency*. For applications involving human input and displays, the latency should not exceed 60 ms, otherwise discontinuities or lags in a response are discernible by a human operator and will affect the response of the operator. Although a frame time of 20 ms is well below this threshold, the actual latency may be increased by sampling stale

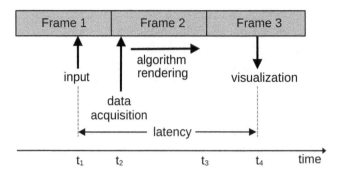

Figure 1.2 Latency.

data (e.g. an input from a previous frame) or the mechanical bandwidth of an actuator or the rendering of a display, as shown in Figure 1.2 for a real-time graphics application.

The input is applied in frame 1 at t_1, but it is not sampled until the start of frame 2 at t_2, the response is computed and the graphics are rendered during frame 2, meeting the 20 ms frame time. However, with real-time graphics, the graphics output is not written to the display (or projector) until frame 3. The acquisition of the image occurs during frame 3 (possibly averaged to the mid-frame time) at t_4. In this example, the latency is not 20 ms, but is closer to 40 ms. There are, in fact, two latencies, the delay between sampling the input and the computed output (t_3-t_2), where the state computed in frame 2 will be used in frame 3 and the delay between the actual pilot input and the cognitive acquisition of the display by the pilot (t_4-t_1). Although increasing the frame rate can reduce the latency, it implies an increase in processor speed and faster data acquisition hardware but cannot eliminate the latency. Considerable care is needed in simulation (and aircraft FCS) to minimise latencies, particularly where they include delays (e.g. additional frames in rendering graphics), which are independent of the user software. In the case of image generation systems, it is very important to clarify the number of frames between rendering the image and its subsequent projection, particularly as some graphics cards require additional frames for post-processing.

1.4 Distributed Computing

In the past, with high unit costs for a computer, emphasis was given to centralised computing with minimal external connections. When external data was connected to a computer, it was often in the form of serial or parallel interfaces. This situation changed with the arrival of the microprocessors used in desktop devices and the subsequent availability of interconnection to a local network, particularly Ethernet, which has become ubiquitous with relatively cheap cards providing data rates in excess of 100 million bits per second (Mbps). Although this data rate is impressive, it is important to bear in mind that if large numbers of users are connected to the network and are making heavy use of it the bandwidth of the network is, in effect, shared between the users, and the data rate between any two users on the network is reduced owing to the loading of the network.

Ethernet has, arguably, revolutionised computing, offering many advantages:

- An Ethernet connection is cheap and simple.
- Large packets of data can be transmitted at relatively high data rates.
- Communication across a network is very reliable.

- Ethernet cables are lightweight and the connectors are small.
- Wireless forms of Ethernet are available for transmission around buildings.
- The signal uses Manchester bi-phase encoding, enabling both the data and the clock to be recovered from two wires.
- The encoding method enables spikes, burst noise and 'stuck-at-one' faults to be detected.
- The packets include a 32-bit cyclic redundancy checksum (CRC); the likelihood that a data error will *not* be detected is of the order of 1 in 2^{32} (1 in 4×10^9) transfers with computation of the CRC implemented in hardware.

Distributed systems offer major advantages in comparison with centralised systems, particularly for applications in simulation:

- If an application is partitioned across n computers, there is a potential speed improvement by a factor of n, although, in practice, the overhead of managing the interconnections reduces this factor.
- Distributed systems are scalable – extra computers can be added to the network to increase processing power or networks can be connected to other networks to increase interconnectivity.
- Conventional software validation can still be applied to individual processors.
- A distributed system can be more fault-tolerant – failures within one computer can be accommodated, possibly by dynamic reconfiguration of the network.
- Standards are published for software interfaces for networks, assuring a high degree of software portability and interoperability if the network hardware is changed.

Nevertheless, distributed systems can also introduce problems:

- Failures of the network can stop all transmissions.
- There is a finite delay in sending data from one computer to another with the potential to introduce latency into computations.
- For the developer, applications have to be redesigned and partitioned in order to operate concurrently, which is not always straightforward with systems with a high sequential content.
- The benefits of parallelism are achieved at the module level rather than the code level.

From the perspective of software design, the major problem is latency or delay in the transmission of data from one computer to another. This delay includes the time to construct the data packet, waiting for access to the network, transmitting the packet on the network and a final delay while the packet is received and copied to the user application. Although these transmissions are likely to be managed by the operating system, in many applications, considerable effort is needed to avoid data becoming 'stale', that is to say ensuring that data used in an algorithm is accessed within some datum of time. This condition is particularly important in real-time systems where the transfers must be *deterministic*, that is to say all transfers are guaranteed to be completed within a finite time, independent of the load on the network or the load on individual computers. However, network latency is difficult to measure; it is the difference between the time measured at the transmitting computer when the message is transmitted and the time measured when the message is received. In practice, the computers on a network are likely to have a local clock to measure time, but there is no guarantee that all these clocks are aligned or synchronised. Consider a series of transmissions for a network of 10 nodes, where each node has to transmit a packet to the other nodes, as shown in Figure 1.3

Each node transmits 9 packets giving a total of 90 packets, although, in practice, not every node may need to send data to every other node. The constraint is that a node only has access to the

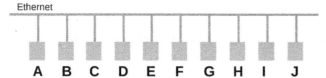

Figure 1.3 The Topology of a Local Network.

network when it is not being used by the other nodes for their transmissions. Ethernet offers a major advantage in this situation, that is, packets may be broadcast over the network, known as multicast transfers. A node waits for the network to be available and then broadcasts its packet. Nodes which require the data can read the incoming packet, while other nodes can ignore or discard the packet. In this example, with multicast transfers, the number of transmissions is reduced from 90 to 10.

However, there are three problems with this simplistic explanation of network transfers:

1) How does the transmitter know the network is available?
2) How does the transmitter know that the message was received?
3) How can the transmitter and receiver detect any errors in the transmission of data over the network?

Assuming messages are transmitted asynchronously, any two nodes may compete for bus access. It is possible that the two nodes may detect that the bus is free at exactly the same time and transmit their messages, corrupting the data sent over the bus, which is known as a *collision*. There are two options to avoid collisions; either the transfers are scheduled to avoid contention for the bus or, alternatively, collisions are detected and the transfers are retried later. With high bus loading, the probability of collisions increases and if the transfers are to be deterministic, it is essential to guarantee the transmissions within a finite time.

Ethernet uses CSMA/CD (carrier sense multiple access with collision detection) – if a node detects a collision during transmission, it waits for a random delay and then tries to retransmit its message. The concern with Ethernet is that, although highly improbable, this process could repeat indefinitely, with two nodes blocking the network as they compete for access. In other words, simply transmitting Ethernet packets will lead to collisions and deterministic transfers cannot be guaranteed.

Three protocols are commonly used in networks to ensure deterministic transfers:

1) *Master-slave*: one node on the network is chosen as the master node. Initially, all the other nodes are passive, reading packets. The master node transmits a packet to a slave node, in effect, giving it ownership of the network. The slave node can then transmit its packets, confident that no other node will attempt to transmit a packet. On completion of its transfers, the slave node transmits a packet to the master node, relinquishing control of the network. The master node has a schedule of transfers and repeats this process for all the slave nodes needing to transmit packets. The method assumes that slave nodes are listening for packets from the master node to avoid delays in changes of bus ownership. The packets transferring ownership of the bus are an overhead.
2) *Token passing*: all nodes have a copy of the schedule of transfers. In a cycle of transfers, the first node transmits its packets. On completion of its transfers, it transmits a token (typically a packet) to the second node in the schedule which transmits its packets, passing on the token to

a third node and so on. Only the node with the token is allowed to transmit. Again, the method assumes that nodes take their token immediately. A variant of the method is that the first packet acts as both a token and a data packet. Care is needed to avoid the situation where nodes towards the end of the chain of token passing experience unacceptable delays.

3) *Time division multiple access*: the cycle of transfers is partitioned into discrete time intervals. Each node is given its time slot and is allowed to transmit packets in this period. At all other times, the node is passive. There are three problems with this method. Firstly, no node must exceed its time slot, otherwise two nodes may transmit at the same time. Secondly, each node needs to monitor time very accurately. In practice, the clocks of each node will drift slowly with respect to the other nodes, so that after a long duration, two clocks may differ by a significant amount. Synchronisation of clocks in distributed system is possible but requires a significant number of extra packet transfers to align local clocks. Thirdly, a node may not need all of its allocated slot, in effect, introducing network delays.

Two questions remain. How does a transmitting node know that its packet was received and what happens if the receiving node detects an error? The problem is resolved in the TCP/IP protocol (Donahoo and Calvert, 2001), which is used throughout the Internet. The protocol is based on the receiver transmitting a reply packet to indicate the success of the transfer together with a timeout to detect a null reply. A certain number of retries are permitted in the case of negative replies or timeouts, before the transfer is deemed to have failed. Clearly, the method works, otherwise the Internet would not be viable. The problem for simulation is that, by its very nature, TCP/IP is nondeterministic. The time to complete a transfer is not guaranteed and the number of retries and the message latency is not known a priori.

An alternative protocol often used in simulation is the User Datagram Protocol (UDP), which is a simpler form of network transfer. With UDP, there is no acknowledgement by the receiver and the sender therefore assumes that all packets are received and that there is no corruption of the data in the packet transmissions. At the receiver, incoming packets are checked for consistency, particularly the packet size and checksum. A packet is read if no errors are detected, otherwise the packet is discarded. UDP protocols are only practical in networks where the error rate is low. There is one further consideration with UDP transfers: the temporal order of packets cannot be guaranteed and packets may not necessarily arrive in the order in which they were transmitted, particularly passing through Ethernet switches. In practice, this event is unlikely to occur in dedicated networks where the bus loading is low. If errors can be ignored, UDP has a minimal overhead and can be used in networks where determinism is essential.

The network protocols are set out and defined in the Open Systems Interconnection (OSI) Reference Model (Anon, 1994). This standardisation ensures that user software is independent of the platform, operating system or network card and there is consistency in the software libraries provided to read and write packets on a network. Taking UDP transfers as an example, the user is required to provide a socket to connect to the network. Once set up, the user provides details of the packet to be read or written. The actual management of the transfer is undertaken by the operating system. To transit a packet, the user software provides:

- the IP address of the destination;
- the port number to be used for the transfers;
- the memory address of the buffer containing the packet to be sent.

The actions are summarised in the following code. A data structure **tx_addr** of type **sockaddr_in** is initialised as follows:

```
memset(&tx_addr, 0, sizeof(tx_addr));
tx_addr.sin_family     = AF_INET;
tx_addr.sin_addr.s_addr = inet_addr(IP_addr);
tx_addr.sin_port       = htons(port);
```

where **IP_addr** is a 32-bit integer holding the destination IP address and **port** is a 32-bit integer holding the port number. The socket is opened by

```
sock = socket(PF_NET, SOCK_DGRAM, IPPROTO_UDP);
```

where **sock** is an integer and a negative result is returned if the socket cannot be opened. The packet is transmitted by:

```
t = sendto(sock, buff, n, 0, (struct sockaddr * &tx_addr, sizeof(tx_addr));
```

where **buff** is an array containing the data to be transferred, **n** is the number of bytes to transfer and **t** is the number of bytes transferred. In the case of an error, **t** is negative. Although this description is simplified, it illustrates the simplicity of Ethernet UDP transfers. A similar function is used to read a UDP packet:

```
r = recvfrom(sock, buff, n 0, (struct sockaddr * &rx_addr, &addr_len));
```

where **r** is the number of bytes received, **sock** is the socket used for reading packets, **buff** is the area of memory where the data is written, **n** is the size of **buff**, **rx_addr** holds the socket information and **addr_len** is a pointer to an integer containing the size of **rx_addr**.

As an example, consider a simulation using five computers, numbered 1–5. The sequence of transfers in any one frame is shown in Table 1.1, where each row shows the progression of time.

Table 1.1 Sequence of Packet Transfers.

Time	Node 1	Node 2	Node 3	Node 4	Node 5
	send pkt1				
		read pkt1	read pkt1	read pkt1	read pkt1
		send pkt2			
	read pkt2		read pkt2	read pkt2	read pkt2
			send pkt3		
	read pkt3	read pkt3		read pkt3	read pkt3
				send pkt4	
	read pkt4	read pkt4	read pkt4		read pkt4
					send pkt5
	read pkt5	read pkt5	read pkt5	read pkt5	

Assuming that node 1 has access to an accurate clock and can start the round of transmissions every 20 ms by broadcasting its packet, the detection of a packet from node 1 can also be used as a start-of-frame signal by the other nodes. On receipt of the packet from node 1, node 2 broadcasts its packet. Similarly, when node 3 has received packets from nodes 1 and 2, it broadcasts its packet, and so on. Although this is a simple protocol, the transmission of each packet is typically less than 100 µs, so that all packet transfers are likely to be completed within 1 ms leaving the rest of the frame for processing and ensuring that the bus loading is low. There are further considerations with this simple protocol:

- The data in the transmitted packet was created during the previous frame.
- One node is responsible for the synchronisation of the frames, implying the need for an accurate clock.
- The network is dedicated, that is to say, there is no other traffic on the network, which would otherwise introduce possible collisions and delays.
- All nodes follow the protocol at all times, reading incoming packets immediately and only broadcasting their packet in their allotted slot.
- If any packet is lost or corrupted, the protocol may fail and the transfers may halt, otherwise a timeout mechanism must be added to the protocol.

The protocol is totally dependent on the reliability of the network. In practice, with a dedicated Ethernet switch and modern Ethernet cards and cables, loss or corruption of data is extremely rare. In addition, in many applications, a corrupted packet could possibly be discarded without a major impact on the simulation.

From the software developer's perspective, while a distributed system offers increased speed, it adds a further layer of complexity. Specific data items used in a simulation can only be stored in one node. Within that node, the data items can be accessed directly with negligible delay. However, for other nodes needing access to these data items, the items are only visible in the transmitted packets and can only be read once a new packet has been received. These constraints place responsibility on the system designer to arrange both the functionality and the storage of data in the most appropriate nodes, in particular to reduce the number of accesses to items via packet transfers. Typically, packets are read and written by copying blocks of data rather than copying individual items.

Note that the protocol just outlined is not the only solution to the interconnection of distributed systems. Other bus systems, including PROFIBUS and CANBUS, are used for real-time applications, although bandwidth reduces with distance between nodes for CANBUS. TCP/IP may also provide a valid solution where the overall bus loading is low. Shared memory systems are also used in distributed systems which allow computers to read and write to dedicated memory shared by all the nodes, where contention for the memory is managed by the shared memory controller. However, the complexity and cost of shared memory systems tend to outweigh their advantages. Finally, it is worth noting that Airbus adapted Ethernet for the AFDX databus used for safety critical applications in the Airbus A380, the Airbus A400M and the Boeing 787 aircraft. AFDX is defined in the ARINC Specification 664, Part 7.

1.5 Processes and Threads

1.5.1 Multi-tasking

While parallelism is explicit in a distributed architecture, there may be cases within an individual computer where the computational tasks of an individual processor can also be executed in parallel. Although an individual processor executes instructions sequentially, if the processor switches between a number of computing tasks during a single frame, it would appear that the tasks have been executed in parallel during that frame.

Consider the three examples of multi-tasking shown in Figure 1.4. In Figure 1.4(a), the three modules are executed sequentially in the order X_1, X_2 and X_3. If all three modules can complete their tasks within the frame, this is the simplest solution. Note that delays in one module will reduce the amount of processing time available to the other modules. All modules share the same code space, data regions, machine registers and system stack.

In Figure 1.4(b), the modules are arranged as processes (Hansen, 1973), which are started and stopped by the operating system and are suspended while waiting for an event. Each process has its own separate code, data, machine registers and stack space. The continuous switching of processes is undertaken by the operating system and care is needed that sufficient time is allocated to the processes to enable them to complete their tasks during each frame. The user may define the priority of processes to ensure the responses to inputs.

An alternative scheme is shown in Figure 1.4(c), which comprises three threads. Threads can be considered as lightweight processes; they are spawned by a parent thread and can share the same code and data, but each thread has its own stack and set of registers. On processors with multiple cores, threads can be allocated to cores, executing code in parallel with the other cores. From the simulation perspective, the main advantage is that threads can be designed to control the synchronisation of threads and the sharing of data. Generally, threads cooperate to complete their computations.

1.5.2 Semaphores

The Posix API enables threads to spawn new threads and to remove completed threads, provides communication between threads and ensures mutual exclusion to shared resources. Threads are particularly useful where computations need to be synchronised, or there are

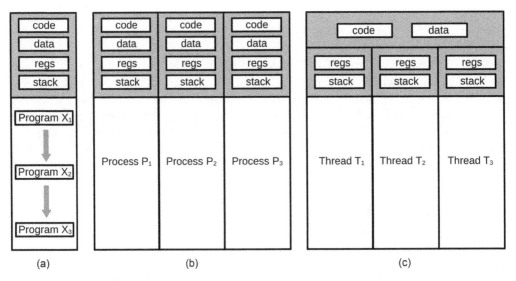

(a)   (b)   (c)

Figure 1.4 Multi-tasking Paradigms.

delays in acquiring data or where common areas of memory are shared. The communication mechanism between threads to control the sequencing of threads and accessing of shared data is provided by *semaphores*, which enable user code to control the sequencing of threads, rather than the operating system. Posix provides two functions to manage semaphores, *sem_post* and *sem_wait*, which are illustrated by the following code, where the semaphore **sem** is represented by an integer:

```
void signal(int *sem)
{
    *sem = *sem + 1;
}
void wait(int *sem)
{
    repeat
    {
        if (*sem > 0)
        {
            *sem = *sem - 1;
            return;
        }
    }
}
```

Although the code appears to be very simple, it is important to understand the significance of these operations, which are provided as Posix functions. The code in the circled sections, which can alter the state of a semaphore, must be atomic, that is, it cannot be interrupted. If an interrupt occurs while the semaphore is being altered in *signal* or *wait* and another thread subsequently modifies the semaphore, the state of the semaphore could become corrupted. This situation is avoided by the coding of *signal* and *wait* in Posix. Rather than disabling and enabling interrupts, the semaphore is modified in a single instruction, which, for many computers, is provided explicitly for that purpose. Note that, if the *wait* fails, the thread cannot continue until the semaphore is signalled by another thread, and the waiting thread is immediately suspended by the operating system.

Synchronisation is where one thread cannot proceed until another thread has reached some point in time. Initially, **s=0** and process A cannot proceed until process B reaches **x**.

```
Process A                    Process B
.                    .
.                    .
.                    .
wait(s)              .
.                    .
.                    .
.            x:      signal(s)
.                    .
.                    .
```

In mutual exclusion, two processes can update the information in a shared buffer without corrupting the data in the buffer, including any pointers. In Posix, the semaphore used for mutual exclusion is referred to as a *mutex*. Initially, `s=1` and between the wait and signal calls, each process has unique access to the content of the shared buffer and can update the structure representing the buffer.

```
Process A                    Process B

wait(s)                      wait(s)
add an item to the buffer    remove an item from the buffer
signal(s)                    signal(s)
```

The producer-consumer problem (Maekawa et al., 1987) is an advanced form of a shared buffer. One thread produces items which are written to a buffer and another thread extracts the items from the buffer. Three semaphores are used. Initially, the semaphore `space = N`, defining the number of items `N` before the buffer overflows, `item` (initially 0) and `buffer` (initially 1) are two semaphores providing mutual exclusion to the item and the shared buffer, respectively. The following fragment shows the insertion and removal of one item in the buffer. If the buffer is full, no further items can be inserted until items have been removed by the consumer thread.

```
Consumer Process         Producer Process
    wait(item)               produce an item
    wait(buffer)             wait(space)
    extract an item          wait(buffer)
    signal(buffer)           insert the item
    signal(space)            signal(buffer)
    consume the item         signal(item)
```

1.5.3 Asynchronous Input

A common example of the use of threads in simulation is where data is entered asynchronously. Simply waiting for data would introduce long delays. One solution is to have one thread acquiring input data and a second thread which only accesses the data when a complete set of data has been acquired. The following example is based on an Airbus Flight Control Unit (FCU), which is connected via an RS-232C serial line at 9600 baud (bits per second). Data is transmitted as a string of ASCII characters when a knob or a selector is turned or a button is pressed. Similarly, the computer can respond by sending an ASCII string to switch lamps on or off or to display values on an LED display. The threads interface uses the Posix interface (IEEE Std 1003, 2017) which provides a well-defined set of functions to create and manage threads. Typically, the strings contain 5–10 characters which take approximately 5–10 ms to transmit at 9600 baud. The following code avoids such delays; when no inputs have been entered, the input thread is dormant, and does not consume any frame time. When a complete string is detected, the characters are removed from the shared buffer and the buffer pointers are updated, typically taking less than a microsecond. An implementation using Posix threads (Lewine, 1991) is shown in Figure 1.5.

During initialisation:

- Two mutexs are initialised for the shared buffers.
- Two semaphores are initialised to manage the states of the threads.
- Two threads are created, one to read serial data from the FCU and one to write serial data to the FCU.

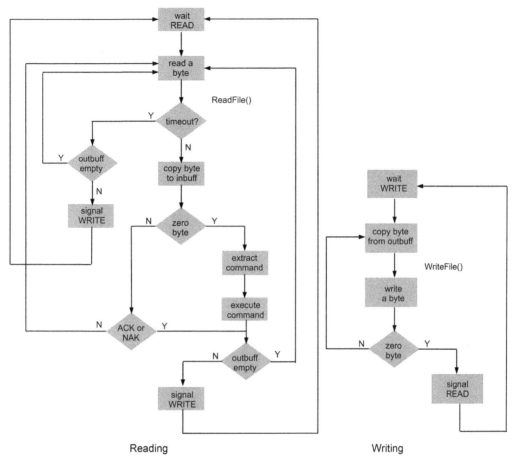

Figure 1.5 Posix Threads to Read Strings from, and Send Strings to, the FCU.

```
pthread_mutex_init(&rxbuf_mutex, NULL);
pthread_mutex_init(&txbuf_mutex, NULL);

if (sem_init(&rxsem, 0, 0) < 0)
{
    printf("sem_init failed rx thread\n");
    exit(1);
}
if (sem_init(&txsem, 0, 0) < 0)
{
    printf("sem_init failed tx thread\n");
    exit(1);
}

if (pthread_create(&port_rxthread, NULL, serial_input, NULL))
{
    printf("Unable to create serial port rx thread\n");
    exit(1);
}
```

```
if (pthread_create(&port_txthread, NULL, serial_output, NULL))
{
    printf("Unable to create serial port tx thread\n");
    exit(1);
}
```

The following code shows the thread reading data from the serial port, with pseudo-code (in italics) to simplify the code and hide unnecessary detail.

```
void *serial_input()
{
    while (1)  /* threads reads one byte from serial input indefinitely */
    {
        sem_wait(&rxsem);

        while (1)
        {
            unsigned char ch;
            DWORD           dwBytesRead = 0;

            Try to read one character from the serial port into ch

            if (dwBytesRead == 0)  /* timeout? */
            {
                pthread_mutex_lock(&txbuf_mutex);
                if (txiptr == txoptr) /* output buffer empty? */
                {
                    pthread_mutex_unlock(&txbuf_mutex);
                    continue;
                }
                else
                {
                    pthread_mutex_unlock(&txbuf_mutex);
                    sem_post(&txsem);
                    break;
                }
            }

            if (dwBytesRead != 1)
            {
                printf("serial port read error 2\n");
                exit(1);
            }

            pthread_mutex_lock(&rxbuf_mutex);
            if (!(ch ==  ACK || ch == NACK))
            {
                Enter the character into the input buffer
            }
            pthread_mutex_unlock(&rxbuf_mutex);
            if (ch == '\0') /* end of string? */
            {
                char v[100];
                unsigned int p = 0;

                pthread_mutex_lock(&rxbuf_mutex);
                while (1)
                {
                    Copy characters from the buffer to v, chx is the last character accessed
                    if (chx == '\0')
                    {
```

```
                    break;
            }
        }
        pthread_mutex_unlock(&rxbuf_mutex);
        Decode(v);
    }
    else if (!(ch == ACK || ch == NACK))
    {
        continue;
    }
    pthread_mutex_lock(&txbuf_mutex);
    if (txiptr == txoptr)
    {
        pthread_mutex_unlock(&txbuf_mutex);
        continue;
    }
    else
    {
        pthread_mutex_unlock(&txbuf_mutex);
        sem_post(&txsem);
        break;
    }
    }
    }
}
```

Although the code appears to be complicated at first sight, the use of a mutex ensures that the circular input buffer can be altered by each thread, safe in the knowledge that it has exclusive access to the buffer. The unconditional loops show that, once initiated, the thread repeats these functions, reading and decoding strings, until it is cancelled.

The function **decode** takes a complete string, terminated with a zero byte and, depending on the specific command, will transmit a string to the FCU in response to the command. In the decode function, as a string is decoded, the appropriate simulator value is updated. This means that the main simulator code simply reads the current values of an FCU variable unhindered by the two threads managing communication with the FCU. Apart from initialisation, all commands are initiated by the FCU, in response to a button being pressed or a knob being turned. The background threads manage the asynchronous reading of incoming strings and the transmission of strings, responding to the commands without introducing any delay during a frame.

1.5.4 Real-time Scheduling

When user processes are executing, they can only be pre-empted when a request is made for system resources. If the request fails, the process is suspended and another process is selected by the scheduler. The one exception to this scheme is the system clock, which interrupts the processor on a regular basis, ensuring a guaranteed point of pre-emption. The system hardware clock, which is not necessarily the clock maintaining the time of day, serves four very important purposes:

1) If a process has not responded for a specific time, the clock interrupt provides a means to investigate or abort a process that is behaving abnormally – this mechanism is often known as a watchdog timer. Strictly, it is the only method to detect a 'rogue' process.

2) It provides further granularity for process scheduling by providing a measure of the number of clock ticks that have occurred while a process has been running or the number of ticks that

have elapsed since a process last ran, avoiding the situation where a process is permanently locked out.

3) If no processes are running, it activates the scheduler on a regular basis, enabling the scheduler to detect if there is a new process to run.

4) With sufficient granularity, it can provide a timing reference for processes.

Note that the processing of these clock interrupts is an overhead, albeit consuming a relatively small amount of the frame time.

1.6 Software Partitioning

The modern flight simulator is likely to be based on a set of computers rather than a single fast computer, where the computers are connected to a high-speed local-area network. Given this framework, the developer is faced with the problem of organising and partitioning the software on a distributed architecture.

Data structures and their variables used in a distributed system fall into three categories: system-wide, global and local. In the overall simulation, *system-wide* variables are shared between several computers. The only mechanism to access these variables is the sharing of packets transmitted over the network. In this sense, the scope of these variables is the packet structures. For example, if the altitude of the aircraft is stored in computer A, computer B can only update its copy of altitude when the aircraft altitude is transmitted by computer A. System variables exist for the life of a program; their storage is allocated when the program is loaded and released when the program is terminated.

Within a computer, there may be several modules and each module may consist of a set of functions (or procedures). These modules may be separately compiled and linked to form an executable program. Variables shared between modules are referred to as *global* variables and, typically, these variables are defined in header files shared by the modules and declared in the module associated with the header file. A module can access these global variables by including the header file in the compilation. Typically, in a C program, these variables have the qualifier *extern*. However, if a variable does not need to be accessed by other modules but is still accessed by several functions of a module, these variables are also global variables, but without the requirement to be defined in header files. In this case, global variables are declared at the start of a program before the functions of a module are declared. Global variables exist for the life of a program; their storage is allocated when the program is loaded and released when the program is terminated.

The other group of data structures are the *local* variables. These variables are defined in a function (and include the parameters of a function) and their scope is strictly limited to the function. Specifically, variables defined in a function are not accessible to other functions. Storage for local variables is created on entry to a function and freed on exit from the function. Normally, local variables exist on a stack, which grows in one direction as variables are pushed onto the stack and reduces as variables are popped from the stack. In high-level languages, the code to manage the stack, pushing and popping variables on entry to and exit from a function, is generated by a compiler and is transparent to the user. In applications where functions are called recursively, that is to say, a function can call itself, a new set of local variables is created each time the function is invoked and deleted on each return from the function.

It is recommended practice that variables are defined in terms of minimal scope. If a variable is only used within a function, it should be a local variable of that function. If a variable is used by two or more functions of a module, it should be defined as a global variable in the module. If a global variable is used by several modules, it should be defined in a shared header file. If a variable is used by several computers, it should be defined in a system-wide header file and accessed via a packet. Consider the example shown in Figure 1.6.

The variable x in computer A is transmitted in a packet to computer B. Module E in computer B has a copy x' of x but x is not visible to computer C. In computer B, modules D and E have access to variable y, if y is defined in a shared header file and declared in module D. y is visible to modules D and E in computer B but not to computers A and C. In computer C, z is visible to functions P, Q and R but variable w is only visible to function R. Variable z is not visible to computers A and B and variable w is not visible to functions P and Q, nor to other modules in computer C, nor to computers A and B.

The partitioning of modules to computers and functions to modules includes the data associated with the modules and functions. The partitioning of data structures and variables across computers is influenced by:

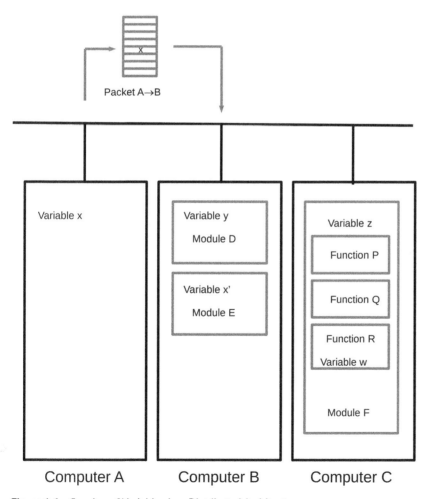

Figure 1.6 Scoping of Variables in a Distributed Architecture.

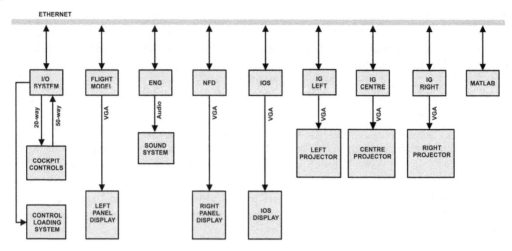

Figure 1.7 A Flight Simulator Configuration.

- the functions which need access to the variables;
- the frequency of access to the variables;
- the storage requirements of the data structures;
- clarity of the code in accessing variables.

Clearly, there is no advantage in accessing a variable in a packet if it is only used within one procedure or one module. There would be an increased overhead to form the packet and to access the variable in the packet. Similarly, if a data structure is large, unless there is an obvious need to include it in a packet transfer, data structures of this form are best restricted to one computer.

One method to partition a large project is to set out all the modules and their variables and then draw a graph with lines between modules to indicate the sharing of variables or their dependence on each other. The intensity of the interconnections is an indication of clustering and modules can be allocated to computers in order to minimise the transfer of data across the network. However, it is often likely that there is a natural partitioning of modules to computers and functions to modules, based on knowledge of the application. Consider the flight simulator configuration shown in Figure 1.7.

This example is taken from an actual engineering flight simulator, comprising nine computers, where the functions of each computer are outlined in Table 1.2. In this example, there is a clear mapping of functions to computers. Depending on the maximum frame time on each computer, some functions could be merged, provided the frame time is not exceeded.

1.7 Simulator Data

Anyone given the task of developing a flight model for a simulator is faced with an immediate problem – where is the data for the aircraft? Some basic data may be in the public domain, for example, brochures issued by the manufacturer or aviation publications (Anon, 2020). Some data may be proprietary, for example, maintenance data issued to an airline or technical manuals, in which case approval is needed to access the data. This data may include basic dimensions and performance data but is likely to lack detailed dimensions, breakdown of masses or

Table 1.2 Computer Functions.

Computer	Function
I/O	Capture digital and analogue inputs and output to the control loading system
Flight model	Aerodynamic model, gear model, FCS and equations of motion
Engine model	Engine model and sound generation
Navigation	Navigation, avionics, FCU panel and radio panels
IOS	Instructor station
IG left	IG left channel (passive)
IG centre	IG centre channel (passive)
IG right	IG right channel (passive)
MATLAB	MATLAB control laws (optional)

moments of inertia, aerodynamic data such as the coefficient of lift or examples of the aircraft response to test inputs.

The major source of data for most manufacturers is flight test data which requires instrumentation of the aircraft to measure accelerations, rates, velocities and attitude etc. The hourly rate of flying prototype aircraft to capture flight data is expensive and the flight tests need to cover the full flight envelope, possibly with hundreds of tests at specific points. Consequently, manufacturers are reluctant to release information into the public domain. In addition, the captured data is processed and organised in a format that can be accessed by a range of developers and users. It is hardly surprising that a manufacturer, investing tens of millions of dollars (and possibly more) in flight test programs, is protective of its data and would seek payment to recover these costs.

For flight simulator companies developing flight simulators for airlines and military organisations, the cost of the data packs for the aircraft, which include aerodynamic data, engine data and avionics data, is included in the price of the simulator and manufacturers are likely to charge several million dollars per aircraft for a data pack. In addition, updates to the aircraft or new data obtained for the aircraft are passed to the simulator manufacturer on a regular basis, in order to keep the simulator software up-to-date.

For other users, needing to develop flight models, but lacking funding to obtain proprietary data packs, the situation is far from clear. Assuming the aircraft is in service, it is possible to undertake flight tests but, of course, the costs of these trials is similar to the costs borne by the manufacturer and the amount of data required depends on the fidelity of the simulator. It is probably fair to say that, apart from simulator manufacturers, the cost of flight trials to acquire data can be prohibitive.

The alternatives are bleak. Data may be available in technical papers or textbooks where the authors have been given access to proprietary information. However, the drawback is that such data is likely to be incomplete. Quite simply, the manufacturer's data may occupy several thousand pages, which cannot be replicated in a chapter of a book or in a journal paper. Moreover, the data needs to include not only the aerodynamic data but also the data to validate the aerodynamic model. Not surprisingly, there are very few examples of the complete data for simulators. An exception is the large repository of NASA reports which are made available to the public via the NASA technical reports server (https://ntrs.nasa.gov). However, there are several drawbacks with using flight data from the NASA archives:

- The catalogue is not ordered in a way that enables users to locate simulation models directly.
- Many of the aircraft covered in the archive are prototype research aircraft (in line with NASA's remit).
- Most of the aircraft are pre-1980.
- Most of the flight models covered are incomplete and lack sufficient data to develop a full flight model.

In passing, it is possibly worth noting that universities and research organisations are mostly excluded from access to simulator data for modern aircraft. There are two exceptions. Firstly, an EU programme Garteur (https://garteur.org) provided data for the Airbus Beluga aircraft to participating organisations. The data is limited to the Beluga aircraft which is used to transport large aircraft components around Europe. Secondly, a generic model of a modern transport aircraft is available under the NASA Technology Transfer Program (https://technology.nasa.gov) to approved participants. The model is provided in Simulink and includes data covering all aspects of flight modelling.

There is one other source of data. Over the last 50 years, numerous papers have been published to derive aerodynamic terms of aircraft. Many of these equations are based on the geometry of an aircraft and its sub-assemblies, such as control surfaces and flaps. By assembling the papers and equations which relate strictly to the geometry of the fuselage, wing, tail, fin and control surfaces etc., it is possible to estimate aerodynamic terms from measurements of the airframe. This approach is the basis of the USAF DATCOM program (Anon, 1979), software developed at the Royal Aircraft Establishment (Mitchell, 1973), entitled 'A Computer Programme to Predict the Stability and Control Characteristics of Subsonic Aircraft' and a textbook by F. O. Smetana (Smetana, 1984) entitled *Computer-assisted Analysis of Aircraft Performance Stability and Control*. In addition, ESDU publications include the estimation of aerofoil characteristics (ESDU, 2006).

Mitchell's software was mostly limited to military organisations and the author is not aware of public domain versions of the software. Smetana's software, developed at North Carolina State University, is in the public domain and is written in FORTRAN IV and has been translated to C by the author. Smetana states clearly that the software is limited to light aircraft configurations of propeller-driven aircraft. His book contains a detailed example of a Cessna-172 aircraft and includes stability analysis but lacks details of engine modelling and undercarriage modelling. Moreover, the validation data is quite limited.

The DATCOM project was developed by the USAF in conjunction with McDonnell Douglas in 1979. It is also a large suite of FORTRAN programs but covers subsonic and supersonic regimes of flight for light aircraft, transport aircraft, military aircraft and missiles. The difficulty with both Smetana's package and DATCOM is the requirement to obtain numerous measurements of the geometry of an aircraft. In principle, once these measurements are entered into the software, it will produce overall dimensions, mass, moments of inertia and a complete set of aerodynamic derivatives. However, it does not generate data for an engine model (performance data needs to be provided) or an undercarriage model. A sample of DATCOM code for a Boeing 737-100 (Roy and Sliwa, 1983) is shown in Example 1.1.

Example 1.1 DATCOM Implementation of a Boeing 737-100

```
CASEID Boeing B-737-100
 $FLTCON WT=115000.,NMACH=1.,MACH(1)=.194,NALT=2.,ALT(1)=1500.,2000.,
         PINF=1967.62,VINF=215.68,TINF=511.57,
         NALPHA=5.,ALSCHD(1)=-2.,0.,1.,2.,4.,GAMMA=0.,RNNUB(1)=1.07E6$
 $OPTINS BLREF=93.0,SREF=1329.9,CBARR=14.3$
```

```
$SYNTHS XW=28.3,ZW=-1.4,ALIW=1.0,XCG=41.3,ZCG=0.0,
 XH=76.6,ZH=6.2,
 XV=71.1,ZV=7.6,
 XVF=66.2,ZVF=13.1,
 VERTUP=.TRUE.$
$BODY NX=14.,
      BNOSE=2.,BTAIL=2.,BLA=20.0,
 X(1)=0.,1.38,4.83,6.90,8.97,13.8,27.6,55.2,
       65.6,69.0,75.9,82.8,89.7,90.4,
 ZU(1)=.69,2.07,3.45,4.38,5.87,6.90,8.28,
       8.28,8.28,8.28,7.94,7.59,7.50,6.9,
 ZL(1)=-.35,-1.73,-3.45,-3.80,-4.14,-4.49,-4.83,
       -4.83,-3.45,-2.76,-0.81,1.04,4.14,6.21,
 R(1)=.34,1.38,2.76,3.45,4.14,5.18,6.21,6.21,
       5.87,5.52,4.14,2.76,.69,0.0,
 S(1)=.55,8.23,28.89,44.31,65.06,92.63,127.81,
       127.81,108.11,95.68,56.88,28.39,3.64,0.11$
$WGPLNF CHRDR=23.8,CHRDTP=4.8,CHRDBP=12.4,
       SSPN=46.9,SSPNOP=31.1,SSPNE=40.0,CHSTAT=.25,TWISTA=0.,TYPE=1.,
       SAVSI=29.,SAVSO=26.0,DHDADI=0.,DHDADO=4.$
$JETPWR NENGSJ=2.0,JEVLOC=-5.2,JIALOC=34.5,JELLOC=15.9,JEALOC=58.0,
       JINLTA=13.4,AIETLJ=-5.$
$VTPLNF CHRDR=15.9,CHRDTP=4.8,SAVSI=33.,
       SSPN=27.6,SSPNOP=0.,SSPNE=20.7,CHSTAT=.25,TWISTA=0.,TYPE=1.$
$HTPLNF CHRDR=12.4,CHRDTP=4.1,
       SSPN=17.6,SSPNE=15.87,CHSTAT=.25,TWISTA=0.,TYPE=1.,
       SAVSI=31.,DHDADI=9.$
$SYMFLP FTYPE=1.,NDELTA=9.,DELTA(1)=-40.,-30.,-20.,-10.,
       0.,10.,20.,30.,40.,SPANFI=0.,SPANFO=14.,CHRDFI=1.72,
       CHRDFO=1.72,NTYPE=1.0,CB=.50,TC=.44,PHETE=.003,PHETEP=.002$
NACA-W-4-0012-25
NACA-H-4-0012-25
DERIV RAD
```

A fragment of the output for a Cessna Citation II aircraft showing several aerodynamic derivatives is given in Table 1.3, which is taken from the spreadsheet output generated by DATCOM. Note that this data is for one specific flight condition in terms of airspeed, altitude and engine setting with the angle of attack varying from $-2°$ to $16°$. These tests would be repeated for the complete flight envelope enabling aerodynamic derivatives to be computed as a function of Mach number and possibly altitude. The output produced by DATCOM can be specified with the input data and several packages have modified the user interface to DATCOM, providing outputs in spreadsheet and MATLAB formats and graphical form using gnuplot.

Software models developed in the NASA LarcSIM (Jackson, 1995) project, JSBSim (http://jsbsim. sourceforge.net) and FlightGear (https://www.flightgear.org), have used data derived from DATCOM. Nevertheless, both the DATCOM package and Smetana's software are not without their problems:

- The user interface is cumbersome and complicated, although a few packages have been developed to simplify the design of user interfaces.
- The output is not in a format that can be used directly in flight modelling.
- The accuracy is variable – in cases where the results are compared with flight data, inaccuracies of 10–20% are reported (Ahmad et al., 2021).

Table 1.3 Aerodynamic Data for a Cessna Citation II.

Alpha	Cladot	CYbeta	CYp	Clbeta	Clp	Clr	CMq	CMadot	CNbeta	CNp	CNr
-2	2.3619	-0.7503	-0.0879	-0.136	-0.4558	0.0295	-16.836	-6.715	0.0573	-0.0002	-0.0984
0	2.405	-0.7503	-0.0965	-0.1301	-0.4649	0.0615	-16.836	-6.8375	0.0573	-0.0154	-0.1002
2	2.486	-0.7503	-0.1055	-0.1242	-0.4724	0.0941	-16.836	-7.0678	0.0573	-0.031	-0.1024
4	2.5321	-0.7503	-0.1147	-0.1182	-0.4695	0.127	-16.836	-7.1989	0.0573	-0.0471	-0.1051
8	2.3822	-0.7503	-0.133	-0.1062	-0.3793	0.1897	-16.836	-6.7729	0.0573	-0.0824	-0.1115
9	2.2906	-0.7503	-0.1362	-0.1033	-0.3392	0.2016	-16.836	-6.5124	0.0573	-0.0908	-0.113
10	2.11	-0.7503	-0.1388	-0.1005	-0.2957	0.2118	-16.836	-5.9988	0.0573	-0.0992	-0.1145
11	1.7501	-0.7503	-0.1408	-0.0977	-0.2473	0.2203	-16.836	-4.9758	0.0573	-0.1073	-0.1157
12	1.4248	-0.7503	-0.142	-0.095	-0.194	0.2269	-16.836	-4.0509	0.0573	-0.1149	-0.1168
13	1.1986	-0.7503	-0.1421	-0.0924	-0.1476	0.2313	-16.836	-3.4078	0.0573	-0.1214	-0.1176
14	0.6856	-0.7503	-0.1437	-0.0898	-0.046	0.2342	-16.836	-1.9491	0.0573	-0.1297	-0.1183
15	-0.2414	-0.7503	-0.1897	-0.0874	0.3134	0.2309	-16.836	0.6862	0.0573	-0.1346	-0.1182
16	-0.7853	-0.7503	-0.0471	-0.0861	0.8667	0.2065	-16.836	2.2327	0.0573	-0.2453	-0.1159

- The learning curve is large – both packages come with extensive documentation but it is necessary to understand the formats and naming conventions.
- It is very easy to make mistakes in data entry and these are not always detected or flagged by the software. If the entered data is incorrect, the output will contain errors, but without validation data it is difficult to locate the sources of errors in the data.
- It is easy to make mistakes in terms of units, for example, confusing metres with feet or degrees with radians.

Despite these reservations, an understandable viewpoint is that slightly inaccurate modelling data is better than no data and, for many investigations, a representative flight model may be acceptable. Certainly, the widespread use and acceptance of DATCOM is an indication of its merit in flight modelling. The documentation provided in both the DATCOM user guide (Anon, 1979) and Smetana's textbook provides full details of the equations used in the software and the various sources used to derive the equations. Users have to decide if the considerable effort to produce aircraft geometric data in the rigid formats required by the packages is worth the benefit of developing models that are otherwise unavailable in the public domain.

1.8 Input and Output

1.8.1 Data Acquisition

A range of devices can be connected to a computer via interfaces which include serial ports, parallel ports, Universal Serial Bus (USB) ports and Integrated Development Environment (IDE), Small Computer System Interface (SCSI), and Serial Advanced Technology Attachment (SATA) hard drive interfaces. The commonality of all these interfaces is that they include registers to access the status of a device, to set up transfers and to transfer data to or from a device. Generally, users are inhibited from accessing hardware registers and transfers are initiated by system calls to the operating system, which is responsible for the transfers, in order to ensure the integrity of transfers.

Consider an example of a serial port reading data at 9600 baud (bits per second). This data rate is approximately 1000 bytes per second or one 8-bit byte per ms, or 20 bytes per frame at 50 fps. Having read one byte from the serial port, the next byte will not be available for approximately 1 ms. There are three options:

1) The processor can simply wait for the next byte to become available, typically by testing a status bit, and then reading the data. This method is known as blocking I/O and introduces the problem that the processor performs no useful operations waiting for a slow transfer. Moreover, if no data is generated, the processor could wait indefinitely.
2) The processor can check to see if a byte is available and, if so, read it, otherwise it can return an error code to indicate that no data is currently available. This method is known as non-blocking I/O and enables the processor to check for data but continue with other processing when no data is available. The processor wastes valuable time checking for data that is not available.
3) The device can be enabled to generate an interrupt when data is available. Once set up, the processor can perform other computations and will only be interrupted when the data is available. The only overhead is stopping to respond to the interrupt, capturing the data and resuming a previous activity.

A similar problem occurs with disk transfers, which have fast data rates but typically are used to transfer large amounts of data in a single transfer. While data is transferred between the disc and memory, which could take several milliseconds, the processor can continue with other processing until the transfer is completed. The method used is known as DMA. The processor accesses the device registers to specify the disc address, the memory address, the size of the transfer and the type and direction of the transfer and initiates the transfer. On completion of the transfer, the device interrupts the processor, which checks on the status of the transfer.

1.8.2 Digital-to-Analogue Conversion

Digital-to-analogue (D/A) conversion enables variables stored in a computer to be output as analogue voltages to drive electrical equipment, such as lamps, motors and actuators, for example, an autothrottle. Figure 1.8 shows a typical D/A configuration.

D/A conversion is relatively straightforward and very fast. The processor writes to the control register to configure the D/A converter, to select a channel and to write a value to be converted. The status register enables the processor to detect if the D/A is busy or if a conversion is completed and to check if any errors occurred in the conversion. The advantage of a D/A is that the bits of an output register are generated by a ladder of resistors, where each resistor has a weighted value that determines the current through the resistor. The currents are added (electrically) and converted to a voltage, as shown in Figure 1.9, which illustrates a 10-bit D/A.

The resistor values produce a current which is weighted to correspond to the bit value of the 10 bits (Horowitz and Hill, 1980). The device is relatively simple, containing 22 resistors and an operational amplifier used for summation. As only two resistor values are used, it is possible to manufacture the device to a very high tolerance. The conversion time depends on the time to write a

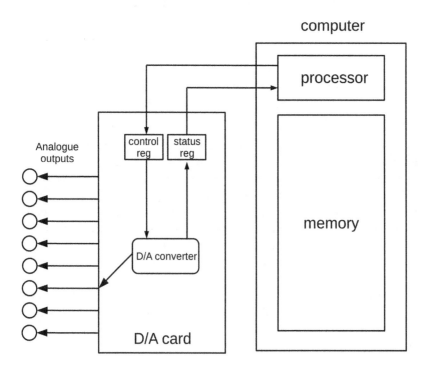

Figure 1.8 Digital-to-Analogue Conversion.

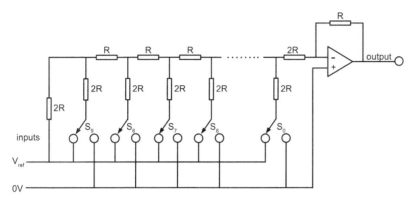

Figure 1.9 A Typical Digital-to-Analogue Converter.

10-bit value to a register and the settling time of the operational amplifier, which is typically less than a microsecond. Extending the device for higher resolutions merely requires the addition of further stages. The output of D/A devices is typically limited to a few milliamps and additional amplification is required to drive lamps and motors.

1.8.3 Analogue-to-Digital Conversion

Flight simulator inputs from the controls, levers, selectors and knobs are connected by potentiometers, LVDT transducers, strain gauges or other forms of sensors, where the signal is an analogue voltage in a defined range. The process of analogue-to-digital conversion (A/D) converts analogue inputs to digital values for use in a computer using dedicated hardware to acquire and convert these signals, which can be read directly by a computer. For A/D conversion, there are several considerations:

- the range of voltage of the inputs – if the input signal is too small, there is a potential loss of resolution and if the signal is too large, it may cause damage to the input circuitry;
- the resolution of the conversion – the number of bits used to represent the voltage;
- the time to convert the input and read it into the computer;
- the linearity of the conversion;
- the filtering of any noise in the signal.

Typically, an A/D converter is interfaced to a computer as shown in Figure 1.10.

The processor writes to the control register to select the input channel, to specify the type of transfer and to initiate the transfer. The input register enables the processor to read the sampled value and to check the status of the A/D, for example, if it has completed the conversion or if an error occurred during conversion. In this example, only eight analogue inputs are shown and they can be switched or connected to the A/D converter. The main limitation with this solution is that the processor is involved with the transfer and has to poll the A/D to detect that the conversion is completed. Alternatively, the A/D may be configured to interrupt the processor on completion of the conversion, to avoid polling. Depending on the A/D device, the number of input channels can vary from 1 to 32 (or more), the resolution can be from 8 to 16 bits and the conversion time can range from a few microseconds to several milliseconds. An improvement is shown in Figure 1.11.

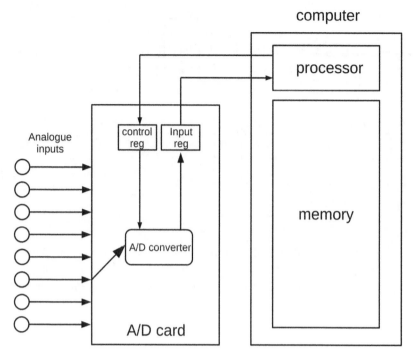

Figure 1.10 Analogue-to-Digital Conversion.

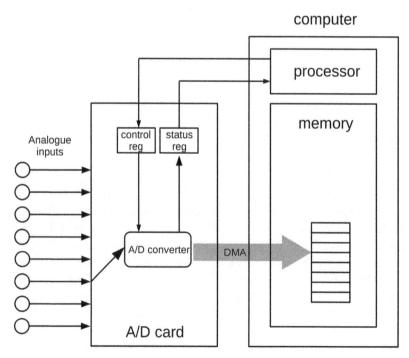

Figure 1.11 Analogue-to-Digital Conversion Using DMA.

Again, only eight analogue inputs are illustrated. The A/D converter outputs are mapped to corresponding memory locations in the processor memory using DMA transfers, so that the processor loading is limited to setting up the A/D transfers. Often, the inputs are sampled autonomously and repetitively, requiring no further intervention by the processor. The sampled values can then be accessed from memory by the simulation software. Normally, the analogue inputs are sampled at the simulator frame rate.

Unlike the D/A, extending an A/D for higher resolution is not straightforward. Three forms of A/D devices are used, with the performance depending on the A/D method:

1) The simplest and slowest form is a counter with a D/A which is initially set to zero. As the counter increments, the voltage output from the D/A rises linearly. The A/D input and the output from the D/A are connected to a comparator which saturates when the D/A ramp voltage exceeds the input voltage, at which point the counter is inhibited. The value in the counter corresponds to the input voltage.
2) A variation is to search for the input voltage by setting or clearing bits in the D/A until the difference between the D/A output and the input is within the voltage corresponding to the least significant bit of the register.
3) The fastest but most expensive device is a flash A/D. The input voltage is applied to a ladder of resistors and operational amplifiers. For n-bit conversion, 2^n stages are required. The 2^n outputs of the operational amplifier are treated as digital inputs to an n-bit encoder to produce an n-bit output. Note that a 16-bit A/D would contain over 64,000 operational amplifiers.

The sequential A/D is slowest as the conversion time depends on the rise time of the ramp voltage, whereas the successive approximation method is guaranteed to convert the input in n cycles for an n-bit A/D and is therefore significantly faster. The flash A/D is the fastest device as it only depends on the settling time of 2^n operational amplifiers, plus the small delay of the encoding; flash A/D is widely used for applications requiring high conversion speeds. However, at frame rates of 50 Hz, with 32 analogue channels, the sampling rate is only 1600 samples per second, which is well within the performance of most A/D devices.

1.8.4 Multiplexing

Although the sampling rates in real-time simulation are relatively low, many flight simulators may have several hundred analogue inputs. Rather than allocating an A/D to each channel, a single high-speed A/D can be used, where the inputs are selected by switching the multiplexor output, as shown in Figure 1.12

The configurations are shown with only eight inputs, for purposes of illustration. In Figure 1.12(a), all A/D devices are selected and the converted values are written, either to local memory (as shown) or memory-mapped in the host computer. In Figure 1.12(b), the multiplexor is addressed to select one of the eight inputs which is converted and read by the host computer.

1.8.5 Encoders

Most potentiometers used for analogue input have a fixed mechanical range. However, there are applications, particularly where the input is derived from a shaft, where the input is angular and continuous. Some potentiometers have a circular track with a small band of insulating material

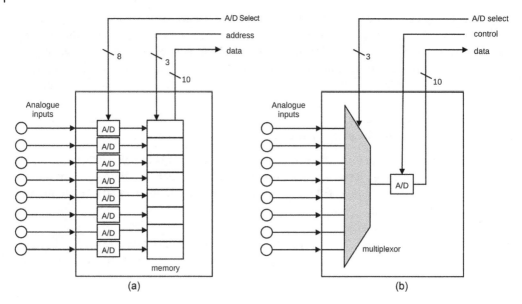

Figure 1.12 Multiplexed Analogue-to-Digital.

between the end points of the track. For example, the track of a potentiometer may range from 1 to 10 KΩ with an infinite resistance when the wiper is positioned over the insulator. An alternative method of measuring continuous angular position, for example, a radio panel knob, is to use a quadrature encoder. The encoder provides two square waveforms, A and B, which enable both motion and the direction of motion to be detected, as shown in Figure 1.13.

The signal is detected by a quadrature decoder where the inputs are A and B and the two digital outputs are the direction and a count enable to detect a change in position. These outputs can be fed to an up–down counter, as shown in Figure 1.14.

Although this arrangement requires one counter per channel, typically with 10–16 bits, it is also possible to detect the change in position and direction of motion in software, requiring only the decoder outputs. The advantage of using quadrature encoders is that the computer input is digital rather than analogue and that the cost of decoders is typically a few dollars. Quadrature decoders are used where the input is a rotary selector or knob, as they avoid additional analogue channels and the ambiguity of detecting an input where the potentiometer is positioned on the insulator.

1.8.6 Digital Input/Output

Digital inputs are used for inputs that can only take the value 0 or 1 (on or off). It is clearly extravagant to allocate an analogue channel to a digital input that can have only two voltages. Similarly, digital outputs for lamps, which are also either 0 or 1, can be represented as digital values rather than analogue values. Typically, an input/output (I/O) card will include digital input and output registers for 16 or 32 bits, or possibly more. The outputs may drive relays or amplifiers as they are likely to be limited to a few milliamps.

The digital inputs of an I/O card will be limited to a voltage range specific to the card. The most common form of input in a simulator is a switch or selector. Rather than providing an active input

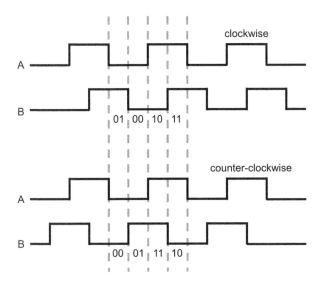

Figure 1.13 Quadrature Encoder Signals.

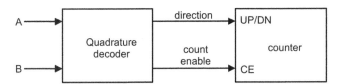

Figure 1.14 Quadrature Decoder.

that switches between 0 V and 5 V, a circuit similar to the one shown in Figure 1.15 is used for digital inputs.

If the switch is open, V_{out} is at the reference voltage V_{cc}. If the switch is closed, V_{out} is at 0 V. The value of the resistor R is chosen so that the circuit only takes a few micro amps when the switch is closed. The main advantage of this arrangement is that if the switch is accidentally grounded, no electrical damage will ensue, avoiding the situation where the supply rail could be connected accidentally to ground.

1.8.7 Signal Conditioning

Depending on the technology of I/O devices, the input range of analogue inputs may be fixed at ±10 V, ±5 V, 0–3.3 V, 0–5 V or even 0–24 V. A problem arises if the inputs are outside these ranges, or alternatively, if they only occupy a small part of a range. In the first case, protection is required to avoid damaging the input circuitry and the signal must be attenuated to the input range. In the other cases, the inputs will only use a fraction of the resolution available, effectively reducing the accuracy of the A/D.

Signal conditioning is provided with analogue inputs to ensure that the inputs extend over the full input range. For example, if a potentiometer is connected to measure the rudder position, but the

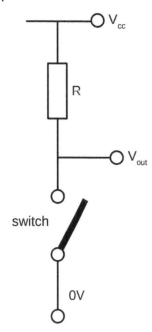

Figure 1.15   Digital Input.

mechanical movement of the assembly only displaces the potentiometer over 40% of its travel, for an input range 0–5 V, it may be found that the left full rudder produces 1.5 V and right full rudder produces 3.5 V. The inputs need to be calibrated to give 0 V for left full rudder and 5 V for right full rudder. Calibration of both the gain and offset provides bias compensation, as shown in the circuit in Figure 1.16 for each analogue input.

The bias is altered by adjusting the variable resistor R_3 where the voltages $\pm V$ are sufficient to ensure the input bias covers the range of possible input values. The gain of the circuit is adjusted by changing the ratio of R_2/R_1 where R_2 is a variable resistor. For each input, the gain and bias are adjusted so that the minimum input is 0 V and the maximum output is 5 V. As the two adjustments affect each other, calibration is an iterative process and is amenable to automation where the voltages between R_5 and R_6 and between R_2 and the negative input of the operational amplifier are provided by D/A outputs. The calibrated values can then be stored and reset at the start of the simulation. The calibration process should be undertaken as part of regular maintenance or following replacement of sensors, partly to ensure the input is correctly calibrated but also to check the polarity of the inputs.

One further consideration, particularly for analogue inputs, is electrical noise. Generally, at low sampling frequencies, there is no likelihood of radio interference or cross-talk from cabling although a circuit board can contain tracks that act as an antenna inducing noise in circuits in close proximity to electrical motors. Noise is mostly picked up from two sources, firstly, mains pick-up, where some component of the AC mains signals leaks into the circuitry and, secondly, poor grounding where there is a small voltage difference between the 0 V lines of different components.

In addition to the 0 V line there is usually a ground line with the cable shield connected to ground. As the currents into an operational amplifier are extremely small, the grounding input of all amplifiers should be common. However, if these connections occur over a relatively long distance (several metres), these grounding points may not all be at the same voltage. Moreover, the 0 V line of digital inputs and analogue inputs should be separated, otherwise interference can occur if these are connected. With 5 V inputs, the least significant bit of a 12-bit analogue input is only 1 mv and, consequently, noise levels above 500 μv are likely to corrupt the input signal. If noise is detected, there are two solutions. Firstly, the grounding of digital and analogue inputs should be checked, isolating (if possible) 0 V lines selectively, in order to identify possible sources of poor grounding. Secondly, a low-pass filter can be added to noisy channels to remove components above 100 Hz. The bandwidth of the filter must be selected to eliminate noise without filtering the actual input signal.

1.8.8   Embedded Systems

Following the development of Arm chipsets for embedded systems (Sloss et al., 2004), several vendors developed small footprint devices including the Raspberry Pi (RPi) and the BeagleBone single-board computers. These devices include USB ports, Ethernet connection, HDMI ports, I/O

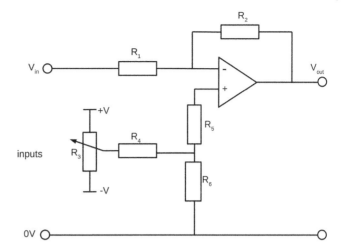

Figure 1.16 Analogue Input Conditioning.

pins and SD-card storage. In addition, there is sufficient processing power and memory capacity to run a Linux operating system. One option is to use these devices as headless embedded systems, where the code is developed to run automatically at power-up in a dedicated system. One other important aspect of these developments is that the cost of an RPi is of the order of $40. With the performance similar to a low-end PC and 40 I/O pins, the RPi can be used as a dedicated I/O interface for real-time simulation.

With the addition of A/D, D/A and digital I/O integrated circuits, the RPi can provide a dedicated controller for analogue and digital I/O at the simulator frame rate. In addition, acquired data can be transmitted in an Ethernet packet, synchronised to the start of every frame, which is read by all the computers connected to the network.

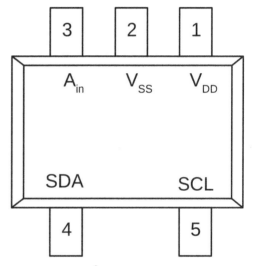

Figure 1.17 An I²C A/D.

The RPi can perform one further function which is the provision of frame timing. The Linux clock is accurate to within a few microseconds and the transmission of a packet containing the system I/O data can signal the start of the simulator-wide frame.

The RPi has a 40-pin header known as the general purpose I/O (GPIO) line. This feature makes the RPi well-suited to hardware applications needing an embedded processor. The I²C (I squared C) protocol was originally developed by Philips Semiconductors in 1982 (Anon, 2012) and has become a worldwide standard for low-cost interfacing. The attraction of I²C is that it only requires four wires, including the supply reference voltage (V_{DD}) and ground (0 V). The other two wires, SDA and SCL, provide the timing, control and data as serial signals between the processor and a device. I²C supports data rates of 100 Kbit/s and 400 Kbit/s. Readers seeking to learn more about I²C signal formats and protocol are referred to the extensive literature on the subject.

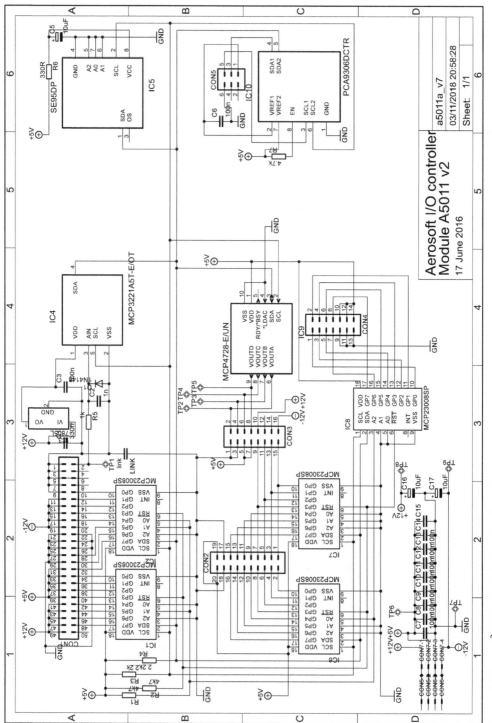

Figure 1.18 An I²C I/O Interface for a Raspberry Pi.

With only four wires between the RPi processor and an I^{2}C device, the hardware interfacing is remarkably straightforward. Many vendors sell a wide range of I^{2}C devices which include motor control, A/D, D/A, serial ports, parallel ports, microcontrollers and so on. For example, the MCP-3221, shown in Figure 1.17, is a 12-bit analogue-to-digital converter manufactured by Microchip Technology Inc (https://www.microchip.com/en-us/product/MCP3221).

Available in surface-mount form, the integrated circuit is approximately 3 mm × 3 mm, V_{DD} is the reference voltage, V_{SS} is 0 V, SDA and SCL are the I^{2}C lines and A_{in} is the analogue input. A slight complication is that the I^{2}C and the RPi reference voltages are 5 V and 3.3 V, respectively; an additional chip is used to convert V_{DD}, SDA and SCL voltage levels between the RPi and the I^{2}C chips. The schematic of a small board containing an A/D for 32 12-bit multiplexed analogue inputs, four 12-bit D/A outputs, 32 digital inputs, and 16 digital outputs is shown in Figure 1.18. The MCP-23008 dual in-line packages are used for digital input and output. Note that, in addition to the four I^{2}C lines, there are three address lines for each IC, enabling up to eight MCP-23008 ICs to be addressed separately. Five of the digital outputs are used to drive two external multiplexors for the 32 analogue inputs.

The simplicity of this solution for an I/O device is remarkable. There are only 10 ICs on the board with minimal wiring owing to the I^{2}C serial protocol. The MCP-3221 A/D sampling rate is over 22,000 samples per second. The MCP-4728 D/A has a settling time of 6 μs. The MCP-23008 I/O expander provides 8-bit digital inputs and outputs and allows the direction of each bit to be set independently.

The programming of I^{2}C devices is straightforward. The I^{2}C protocol must be enabled and the *i2c-tools* package, which can be downloaded for most distributions, provides the drivers for RPi I^{2}C devices. The code to sample a single analogue input channel is shown here, to illustrate the basic principles of transferring data via I^{2}C devices:

```
void Sample_ADC(unsigned int chn)
{
    unsigned char            inbuf[2];
    unsigned char            outbuf[2];
    struct i2c_rdwr_ioctl_data packets;
    struct i2c_msg           messages[2];

    outbuf[0] = 9;  /* reg 9 = channel no for analogue mux */
    outbuf[1] = (unsigned char) chn;

    messages[0].addr = MUX_ADR;
    messages[0].flags = 0;
    messages[0].len = 2;
    messages[0].buf = outbuf;

    packets.msgs = messages;
    packets.nmsgs = 1;

    if (ioctl(i2c, I2C_RDWR, &packets) < 0)
    {
        printf("Unable to set the MUX dir register\n");
        exit(1);
    }
    messages[0].addr = ADC_ADR;
    messages[0].flags = I2C_M_RD;
    messages[0].len = 2;
    messages[0].buf = inbuf;
```

```
    packets.msgs = messages;
    packets.nmsgs = 1;

    if (ioctl(i2c, I2C_RDWR, &packets) < 0)
    {
        printf("unable to read ADC ch=%d\n", chn);
        exit(1);
    }
    ADC[chn] = (((unsigned int) inbuf[0] & 0xf) << 8) + (unsigned int) inbuf[1];
}
```

Each MCP-23008 has a unique address and **MUX_ADDR=0x23**. The system function `ioctl` is used to access the I^2C devices with three arguments: the file descriptor, the I/O request and a pointer to the data. The first `ioctl` call sets the multiplexor to one of the 32 channels. The second `ioctl` call reads the 12-bit data from two registers (`inbuf`) which is written to an array holding 32-bit values. Note that the `ioctl` calls are checked and if errors are detected, the software is terminated, as further I^2C transfers are likely to fail. In this example, the initialisation of the I^2C devices is omitted.

The use of a dedicated embedded system for I/O transfers has many advantages. Although the I^2C polling is an overhead for the RPi, it can easily sustain the frame rate. For the other computers, the overhead is one packet transfer; the raw data sampled by the RPi is read in a small Ethernet packet broadcast by the RPi at the start of each frame. The I/O board is sparsely populated and approximately three inches square with a small connector to four GPIO pins. The main limitations with I^2C are that it is only possible to directly address 128 I^2C devices, the data rate is nominally limited to 400 Kbit/s and the interconnection distance is limited to a few feet. However, none of these limitations apply to real-time flight simulation and expansion of the I/O system using additional boards or additional RPi computers is straightforward.

1.8.9 USB Interfacing

In recent years, many vendors have interfaced their equipment via USB ports rather than analogue or digital inputs, partly because USB ports are commonly provided on modern computers but also because USB simplifies the I/O interface. The technology of USB has advanced, with data rates for version 1.0 of 1.5 Mbits/s increasing to 5 Gbits/s with version 3.0. The acquisition of analogue and digital inputs is implemented in external hardware, reducing the loading on the host computer to acquire input data.

Although drivers may be available to interface to commercial equipment, Linux provides generic USB interface libraries; for example, a system header file `<linux/joystick.h>` is used for access to games devices via USB ports. One example of simulator hardware is the ThrustMaster™ set of controls, requiring three USB inputs for the side-stick, rudder and throttle inputs. The following code opens an input channel for the first USB port:

```
js0 = open("/dev/input/js0", O_RDONLY | O_NONBLOCK);

if (js0 < 1)
{
    printf("Unable to open joystick\n");
    exit(-1);
}
```

The USB input is opened for reading from port **/dev/input/js0** in a non-blocking mode, that is, incoming data can be read without waiting for the data to be generated. If data is available, it can be read, otherwise the code returns immediately, with a status indicating that no data is available. Linux system **ioctl** calls are used to access USB ports. For example, to check the initial status of a side-stick, the following code obtains the name of the device, its version number, the number of axes and the number of buttons.

```
ioctl(js0, JSIOCGNAME(sizeof(name)), name);
ioctl(js0, JSIOCGVERSION, &version);
printf("%s: version %d\n", name, version);
ioctl(js0, JSIOCGAXES, &naxes);
printf("%d axes\n", naxes);
ioctl(js0, JSIOCGBUTTONS, &nbuttons);
printf("%d buttons\n", nbuttons);
```

One reason for accessing the device name is that control inputs may be connected arbitrarily to the USB ports; identifying the device name enables a dynamic mapping to be used between the device channels and USB ports, to ensure that the correct channel is accessed.

The main function is to read characters from the device and this is implemented in the following code:

```
int read_event(int fd, struct js_event *event)
{
    ssize_t bytes;

    bytes = read(fd, event, sizeof(*event));

    if (bytes > 0)
    {
        return 0;
    }
    else
    {
        return -1;   /* probably no event */
    }
}
```

where **event** is a pointer to a structure holding the result of the operation. If the read operation succeeds, 0 is returned, otherwise −1 is returned, indicating that the read failed. The following fragment of code illustrates access to the digital and analogue inputs, for the first two inputs.

```
void IOLib_UpdateIO(unsigned char DigitalOutputA, unsigned char DigitalOutputB)
{
    int e0, e1, e2;

    do
    {
        e0 = read_event(js0, &event);
        if (e0 == 0)
        {
```

```
switch (event.type)
{
  case JS_EVENT_BUTTON:
    switch (event.number)    /* joystick buttons/switches */
    {
      case 0:               /* A7 stick rear trigger (auto-pilot disconnect) */
        IOLib_DigitalDataA = (event.value) ? IOLib_DigitalDataA | BIT7 :
                                              IOLib_DigitalDataA & ~BIT7;
      break;
      case 1:               /* A6 stick red button */
        IOLib_DigitalDataA = (event.value) ? IOLib_DigitalDataA | BIT6 :
                                              IOLib_DigitalDataA & ~BIT6;
      break;
      .
      .
      .
    }
    break;

  case JS_EVENT_AXIS:
    switch (event.number)       /* joystick analogue inputs */
    {
      case 0:                   /* P01 aileron */
        IOLib_AnalogueData[0] = convert(event.value);
        break;
      case 1:                   /* P02 elevator */
        IOLib_AnalogueData[1] = convert(event.value);
        break;
      .
      .
      .
    }
    break;
```

This code is repeated for the USB ports **JS1** and **JS2**, where the function **convert** is used to acquire analogue data, converting the inputs to normalised (±1.0) values for use in the simulation. The simplicity of the interface facilitates quick adaptation of USB devices, while the overhead of reading inputs is of the order of a few microseconds.

References

Ahmad, M., Hussain, Z. L., Shah, S. I. A., and Shams, T. A. (2021), Estimation of Stability Parameters for Wide Body Aircraft Using Computational Techniques, *Applied Sciences*, Vol. 11, No. 2087.

Allen, L. (1993), *Evolution of Flight Simulation, AIAA Conf. Flight Simulation and Technologies*, AIAA-93-3545-CP, Monterey.

Allerton, D. J. (1996), Avionics, Systems Design and Simulation, *The Aeronautical Journal*, Vol. 100, No. 1000, pp. 439–48.

Allerton, D. J. (2000), Flight Simulation – Past, Present and Future, *The Aeronautical Journal*, Vol. 104, No. 1042, pp. 651–63.

Allerton, D. J. (2010), The Impact of Flight Simulation in Aerospace – A UK Perspective, *The Aeronautical Journal*, Vol. 106, No. 1065, pp. 607–18.

Anon. (1979), *The USAF Stability and Control Datcom*, User's Manual, AFFDL-TR-79-3032, Vol. 1, McDonnel Douglas, St Louis.

Anon. (1994), *Information Technology – Open Systems Interconnection – Basic Reference Model*, ISO/IEC 7498-1.

Anon. (2012), *I^2C-bus Specification and User Manual, Rev. 5*, available from https://www.nxp.com/docs/en/user-guide/UM10204.pdf (accessed 4 June 2022).

Anon. (2020), *Jane's All the World's Aircraft: In Service Yearbook,* 2020/2021 Edition, Jane's Group.

Burns, A. and Wellings, A. (2001), *Real-time Systems and Programming Languages*, Addison Wesley.

Cohen, L. L. (1955), *Report of the Court of Inquiry into the Accidents to the Comet Aircraft G-ALYP on 10 January 1954 and Comet G-ALYY on 8 April 1954*, Ministry of Transport and Civil Aviation (UK).

Donahoo, M. J. and Calvert, K. L. (2001), *TCP/IP Sockets in C*, Elsevier Sciences.

ESDU (2006), *Computer Program for Estimation of Aerofoil Characteristics at Subcritical Speeds: Lift-curve Slope, Zero-lift Incidence and Pitching Moment, Aerodynamic Centre and Drag Polar Minimum*, Item 06020, Engineering Sciences Data, London.

Hansen, P. B. (1973), *Operating System Principles*, Prentice-Hall.

Horowitz, P. and Hill, W. (1980), *The Art of Electronics*, Cambridge University Press.

IEEE Std 1003 (2017), *The Open Group Technical Standard Base Specifications*, Issue 7, IEEE.

Jacazio, G. and Balossini, G. (2005), Real-time Loading Actuator Control for an Advanced Aerospace Test Rig, *Journal of Systems and Control Engineering*, Vol. 1, No. 2, pp. 199–210.

Jackson, E. B. (1995), *Manual for a Workstation-based Generic Flight Simulation Program (Larcsim), Version 1.4*, NASA TM-110164.

Korn, G. A. and Korn, T. M. (1965), *Electronic Analog and Hybrid Computers*, McGraw-Hill.

Lewine, D. (1991), *POSIX Programmer's Guide*, O'Reilly and Associates Inc.

Link, E. A. (1930), *Combination Training Device for Student Aviators and Student Entertainment Apparatus*, US Patent Specification 1, Serial No. 825462.

Maekawa, M., Oldehoeft, A. E., and Oldehoeft, R. R. (1987), *Operating Systems - Advanced Concepts*, Benjamin Cummings.

McRuer, D. T. (1995), *Pilot-Induced Oscillations and Human Dynamic Behaviour*, NASA CR-4683.

Mitchell, C. G. B. (1973), *A Computer Programme to Predict the Stability and Control Characteristics of Subsonic Aircraft*, TR 73079, Royal Aircraft Establishment.

Moore, G. E. (1965), Cramming More Components onto Integrated Circuits, *Electronics*, Vol. 38, No. 8.

Roy, M. and Sliwa, S. M. (1983), *A Computer Program for Obtaining Configuration Plots from Digital Datcom Input Data*, NASA TM-84639.

Sloss, A. N., Symes, D., and Wright, C. (2004), *ARM System Developer's Guide*, Elsevier.

Smetana, F. O. (1984), *Computer-assisted Analysis of Aircraft Performance Stability and Control*, McGraw-Hill.

White, A. D. (1989), G-seat Heave Motion Cueing for Improved Handling in Helicopters, *AIAA Conf. Flight Simulation Technologies*, Boston.

Wilkinson, B. (1996), *Computer Architecture Design and Performance*, Prentice-Hall.

2

Software Methods in Simulation

There are two methods in software design. One is to make the program so simple, there are obviously no errors. The other is to make it so complicated, there are no obvious errors.

Tony Hoare

Software used in flight simulators differs in many ways from software developed for commercial or scientific applications:

- Much of the underlying software is implemented as linear sequences of equations, with relatively few data structures.
- The data for models may be supplied by manufacturers in graphical or tabular form and tools are needed to transform this data to code to be used without loss of accuracy.
- The emphasis is on real-time performance, typically provided by distributed architectures, where the simulator data is shared between multiple processors with data passed in packets transmitted over a local network.
- Aircraft systems are often defined as transfer functions, and techniques to simulate transfer functions are commonplace in simulator development.
- Many of the equations are given in the form of differential equations, which are solved by numerical methods in real-time simulation (Benyon, 1968).

2.1 The Laplace Transform

By understanding the physical relationships of electrical and mechanical components and their interconnections as a network, it is possible to derive a mathematical model of a system. For many systems, it is also possible to describe a model as a set of differential equations. The problem of describing a system in a notation of differential equations is that it is far from straightforward to analyse the system behaviour or to simulate the system.

The Laplace transform provides a short-hand notation to describe differential equations together with a method to allow algebraic manipulation of the equations. The derivation of Laplace transformations is covered in most textbooks on control theory and is treated in this book as an accepted practice. The Laplace transformations provide a method to transform signals and systems, defined in terms of time, to polynomials in the s operator. The method also provides the inverse

Flight Simulation Software: Design, Development and Testing, First Edition. David Allerton.
© 2023 John Wiley & Sons Ltd. Published 2023 by John Wiley & Sons Ltd.
Companion Website: www.wiley.com/go/flightsimulationsoftware

transformation from polynomials defined in *s* to an equivalent signal with respect to time. The Laplace transformation offers many advantages in the design and analysis of systems:

- Algebraic manipulation can be applied to both variables and the *s* operator.
- The transformations include standard forms of inputs.
- Transfer functions can be written as polynomials in *s*.
- Networks of elements can be simplified by algebraic manipulation.
- Analysis of transfer functions provides insight into the stability, frequency response and time response of a system.
- Transfer functions can be analysed and modified to achieve a desired response.
- Tables of transformations and inverse transformations are published.
- Many systems are defined as transfer functions or polynomials in *s*.
- Controllers can also be designed as transfer functions.

The power of being able to transform a system of equations defined as a function of time to a polynomial in *s* cannot be overstated. One condition of the Laplace transform is that variables that are a function of time are differentiable. A second condition is that the Laplace transform of the input $u(t)$ exists. However, typical inputs used to analyse a system will include steps, ramps, sine waves and impulses for which the Laplace transforms exist.

There are three aspects of the Laplace transform which provide essential aids in system design:

1) The roots of the numerator and denominator polynomials of a transfer function provide insight into the stability, response and robustness of a system.
2) For analysis in the frequency domain, the *s* operator can be replaced by $j\omega$ where $j = \sqrt{-1}$. This conversion produces a complex number as a function of frequency ω. The magnitude of the complex number gives the system gain and the direction of the complex number gives the phase shift, enabling a Bode plot (Franklin et al., 1994) to be produced to determine gain and phase of a system as a function of frequency.
3) As the *s* operator effectively denotes d/dt or rate of change, if a system is in a steady-state condition, the variables are unchanging and consequently, setting $s = 0$ in the transfer function gives the steady-state gain of a system.

As an example, consider the following differential equation, where $y = 0$ and $\dfrac{dy}{dt} = 0$ initially.

$$\frac{d^2 y(t)}{dt^2} + \frac{5\,dy(t)}{dt} + 6y(t) = 4 \tag{2.1}$$

Applying the Laplace transform to each term

$$s^2 y(s) + 5sy(s) + 6y(s) = \frac{4}{s} \tag{2.2}$$

where the Laplace transform of a unit step input is $1/s$
Solving for $y(s)$

$$y = \frac{4}{s(s^2 + 5s + 6)} = \frac{4}{s(s+3)(s+2)} \tag{2.3}$$

Applying partial fractions

$$y(s) = \frac{2/3}{s} + \frac{4/3}{(s+3)} - \frac{2}{(s+2)} \tag{2.4}$$

Applying the inverse Laplace transform

$$y(t) = \left(\frac{2}{3} + \frac{4e^{-3t}}{3} - 2e^{-2t} \right) \tag{2.5}$$

Note that, although these techniques enable an equation to be solved, solving an equation as a function of time is rarely useful in simulation, whereas being able to transform an equation from the time domain to a polynomial in s, using the method outlined earlier, is a particularly powerful technique.

Transfer functions in s are commonplace in the design and simulation of most systems. Numerous tools exist to analyse systems defined as transfer functions and the connection of systems defined as transfer functions or the addition of a controller also defined as a transfer function simply transforms one transfer function to another. These concepts are explored in more detail in Chapter 4, where transfer functions are used in the design of controllers.

2.2 Transfer Functions

In a system of linear components, it is possible to draw the system as a block diagram, reduce the block diagram and derive the transfer function which represents the system in terms of the Laplace operator s. It is important to bear in mind that the components of a system must be linear. In electrical systems, this precludes transistors, diodes and thyristors. Similarly, in mechanical systems, backlash in gear trains or stiction cannot be realised as linear elements.

Consider the circuit shown in Figure 2.1,

where

$$\frac{v_{out}}{v_{in}} = \frac{R_2}{R_1 + R_2} \tag{2.6}$$

If R_1 is a resistor of impedance R and R_2 is a capacitor of impedance C, or in a mechanical system, if R_1 is a damper D and R_2 is a mass M, then

$$\frac{v_{out}}{v_{in}} = \frac{\dfrac{1}{sC}}{R + \dfrac{1}{sC}} = \frac{\dfrac{1}{RC}}{s + \dfrac{1}{RC}} = \frac{a}{s+a} \tag{2.7}$$

where $a = \dfrac{1}{RC}$ for an electric circuit and $a = \dfrac{1}{DM}$ for a mechanical system. From Laplace tables, this transfer function corresponds to a decaying exponential given by

$$ae^{-at} \tag{2.8}$$

For a given system, the first step is to construct the network defining the connections of each component, substituting the component values and its associated Laplace transform term. The network can

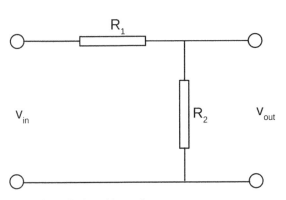

Figure 2.1 Resistor Network.

Figure 2.2   Reduction of Serial Components.

then be simplified and reduced, recalling that algebraic manipulation of s is permitted. Two transfer functions, G_1 and G_2, in series can be reduced to a single transfer function G_1G_2, as shown in Figure 2.2.

Similarly, the addition of two transfer functions G_1 and G_2 in parallel can be reduced to a single transfer function $G_1 + G_2$ as shown in Figure 2.3.

One other common configuration is the feedback arrangement shown in Figure 2.4, where the forward path transfer function G_1 and the feedback path transfer function G_2 are reduced to a single transfer function $\dfrac{G_1}{1+G_1G_2}$.

Readers involved in system design are referred to the numerous textbooks covering control system analysis.

It is important to appreciate the power of the Laplace transform and the ease with which it can be used to describe systems and identify the behaviour of linear systems. The main limitation is that not all systems are linear, although it is often possible to linearise a system about some operating point in order to analyse its behaviour near to that point. Clearly, the dynamic response of an aircraft in the cruise at 35,000 ft at 450 Kt is very different from the dynamic response in the approach at 130 Kt with the undercarriage down and the flaps deployed. This is an important consideration, because it is very tempting to derive a transfer function, only to find that using it outside the range where the linearisation was applied produces unexpected results.

A differential equation occurring commonly in engineering and science is the second-order equation given by

$$\frac{d^2}{dt^2}y(t) + 2\zeta\omega_n\frac{dy}{dt}y(t) + \omega_n^2 y(t) = K\omega_n^2 u(t) \qquad (2.9)$$

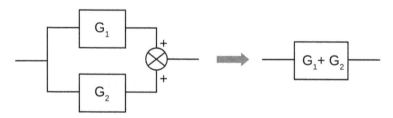

Figure 2.3   Reduction of Parallel Components.

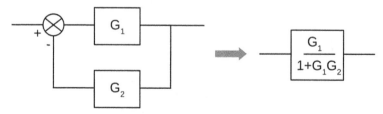

Figure 2.4   Reduction of Feedback Paths.

where ζ is the damping factor, ω_n is the natural frequency, K is a constant and $u(t)$ is the input or driving function. If $\zeta > 1$, the system is overdamped, if $\zeta < 1$, the system is underdamped and if $\zeta = 1$, the system is critically damped. For the underdamped case, it can be shown that the response to a unit step input ($u = 1, t \geq 0$) is given by

$$y(t) = K\left[1 - e^{-\zeta\omega_n t}\left(\cos\omega_d t + \frac{\zeta\omega_n}{\omega_d}\sin\omega_d t\right)\right] \tag{2.10}$$

where the damping frequency is

$$\omega_d = \omega_n\sqrt{1-\zeta^2} \tag{2.11}$$

Setting $K = 1$, $\omega_n = 5$ and $\zeta = 0.1$, the underdamped response is shown in Figure 2.5.

In this example, the input is a step input. The transfer function in s can be reordered so that it is defined in terms of $1/s$, enabling the transfer function to be represented as a set of integrations. From equation 2.9, $y(s)$ can be rewritten as follows.

$$s^2 y(s) + 2\zeta\omega_n s y(s) + \omega_n^2 y(s) = K\omega_n^2 u(s) \tag{2.12}$$

Giving the transfer function

$$\frac{y(s)}{u(s)} = \frac{K\omega_n^2}{s^2 + 2\zeta\omega_n s + \omega_n^2} \tag{2.13}$$

Rearranging the equation and dividing both sides by s

$$sy + 2\zeta\omega_n y = \frac{1}{s}\left(K\omega_n^2 u - \omega_n^2 y\right) \tag{2.14}$$

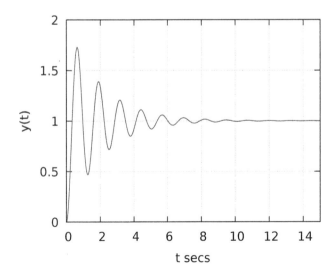

Figure 2.5 Second-order System Response.

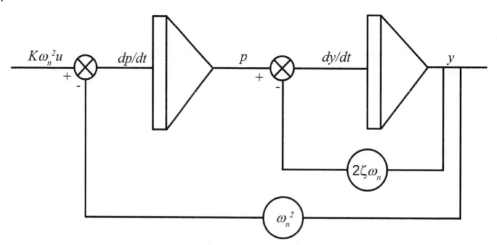

Figure 2.6 Block Diagram of a Second-order System.

Again, dividing both sides by s

$$y = \frac{1}{s}\left[\frac{1}{s}\left(K\omega_n^2 u - \omega_n^2 y\right) - 2\zeta\omega_n y\right] \tag{2.15}$$

Let $pdot = K\omega_n^2 u - \omega_n^2 y$, then

$$p = \frac{1}{s}\left[pdot\right] \tag{2.16}$$

$$y = \frac{1}{s}\left[p - 2\zeta\omega_n y\right] \tag{2.17}$$

The system can be drawn in block diagram form as shown in Figure 2.6.

Assume that prior to $t = 0$, $u = 0$ and $y = 0$, therefore $dp/dt = 0$, $p = 0$ and $dy/dt = 0$; the system is at rest, defining the initial conditions of the two integrators. At time $t = 0$, $u = 1$, the system is excited with a step input. Using a library function **integrate(double y, double p)**, which integrates an integral **y** with the integrand **p**, the code to compute **y** as a function of time is shown in Example 2.1 (the procedure **integrate** is omitted).

Example 2.1   Simulation of a Second-order System

```
#include <stdio.h>
#include <stdlib.h>

#define StepLength 0.01

int main(int argc, char *argv[])
{
    double t    = 0.0;
    double k    = 1.0;
    double wn   = 5.0;
    double zeta = 0.1;
    double u    = 1.0;
```

```
    double y     = 0.0;
    double p     = 0.0;
    int i;

    for (i=1; i<=1500; i+=1)
    {
        double pdot = k * wn * wn * u - wn * wn * y;
        double ydot = p - 2.0 * zeta * wn * y;
        p = integrate(p, pdot);
        y = integrate(y, ydot);
        printf("%f %f\n", t, y);
        t += StepLength;
    }

    return 0;
}
```

There are several points to note from this program:

- pdot and ydot are evaluated before the integrals are computed so that the values p and y are defined at the start of the current step, rather than using values computed during the step.
- The integration step length is 0.01s.
- The initial values of y and p must be known – in this example, y = 0 and p = 0 at time t = 0.
- The simulation runs for 1500 steps, which corresponds to 15 seconds with a step length of 0.01.
- The variable p is not a direct function of y and is introduced as an intermediate variable, to simplify the coding.

The simplicity of this method is appealing. The transfer function is reordered algebraically in terms of $1/s$ operators, enabling the integral and integrand terms to be substituted as arguments of an integration function. Care is needed to select an appropriate integration method and to choose an acceptably small step length to minimise numerical errors in the computation; these considerations are covered in Section 2.5. Usually, the step length is based on the frame rate but significant errors can be introduced if the step length is too coarse.

Another method, known as the Tustin Transform (Van de Vegte, 1990), can also be used to implement transfer functions and is based on a bilinear transform; the Laplace operator s is replaced by the sampled-data operator Z, where

$$s = \frac{2}{T}\left(\frac{Z-1}{Z+1}\right) \tag{2.18}$$

and T is the sampling interval. Although manual substitution is possible, the substitution is best implemented by means of an algebraic manipulation tool, to avoid the introduction of errors. The numerator and denominator in s are replaced by their equivalent polynomials in Z, where

$$\frac{y(Z)}{x(Z)} = \frac{b_0 + b_1 Z^{-1} + b_2 Z^{-2} + \cdots}{a_0 + a_1 Z^{-1} + a_2 Z^{-2} + \cdots} \tag{2.19}$$

Cross-multiplying the equation and ignoring the terms after Z^{-2}, where Z^{-n} is a difference operator for the nth difference of a variable, then

$$a_0 y_k + a_1 y_{k-1} + a_2 y_{k-2} = b_0 x_k + b_1 x_{k-1} + b_2 x_{k-2} \tag{2.20}$$

giving

$$y_k = \frac{-a_1 y_{k-1} - a_2 y_{k-2} + b_0 x_k + b_1 x_{k-1} + b_2 x_{k-2}}{a_0} \qquad (2.21)$$

Using the transfer function from the example given by equation 2.13, with a step length $T = 0.01$ s,

$$\frac{y(Z)}{u(Z)} = \frac{1 + 2Z^{-1} + Z^{-2}}{1609 - 3198Z^{-1} + 1593Z^{-2}} \qquad (2.22)$$

giving

$$y_k = \frac{3198 y_{k-1} - 1593 y_{k-2} + u_k + 2u_{k-1} + u_{k-2}}{1609} \qquad (2.23)$$

where k is the current time step. For a unit step input, at time $t = 0$, $u_k = 1$, $u_{k-1} = 0$, $u_{k-2} = 0$, $y_k = 0$, $y_{k-1} = 0$ and $y_{k-2} = 0$. The transfer function, with $K = 1$, $\omega_n = 5$ and $T = 0.01$, is given by

$$\frac{25}{s^2 + s + 25} \qquad (2.24)$$

While the coefficients can be derived analytically, as shown, packages such as Octave also provide a method to convert transfer functions in s to polynomials in Z. For example, the following Octave statements produce the coefficients for **numz** and **denomz** for the discrete transfer function:

```
nums = [25]
denoms = [1 1 25]
T = 0.01
[numz, denomz] = c2dm(nums, denoms, T, 'tustin')
```

Simulation of the transfer function in equation 2.15 using the Tustin Transform method for 1500 steps (15 s) is shown in Example 2.2.

Example 2.2   Tustin Transform

```
#include <stdio.h>
#include <stdlib.h>

#define StepLength 0.01

int main(int argc, char *argv[])
{
    double t     = 0.0;
    double k     = 1.0;
    double wn    = 5.0;
    double uk    = 0.0;
    double uk1   = 0.0;
    double uk2   = 0.0;
    double yk    = 0.0;
    double yk1   = 0.0;
    double yk2   = 0.0;
    int     i;
```

```
for (i=1; i<=1500; i+=1)
{
    uk = 1.0;
    yk = (3198.0 * yk1 - 1593.0 * yk2 + uk + 2.0 * uk1 + uk2) / 1609.0;
    printf("%f %f\n", t, yk);
    uk2 = uk1;
    uk1 = uk;
    yk2 = yk1;
    yk1 = yk;
    t += StepLength;
}

return 0;
}
```

Note that, as the computation proceeds, the previous inputs u_{k-1} (**uk1**) and u_{k-2} (**uk2**) and the previous outputs y_{k-1} (**yk1**) and y_{k-2} (**yk2**) are updated, ready for the next iteration.

For this example of an underdamped response to a step input, equation 2.10 gives the analytic solution, equation 2.15 gives the implementation using integration, and equation 2.23 defines the computation using the Tustin Transform. The relative errors of the integration method and the Tustin Transform methods are shown in Figure 2.7.

Although the results imply that the integration method has an error of the order 5%, whereas the Tustin method has an error of the order 1%, for this particular example it is important to bear in mind that:

- a first-order integration method was used – higher-order methods are far more accurate;
- the Tustin method uses the trapezoidal rule with an accuracy $O(h^2)$;
- the Tustin method is truncated after three terms in the numerator and denominator.

Notwithstanding these reservations, both methods provide a semi-automated approach to converting transfer functions to equations that can be readily formulated as code and are suited to the simulation of transfer functions of high order.

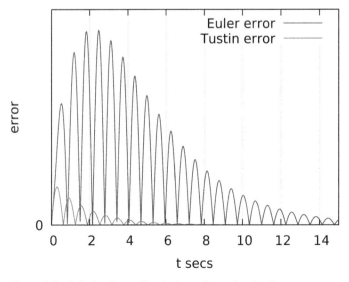

Figure 2.7 Relative Errors Simulating a Second-order System.

2.3 Discrete-event Systems

For many systems, it is essential to know the system behaviour at all times and with sufficient resolution that the system appears continuous, even over short intervals of time. Alternatively, there are also systems, known as discrete-event systems, where it is sufficient to solve the equations over relatively coarse intervals, where a change may only apply when an event occurs; at all other times, the system remains in a constant state. The attraction of discrete-event systems is that a simulation can proceed at a fast pace as entities in the simulation only change state when an event occurs and they are often used for 'faster than real-time' studies. The drawback is that the level-of-detail of the model is reduced and integrating the simulation with real-time models can prove to be difficult.

Consider the simulation of air traffic. The airspace is populated with different types of aircraft, which arrive at and depart from airports and pass through the airspace. Events in the simulation might include taxiing, arrivals, departures, reaching waypoints, traffic conflicts detected by traffic collision avoidance systems (TCAS) and interventions by air traffic controllers. Generally, there is no dynamics in the simulation and, in the case of an air traffic simulation, aircraft would follow trajectories, possibly derived from flight plans or a database of aircraft performance, for example, the Base of Aircraft Data (BADA) database (Eurocontrol, 2010).

In many discrete-event simulations, although specific values may not be known, the statistical properties of entities may be available or can be reasonably estimated, for example, the number of departures from an airport as a function of time. It may be possible to obtain data from flight plans, airport schedules and air traffic communications etc. which would provide the data for a simulation. In practice, such data is likely to be based on statistical data rather than raw data, for example, the number of departures per hour or airways routing altitudes through the airspace. Ideally, the simulation should be based on typical data which, for example, would allow the air traffic density to be increased realistically, in a study of the response of air traffic controllers to high densities of air traffic.

In simulations where the behaviour of entities is estimated, random number generators are used in the estimation. Although random events such as the lottery, rolling dice or a roulette wheel produce truly random events (the next sequence is totally independent of the previous sequence), they are not applicable to computer simulations. An algorithm to compute a random number in a given range should have the following properties:

- a uniform distribution;
- independent output over several dimensions;
- repeatability;
- fast computation time;
- a very long period before any repetition occurs.

Such numbers, which are computer-generated, are often referred to as pseudo-random numbers. The first property implies that the randomness can be measured and should compare favourably with true random events. If successive random numbers are used in different dimensions, the cross-correlation between these numbers should be zero. The requirement of repeatability is particularly important; if a test is run with a set of random numbers, an identical sequence of numbers can be generated in further tests. Although this seems to contradict the concept of a random number, it means that the generator can be reset to repeat a previous random sequence. The requirement for a fast method implies simple arithmetic to generate a random number. Finally, if a sequence does repeat, this period should be sufficiently long that it will not affect, or bias, any simulation.

Random number generators are provided on most computers. Example 2.3 prints five random numbers in the range 0–49 using the default C library (stdlib.h).

Example 2.3   Random Number Generation in C

```c
#include <stdio.h>
#include <stdlib.h>

int main()
{
    int i;

    srand(12345);

    for (i=1; i<=5; i+=1)
    {
        printf("%d\n", rand() % 50);
    }
    return 0;
}
```

The function **srand** initialises the generator so that the same random sequence will be generated if the program is rerun with an argument 12345. The function **rand** returns an integer in the range 0–32,767. The important question is: how random are these numbers? There are numerous papers on the randomness of pseudo-random numbers and methods to compute and analyse random numbers (Knuth, 1981). Bear in mind that there is a further question: just how random does an algorithm need to be? In some applications, any bias in the random numbers generated may introduce unacceptable errors but, for many applications, a basic pseudo-random sequence may be adequate. Readers concerned about this issue are advised to consult the extensive sources on the subject in the literature.

In the above-mentioned pseudo-random number generator, each number should only appear once in the list of 32,768 numbers, which will repeat after 32,768 invocations. This characteristic is common to methods known as linear congruential methods. In their paper, 'Random Number Generators: Good Ones are Hard to Find', Park and Miller (1988) give a linear congruential method which they suggest is the minimum standard for a pseudo-random number generator. The following fragment of code shows a pseudo-random number generator based on an algorithm by Cooper (1990):

```c
#define MODULUS     2147483647
#define MULTIPLIER  16807
#define QUOTIENT    127773
#define REMAINDER   2836

int seed;

int Generate()
{
    int hi  = seed / QUOTIENT;
    int low = seed % QUOTIENT;
    int t   = (MULTIPLIER * low) - (REMAINDER * hi);

    seed = (t > 0) ? t : t + MODULUS;
    return seed;
}
```

Note the computation needs only two integer multiplications, two integer divisions, one addition and one subtraction. The value 2,147,483,647 is $2^{31}-1$, giving a range 0–2,147,483,647. The output for the first 100 random numbers in the range 0–999 is shown in Figure 2.8(a).

However, trying to judge the randomness of a plot of this form is almost meaningless. Figure 2.8(b) shows the distributed frequency of one million generations, that is, the number of times each random number is generated. Ideally, each number should have been generated 1000 times. From inspection, the average is approximately 1000 and the variance is less than 50 which, for many applications, would be adequate. One word of caution: the least significant bits of numbers generated by linear congruential methods are generally less random than their high-order bits. For example, for random numbers in the range 0–99, it is better to generate floating-point values in the range 0–1.0 and then use `(int)` `(100.0 * random())` rather than `(int)` `(1000000.0 * random())` `% 100`. In the latter case, the number has effectively been shifted many places to the left, introducing zeros in the least significant bits prior to the modulo operation.

One further application of pseudo-random number generators is to generate a set of random values with specific statistical properties. For example, a Gaussian distribution has a probability density function of a normal (Gaussian) distribution given by

$$p(y) = \frac{1}{\sqrt{2\pi}} e^{\frac{-y^2}{2}} \tag{2.25}$$

The Box–Muller method (Thomas et al., 2007) can be used to transform two random numbers x_1 and x_2 in the range 0–1 to two random numbers y_1 and y_2, which satisfy a Gaussian distribution, where

$$y_1 = \sqrt{-2\ln(x_1)} \cos 2\pi x_2 \tag{2.26}$$

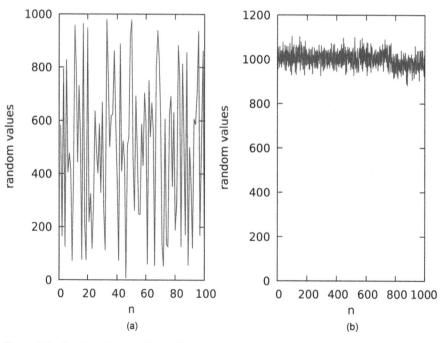

Figure 2.8 Random Number Generation.

$$y_2 = \sqrt{-2\ln(x_1)}\, \sin 2\pi x_2 \tag{2.27}$$

The equation can be further simplified if the values x_1 and x_2 are constrained to lie within the unit circle, then

$$R = x_1^2 + x_2^2 \tag{2.28}$$

where R is a uniform deviate and the angle of (x_1, x_2) with respect to the x_2 axis is the random angle $2\pi x_2$. This method avoids computation of the sine and cosine terms which can be replaced by $v_1/\sqrt{R}$ and $v_2/\sqrt{R}$, respectively. Code to generate a Gaussian distribution using the Box–Muller method is given in Example 2.4.

Example 2.4 Generation of Gaussian Random Numbers

```
float Gaussian(float mean, float std) /* Box-Muller. */
{
    if (odd)
    {
        odd = false;
        return a2;
    }
    else
    {
        float u1;
        float u2;
        float r;
        float d;

        do
        {
            u1 = 2.0 * Generate() - 1.0;
            u2 = 2.0 * Generate() - 1.0;
            r = u1 * u1 + u2 * u2;
        } while (r < 0.0001 || r > 1.0);

        d = sqrt(-2.0 * log(r) / r);
        a1 = u1 * d * std + mean;
        a2 = u2 * d * std + mean;
        odd = true;
        return a1;
    }
}
```

where the function **Generate** produces a random number in the range 0–1. In this example, two values (**a1** and **a2**) are produced for each call of the function **Gaussian** and an odd/even flag is used to either compute two more values or return the second value from the previous call. Note that the random values **u1** and **u2** are transformed to the range ±1 and are discarded (the computation is repeated) for very small values of **r**, to avoid overflow in the subsequent division, or if **r** is greater than 1 (as a result of rounding errors). The mean and standard deviation values are passed as arguments to the function. Figure 2.9 shows the probability distribution for one million random numbers where the distribution is Gaussian, the mean is 500 and the standard deviation is 200.

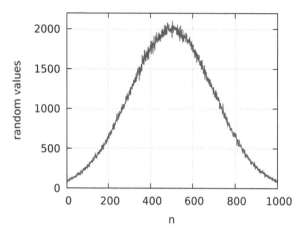

Figure 2.9 Gaussian Distribution of Random Numbers.

For a long time, discrete-event simulation was ignored in flight simulation but this was at a time when simulation focused on the training of flight crews in a single aircraft during a specific training exercise. The instructor was able to fulfil the role of the air traffic controller communicating with the pilot. However, with modern aircraft, with TCAS, datalink messages and busy airspace, the introduction of computer-generated traffic and computer-generated voice messages, which correspond closely to the air traffic and flight conditions, flight simulators nowadays incorporate discrete-event simulators, particularly for air traffic, communications and ground traffic. These simulators are complex but must not introduce unnatural effects, for example, an aircraft contravening a clearance or a ground vehicle moving to an unauthorised part of the airfield. Moreover, in a flight training simulator, it can be useful to be able to repeat an exact set of events or, alternatively, to repeat an exercise with a different set of events. Considerable care is needed to integrate discrete-event simulators with continuous system simulators, particularly to ensure synchronisation of the two simulators.

2.4 Data Fitting

2.4.1 Data Sources

Consider the development of a model of a petrol-driven motor car engine, where the inputs to the model are the fuel in the petrol tank, the car speed and the position of the accelerator pedal and the outputs are the engine RPM and thrust. It would be possible to derive the model from the principles of thermodynamics, but there would be numerous assumptions: the change of viscosity of engine oil with engine temperature, the detonation properties of petrol, losses due to leakage from engine wear, friction of the various engine components and so on. Having produced the model, based on such assumptions, if the characteristics of the model differ from the actual car, it may prove difficult to isolate and correct the sources of any inconsistencies. An alternative approach is to instrument the car and its engine and carburettor and take measurements of thrust and RPM as a function of the accelerometer pedal position and fuel flow at various speeds. However, this latter approach introduces four problems:

- In order to provide a sufficiently accurate model, hundreds or possibly thousands of readings may be needed to cover the range of all possible inputs and outputs.

- The accuracy and resolution of the sensors may affect the fidelity of the model.
- The data has to be provided in a form that can be implemented as a model by the developers.
- The data cannot necessarily be adapted to different engines.

The advantage of basing a model on data is that it is possible to compare the relationships of the simulated model against the measured system and, if there is a close match, there is a high degree of confidence that the system behaviour has been captured in the model.

The basis of modern flight simulator models is the data pack. The aircraft manufacturer will provide the simulator developer with aerodynamic data, derived from numerous flight tests, to enable the developer to construct an aerodynamic model. Similarly, the engine manufacturer will supply engine data to enable engine models to be developed and avionics companies may provide data to enable avionics equipment to be modelled accurately. Understandably, the manufacturers will charge for this data and these costs, which are typically in excess of $1 million, are included in the sale price of a flight simulator. The data pack also provides the basis for qualification of a simulator, enabling the regulator to compare the accuracy of the simulator models with data provided in the data pack.

An aircraft manufacturer will instrument an aircraft, conduct numerous flight tests, acquire and analyse the flight data, producing the data in a machine-readable format for the customer. These activities are labour-intensive and consequently expensive. For older aircraft and smaller aircraft, the data may simply not exist or the cost to the manufacturer of undertaking flight trials may be prohibitive. It is a recurring observation in the flight simulation industry that a simulator is only as good as its data.

There is a further important consideration with the acquisition of aircraft data in flight testing. If a variable is a function of two variables, it is necessary to set one of the variables to a constant value, while measurements are taken for different values of the second variable. These tests are known as steady-state or static tests. For example, aerodynamic data may be obtained from wind-tunnel tests with a scaled model, where the wind speed of the tunnel may be held constant while forces or torques are measured as a function of the angle of attack. Such conditions may be difficult to recreate in airborne tests, for example, the airspeed may change during the subsequent manoeuvre. One other approach is to take measurements during dynamic manoeuvring of the aircraft, applying parameter estimation methods to derive the aerodynamic terms (Maine and Iliff, 1975). However, such methods are limited to matching the data generated by a model of equivalent complexity against the aircraft data. There is an additional problem with airborne flight tests; the acquisition of data throughout the aircraft flight envelope may be impractical or possibly dangerous, for example, measuring the effects of a tyre burst during high-speed taxiing.

Data provided by a manufacturer is nowadays provided in electronic format, in the form of tabular data or graph plots, where data provided as a function of one variable is simply a table of two columns or a single plot. For functions of two variables, sets of tables are provided for different settings of one of the variables or several plots are provided on an individual graph.

Consider the plot of the coefficient of drag (C_d) against the coefficient of lift (C_l) for four Mach values, shown in Figure 2.10 for a military jet trainer. There are several significant points with the data provided in this form:

- The data is a nonlinear function of the coefficient of lift and the Mach number – the curves are given for four specific values of Mach number: 0.7, 0.75 and 0.765 with the fourth curve applicable to Mach numbers in the range 0–0.65.
- No specific data points are plotted – the plots are relatively smooth and it is not clear if the data has been interpolated from raw data.

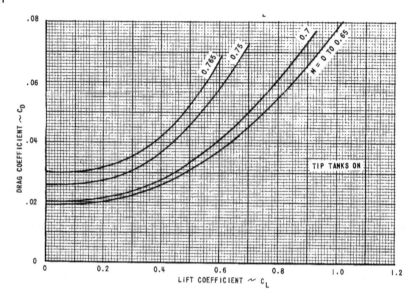

Figure 2.10 C_d Curve versus C_l at Different Mach Numbers.

- The characteristics of the four curves are slightly different – from inspection, it is unlikely that a single equation applies to all four curves.
- This data only applies to the aircraft with tip tanks fitted – there is another set of data for tip tanks removed – strictly, this data is a function of three variables.

Clearly, data can be captured from this graph. For example, $C_d = 0.0335$ at $C_l = 0.5$ for $M = 0.7$. However, values offset from these four curves must be interpolated or extrapolated. Consider also the data shown in Figure 2.11, where C_{lp} is a function of Mach number.

From inspection, is it clear that the data changes very little from $M = 0$ to $M = 0.4$ but that the changes around $M = 0.7$ are significant and it is therefore likely that additional measurements would have been taken between $M = 0.5$ and $M = 0.8$ and close to $M = 0.7$. Graphs of this form are typical of the data produced from aircraft flight tests.

2.4.2 Least-squares Method

One method to simplify data is to assume that it can be represented by a polynomial and attempt to fit a polynomial of order n to the points, where the polynomial coefficients are chosen to minimise the distance between the polynomial and the tabulated values. Numerous methods exist in the literature to derive a polynomial that fits the data closely. Most popular methods include least-squares curve fitting (Griffiths and Smith, 1991) and Chebyshev polynomials. One problem, particularly with Chebyshev polynomials, is to assume that a polynomial of high order will provide a more accurate fit than a polynomial of low order. In practice, a high-order polynomial may contain high-frequency components, producing a line that passes close to the points but containing oscillations, which bear no resemblance to the characteristics of the actual data. Alternatively, a fit may produce a polynomial that is close to all the points but does not pass through any of the points, even though specific points were measured during data acquisition.

The general form of the least-squares method is the solution of a set of simultaneous linear equations given by

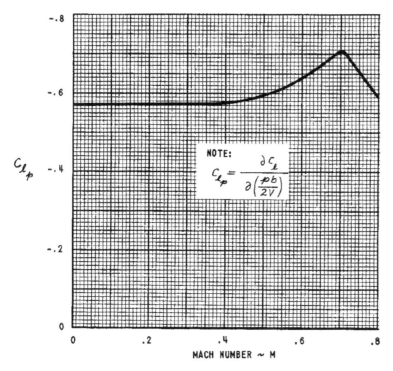

Figure 2.11 C_{l_p} Versus Mach Number.

$$Ma = b \qquad (2.29)$$

where

$$
\begin{bmatrix}
n & \sum_{i=1}^{n} x_i & \cdots & \sum_{i=1}^{n} x_i^k \\
\sum_{i=1}^{n} x_i & \sum_{i=1}^{n} x_i^2 & \cdots & \sum_{i=1}^{n} x_i^{k+1} \\
\vdots & \vdots & \vdots & \vdots \\
\sum_{i=1}^{n} x_i^k & \sum_{i=1}^{n} x_i^{k+1} & \cdots & \sum_{i=1}^{n} x_i^{2k}
\end{bmatrix}
\begin{bmatrix}
a_0 \\
a_1 \\
\vdots \\
a_k
\end{bmatrix}
=
\begin{bmatrix}
\sum_{i=1}^{n} y_i \\
\sum_{i=1}^{n} x_i y_i \\
\vdots \\
\sum_{i=1}^{n} x_i^k y_i
\end{bmatrix}
\qquad (2.30)
$$

This set of equations is referred to as the *Normal Equations* and the solution of these equations is computed by pre-multiplying both sides by the inverse of M, giving

$$a = M^{-1}b \qquad (2.31)$$

For a first-order least-squares fit

$$
\begin{bmatrix}
n & \sum_{i=1}^{n} x_i \\
\sum_{i=1}^{n} x_i & \sum_{i=1}^{n} x_i^2
\end{bmatrix}
\begin{bmatrix}
a_0 \\
a_1
\end{bmatrix}
=
\begin{bmatrix}
\sum_{i=1}^{n} y_i \\
\sum_{i=1}^{n} x_i y_i
\end{bmatrix}
\qquad (2.32)
$$

For a second-order least-squares fit

$$\begin{bmatrix} n & \sum\limits_{i=1}^{n} x_i & \sum\limits_{i=1}^{n} x_i^2 \\ \sum\limits_{i=1}^{n} x_i & \sum\limits_{i=1}^{n} x_i^2 & \sum\limits_{i=1}^{n} x_i^3 \\ \sum\limits_{i=1}^{n} x_i^2 & \sum\limits_{i=1}^{n} x_i^3 & \sum\limits_{i=1}^{n} x_i^4 \end{bmatrix} \begin{bmatrix} a_0 \\ a_1 \\ a_2 \end{bmatrix} = \begin{bmatrix} \sum\limits_{i=1}^{n} y_i \\ \sum\limits_{i=1}^{n} x_i y_i \\ \sum\limits_{i=1}^{n} x_i y_i^2 \end{bmatrix} \qquad (2.33)$$

where the polynomial to the least-squares fit is given by

$$y = a_0 + a_1 x + \ldots + a_k x^k \qquad (2.34)$$

and k is the order of the polynomial and n is the number of data points.

The *cglm* matrix package (a variant of the *glm* package) is open software and provides a useful set of matrix functions in C for two-dimensional and three-dimensional matrix operations and is described in Section 4.3.2. The functions `glm_mat2_inv(a,b)` and `glm_mat3_inv(a,b)` return the inverse of the matrix **a** in the matrix **b**, where **a** and **b** are square matrices of type `mat2` and `mat3` for two- and three-dimensional matrices, respectively. The functions `glm_mat2_mulv(a,b,c)` and `glm_mat3_mulv(a,b,c)` return the vector **c** formed by multiplying the matrix **a** by the vector **b**, for two- and three-dimensional matrices and vectors, respectively. The following fragment of code solves the equation

$$x = s^{-1} b \qquad (2.35)$$

```c
int main(int argc, char *argv[])
{
    const float xval[n] = { 5.93506,  6.90909,  7.90476,  8.90043,  9.91775,
                            10.8918,  11.8874,  12.9048,  13.9004 };
    const float yval[n] = { 0.731538, 0.827385, 0.902769, 0.964154, 1.00938,
                            1.04600,  1.07400,  1.09015,  1.09877 };

    float sumx    = 0.0;
    float sumy    = 0.0;
    float sumxy   = 0.0;
    float sumx2y  = 0.0;
    float sumx2   = 0.0;
    float sumx3   = 0.0;
    float sumx4   = 0.0;

    int j;
    mat3 s;          /* normal equations A matrix Ax = y */
    mat3 sinv;       /* inverse of A */
    vec3 x, b;

    for (j=0; j<n; j+=1)
    {
        float x = xval[j];
        float y = yval[j];

        sumx  += x;
        sumy  += y;
        sumxy += x * y;
        sumx2y += x * x * y;
```

```
        sumx2 += x * x;
        sumx3 += x * x * x;
        sumx4 += x * x * x * x;
    }

    s[0][0] = n;
    s[0][1] = sumx;
    s[0][2] = sumx2;
    s[1][0] = s[0][1];
    s[1][1] = s[0][2];
    s[1][2] = sumx3;
    s[2][0] = s[1][1];
    s[2][1] = s[1][2];
    s[2][2] = sumx4;

    b[0] = sumy;
    b[1] = sumxy;
    b[2] = sumx2y;

    glm_mat3_inv(s, sinv);        /* sinv = s^-1 */
    glm_mat3_mulv(sinv, b, x);    /* x = sinv * y */

    printf("a=%f b=%f c=%f\n", x[0], x[1], x[2]);
}
```

where **s** and **sinv** (the inverse of **s**) are three-dimensional matrices and **x** and **b** are two three-dimensional vectors. The variables needed to populate the *Normal Equations* are computed in the for-loop, where the arrays **xval** and **yval** contain **n** tabulated values of the x and y ordinates, **sumx** is Σx, **sumy** is Σy, **sumxy** is Σxy, **sumx2y** is $\Sigma x^2 y$, **sumx2** is Σx^2, **sumx3** is Σx^3 and **sumx4** is Σx^4. The solution of the equation is given by the last two lines, where the function **glm_mat3_inv** inverts the matrix **s** and copies the result to the matrix **sinv** and the function **glm_mat3_mulv** multiplies the matrix **sinv** by the vector **b**, copying the result to the vector **x**. The coefficients of the equation $y = a + bx + cx^2$ are given by **x[0]**, **x[1]** and **x[2]**, respectively.

As an example, the plots shown in Figure 2.12 are taken from a report for the T-33 aircraft, showing the coefficient of lift versus angle of attack (Hall and Huber, 1970). The data values in **xval** and **yval** are digitised from Figure 2.12(a), where **n** = 9. Figure 2.12(b) shows a first-order fit, given by $C_l = 0.0154397 + 0.097583\alpha$ for $\alpha < 6°$ and a second-order fit, given by $C_l = -0.026337 + 0.164650\alpha - 0.006046\alpha^2$ for $\alpha > 6°$. Note the extrapolation beyond 14°.

2.4.3 Spline Methods

An attraction of the least-squares method is the simplicity of the computation and the reduction of a set of data points to an equivalent polynomial, which can be computed to evaluate a data point. However, the polynomial may not pass through any of the acquired data points. In flight testing, data is often acquired at convenient points or where it is felt that additional data points would be useful. The spacing and number of points vary with individual tests and there is no guarantee that data is acquired at regular intervals. An alternative approach is the use of spline fitting (Ahlberg et al., 1967), which computes a polynomial that passes smoothly through the previous point, the current point and the subsequent point for all the tabulated points.

If a polynomial is defined for the interval between two points, then interpolation is simply computation of the polynomial. For cubic splines, a third-order or cubic polynomial is derived of the form

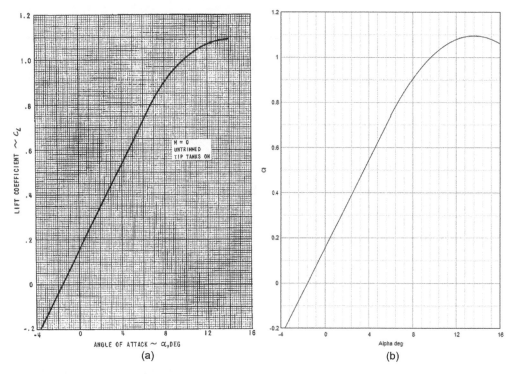

Figure 2.12 Comparison of Flight Data with a Least-squares Fit.

$$a_i x^3 + b_i x^2 + c_i x + d_i \tag{2.36}$$

where $1 < i < n-1$ and n is the number of points.

A cubic spline is the solution of the equation to determine the four coefficients a, b, c and d for each interval and is shown in Figure 2.13. The data points x_0, y_0 to x_3, y_3 are joined by three lines, in this example. The polynomial f_0 is applied from x_0 to x_1, the polynomial f_1 is applied from x_1 to x_2 and the polynomial f_2 is applied from x_2 to x_3. At the intermediate points, known as knots, the first and second derivatives of the two joining curves are identical. The requirement is to derive $n-1$ equations for the n points, which satisfy the criteria that the two adjacent curves pass through the point and that the first and second derivatives of the polynomials are identical at the common point.

Taking the points at $f_i(x)$ and $f_{i+1}(x)$ gives $2n-2$ equations. The first derivative is continuous, so that $\dot{f}_i(x) = \dot{f}_{i+1}(x)$, giving $n-2$ equations. Similarly, the second derivative is continuous, so that $\ddot{f}_i(x) = \ddot{f}_{i+1}(x)$, giving $n-2$ equations. This is a total of $4n-6$ equations in $4n-4$ unknowns. If the assumption is made that the second derivative of the first and last points is zero (the gradient of the curve at the start and end points is not changing), which is known as the *natural* cubic spline, then the set of equations is soluble. Alternatively, the first derivative at the first and end points can be computed from the gradient between the first two or last two points, respectively.

Note that care is needed with two conventions. Firstly, a list of n points can be numbered from 1 to n or from 0 to $n-1$ and, secondly, a fence containing n posts will contain $n-1$ panels. In other words, the numbering must be consistent and any algorithm will contain both data points and functions between points.

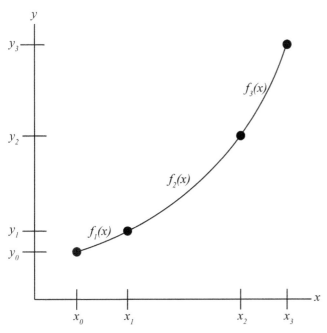

Figure 2.13 A Cubic Spline.

The set of polynomials forming the cubic spline equations is given by

$$S_0(x) = a_0 x^3 + b_0 x^2 + c_0 x + d_0 \tag{2.37}$$

$$S_1(x) = a_1 x^3 + b_1 x^2 + c_1 x + d_1 \tag{2.38}$$

$$\vdots$$

$$S_{n-1}(x) = a_{n-1} x^3 + b_{n-1} x^2 + c_{n-1} x + d_{n-1} \tag{2.39}$$

where

$$S_{i-1}(x) = S_i(x) \tag{2.40}$$

$$\dot{S}_{i-1}(x) = \dot{S}_i(x) \tag{2.41}$$

$$\ddot{S}_{i-1}(x) = \ddot{S}_i(x) \tag{2.42}$$

$$\ddot{S}(x_0) = \ddot{S}(x_{n-1}) = 0 \tag{2.43}$$

and the n data points are given by the arrays $[x_0, x_1, x_2, \cdots, x_{n-2}, x_{n-1}]$ and $[y_0, y_1, y_2, \cdots, y_{n-2}, y_{n-1}]$.

It can be shown that the natural cubic spline can be represented by the following set of linear equations:

$$
\begin{bmatrix}
u_1 & h_1 & & & & & \\
h_1 & u_2 & h_2 & & & & \\
& h_2 & u_3 & h_3 & & & \\
& & \ddots & \ddots & \ddots & & \\
& & & h_{n-3} & u_{n-2} & h_{n-2} \\
& & & & h_{n-2} & u_{n-1}
\end{bmatrix}
\begin{bmatrix}
z_1 \\ z_2 \\ z_3 \\ \vdots \\ z_{n-2} \\ z_{n-1}
\end{bmatrix}
=
\begin{bmatrix}
v_1 \\ v_2 \\ v_3 \\ \vdots \\ v_{n-2} \\ v_{n-1}
\end{bmatrix}
\tag{2.44}
$$

where

$$
h_i = x_{i+1} - x_i \tag{2.45}
$$

$$
b_i = \frac{y_{i+1} - y_i}{h_i} \tag{2.46}
$$

$$
u_i = 2\left(h_{i-1} + h_i\right) \tag{2.47}
$$

$$
v_i = 6\left(b_i - b_{i-1}\right) \tag{2.48}
$$

for $i = 1, \cdots, n-1$.

Note that, if the original set contains n data points, then the dimension of the matrix is $[1{:}n{-}2,$ $1{:}n{-}2]$ and z_0 and z_n are set to zero after the set of equations are solved for z_1 to z_{n-1}. Note also that the square matrix is tridiagonal.

Solving these equations for z produces the second-order derivatives needed to compute the polynomials for each interval, which are given by

$$
y = \frac{z_{i+1}}{6h_i}\left(x - x_i\right)^3 + \frac{z_i}{6h_i}\left(x_{i+1} - x\right)^3 + \left(\frac{y_{i+1}}{h_i} - \frac{z_{i+1}}{6}h_i\right)\left(x - x_i\right) + \left(\frac{y_i}{h_i} - \frac{h_i}{6}z_i\right)\left(x_{i+1} - x\right) \tag{2.49}
$$

The x value is used to search the table of the x data points to locate the value of z_i to be used in the expression above. A binary search method can be used to locate the appropriate interval. The code to generate a spline fit for a set of points is as follows:

```
void Generate_spline(double x[], double y[], int np, double f[])
{
    double a[MaxPts][MaxPts];
    double b[MaxPts];
    double dx[MaxPts];
    double dy[MaxPts];
    double v[MaxPts];
    double u[MaxPts];

    int n = np - 1;
    int i;

    for (i=1; i<=n-1; i+=1)
    {
        int j;
```

```
        for (j=1; j<=n-1; j+=1)
        {
            a[i][j] = 0.0;
        }
    }

    for (i=0; i<=n-1; i+=1)
    {
        dx[i] = x[i+1] - x[i];
        dy[i] = y[i+1] - y[i];
        b[i] = (dy[i] / dx[i]);
    }

    for (i=1; i<=n-1; i+=1)
    {
        v[i] = 2.0 * (dx[i-1] + dx[i]);
        a[i][i] = v[i];
        u[i] = 6.0 * (b[i] - b[i-1]);
    }

    for (i=1; i<=n-1; i+=1)
    {
        a[i][i+1] = dx[i];
        a[i][i-1] = dx[i-1];
    }

    TriDiagonal(n-1, a, f, u);
    f[0] = 0.0;
    f[n] = 0.0;
}
```

The arrays **x** and **y** contain **np** tabulated data points and the second derivatives are returned in the array **f**. The function **Tridiagonal** solves the set of simultaneous equations to derive the second derivative terms, as follows:

```
void TriDiagonal(int n, double m[MaxPts][MaxPts], double x[MaxPts], double
b[MaxPts])
{
    double a[MaxPts];
    double d[MaxPts];
    double c[MaxPts];
    double dnew[MaxPts];
    double bnew[MaxPts];
    int    i;

    for (i=1; i<=n; i+=1)
    {
        d[i] = m[i][i];
    }

    for (i=1; i<=n-1; i+=1)
    {
        a[i] = m[i+1][i];
        c[i] = m[i][i+1];
    }

    bnew[1] = b[1];
```

```
    dnew[1] = d[1];

    for (i=2; i<=n; i+=1)
    {
        double m = a[i-1] / d[i-1];

        dnew[i] = d[i] - m * c[i-1];
        bnew[i] = b[i] - m * b[i-1];
    }

    x[n] = bnew[n] / dnew[n];
    for (i=n-1; i>=1; i-=1)
    {
        x[i] = (bnew[i] - c[i] * x[i+1]) / dnew[i];
    }
}
```

The function **Generate_spline** is executed offline to produce the tables containing the x, y and $\ddot{y}$ terms. At run-time, a function **spline** is used to derive the interpolated values from these three tables. A simple data structure is used to define these spline tables as follows:

```
#define MaxCurves 20
#define MaxPts    100

typedef struct
{
    unsigned int np;
    float        xmin;
    float        xmax;
    float        z;
    float        xa[MaxPts];
    float        ya[MaxPts];
    float        y2[MaxPts];
} SData;

typedef struct
{
    unsigned int ncurves;
    SData        curves[MaxCurves];
} SplineData;
```

For example, the data structure for the data shown in Figure 2.11 is as follows:

```
const SplineData Cl_p_off =
{  1,
    {
        { 9, -0.000546,  0.800291,  0.000000,
            { -0.000546,  0.300782,  0.399419,  0.498927,  0.599220,  0.651917,
              0.699161,  0.735382,  0.800291 },
            { -0.428262, -0.429894, -0.430408, -0.443817, -0.475485,
              -0.501019, -0.528613, -0.501290, -0.443302 },
            {  0.000000,  0.172835, -1.389197, -2.449801,  0.268113, -15.474355,
              53.548361, -4.672245,  0.000000 }  }
    }
};
```

In this example, there is only one curve, represented by nine data points in the range $-0.000\,546$–$0.800\,291$. The three rows define the *x* values, the *y* values and the second derivative values, respectively. The code for **spline** uses a binary search method to locate the tabulated ordinates nearest to **x**, as shown in the following code:

```
float spline(float xa[], float ya[], float f[], int n, float x)
{
    int xlo = 0;
    int xhi = n-1;
    int m;
    float dx;

    while ((xhi - xlo) > 1)
    {
        m = (xlo + xhi) / 2;
        if (x > xa[m])
        {
          xlo = m;
        }
        else
        {
          xhi = m;
        }
    }
    dx = xa[xhi] - xa[xlo];

    return f[xlo] * pow(xa[xhi] - x, 3.0) / (6.0 * dx) +
           f[xhi] * pow(x - xa[xlo], 3.0) / (6.0 * dx) +
           (ya[xlo] / dx - f[xlo] * dx / 6.0) * (xa[xhi] - x) +
           (ya[xhi] / dx - f[xhi] * dx / 6.0) * (x - xa[xlo]);
}
```

The arrays **xa**, **ya**, and **f** hold the *x*, *y* and second derivative tabulated values, **n** is the number of values and **x** is the value to be interpolated. The binary search reduces the number of operations to $O(\log_2 n)$ and the simplicity of the method is evident from the last four lines of the function.

Figure 2.14(a) shows the original plot for C_{lp} versus Mach number for the T-33 aircraft. The resultant plot, using a cubic spline fit, is shown in Figure 2.14(b). Note that considerable care is needed with data entry, particularly as, in this case, the thickness of the line in the original plot is equivalent to 0.01. In addition, the selection of data points should reflect the curvature of the plotted line, thus aiding the spline fit, which is only capable of providing a curve fit to the data points provided.

The method is readily adapted for multiple curves as shown in the following code.

```
float Lookup1(SplineData s, unsigned int n, float x)
{
    SData *p = &s.curves[n];

    return splint(p->xa, p->ya, p->y2, p->np, x);
}

float Lookup2(SplineData s, float var1, float var2)
{
    int    i;
    int    cn = -1;
    float  y1;
```

```
    float y2;
    float z1;
    float z2;

    for (i=0; i<s.ncurves-1; i+=1)
    {
        if (var2 < s.curves[i+1].z)
        {
            cn = i;
            break;
        }
    }
    if (cn < 0)
    {
        cn = s.ncurves - 2;
    }

    z1 = s.curves[cn].z;
    z2 = s.curves[cn+1].z;
    y1 = Lookup1(s, cn, var1);
    y2 = Lookup1(s, cn + 1, var1);
    return y1 + (var2 - z1) * (y2 - y1) / (z2 - z1);
}
```

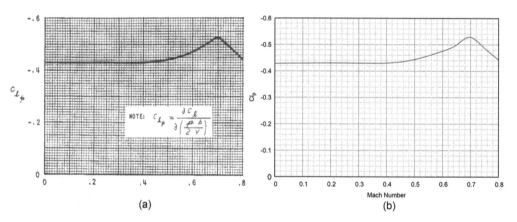

Figure 2.14 Spline Interpolation for a Single Curve.

The function **Lookup1** is used for a single curve whereas the function **Lookup2** interpolates the values for two curves, where **var1** is the x ordinate and **var2** is the z ordinate, that is, the value between two curves. The function searches for the two curves adjacent to **var2**, interpolates each curve using the value **var1** and then computes a linear interpolation of the two points. Figure 2.15 shows a spline fit for multiple curves for C_{lp} for Mach values in the range 0–0.8 and a coefficient of lift C_l in the range 0–1 in steps of 0.2 shown in the original plot in Figure 2.15(a). Note that an additional curve for $C_l = 0.5$ has been interpolated from the tabulated data in Figure 2.15(b).

2.4.4 Extrapolation

For both the least-squares and the cubic splines methods, the values are interpolated at run-time from the tabulated points. Points outside the range of the tabulated values must be extrapolated.

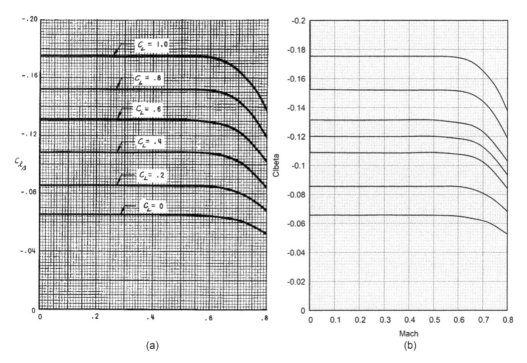

Figure 2.15 Spline Interpolation for Multiple Curves.

Bear in mind that errors in interpolation are relatively small, whereas errors in extrapolation increase with the distance outside the range of the tabulated points. A commonly used method is the Neville algorithm (Neville, 1934), which is based on constructing a table using tabulated points, as follows:

x_0	y_0				
		$P_{01}(x)$			
x_1	y_1		$P_{012}(x)$		
		$P_{12}(x)$		$P_{0123}(x)$	
x_2	y_2		$P_{123}(x)$		$P_{01234}(x)$
		$P_{23}(x)$		$P_{1234}(x)$	
X_3	Y_3		$P_{234}(x)$		
...	...	...	...	...	...
x_k	y_k	$P_{k-1,k}(x)$	$P_{k-2,k-1,\,k}(x)$		

where

$$P_{01}(x) = \frac{(x - x_1) P_0(x) - (x - x_0) P_1(x)}{x_0 - x_1} \tag{2.50}$$

$$P_{02}(x) = \frac{(x - x_2) P_0(x) - (x - x_0) P_2(x)}{x_0 - x_2} \tag{2.51}$$

$$P_{012}(x) = \frac{(x - x_2) P_{01}(x) - (x - x_1) P_{02}(x)}{x_1 - x_2} \tag{2.52}$$

The code for the Neville method is as follows:

```
float Neville (int n, float xa[], float ya[], float x)
{
    unsigned int i;
    unsigned int j;
    float     f[100];

    for (i=0; i<n; i+=1)
    {
        f[i] = ya[i];
    }

    for (j=1; j<n; j+=1)
    {
        for (i=n-1; i>=j; i-=1)
        {
            f[i] = ((x - xa[i-j]) * f[i] - (x - xa[i]) * f[i-1]) / (xa[i] - xa[i-j]);
        }
    }

    return f[n-1];
}
```

where the arrays **xa** and **ya** contain **n** tabulated x and y values, respectively, and **x** is the value to be extrapolated. Note the nested for-loops, implying that the number of operations are $O(n^2)$.

2.4.5 Observations on Data Fitting

In cases where a linear or quadratic relationship is implied by the data, a first- or second-order least-squares method may produce an acceptable fit to the data, with the proviso that the fit may not necessarily pass through the data points. However, once converted to a polynomial form, the data points can be represented by one or two coefficients. Otherwise a cubic spline method can be used based on the tabulated data points and offline computation of the second-order derivative terms. The values do not need to be tabulated at regular intervals and, at run-time, computation of values reduces to applying a cubic polynomial using the coefficients extracted from the tabulated data. Care is needed to capture the appropriate number of points to ensure the consistency of the fit. One further limitation with spline fitting is that care is also needed with straight line segments; it is always advisable to check the resultant curves to ensure that no distortion has been added in satisfying the spline curvature requirements. A worst-case condition for spline fitting is a pulse train but, of course, such waveforms are straightforward to generate without recourse to splines.

2.5 Numerical Methods

In computing, algebraic and trigonometric expressions are implemented to the accuracy of the word length of the computer. However, it is significantly more difficult to implement differentiation. In differentiating a variable with respect to time, any noise added to a variable will also be differentiated, amplifying the noise. In other words, differentiation can introduce noise into a simulation which is a by-product of the differentiation process rather than a characteristic of the system being simulated. The solution, first realised in analogue computing, is not to use differentiation but to use integration. The integration symbol shown in Figure 2.16 can be thought of as producing the integral of x with respect to time. Equally, if the output of the integrator is p, the input must be dp/dt, that is the first derivative of p with respect to time. Note that, in both cases, the initial value of the integrator output must be known.

Consider the simplified car suspension system shown in Figure 2.17, where M is the mass of the car, D is the damping of the shock absorbers and K is the spring stiffness of the suspension system. The dynamics is given by the following differential equation:

$$M\frac{d^2x(t)}{dt^2} = -D\frac{dx}{dt} - Kx(t) \tag{2.53}$$

In block diagram form, the system can be represented using two integrators as shown in Figure 2.18.

In order to simulate the dynamics, an integration function is required that computes $\int x\,dt$. In simple terms, this is the area under the curve of x against time. The function can be approximated by thin strips of width δt, with the area computed by summing the areas of the individual strips. In systems with a constant update rate, where δt is the interval between updates, the change in

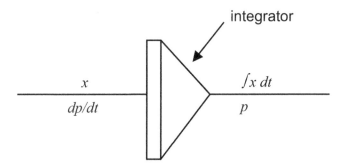

Figure 2.16 An Integration Operator.

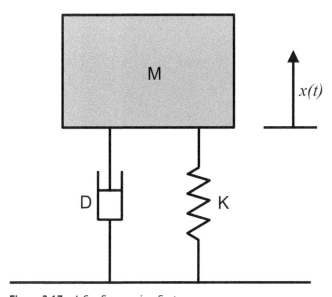

Figure 2.17 A Car Suspension System.

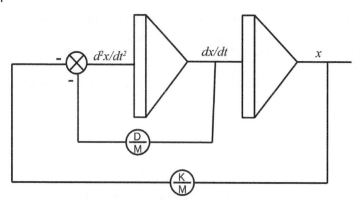

Figure 2.18 Block Diagram of a Car Suspension System.

integral is simply $x\delta t$. The problem is that the value of x is likely to change during the interval. In other words, representing these strips as rectangles is a poor approximation to integration.

A commonly used integration algorithm, known as Euler's method, is illustrated in Figure 2.19. Between x_0 and x_1, the increase in the area under the curve is hy_0 plus the area of the triangle ABC. The gradient of the tangent to the curve at $A = \dfrac{dy}{dx} = f(x_0) = (y_1 - y_0)/h$, giving

$$y_1 = y_0 + hf(x_0) \tag{2.54}$$

In other words, knowing the values of x_0 and y_0 from the previous step, it is possible to predict the value of y_1. The method has the advantage that it is simple to compute and only depends on values of the previous step. The drawback is that the true position of the curve at x_1 is y'_1 rather than y_1 with an error e. Computation of the next step progresses in the same way except that the subsequent computation of y_1 contains the error from the previous step. In other words, not only is there a numerical error with this method but, more importantly, the error accumulates with time. One way to reduce this error is to reduce the step length h. Although this reduces the growth in the error, it does not eliminate the error and, moreover, the number of computations per frame is increased.

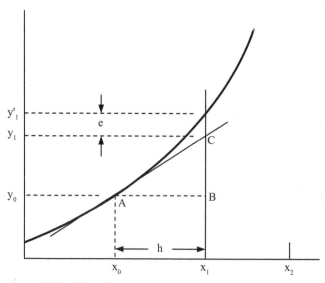

Figure 2.19 Euler's First-order Method.

To illustrate the accuracy of Euler's method, the method can be applied to an equation with a known solution, for example, $y = \dfrac{dy}{dx}$, which has the solution $y = e^x$. In equation 2.54, $f(x_0) = y_0$, giving

$$y_1 = y_0 + hy_0 \qquad (2.55)$$

Computing this solution for 500 steps with $h = 0.01$ gives the results shown in Table 2.1. Even after 10 steps, the error is 0.000549, and after 500 steps, $x = 5$ and the error has increased to 3.64.

Another useful test is the harmonic equation $\dfrac{d^2y}{dx^2} = -y$. The solution is a sine wave of unity amplitude or a unit circle as a phase-plane plot. Any errors, accumulating from the double integration, will be evident from variations in the radius of the circle in the phase-plane plot, as shown in Example 2.5.

Table 2.1 Comparison of Euler's Method.

x	e^x	Euler's method
0.000000	1.000000	1.000000
0.010000	1.010000	1.010050
0.020000	1.020100	1.020201
0.030000	1.030301	1.030455
0.040000	1.040604	1.040811
0.050000	1.051010	1.051271
0.060000	1.061520	1.061837
0.070000	1.072135	1.072508
0.080000	1.082857	1.083287
0.090000	1.093685	1.094174
0.100000	1.104622	1.105171
1.000000	2.704814	2.718282
2.000000	7.316018	7.389056
3.000000	19.788466	20.085537
4.000000	53.524117	54.598150
5.000000	144.772772	148.413159

Example 2.5 Phase-plane Plot of the Harmonic Equation

```
#include <stdio.h>
#include <stdlib.h>
#include <math.h>

#define h 0.01    /* step length */

float Integrate(float y, float x)   /* 1st order Euler integration */
{
    return y + h * x;
```

```
}

int main(int argc, char *argv[])
{
    double y    = 1.0;    /* initial values */
    double ydot = 0.0;
    double t    = 0.0;    /* start from time t=0 */

    do
    {
        double oldydot = ydot;  /* ydot from previous frame */
        printf("%f %f\n", y, ydot);
        ydot = integrate(ydot, -y);
        y = integrate(y, oldydot);
        t += h;
    } while (t <= 100.0);
}
```

Notice that **ydot** is updated during the first integration and therefore should not be used in the second integration as it differs from the value of **ydot** at the start of the step. The variable **oldydot** is used in the second integration instead. The plot of *y* versus *dy/dt* is shown in Figure 2.20, where the growth of error in the simulation, resulting from the integration method, is shown clearly.

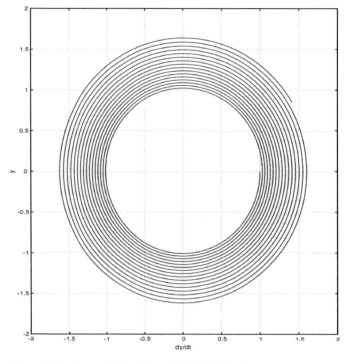

Figure 2.20 Accumulating Error of Euler's Method.

The source of the errors in Euler's method is evident from the expansion of the Taylor series:

$$f(x+h) = f(x) + hf'(x) + \frac{h^2}{2!}f''(x) + \frac{h^3}{3!}f'''(x) + \cdots \tag{2.56}$$

Euler's method corresponds to the Taylor series truncated after two terms, implying that the error is of the order h. This is not surprising, as Euler's method only uses information at the start of the current frame, ignoring previous values of y and dy/dt. However, values of $f(x)$ can be obtained by finite differences of previous values of $f(x)$. This is the basis of the Adams–Bashforth multistep methods which are commonly used in simulation (Cardullo et al., 1991). The coefficients used in these methods are derived by solving the interpolation equations of polynomials that pass through the few previous points. These equations are not derived here but are simply stated because they are covered widely in the literature on numerical methods for differential equations (Ortega and Poole, 1981).

The second-order Adams–Bashforth equation, so called because the error is of the order h^2, is given by

$$y_{k+1} = y_k + \frac{h}{2}(3f_k - f_{k-1}) \tag{2.57}$$

There are two other considerations with multistep methods. Firstly, equation 2.57 includes the term f_{k-1} which will not have been computed for the very first step. Consequently, a starter formula is needed, in order to generate the derivative f_0 at step $k = 0$, for use at step $k = 1$. It is necessary to use a starter formula of similar accuracy, otherwise the errors generated in the first few steps will promulgate in the Adams–Bashforth computations. Secondly, the variables f and y need to be represented as a structure, which is updated at the end of each step in order to refresh the previous values, that is, $f_{k-3} = f_{k-2}, f_{k-2} = f_{k-1}, f_{k-1} = f_k$ (and similarly for the y terms) depending on the number of terms.

The integration algorithm for second-order Adams–Bashforth is as follows:

```
double integrate_AB2(double y, double f0, double f1)
{
    return y + 0.5 * StepLength * (3.0 * f0 - f1);
}
```

where `f0` is the current integrand and `f1` is the integrand from the previous step. The code from the previous example of an underdamped second-order system is modified, firstly to include Euler's trapezoidal method as a starter formula and, secondly, to update the integrand values at the end of each step. The Euler trapezoidal method, which is used for the first step of the Adams–Bashforth method – also known as Heun's method – is given by

$$y_{k+1} = y_k + \frac{h}{2}\left[f(x_k, y_k) + f(x_{k+1}, y_k + hf(x_k, y_k))\right] \tag{2.58}$$

and has an error $O(h^2)$ and is shown in Figure 2.21.

Using Euler integration, y_1 is computed using the gradient at x_0. The gradient at y_1 is computed and the average gradient is used to provide a more accurate estimation of y_1. This process is repeated to derive a new value of y_1 and a new gradient at y_1 until changes in the gradient at y_1 are negligible. The solution of equation 2.15 for 15 s, using the second-order Adams–Bashforth

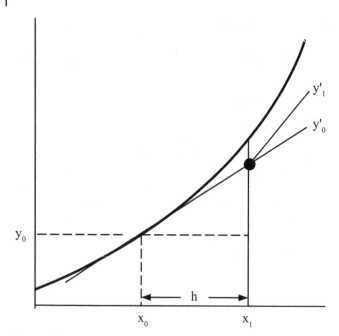

Figure 2.21 Euler's Trapezoidal Method.

method with Euler's trapezoidal method as a starter formula, is as follows, with annotation added in italics:

```
for (i=0; i<=1500; i+=1)   15 secs with step length =0.01
{
    double pdot = k * wn * wn * u - wn * wn * y;  compute derivatives at the start of the step
    double ydot = p - 2.0 * zeta * wn * y;

    if (i == 0)  only applies for the first step
    {
        double p0 = p;  initial values of p and y
        double y0 = y;
        double p1 = p0 + StepLength * pdot0;  Euler integration for one step
        double y1 = y0 + StepLength * ydot0;

        pdot0 = pdot;   retain dp/dt and dy/dt at the start of the step
        ydot0 = ydot;

        while (1)
        {
            double ptmp = pdot;    remember pdot and ydot before updating
            double ytmp = ydot;

            pdot = k * wn * wn * u - wn * wn * y1; update pdot and ydot at end of step
            ydot = p1 - 2.0 * zeta * wn * y1;

            p1 = p0 + 0.5 * StepLength * (pdot0 + pdot);  Euler trapezoidal integration
            y1 = y0 + 0.5 * StepLength * (ydot0 + ydot);

            if (fabs(pdot-ptmp) < 0.000001 && fabs(ydot-ytmp) < 0.000001)
            {
```

```
                    break;    stop if pdot and ydot have converged
                }
                ptmp = pdot;    otherwise remember last pdot and ydot and repeat
                ytmp = ydot;
            }
        }
    else
    {
            p = integrate_AB2(p, pdot, pdot0);    2nd order Adams-Bashforth
            y = integrate_AB2(y, ydot, ydot0);
    }

        pdot0 = pdot;    update previous pdot0 and ydot0 for next step
        ydot0 = ydot;

        t += StepLength;
}
```

Errors in the Adams–Bashforth method are $O(h^2)$ and, in this example, error values are reduced by a factor of approximately 20 in comparison with Euler's method, as shown in Figure 2.22. An essential point to note is the importance of the accuracy of the initial values of **pdot** and **ydot**, otherwise these errors will reduce the accuracy of the Adams–Bashforth algorithm.

The fourth-order Runge–Kutta method, which is used in many scientific applications, is given by

$$y_{k+1} = y_k + \frac{h}{6}\left(F_1 + F_2 + F_3 + F_4 \right) \tag{2.59}$$

where

$$F_1 = f\left(x_k, y_k \right) \tag{2.60}$$

$$F_2 = f\left(x_k + \frac{h}{2}, y_k + \frac{h}{2}F_1 \right) \tag{2.61}$$

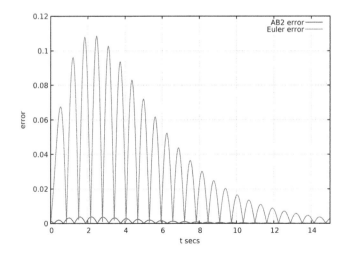

Figure 2.22   Comparison of Adams–Bashforth and Euler Methods.

$$F_3 = f\left(x_k + \frac{h}{2}, y_k + \frac{h}{2}F_2\right) \tag{2.62}$$

$$F_4 = f\left(x_{k+1}, y_k + hF_3\right) \tag{2.63}$$

While the method gives an accuracy $O(h^4)$, the need to compute values at the mid-point of the step means that the equations must be updated at twice the frame rate, in a fixed frame rate simulation.

In simulation, accuracy can be increased either by reducing the step length or, alternatively, by increasing the order of the method. Extending the Adams–Bashforth method to fourth order (fitting a polynomial to four points), for example, gives an error $O(h^4)$ with the following equation:

$$y_{k+1} = y_k + \frac{h}{24}\left(55f_k - 59f_{k-1} + 37f_{k-2} - 9f_{k-3}\right) \tag{2.64}$$

Note that, as before, previous values of the integrand must be retained and updated at each step and that a starter formula is needed for the first three points. One further implication for simulation is that if the simulation is repositioned or reset in some way, unless the previous integrand values are stored and can be reinstated, then the starting formula must be applied to the subsequent three steps.

2.6 Numerical Stability and Accuracy

2.6.1 Numerical Stability

There are two primary sources of error in numerical integration. Truncation errors are caused by omission of higher terms in the algorithm. Rounding errors arise from the word length of the processor's internal registers. Figure 2.23 shows three simulations of the second-order system given in equation 2.15.

All three simulations solve the same equations and all three use Euler's method. The only difference between the three methods is the step length. In Figure 2.23(a), the plot is close to the analytic solution using a step length of 0.01. In Figure 2.23(b), the step length is increased to 0.02 and the difference in the solution is evident; the system appears to be less stable than in Figure 2.23(a). In

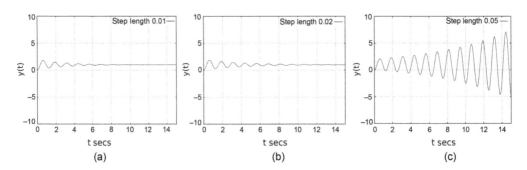

Figure 2.23 Effects of Step Length in Numerical Methods.

Figure 2.23(c), the step length is further increased to 0.05 and the numerical instability (Higham, 1996) is shown clearly.

Numerical integration at a fixed step length is a sampled-data system and Shannon's sampling theorem states that to recover a signal, it must be sampled at least twice the highest frequency of the signal (Betts, 1970). This consideration is very important in real-time simulation, where, if the frame rate is 50 fps, the simulated model must not contain fundamental frequencies above 25 Hz. In the case of modelling an agile fighter aircraft and its flight control systems, this is unlikely to be the case and it would be necessary to increase the frame rate. Using a higher-order algorithm will not change the sampling frequency. This situation, where instability is introduced by numerical integration, is not uncommon in flight simulation and different parts of the models may be solved at different iteration rates, according to the time constants of the systems modelled.

The dilemma for the simulator developer, if instabilities are detected, is to establish whether they are inherent in the simulated system or are caused by a low sampling rate or possibly by a programming error; this situation is particularly difficult to resolve in the simulation of nonlinear systems. An ad hoc solution is to determine the main time constants of the system being simulated (possibly by linearising the system and extracting the eigenvalues) and then simulating the equations with a known solution and similar natural frequencies in order to establish the effect of step length on the overall simulation.

Consider the first-order equation

$$\frac{dy}{dt} = \lambda y \tag{2.65}$$

which has the solution

$$y = \lambda e^t \tag{2.66}$$

Applying Euler's method

$$y_{n+1} = y_n + h \dot{y}_n \tag{2.67}$$

Substituting $\dot{y}_n = \lambda y_n$

$$y_{n+1} = y_n + h\lambda y_n = y_n(1 + h\lambda) \tag{2.68}$$

After n steps from y_0,

$$y_{n+1} = y_0 + (1 + h\lambda)^n \tag{2.69}$$

The solution is stable if $|1 + h\lambda| < 1$, as shown in Figure 2.24. For real values of λ, $-2 < h\lambda < 0$ or if λ is negative, $h \leq 2/\lambda$.

There are numerous integration methods described in the literature, based on predictor–corrector methods (Howe, 1991). Although Euler's method is often discounted for its poor accuracy, the implementation is uncomplicated and compact. The Adams–Bashforth method is in the class of multistep methods and can provide high accuracy, with the penalty that initial derivatives must be computed using a starter formula and variables must be represented by the current and previous values.

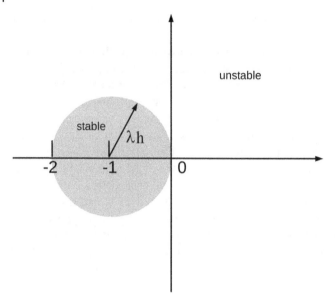

Figure 2.24 Stability of Euler's Method.

There is one further observation that is significant in simulation. If the simulation is open-loop, the numerical errors will grow unbounded and care is needed to select an algorithm with appropriate error growth. However, in applications where there is feedback, the errors are included with the system dynamics and therefore may be reduced to zero as part of any feedback loops. For example, a pilot approaching a runway may have a position offset error of 100 m, as a result of an accumulation of numerical errors. However, the pilot will still be able to line up on the runway centre line, eliminating both piloting errors and numerical errors. Nevertheless, considerable attention should be given to the selection of numerical methods used in a real-time simulation (Cardullo et al., 1991) because of the potential to introduce extraneous instability.

2.6.2 Numerical Accuracy

Versions of C supported by the gcc compiler provide both single-precision and double-precision floating-point arithmetic for *float* and *double* types, respectively. The provision of both types can be traced back two or more decades, when memory capacity was limited and double-precision arithmetic was significantly slower. Nowadays, memory capacity is not a significant consideration and the difference in processing speed between 32-bit and 64-bit floating-point operations is negligible.

Floating-point numbers are used for variables, which can take any values, and generally manufacturers of integrated circuits for processors comply with the IEEE standard (IEEE, 2019) for floating-point numbers, which contain an exponent and a mantissa as shown in Figure 2.25.

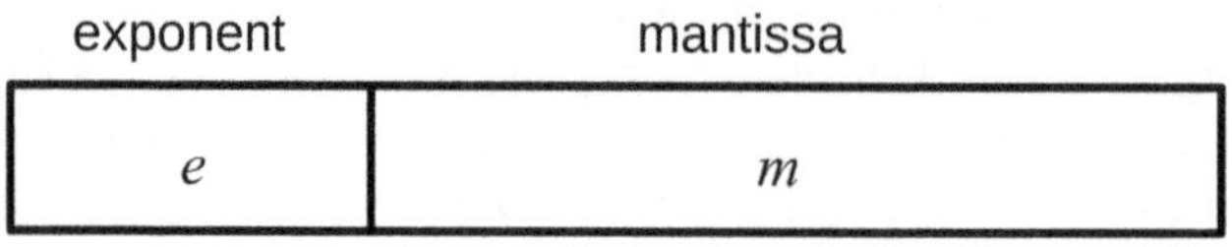

exponent	mantissa
e	m

Figure 2.25 Representation of a Floating-point Number.

For 32-bit arithmetic, e is 8 bits and m is 24 bits. For 64 bits, e is 11 bits and m is 53 bits. The number is given by $2^e \times m$, where m is a fraction in the range from -1 to $+1$. The exponent is a signed integer giving a range of approximately $\pm 2^{127}$ (1.7×10^{38}) for 32-bit floating-point and $\pm 2^{1023}$ for 64-bit arithmetic. Note that the exponent and mantissa can represent positive or negative numbers. With a negative exponent, the smallest positive number is 2^{-127} for 32-bit arithmetic and 2^{-1023} for 64-bit arithmetic.

In order to retain as much precision as possible, the mantissa is normalised, that is, it is shifted left until the most significant bit is a one. Shifting a binary number one place left doubles it, so that each time the mantissa is shifted left, the exponent is decremented. The reason for normalisation is to avoid discarding the least significant bits during addition, subtraction, multiplication and division. Thankfully, the user does not need to be aware of any of these conventions – they are all transparent and managed by the processor arithmetic unit. However, the user does need to appreciate the limitations of these number formats in terms of range and resolution. Even with floating-point arithmetic, numbers can exceed the maximum range, known as *overflow*, particularly as a result of the multiplication of two large numbers or the division of a large number by a small number. Similarly, precision can be lost, known as *underflow*, in the multiplication of two small numbers or the division of a small number by a large number.

On some computers, overflow and underflow may cause an exception error. On other computers, the processor may continue with potentially corrupted values. In addition, special bit patterns can be used to represent ±infinity and 'not a number' (nan) where a number exceeds the largest or smallest range. Variables, which are not explicitly initialised, may be set to zero or –infinity by some compilers. In the latter case, significant errors can occur during subsequent arithmetic. If a and b are very small fractions and c and d are very large numbers, the order of multiplication of the expression $a \times b \times c \times d$ may occur in the order $((ab)c)d$ or $(cd)ba$. The former may result in underflow of ab and the latter may result in overflow of cd, whereas $a(bc)d$ may produce a valid result. Particular care is needed in expressions involving trigonometric terms, where angles are in the range $\pm \pi$ and sine and cosine values are in the range ± 1. Multiplication of trigonometric terms can lead to very small values and underflow and potential loss of precision.

The most important choice is usually whether to use 32-bit floating-point arithmetic or 64-bit floating-point arithmetic. Inevitably, in a real-time system, reading/writing 64-bit values from/to memory and 64-bit arithmetic operations will be slower than 32-bit operations. Data sheets for a processor provide some insight into the relative speed of the two options, and analysis of the software can indicate the worst-case number of memory accesses, additions, subtractions, multiplications, divisions and mathematic functions, per frame. If the number of memory accesses and arithmetic operations can be estimated, it is not difficult to run tests to measure the relative speed of using 32-bit floating-point versus 64-bit floating-point. It is also important to bear in mind three points:

1) An increase in the packet size of data transferred between computers in a distributed system will increase the time allocated to packet transfers.
2) Not all variables justify the use of 64-bit arithmetic. For example, the coefficient of lift will be in range ± 2 and the lift slope curve is itself an approximation, and an error less than 10^{-3} is probably better than the accuracy of the supplied data.
3) Mixing single- and double-precision variables in arithmetic expressions can reduce the overall resolution of an expression to single precision. The compiler may apply implicit coercion between types, without informing the user. The use of brackets to ensure the precedence (and order) of operations and explicit type coercion is essential when single- and double-length arithmetic is mixed.

The decision to use 32-bit or 64-bit arithmetic is totally at the discretion of the user. Often, it is simpler to opt to use either 32-bit or 64-bit throughout and thereby minimise the need for coercion between different floating-point types. The decision is also influenced by the application. For example, the use of 32-bit arithmetic for latitude and longitude can limit position accuracy to 50 cm, justifying the use of 64-bit arithmetic. Similarly, orbital prediction of inter-planetary missions, where distance is given to the nearest metre, would limit distance to just over two million km in 32-bit floating-point computations. Some simulations result in the damping out of dynamics. With a lower resolution of arithmetic, a value may go to zero (from the loss of the least significant bits), whereas damped oscillations may continue for much longer with higher precision arithmetic.

2.7 Timing Analysis

One method to time the execution of real-time code is to read the system clock before and after execution of a section of code, either once, if the clock has sufficient granularity, or possibly one million times to determine the average execution time. However, there may be a significant variance in the timing measurements as the operating system may run other processes between the timing measurements. In addition, profile counting can be used to determine the number of times a section of code is executed.

The duration of a typical frame in a real-time simulation comprises: the time to execute the current cycle of the simulation software, the overhead of any operating system functions and a small margin to allow for frame-to-frame variation. Frames can be timed for worst-case conditions, in order to ensure that all operations can be accommodated within the frame. The margin, which is used to ensure that the frame time limit is never exceeded, depends on the application and the variance of its frame times. If the margin is too generous, valuable frame time is lost and, similarly, if it is too small, the computations during a frame may exceed the allocated frame-time bound and the simulation may 'slip' by one complete frame. In other words, although the time for 50 frames may have elapsed, only 48 or 49 frames have actually been computed. Such situations can either cause the simulator to appear to run slowly or introduce a delay, which may accumulate; both situations are unacceptable. The latter condition may occur in a distributed system because incoming packets are buffered. In such cases, the visual system is using flight data from several frames ago, giving the effect of severe latency.

It is possible to evaluate the time needed for each instruction by reference to the data sheets for the processor. However, this is a very time-consuming exercise and, in the case of branches in the code, can determine the number of cycles in each branch to identify worst-case paths through the code. If the code is changed, this analysis may need to be repeated.

An alternative approach is to use the hardware clock of a computer to measure the time taken to execute a fragment of code. Bear in mind, that if the speed of instructions is to be measured to a resolution of nanoseconds, the clock must have a sufficiently fast clock speed and resolution to provide such measurements. Alternatively, for a clock resolution of microseconds, the code could be timed 1000 times. The limitation with this latter approach is that the execution times are averaged and excursions from the average time may be hidden. There is one further problem known as the probe effect. The addition of code to measure the execution time is included in the measurement but also may change the behaviour of the code in terms of synchronisation with other code, giving misleading insight into the performance of the code being tested.

The system function `gettimeofday(&tv, NULL)` on Linux platforms returns the system time in the structure `tv` of type `timeval` (defined in the system header file `time.h`) which has two fields, the time since the start of Linux in seconds and the number of microseconds since the start of the current second. A second call to **gettimeofday** enables the elapsed time in microseconds between successive calls to be computed.

An example of measuring the time to compute engine thrust as a function of EPR and Mach number using 2D splines is shown in Example 2.6.

Example 2.6   Timing of Thrust Computation

```
int main(int argc, char *argv[])
{
    unsigned int   i;
    struct timeval tv1;
    struct timeval tv2;
    int            time[1001];

    for (i=1; i<=1000; i+=1)
    {
        int j;

        gettimeofday(&tv1, NULL);
        for (j=1; j<=1000; j+=1)
        {
            float t = SplineLib_Lookup2(Net_thrust_table, 1.1, 0.7);
        }
        gettimeofday(&tv2, NULL);
        time[i] = (tv2.tv_sec - tv1.tv_sec) * 1000000 + tv2.tv_usec - tv1.tv_usec;
    }

    for (i=1; i<=1000; i+=1)
    {
        printf("%f %f\n", (float) i, (float) time[i] / 1000.0);
    }

    return 0;
}
```

The function **SplineLib_Lookup2** is called 1000 times and the elapsed time between the two calls of **gettimeofday** (divided by 1000) gives the time in microseconds, as shown in Figure 2.26.

These tests were run on a Raspberry Pi Model 4B with a clock speed of 1.5 GHz or cycle time of 0.67 ns. Although there is some noise in the measurements, the spline lookup, albeit for one specific value, takes between 9.4 and 9.6 μs. However, care is needed in interpreting results from sampling the system clock:

- The timing of very short sequences may exceed the granularity of the clock, such that variations in the measurements are noise and obscure useful measurements.
- There is an overhead from executing the for-loop and the calls to read the clock, which will be included in the measurement.
- Executing these tests in a for-loop 1000 times, or possibly more, produces an average time and may miss cases where the elapsed time is excessive.

Typical frame timing of a simulation is shown in Figure 2.27, also for the Raspberry Pi Model 4B.

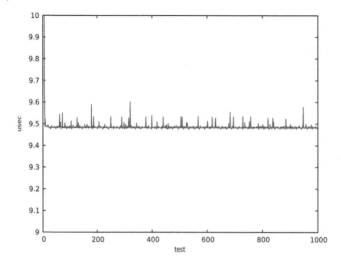

Figure 2.26   Variation in Timing Measurements.

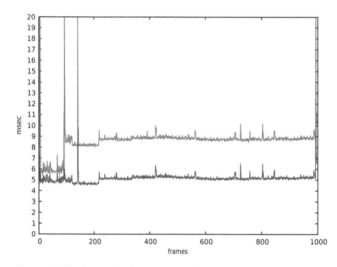

Figure 2.27   Frame Timing Measurements.

Timing was applied during the first 1000 frames (20 s), where the lower trace shows the graphics rendering and the upper trace shows the overall frame computations, including the graphics. The initial spike is caused by the loading of simulator files and the second spike occurs when the simulator is reset and reloads a saved file. The graphics rendering is 5–6 ms and the overall frame is 8–10 ms, giving a considerable margin.

While the use of hardware clocks can provide insight into the performance of real-time applications, considerable care is needed to avoid including clock errors in the measurement and to ensure that the accuracy of measurements matches the number of samples. The important point to bear in mind is that it is both the number of times the code is executed, which can be obtained from profiling, *and* the time to execute the code that determine the frame time occupied by the code. Timing analysis can identify sections of code where further optimisation can significantly reduce the frame loading. Finally, care is also needed with compiler optimisation settings, which will affect the performance of the compiled code.

2.8   Simulation Packages

During the 1980s, several vendors developed packages to capture and simulate engineering systems, particularly in electronics and control system design. This growth coincided with advances in circuit board design and microelectronics, as well as control theory, with new developments in multivariable control, adaptive control, optimal control, H-infinity methods, fuzzy logic, predictive control, neural nets and so on. Of all the packages developed, probably MATLAB is the most well-known (Klee, 2007). Although developers had been able to write simulation models in conventional programming languages, most applications required specialist libraries for matrix operations and visualisation and forced developers to learn the intricacies of a specific programming language. By contrast, MATLAB supports vectors and matrices directly, relaxes and simplifies the programming syntax and incorporates various forms of plotting and visualisation. A limitation of MATLAB is that the user interface is a textual description of a system but this restriction was overcome with the introduction of Simulink, which enabled systems to be captured in a block diagram format and visualised directly, albeit with an underlying structure based on MATLAB.

The forgoing description of MATLAB is not intended as an advertisement for MATLAB. Quite simply, MATLAB is widely used in university teaching in science and engineering and in industry. In addition, its widespread use has attracted developers interested in more specialist applications and sets of 'toolboxes' have been added to MATLAB covering signal processing, statistics, fluid modelling and so on. This is not to say that MATLAB is without its problems. Firstly, it is a commercial product with implications of pricing and licensing. Secondly, it is interpretive; the system description is extracted and constructed in an internal format where each operation is executed by applying library functions, which has the impact of slowing the performance of MATLAB, which poses problems for real-time applications. The solution has been to provide further tools to optimise the code generation to provide speed improvements of orders of magnitude or, alternatively, to run the MATLAB code on custom hardware devices. The downside of both approaches is the additional effort required to integrate a MATLAB model with real-time systems.

Concurrent with the take-up of MATLAB, there has been a growing movement in open systems, which is characterised by the Linux operating system. The attraction of this initiative is that all the source code is available in the public domain and users are both allowed and encouraged to modify and improve the software. Most Linux implementations are free and this positive attitude towards open software has encouraged the growth in development of open applications for Linux (and to some degree Windows) platforms. The Octave package (Eaton, 2020) epitomises these developments. It provides a high degree of compatibility with MATLAB and is freely available for various platforms. Similarly, Scilab affords many of the facilities found in Simulink. This book is not intended as a forum to discuss the merits of open software but all the software covered in this book is freely available; and is therefore more closely aligned to Octave than MATLAB.

Example 2.7 shows two programs written in C and MATLAB. An array is initialised with 1000 random numbers and then copied to a second array and sorted in ascending order. Note that the method of ordering is deliberately simple and is merely used to illustrate memory reference accesses in C and MATLAB. Readers interested in sorting algorithms are advised to review the Quicksort algorithm (Hoare, 1962) or similar faster sorting methods. Running the software on a desktop PC with a 3.3 GHz processor, the C version runs in 6.31 s while the MATLAB version takes 34.47 s, a factor of over five times slower. Surprisingly, the Octave version takes 3691 s, which is over 500 times slower.

Example 2.7   Coding in C and MATLAB

```c
#include <stdio.h>
#include <stdlib.h>
#include <math.h>
#include <GLFW/glfw3.h>

int main(int argc, char *argv[])
{
    unsigned int i, j, k;
    double        t1;
    double        t2;
    int           a[1001];
    int           b[1001];

    if (!glfwInit())
        exit(EXIT_FAILURE);

    for (i=1; i<=1000; i+=1)
    {
        a[i] = rand();
    }

    t1 = glfwGetTime();

    for (k=1; k<=10000; k+=1)
    {
        for (i=1; i<=1000; i+=1)
        {
            b[i] = a[i];
        }

        for (i=1; i<=999; i+=1)
        {
            for (j=i+1; j<=1000; j+=1)
            {
                if (b[j] <= b[i])
                {
                    int t = b[j];
                    b[j] = b[i];
                    b[i] = t;
                }
            }
        }
    }

    t2 = glfwGetTime();

    printf("%f secs\n", (t2 - t1));
    fflush(stdout);

    return 0;
}
```

```matlab
a=zeros(1000, 1);
b=zeros(1000, 1);

for i = 1:1000
  a(i) = rand() * 10000;
end

tic()

for k = 1:10000
  for I = 1:1000
    b(i) = a(i);
  end

  for i = 1:1000
    for j = i+1:1000
      if b(j) < b(i)
        t = b(j);
        b(j) = b(i);
        b(i) = t;
      end
    end
  end
end

toc()
```

These figures are hardly surprising. MATLAB is designed as a tool to analyse system behaviour rather than providing an efficient implementation. Its strength is the ease of creating models, providing detailed analysis of system stability, robustness and performance and visualising the results in a wide range of formats. Tools such as MATLAB are primarily prototyping design tools. Results from such studies provide valuable insight into the software design of actual algorithms, which may be integrated with hardware, prior to installation of the code and hardware in a production system.

Nevertheless, there may be times when it is necessary to interface a MATLAB module with an external simulator and where the benefits of direct integration outweigh the effort needed to rewrite the software in another programming language. Of course, the requirements for real-time simulation covered in Section 1.3 are still valid. Any MATLAB code should be tested to ensure it meets the frame update rate under all conditions.

Both MATLAB and Octave provide software libraries in the form of a 'wrapper' between the MATLAB code and the real-time system code. For MATLAB, the wrapper contains function calls to access data from the simulator and return results to the simulator, where the actual computation is performed in MATLAB code. In the flight simulator outlined in Chapter 1, the real-time computing is partitioned across a distributed set of PCs. For one application, flight control functions were developed in MATLAB and integrated, as shown in Figure 2.28.

The MATLAB module M is developed to run in a MATLAB environment with the host computer connected via Ethernet to the network. Packets transmitted by the other computers can be read by the module M and results of the computation by the module M are broadcast in a packet to the other computers. An additional complication is that the simulation is required to run with the module M either active or inactive (disconnected from the network), requiring the module to join and leave the network without disrupting the protocol.

MATLAB MEX files provide an extension to MATLAB to access functions developed in C/C^{++}, which can be called from MATLAB. A custom MEX function was written in C to manage the transfer of UDP packets, synchronise with the simulator frame rate and comply with the network protocol. A MATLAB script Gnumex is available from http://gnumex.sourceforge.net, which automates configuring MATLAB to recognise a specific compiler, in this case the msys2 MinGw 64-bit C compiler (https://www.msys2.org). One particular problem with using a wrapper between different languages is the conventions used for parameter passing, particularly the number of parameters and their associated storage. The function *mexFunction()* provides a mechanism to set

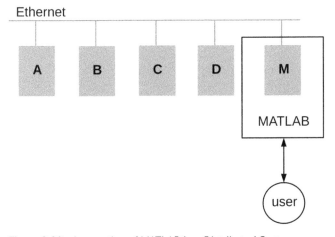

Figure 2.28 Integration of MATLAB in a Distributed System.

the number of inputs and outputs used by a function. The input and output parameters ***plhs** and ***prhs** are MATLAB *mxArrays*, which are used to populate the inputs and outputs of the functions. Note that MEX files can only pass data to and from a MATLAB script via these arrays.

Four functions were written to manage UDP transfers: **UDP_Start(port)** to establish a connection to a specific UDP port, **UDP_Stop()** to terminate a UDP connection, **UDP_Send()** to broadcast a UDP packet and **UDP_Recv()** to read an incoming UDP packet. In addition, two functions, **UDP_Data_Set()** and **UDP_Data_Get()**, were written to form data in a packet and to read data from a packet, respectively, in a specified format.

The function **mexFunction()** is shown in Example 2.8, where **nlhs** is the number of return arguments, ***plhs** is an array of pointers to the return arguments, **nrhs** is the number of input arguments and ***phrs** is an array of pointers to the input arguments. The switch calls the functions corresponding to the **prhs[0]** element. The functions to manage the UDP transfers are based on the functions used by the other computers in the network, described in Section 1.4.

Example 2.8   mexFunction

```c
void mexFunction (int nlhs, mxArray *plhs[], int nrhs, const mxArray *prhs[])
{
    int functionId;
    int rv = 0;
    unsigned short port = 0;
    double *inMatrix;          /* 1xN input matrix */
    mwSize ncols;              /* size of input matrix */
    double *outMatrix;         /* output matrix - holds data from simulator */

    functionId = (int)mxGetScalar(prhs[0]);

    switch(functionId)
    {
        case MEX_UDP_START:
            port = (unsigned short) mxGetScalar(prhs[1]);
            rv = UDP_Start(port);
            break;
        case MEX_UDP_STOP:
            UDP_Stop();
            break;
        case MEX_UDP_SEND:
            rv = UDP_Send();
            break;
        case MEX_UDP_RECV:
            rv = UDP_Recv();
            break;
        case MEX_UDP_DATASET:
            inMatrix = mxGetPr(prhs[1]);
            ncols = mxGetN(prhs[1]);
            UDP_Data_Set(inMatrix);
            break;
        case MEX_UDP_DATAGET:
            plhs[0]= mxCreateDoubleMatrix(1, (mwSize)MEX_OUTPUT_DATANUM,mxREAL);
            outMatrix = mxGetPr(plhs[0]);
            UDP_Data_Get(outMatrix);
            break;
    }
}
```

A template of the MATLAB program to implement simple flight control laws is shown in Example 2.9.

Example 2.9   An Example of a MATLAB Interface

```
fslink(fslink_open, fslink_defaultport);

while done == 0
    fslink(fslink_recv);
    DIN = fslink(fslink_dataget);
    matlabMode = DIN(fslink_Active);

    U     = DIN(fslink_U);          % U
    Udot  = DIN(fslink_Udot);       % Udot
    H     = DIN(fslink_Altitude);   % Altitude
    pitch = DIN(fslink_Pitch);      % Pitch
    alpha = DIN(fslink_Alpha);      % Alpha
    q     = DIN(fslink_Q);          % Q
    Vd    = DIN(fslink_Vd);         % Vd
    de    = DIN(fslink_Elevator);   % elevator
    da    = DIN(fslink_Aileron);    % aileron
    dr    = DIN(fslink_Rudder);     % rudder
    dt    = DIN(fslink_Throttle);   % throttle


    %
    % CUSTOM FLIGHT CONTROL CODE STARTS HERE[1]
    %

    DOUT = [da, de, dr, dt];
    fslink(fslink_dataset, DOUT);
    fslink(fslink_send);
end

fslink(fslink_close);
clear fslink;
```

For reasons of brevity, the code to implement the specific network protocol is omitted. The Boolean variable **done** indicates that the protocol is active. The program reads the simulator variables, including **de** (elevator), **da** (aileron), **dr** (rudder) and **dt** (throttle), and returns these four variables to the simulation. The definition of aircraft variables is given in a MATLAB file *simctrlconstants.m*. In the program, a flight control law can be added in the commented section, for example, to control altitude or airspeed.

Design tools such as MATLAB and Octave are particularly useful in the design of flight control laws. If the software is organised so that simulated inputs and outputs can be used to develop a control law, the software can be analysed in terms of stability, or frequency response, or accuracy and then, using the same software framework, can be integrated directly with the simulator software. Alternatively, once the prototype MATLAB is developed, it can provide a template for manual conversion to another programming language or automated translation tools can be used to generate much faster code than basic MATLAB. Nevertheless, the developer must take care to ensure the correctness of the interface and that the performance of the code meets the real-time constraints.

---

1 The user-specific code is inserted after acquiring the inputs.

References

Ahlberg J. H., Nilson E. N. and Walsh J. L. (1967), *The Theory of Splines and Their Applications*, Academic Press.

Benyon P. K. (1968), A Review of Numerical Methods for Simulation, *Simulation*, Vol. 11, No. 5, pp. 219–238.

Betts J. A. (1970), *Signal Processing, Modulation and Noise*, Hodder and Stoughton, London.

Cardullo F. M., Kaczmarck B. and Waycechowsky B. J. (1991), A Comparison of Several Numerical Integration Algorithms Employed in Real-time Simulation, *AIAA Flight Simulation Technologies Conference*, New Orleans.

Cooper D. (1990), *Oh My! Modula-2!*, W. W. Norton & Company, Inc.

Eaton J. W. (2020), *GNU Octave, Version 6.4.0*, https://octave.org/doc/v6.4.0.

Eurocontrol Experimental Centre (2010), *User Manual for the Base of Aircraft DATA (BADA) Revision 3.8*, *EEC Technical/Scientific Report No. 2010-003*.

Franklin G. F., Powell J. D. and Emami-Naeini A. (1994), *Feedback Control of Dynamic Systems*, third edition, Addison-Wesley.

Griffiths D. V. and Smith I. M. (1991), *Numerical Methods for Engineers*, Blackwell Scientific Publications, Oxford.

Hall G. W. and Huber R. W. (1970), *System Description and Performance Data for the USAF/CAL Variable Stability T-33 Airplane*, CAL report No. BM-2821-F-2, AFFDL-TR-70-71.

Higham N. J. (1996), *Accuracy and Stability of Numerical Algorithms*, Society for Industrial and Applied Mathematics, Philadelphia Soc.

Hoare C. A. R. (1962), Quicksort, *The Computer Journal*, Vol. 5, No. 1, pp. 10–16.

Howe R. M. (1991), A New Family of Real-time Predictor-corrector Integration Algorithms, *Simulation*, Vol. 57, No. 3, pp. 177–186.

IEEE (2019), IEEE Standard for Floating-Point Arithmetic, in *IEEE Std 754-2019 (Revision of IEEE 754-2008)*, doi: 10.1109/IEEESTD.2019.8766229.

Klee H. (2007), *Simulation of Dynamic Systems with MATLAB and Simulink*, CRC Press, London.

Knuth D. E. (1981), *Semi-numerical Algorithms*, second edition, *The Art of Computer Programming*, Addison-Wesley.

Maine R. E. and Iliff K. W. (1975), *A FORTRAN Program for Determining the Aircraft Stability and Control Derivatives from Flight Data*, NASA TN-D-7831.

Neville E. H. (1934), Iterative Interpolation, *The Journal of the Indian Mathematical Society*, No. 20, pp. 87–120.

Ortega J. M. and Poole W. G. (1981), *Numerical Methods for Differential Equations*, Pitman Publishing, Melbourne.

Park S. K. and Miller K. W. (1988), Random Number Generators: Good Ones are Hard to Find, *Communications of the ACM*, Vol. 32, No. 10, pp. 1192–1201.

Thomas D. B., Luk W., Leong P. H. W. and Villasenor J. D. (2007), Gaussian Random Number Generators, *ACM Computing Surveys*, Vol. 39, No. 4, article 11.

Van de Vegte J. (1990), *Feedback Control Systems*, Prentice-Hall, Englewood Cliffs, NJ.

3

Aircraft Equations of Motion

If anybody ever tells you anything about an aeroplane which is so bloody complicated you can't *understand it, take it from me: it's all balls.*

R. J. Mitchell

3.1 Atmospheric Model

3.1.1 The Atmosphere

Most flight simulation covers flight in the troposphere, which is the lowest layer of the atmosphere from the surface of the earth up to an altitude of approximately 18,000 m. In the troposphere, Boyle's law applies in terms of the relationships between pressure, temperature and density. As pressure varies with altitude, an altimeter, which is effectively a barometer, measures altitude. The temperature of the air entering an engine affects engine performance for both piston-engine aircraft and jet-engine aircraft. The density of air is used in the computation of lift, drag and other aerodynamic terms, affecting aircraft performance. Consequently, accurate modelling of the atmosphere is a critical component of flight simulation. The equations covered in this chapter are mostly based on the *US Standard Atmosphere* (Anon, 1976).

The International Standard Atmosphere (ISA), which corresponds to a temperature of 15°C at sea level, a pressure of 101,325 Pa and an air density of 1.225 kgm^{-3}, provides a set of reference values. These variables, which define weather conditions, vary over the surface of the earth and with altitude and time. In practice, temperature and pressure are normally set to default (ISA) or specific values in simulation, to define local weather conditions.

Gravity also affects the atmosphere, holding the troposphere in place over the surface of the earth and variations in gravity can be estimated with reasonable accuracy from published data. A detailed model of gravitational acceleration is covered in Chapter 5. In many simulators, gravity is defined as a constant in computations of the atmosphere, varying only in distance from the centre of the earth. Note that the geometric altitude of an aircraft is not the same as the geopotential altitude which takes account of the gravitational force on the air particles. The geopotential altitude h is given by

$$h = \frac{R_0 z}{R_0 + z} \tag{3.1}$$

Flight Simulation Software: Design, Development and Testing, First Edition. David Allerton.
© 2023 John Wiley & Sons Ltd. Published 2023 by John Wiley & Sons Ltd.
Companion Website: www.wiley.com/go/flightsimulationsoftware

where R_0 is the radius of the surface of the earth (6,356,766 m) and z is the geometric altitude. For example, at an altitude of 11,000 m, $h = 10,813$ m, a difference of 1.7% and, strictly, h should be used in any computations involving temperature, pressure or density rather than z.

Up to 11 km, the variation of temperature with altitude, which is known as the temperature lapse T_L, is a constant, enabling temperature to be computed from the geopotential altitude.

$$T = T_0 - T_L h \tag{3.2}$$

where T is temperature (K), T_0 is the temperature at sea level (K), T_L is the temperature lapse $0.0065°m^{-1}$ and h is the geopotential altitude. For example, if the temperature on the surface is 15°C (288.15K), the temperature is −4.8°C at 10,000 ft (3048 m). From 11 km to 20 km, the temperature is constant at 216.65K. The computation of temperature, which is relative to the temperature on the surface of the earth, normally assumes dry air.

Air pressure is the mass of air molecules per unit area in Pascals (Pa) and varies with location and time. In aviation, air pressure at a location is measured regularly and broadcast by air traffic services to enable altimeters to be set accurately for all aircraft operating within that area. In simulation, the setting of air pressure ensures that operation of altimeters is identical to aircraft operations and altimeters contain an adjustment to set the altitude indicated by an altimeter, as described in Section 5.3.3.

Below 11 km, the computation of air pressure is a function of temperature, given by

$$P = P_0 \left(\frac{T_0}{T}\right)^{\frac{GMR}{T_L}} \tag{3.3}$$

where P is pressure (Pa) and the constant GMR has the following value:

$$GMR = \frac{g_0 M_0}{R_s} \tag{3.4}$$

where g_0 is the acceleration of gravity at a latitude of 45°, P_0 is the pressure at g_0, M_0 is a constant for the mean molecular weight of air (28.9644 kg/kmol) and R_s is the gas constant (8314.32 Nm/kmol/K). From 11,000 m to 20,000 m, air pressure is given by

$$P = P_{11}e^{\frac{-GMR(h-11,000)}{T_{11}}} \tag{3.5}$$

where P_{11} and T_{11} are the pressure and temperature at 11,000 m (22,631.7 Pa and 216.65K, respectively).

Usually, the true altitude of an aircraft (the geometric height) is known in a simulator and the pressure altitude is computed to take account of the variation in air pressure with altitude. Pressure is also used in the computation of air density. Knowing pressure and temperature, air density can be computed from the gas law.

$$\rho = \frac{P}{T} \cdot \frac{M_0}{R_s} \tag{3.6}$$

Air density reduces with altitude and is used in the computation of aerodynamic forces and propulsive forces, including propellers.

The speed of sound, which depends on air temperature, is used in the computation of Mach number and is given by

$$a = \sqrt{\gamma T \frac{R_s}{M_0}} \tag{3.7}$$

where a is the speed of sound (ms^{-1}) and $\gamma = 1.4$. Note that the speed of sound reduces with temperature and therefore with increasing altitude. The computation of Mach number is as follows:

$$M = \frac{V}{a} \tag{3.8}$$

where M is the Mach number and V is the true airspeed of the aircraft (ms^{-1}).

The code fragment in Example 3.1 computes temperature, pressure and air density.

Example 3.1   Computation of temperature, pressure and air density

```
void Weather_WeatherModel(bool turbulence, float Pz, float U)
{
    const double Rs       =    8314.32;     /* Nm/(kmol K), gas constant */
    const double M0       =    28.9644;     /* kg/kmol, mean molecular weight of air */
    const double g0       =    9.80665;     /* m/s^2, accn of gravity at 45 deg lat. */
    const double r0       = 6356766.0;      /* m, Earth radius at g0 */
    const double P0       =   101325.0;     /* Pa, pressure at g0 */
    const double T0       =    288.15;      /* K, standard sea-level temperature */
    const double Gamma    =      1.40;      /* gas constant */
    const double GMR      = g0 * M0 / Rs;   /* a constant term */
    const double TLAPSE   = -0.0065;        /* deg/m, temperature lapse*/
    const double T11      = 216.65;         /* temperature at 11,000 m */
    const double P11      = P0 * 0.2233631; /* pressure at 11,000 m */
    double Z              = (double) -Pz;   /* geomteric altitude */
    double H              = r0 * Z / (r0 + Z); /* geopotential altitude */
    double T, P;

    if (H < 11000.0)
    {
        T = T0 + TLAPSE * H;
        P = P0 * pow(T0 / T, GMR / TLAPSE);
    }
    else
    {
        T = T11;
        P = P11 * exp(-GMR * (H - 11000.0) / T11);
    }

    Weather_Temperature    = (float) T;
    Weather_Pressure       = (float) P;
    Weather_Rho            = (float) ((P / T) * (M0 / Rs));
    Weather_DensityRatio   = Weather_Rho / 1.225;
    Weather_SpeedOfSound   = (float) sqrt((Gamma * Rs / M0) * T);
    Weather_PressureAltitude = (float) (P * 0.00029536); /* In Hg */
}
```

Flight data acquired in flight tests is often referenced to specific flight conditions (ambient temperatures and pressures) and care is needed with performance data to ensure that it is correctly referenced to the ambient conditions. A typical example in a simulation session is to set a high ground temperature, so that the ground roll on take-off will be extended considerably as a result of the increase in air temperature.

3.1.2 Wind

Often, wind is introduced in simulation to increase the pilot workload during an approach or in a navigation exercise. The magnitude and direction of wind vary over the surface of the earth and with altitude and time. For many simulators, it is not necessary to replicate actual wind flows (assuming the flow data exists). Nevertheless, wind data can be downloaded from the National Oceanic and Atmosphere Administration (NOAA) which provides historical wind data on a worldwide basis (https://www.ncdc.noaa.gov/cdo-web). One reason for considering the use of such data is that specific weather conditions can be replayed in a simulator to ensure that the weather conditions are identical over a series of tests. Quite often, a single wind setting is used in a simulator setting, for example, to provide cross-wind conditions during an approach or in a holding pattern. One important consideration with wind is that it should be introduced gradually to avoid inducing an abrupt or unnatural disturbance to the aircraft dynamics.

Note that a 'cube of weather', defined at 1 km intervals over an area 100 km × 100 km up to an altitude of 11 km, comprises over 10^5 data points of wind speed and direction. During simulation, as the aircraft flies through this region, the wind vector is interpolated to avoid abrupt changes at 1 km boundaries. The distribution of wind vectors over this volume should reflect the characteristics of weather patterns, which may justify the use of historical data.

3.1.3 Turbulence

Several models of clean air turbulence exist in the public domain. The most well-known version is the Dryden model (Beal, 1993). In addition, a turbulence model was also developed at the Royal Aircraft Establishment (RAE) Bedford (Tomlinson, 1975). These models are based on passing white noise through linear filters to match the power spectral density with actual measurements from turbulence. With linear turbulence models, three components (u, v, w) are added to the aircraft velocities to disturb the aircraft motion. If the gradients of gusts are extracted, the three additional rotational components can also be derived and added to the aircraft angular rates. The formulation of these models are too detailed to describe and the reader is referred to the extensive list of publications covering the modelling of turbulence.

A major problem with turbulence models is the number of parameters that need to be defined to produce representative turbulence. It is not simply a case of defining clean air as zero and very severe turbulence as 100%. In addition to the mean and variance of components, intermittency is a significant (and also random) characteristic. Furthermore, the aircraft dimensions, particularly the tail arm and wingspan, will also affect the turbulence components added to the aircraft dynamics.

Figure 3.1 shows the response of a Boeing 747-100 at 10,000 ft and 300 Kt to severe turbulence using the RAE model. Despite the autopilot being engaged, there is a change of altitude of approximately 70 ft and bank angle changes of approximately 5°. The severity of the turbulence is indicated by the change in indicated airspeed of almost 20 Kt.

Wake turbulence differs from clean air turbulence as it is generated by disturbance to the air flow caused by another aircraft and its engines. Wake turbulence and wake upset have become important

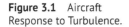

Figure 3.1 Aircraft Response to Turbulence.

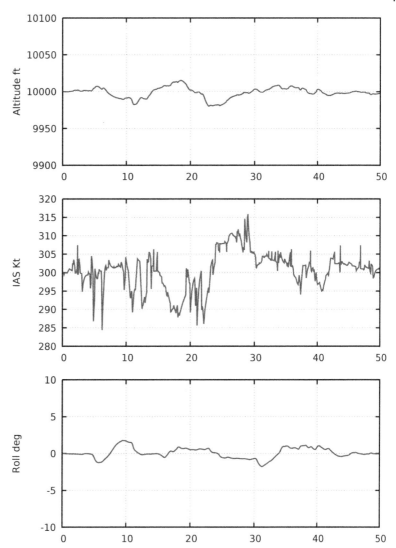

topics in flight training following a number of fatal accidents resulting from wake turbulence encounters. The modelling of wake turbulence introduces an interesting dilemma in flight simulation. The purpose of a wake turbulence model is to expose pilots to wake turbulence events during training and to appreciate (and practise) the appropriate corrective action. The dilemma is just how far the turbulence upset should accurately model wake turbulence; in particular to model the rotary components of the wake and its persistence. In many airline simulators, the wake is represented as an expanding spiral so that an aircraft may encounter a rotary flow at any point of the circular flow and at any point along the spiral, which represents the expanding and decaying turbulence.

An alternative approach, and one developed by a group at the University of Sheffield, is to use computational fluid dynamics (CFD) to model turbulent flows. Using LES methods (Spence et al., 2007), the flow field emanating from an aircraft and its engines can be generated, representing a physical volume of 3D flows over a given distance. The attraction of this method is that both the rotational components and longitudinal components (including Crow instabilities) are modelled with a high degree of accuracy. The drawback is the time (several days) needed to compute the flows for a single wake, at one specific altitude and airspeed, and the large volume of flow data generated. Currently, there is no standard for the requirement to provide turbulence in Level-D flight simulators and, moreover, there is very little data to validate turbulence models. The Sheffield group replicated the Airbus A300 accident in Belle Harbor, New York, in 2001 (Anon, 2004) and reproduced structural loading conditions leading to the departure of the tail, which accorded closely with flight data recorded in the accident.

3.1.4 Wind Shear

Wind shear conditions normally occur during thunderstorms, where there are significant down draughts from clouds near to an airfield. The vertical downdraughts fan out towards the ground and are disruptive to aircraft movements near an airfield and particularly dangerous for aircraft encountering wind shear on approach to a runway, where this category of wind shear is also referred to as a microburst, as shown in Figure 3.2.

Microburst events can have downdraughts up to a maximum of 60–70 ft/s over a region of 1000–10,000 ft diameter (Roberts and Wan, 1985). An aircraft on approach encounters a headwind increasing the indicated airspeed and the power is reduced to maintain the ILS reference speed. Passing through the headwind components, particularly if a tailwind component is encountered, the indicated airspeed drops rapidly (possibly accompanied by the downdraught increasing the rate of descent). At this point, the aircraft is low (below the glide path) and slow (below the approach reference speed) and near the ground. Flight crews are trained to detect and recover from such encounters.

In simulation, microburst events can be introduced during an approach. Mil. Spec. MIL-F-8785C (Moorhouse and Woodcock, 1982) gives the following equation for the magnitude of wind shear as a function of altitude above the ground:

$$u_w = W_{20} \frac{\ln\left(\dfrac{h}{z_0}\right)}{\ln\left(\dfrac{20}{z_0}\right)} \tag{3.9}$$

Figure 3.2 Microburst Conditions.

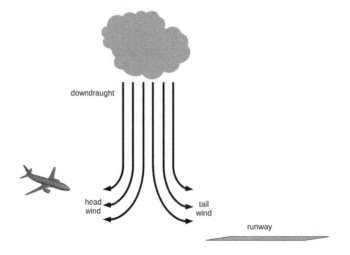

where u_w is the wind speed (Kt), W_{20} is the wind speed at 20 ft (the severity of the wind shear), h is the aircraft altitude (ft) and z_0 is 0.15 for approaches above 140 Kt and 2.0 otherwise.

3.2 Axes

Generally, the axes used in flight simulation are chosen for their suitability for the computations of the variables associated with the axes and their transformations. Although this convention appears to complicate computations, implementation of the transformations between these primary axes is generally straightforward. For example, applying the toe brakes to the wheels, the forces act on each wheel, which is connected to an oleo, which is connected to the airframe, which is moving over the surface of the earth. It is simplest to compute the braking force in the axes of the wheel, which is transformed to derive the resultant force on the airframe, which enables the aircraft motion to be derived.

3.2.1 Body Axes

The body axes are defined at the centre of gravity (C/G) of the aircraft, where u is positive forwards towards the nose of the aircraft, v is perpendicular to u and positive in the direction of the starboard wing, and w is perpendicular to u and v and positive downwards from the airframe, as shown in Figure 3.3.

Note that the location of the C/G varies with the distribution of mass of the aircraft. If the body axes are labelled x, y and z, as shown in Figure 3.3, then the forces computed are X, Y and Z, which enable the linear velocities u, v and w to be computed. Similarly, the moments L, M and N about the x, y and z axes, respectively, enable the body angular rates p, q and r to be computed.

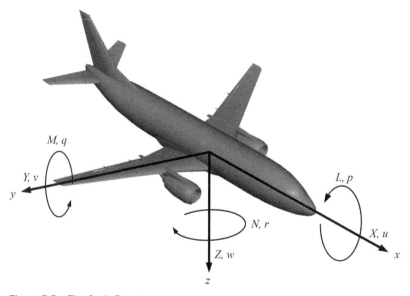

Figure 3.3   The Body Frame.

The angle of attack α and the angle of sideslip β and their derivatives are computed in the body frame and are defined as follows (McFarland, 1975):

$$\alpha = \tan^{-1}\left(\frac{w}{u}\right) \tag{3.10}$$

$$\beta = \tan^{-1}\left(\frac{v}{\sqrt{u^2 + w^2}}\right) \tag{3.11}$$

$$\dot{\alpha} = \frac{u\dot{w} - w\dot{u}}{u^2 + w^2} \tag{3.12}$$

$$\dot{\beta} = \frac{\dot{v}\left(u^2 + w^2\right) - v\left(u\dot{u} + w\dot{w}\right)}{\left(u^2 + v^2 + w^2\right)\sqrt{u^2 + w^2}} \tag{3.13}$$

where the terms for $\dot{\alpha}$ and $\dot{\beta}$ assume small angles ($\tan\alpha \approx \alpha$ and $\tan\beta \approx \beta$). In addition, u is assumed to be positive. Note that these terms are derived from the body frame variables u, v and w and their first derivatives (accelerations). The following code fragment computes α, $\beta, \dot{\alpha}$ and $\dot{\beta}$:

```
uw2              = Model_U * Model_U + Model_W * Model_W;
Vc2              = Model_U * Model_U + Model_V * Model_V + Model_W * Model_W;
Model_Alpha      = atan2(Model_W,  Model_U);
Model_Beta       = atan2(Model_V,  sqrt(uw2));
Model_AlphaDot = (Model_U * Model_WDot - Model_W * Model_UDot) / uw2;
Model_BetaDot  = (uw2 * Model_VDot - Model_V * (Model_U * Model_UDot +
                 Model_W * Model_WDot)) / (VC2 * sqrt(uw2));
```

Two functions are provided to compute $\tan^{-1}(y/x)$ in the *gcc* run-time library *math.h*, **atan(y/x)** and **atan2(y,x)**. Care is needed with the function **atan** to avoid a 'division by zero' exception and to ensure that the result is computed for the correct quadrant. With the function **atan2**, the exception is avoided and the result is correctly computed for each quadrant (in the range $-\pi$ to $+\pi$). Example 3.2 verifies the results for all nine combinations of x and y, taking the values $-1, 0$ and $+1$. An alternative solution to avoid division by zero is to set $\alpha = 0$ for $u < u_{min}$, where u_{min} is an airspeed well below flying conditions (e.g. 5 Kt).

Example 3.2   Verification of the Inverse Tan Function

```
#include <stdio.h>
#include <stdlib.h>
#include <math.h>

#define ONERAD (180.0 / M_PI)

int main ()
{
    int    i, j;
    double x, y;

    for (i=-1; i<=1; i+=1)
    {
        x = (double) i;
```

```
      for (j=-1; j<=1; j+=1)
      {
            y = (double) j;
            printf("atan2(%d, %d)=%f\n", i, j, atan2(x, y) * ONERAD);
      }
   }
   return 0;

}
```

As the engines and undercarriage assemblies are attached to the airframe, these forces and moments are most conveniently computed in the body frame, and derivations are described in Sections 3.4 and 3.5, respectively.

3.2.2 Stability Axes

The forces and moments associated with the aerodynamics of an aircraft are usually referenced to a frame rotated about the Y axis by the angle of attack α and about the Z axis by the angle of sideslip β. These axes are commonly known as the stability axes or wind axes.

Often, data acquired from wind-tunnel data or data that is a function of the angle of attack or angle of sideslip is defined and measured with respect to the stability axes. Similarly, aerodynamic data provided in data packages is normally referenced to stability axes. It is important to understand the difference between the attitude of an aircraft (its orientation) and the flight path (its trajectory in space). In particular an aircraft does not necessarily fly in the direction it is pointing. In the vertical plane, the flight path is given by an offset from the longitudinal body axis of the aircraft defined by the angle of attack α, as shown in Figure 3.4. Similarly, in the horizontal plane, for an aircraft with an angle of sideslip β, the flight path is offset from the longitudinal body axis as shown in Figure 3.5.

3.2.3 Local Frame

Thus far, the use of body axes and stability axes have provided reference frames for forces, accelerations and velocities, both linear and rotational. What they do not provide is the position and orientation (known as attitude) of the simulated aircraft in space. A commonly used local frame, which defines the aircraft position and attitude, is the Euler frame (also known as a 'flat-earth' frame). This frame provides a right-handed orthogonal plane parallel to the tangent of the surface of the earth, with one axis pointing to true north, a second axis pointing to the east and a third axis pointing vertically downwards, to the centre of the earth. The usefulness of the Euler frame is that over a range of 100 nm or so it is a reasonable assumption that the earth is more or less flat (errors

Figure 3.4 The Angle of Attack.

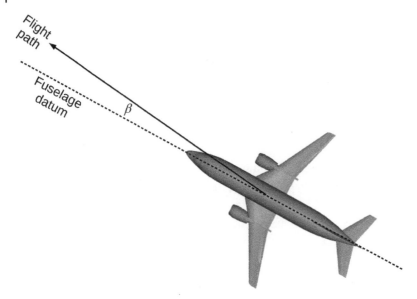

Figure 3.5 The Angle of Sideslip.

arising from curvature of the surface of the earth can be ignored) and an origin can be defined so that the position of an aircraft is given by the offset from that origin and an altitude above the earth. In applications in engineering flight simulation involving exercises in the vicinity of an airport (or set of airports), the Euler frame is adequate. In practice, an aircraft flying level is following a curved path and, over long distances, the rotation of the earth needs to be included in the computation of position. These considerations are covered in Section 3.5.

The aircraft attitude is defined by the angles of pitch, roll and yaw. Pitch (θ) is the angle of the longitudinal axis of the aircraft, with respect to the distant horizon, in the range $-\pi/2$ to $+\pi/2$. When the aircraft is flying vertically upwards, the pitch angle is +90° and when the aircraft is flying vertically downwards, the pitch angle is −90°. Roll (ϕ) is the angle of bank, in the range $-\pi$ to $+\pi$, where the roll angle is 0° with wings level, 45° if the aircraft is banked 45° to the horizon and −180° if the aircraft is inverted. Yaw (ψ) is the aircraft heading, in the range $-\pi$ to $+\pi$, where yaw angles are 0°, 90°, 180° and −90° if the aircraft is pointing to true north, east, south and west, respectively. Note that for both roll and yaw angles, +180° and −180° are equivalent. In the body frame, the linear and angular velocities are u, v, w, p, q and r. In the Euler frame, the corresponding variables are V_N, V_E and V_d, $\dot{\theta}$, $\dot{\phi}$ and $\dot{\psi}$, respectively, where the suffixes N, E and d denote north, east and down, respectively.

The following equations are used to transform angular rates in the body frame to Euler frame:

$$\dot{\phi} = p + \left(q \sin\phi + r \cos\phi \right) \tan\theta \tag{3.14}$$

$$\dot{\theta} = q \cos\phi - r \sin\phi \tag{3.15}$$

$$\dot{\psi} = \left(q \sin\phi + r \cos\phi \right) \sec\theta \tag{3.16}$$

or, in matrix form

$$\begin{bmatrix} \dot{\phi} \\ \dot{\theta} \\ \dot{\psi} \end{bmatrix} = \begin{bmatrix} 1 & \sin\phi\tan\theta & \cos\phi\tan\theta \\ 0 & \cos\phi & -\sin\phi \\ 0 & \sin\phi\sec\theta & \cos\phi\sec\theta \end{bmatrix} \begin{bmatrix} p \\ q \\ r \end{bmatrix} \tag{3.17}$$

Similarly, to transform angular rates from the Euler frame to the body frame:

$$p = \dot{\phi} - \dot{\psi} \sin \theta \tag{3.18}$$

$$q = \dot{\theta} \cos\theta \cos\phi + \dot{\psi} \cos\theta \sin\phi \tag{3.19}$$

$$r = \dot{\psi} \cos\theta \cos\phi - \dot{\theta} \sin\phi \tag{3.20}$$

or, in matrix form

$$\begin{bmatrix} p \\ q \\ r \end{bmatrix} = \begin{bmatrix} 1 & 0 & -\sin\phi \\ 0 & \cos\phi & \cos\theta\sin \\ 0 & -\sin\phi & \cos\theta\cos\phi \end{bmatrix} \begin{bmatrix} \dot{\phi} \\ \dot{\theta} \\ \dot{\psi} \end{bmatrix} \tag{3.21}$$

Although these equations provide a simple mechanism to transform between body rates and Euler frame rates, there is an obvious problem that these equations include the terms $\tan\theta$ and $\sec\theta$. In cases where θ approaches $\pm 90°$, these values are infinite and will probably cause a run-time exception. The use of quaternions, described in Section 3.3, avoids this situation completely. A simpler method is to limit the pitch angle, for example, $\tan 89.9° = 572.957$, and this value can be used in the equations until the absolute value of pitch is less than 89.9°. This condition only applies if an aircraft is flying vertically (upwards or downwards), and in conventional flight will only occur for a relatively short duration and it is unlikely that this simplification would be noticed in a flight simulator, particularly as it only applies during aerobatic manoeuvres such as a loop. Note one further condition that is not immediately obvious; if an aircraft commences a loop, then, as it pitches through $+90°$, the heading changes by 180° as the aircraft is now pointing in the opposite direction and the bank angle also changes by 180° as the aircraft is now inverted. A similar reversal occurs as the aircraft pitches through $-90°$.

To transform body frame velocities u, v and w to Euler frame velocities V_N, V_E and V_d:

$$V_N = u\cos\theta\cos\psi + v\left(\sin\phi\sin\theta\cos\psi - \cos\phi\sin\psi\right) + w\left(\cos\phi\sin\theta\cos\psi + \sin\phi\sin\psi\right) \tag{3.22}$$

$$V_E = u\cos\theta\sin\psi + v\left(\sin\phi\sin\theta\sin\psi + \cos\phi\cos\psi\right) + w\left(\cos\phi\sin\theta\sin\psi - \sin\phi\cos\psi\right) \tag{3.23}$$

$$V_d = -u\sin\theta + v\sin\phi\cos\theta + w\cos\phi\cos\theta \tag{3.24}$$

or, in matrix form

$$\begin{bmatrix} V_N \\ V_E \\ V_d \end{bmatrix} = \begin{bmatrix} \cos\theta\cos\psi & \sin\phi\sin\theta\cos\psi - \cos\phi\sin\psi & \cos\phi\sin\theta\cos\psi + \sin\phi\sin\psi \\ \cos\theta\sin\psi & \sin\phi\sin\theta\sin\psi + \cos\phi\cos\psi & \cos\phi\sin\theta\sin\psi - \sin\phi\cos\psi \\ -\sin\theta & \sin\phi\cos\theta & \cos\phi\cos\theta \end{bmatrix} \begin{bmatrix} u \\ v \\ w \end{bmatrix} \tag{3.25}$$

A matrix transforming vectors between two frames, as shown, is referred to as a direction cosine matrix (DCM). Note that the matrix elements are only functions of the three Euler angles and that there are no singularities. A DCM can be used to transform forces, accelerations, velocities and position between frames. Although the reverse transformation implies matrix inversion, in the case of a DCM, this is implemented by using the transpose of the DCM. For example, to transform a vector from the body frame to the Euler frame:

$$\begin{bmatrix} u \\ v \\ w \end{bmatrix} = \begin{bmatrix} \cos\theta\cos\psi & \cos\theta\sin\psi & -\sin\theta \\ \sin\phi\sin\theta\cos\psi - \cos\phi\sin\psi & \sin\phi\sin\theta\sin\psi + \cos\phi\cos\psi & \sin\phi\cos\theta \\ \cos\phi\sin\theta\cos\psi + \sin\phi\sin\psi & \cos\phi\sin\theta\sin\psi - \sin\phi\cos\psi & \cos\phi\cos\theta \end{bmatrix} \begin{bmatrix} V_N \\ V_E \\ V_d \end{bmatrix} \tag{3.26}$$

The aircraft position (relative to an initial position) is computed by integrating V_N, V_E and V_d and the aircraft attitude is computed by integrating $\dot{\theta}$, $\dot{\phi}$ *and* $\dot{\psi}$ and constraining these angles to the ranges from $-\pi/2$ to $+\pi/2$, $-\pi$ to $+\pi$ and $-\pi$ to $+\pi$, respectively. In many applications, the use of the Euler frame is adequate to model aircraft motion. Indeed, navigation over several hundred nautical miles is possible if the origin of the Euler frame is regularly updated, so that any errors are relative to the current origin. For example, a large region of terrain can be represented as a set of tiles, where the local origin is reset as the aircraft moves from tile to tile, although methods of this form are not needed if the full equations, covered in Section 3.5, are used.

3.2.4 Earth-centred Earth-fixed Frame

The origin of the earth-centred earth-fixed (ECEF) frame is the centre of the earth. This frame ignores the rotation of the earth and generally, when used in flight simulation, the earth is assumed to be a perfect sphere and gravity is assumed to be constant. These assumptions reflect the common use of flight simulation in both flight training and engineering flight simulation. For the majority of applications, errors resulting from variation in gravity or the shape or rotation of the earth are negligible. That is not to say that these effects should always be ignored and these aspects are covered in Section 3.5 and Chapter 5.

In the ECEF frame, meridians of longitude are defined in the range $\pm\pi$ with the $0°$ meridian of longitude passing through Greenwich in London. Latitude is defined in the range $\pm\pi/2$ with latitude positive in the northern hemisphere. The latitude of the equatorial plane is $0°$ with latitudes of $90°$ and $-90°$ for the north and south poles, respectively.

Latitude and longitude are computed as follows:

$$\dot{\lambda} = \frac{V_n}{R+h} \tag{3.27}$$

$$\lambda = \int \dot{\lambda}\, dt \tag{3.28}$$

$$\dot{\mu} = \frac{V_E \sec \lambda}{R+h} \tag{3.29}$$

$$\mu = \int \dot{\mu}\, dt \tag{3.30}$$

where R is the radius of the earth and h is the altitude above mean sea level. Note that latitude and longitude must be initialised and there is a potential singularity flying over the poles ($\lambda = \pm 90°$). Algorithms to avoid the problem of this singularity are outlined in Chapter 5, although, in practice, few simulator exercises involve flying over the poles.

3.2.5 Rotating Earth Frame

Two additional forces are omitted in flat-earth models. Firstly, if an aircraft is flying at a constant altitude, its motion is parallel to the surface of the earth directly below the aircraft. In an inertial frame, after a short time, this motion is no longer parallel to the surface of the earth and the altitude must therefore be changing. The centripetal force resulting from the velocity of the aircraft should be included in the equations of motion. Secondly, as an aircraft flies, the earth is slowly rotating. An aircraft flying along a meridian of longitude, for example, will be displaced from that meridian over time, if the terms for rotation of the earth are included. Strictly, this motion does not arise from a force, it results from the rotation of two frames with respect to each other and is an apparent rotation seen in the observer's frame, which is known as the Coriolis effect.

An example of Coriolis is the motion of the moon, which rotates about the earth by 360° every 27.322 days or 655.728 hours. In an hour, the moon rotates about the earth by only 0.549°. During that hour, the earth will rotate by 15° (360° per day) about its axis and the moon (and distant stars) will appear to have travelled some 15° across the night sky. The apparent motion of the moon, as seen from the earth, results not from the motion of the moon, but from the rotation of the observer's frame, that is, the rotating earth.

Coriolis acceleration occurs between any two frames which rotate with respect to each other and, for the earth frame, is given by $a = 2\Omega \times V$ where a is acceleration, Ω is the earth's rotation rate and V is the velocity of the observer's platform, where these variables are vectors and the $\times$ symbol denotes the vector cross operator. The centripetal term implies a change in attitude of the platform to keep it continually parallel to the surface of the earth. Both Coriolis acceleration and centripetal acceleration are covered in Section 3.5 in the equations of motion and in Section 5.8 on inertial navigation systems. It is important to bear in mind that, in comparison with gravitational, aerodynamic and propulsive accelerations, these are relatively small accelerations and mostly apply to missile guidance and long-range aircraft navigation and, consequently, are omitted from most simulations.

3.3 Quaternions

In Section 3.2, potential singularities were observed in the transformations between axes. Rather than using the three rotation variables pitch, roll and yaw, as defined for direction cosines, a four-parameter method is used, which avoids any singularities, while also providing a means of deriving the DCM from the quaternions and vice-versa. Quaternions are commonly used in military flight simulation in the equations of motion for aerobatic manoeuvres. The four quaternion parameters are defined as follows:

$$e_0 = \cos\frac{\psi}{2}\cos\frac{\theta}{2}\cos\frac{\phi}{2} + \sin\frac{\psi}{2}\sin\frac{\theta}{2}\sin\frac{\phi}{2} \tag{3.31}$$

$$e_1 = \cos\frac{\psi}{2}\cos\frac{\theta}{2}\sin\frac{\phi}{2} - \sin\frac{\psi}{2}\sin\frac{\theta}{2}\cos\frac{\phi}{2} \tag{3.32}$$

$$e_2 = \cos\frac{\psi}{2}\sin\frac{\theta}{2}\cos\frac{\phi}{2} + \sin\frac{\psi}{2}\cos\frac{\theta}{2}\sin\frac{\phi}{2} \tag{3.33}$$

$$e_3 = \sin\frac{\psi}{2}\cos\frac{\theta}{2}\cos\frac{\phi}{2} - \cos\frac{\psi}{2}\sin\frac{\theta}{2}\sin\frac{\phi}{2} \tag{3.34}$$

The quaternions are initialised in the following procedure `SetQuaternions`:

```
void Model_SetQuaternions()
{
    float p, r, y;
    float sp, cp, sr, cr, sy, cy;

    p       = Model_Pitch * 0.5;
    r       = Model_Roll * 0.5;
    y       = Model_Yaw * 0.5;
    sp      = sin(p);
```

```
cp       = cos(p);
sr       = sin(r);
cr       = cos(r);
sy       = sin(y);
cy       = cos(y);
Model_e0 = cr * cp * cy + sr * sp * sy;
Model_e1 = sr * cp * cy - cr * sp * sy;
Model_e2 = cr * sp * cy + sr * cp * sy;
Model_e3 = cr * cp * sy - sr * sp * cy;
}
```

The quaternion parameters are initialised from the Euler angles at the start of the simulation or if the aircraft is repositioned or if the aircraft state is modified, other than by the equations of motion. In the four-parameter method, the quaternion parameters are continually adjusted to satisfy the following constraint:

$$e_0^2 + e_1^2 + e_2^2 + e_3^2 = 1 \tag{3.35}$$

The quaternion rates are derived from the body rates p, q and r and the current four quaternions as follows:

$$\dot{e}_0 = -\frac{1}{2}\left(e_1 p + e_2 q + e_3 r\right) \tag{3.36}$$

$$\dot{e}_1 = \frac{1}{2}\left(e_0 p + e_2 r - e_3 q\right) \tag{3.37}$$

$$\dot{e}_2 = \frac{1}{2}\left(e_0 q + e_3 p - e_1 r\right) \tag{3.38}$$

$$\dot{e}_3 = \frac{1}{2}\left(e_0 r + e_1 q - e_2 p\right) \tag{3.39}$$

These four quaternion rates are integrated to derive the quaternions:

$$e_0 = \int \dot{e}_0 \, dt \tag{3.40}$$

$$e_1 = \int \dot{e}_1 \, dt \tag{3.41}$$

$$e_2 = \int \dot{e}_2 \, dt \tag{3.42}$$

$$e_3 = \int \dot{e}_3 \, dt \tag{3.43}$$

One further attraction of this method is that the DCM can be derived directly from the four quaternions.

$$\begin{bmatrix} a_{11} & a_{12} & a_{13} \\ a_{21} & a_{22} & a_{23} \\ a_{31} & a_{32} & a_{33} \end{bmatrix} = \begin{bmatrix} e_0^2 + e_1^2 - e_2^2 - e_3^2 & 2(e_1 e_2 - e_0 e3) & 2(e_0 e_2 + e_1 e_3) \\ 2(e_1 e_2 + e_0 e_3) & e_0^2 - e_1^2 + e_2^2 - e_3^2 & 2(e_2 e_3 - e_0 e_1) \\ 2(e_1 e_3 - e_0 e_2) & 2(e_2 e_3 + e_0 e_1) & e_0^2 - e_1^2 - e_2^2 + e_3^2 \end{bmatrix} \tag{3.44}$$

The nine scalar values of the DCM are set as follows:

```
void Model_SetDCM()
{
    float e00, e11, e22, e33;

    e00       = Model_e0 * Model_e0;
    e11       = Model_e1 * Model_e1;
    e22       = Model_e2 * Model_e2;
    e33       = Model_e3 * Model_e3;
    Model_A11 = e00 + e11 - e22 - e33;
    Model_A12 = 2.0 * (Model_e1 * Model_e2 - Model_e0 * Model_e3);
    Model_A13 = 2.0 * (Model_e0 * Model_e2 + Model_e1 * Model_e3);
    Model_A21 = 2.0 * (Model_e1 * Model_e2 + Model_e0 * Model_e3);
    Model_A22 = e00 - e11 + e22 - e33;
    Model_A23 = 2.0 * (Model_e2 * Model_e3 - Model_e0 * Model_e1);
    Model_A31 = 2.0 * (Model_e1 * Model_e3 - Model_e0 * Model_e2);
    Model_A32 = 2.0 * (Model_e2 * Model_e3 + Model_e0 * Model_e1);
    Model_A33 = e00 - e11 - e22 + e33;
}
```

Similarly, the Euler angles can also be derived from the DCM, where

$$\theta = \sin^{-1}\left(-a_{31}\right) \tag{3.45}$$

$$\phi = \tan^{-1}\left(\frac{a_{32}}{a_{33}}\right) \tag{3.46}$$

$$\psi = \tan^{-1}\left(\frac{a_{21}}{a_{11}}\right) \tag{3.47}$$

The DCM is formed from the quaternions and the computation of the three Euler angles is as follows:

```
Model_Pitch = asin(-Model_A31);
Model_Roll  = atan2(Model_A32, Model_A33);
Model_Yaw   = atan2(Model_A21, Model_A11);
```

It is necessary to ensure that the constraint in equation 3.35 is applied at all times, which is achieved by introducing an extra term $k\lambda$ in equations 3.36–3.39, where $kh \leq 1$ and h is the simulation step length. This method is known as the *method of algebraic constraint*, where λ is given by

$$\lambda = 1 - \left(e_0^2 + e_1^2 + e_2^2 + e_3^2\right) \tag{3.48}$$

λ is adjusted towards zero during each frame and added to the computation of the quaternion rates as follows:

$$e_0' = -\frac{1}{2}\left(e_1 p + e_2 q + e_3 r\right) + k\lambda e_0 \tag{3.49}$$

$$e_1' = \frac{1}{2}\left(e_0 p + e_2 r - e_3 q\right) + k\lambda e_1 \tag{3.50}$$

$$e'_2 = \frac{1}{2}\left(e_0 q + e_3 p - e_1 r\right) + k\lambda e_2 \tag{3.51}$$

$$e'_3 = \frac{1}{2}\left(e_0 r + e_1 q - e_2 p\right) + k\lambda e_3 \tag{3.52}$$

The code to compute λ and the quaternion parameters is as follows:

```
Lambda = 1.0 - (Model_e0 * Model_e0 + Model_e1 * Model_e1 +
                Model_e2 * Model_e2 + Model_e3 * Model_e3);
e0Dot = -0.5 * (Model_e1 * Model_P + Model_e2 * Model_Q + Model_e3 * Model_R) +
         Lambda * Model_e0;
e1Dot = 0.5 * (Model_e0 * Model_P + Model_e2 * Model_R - Model_e3 * Model_Q) +
         Lambda * Model_e1;
e2Dot = 0.5 * (Model_e0 * Model_Q + Model_e3 * Model_P - Model_e1 * Model_R) +
         Lambda * Model_e2;
e3Dot = 0.5 * (Model_e0 * Model_R + Model_e1 * Model_Q - Model_e2 * Model_P) +
         Lambda * Model_e3;
Model_e0 = Maths_Integrate(Model_e0, e0Dot);
Model_e1 = Maths_Integrate(Model_e1, e1Dot);
Model_e2 = Maths_Integrate(Model_e2, e2Dot);
Model_e3 = Maths_Integrate(Model_e3, e3Dot);
```

The quaternions and their rates are initialised whenever the simulation is reset or initialised. In addition, the convergence of λ must ensure that the numeric constraint is valid at all times. The advantage in simulation is that the use of quaternions avoids completely the problems associated with the singularities in the Euler equations, at the expense of a small increase in the computation of the equations of motion.

3.4 Aerodynamics

The implementation of an aerodynamic model is based on data estimated for aerodynamic variables, or equations from plots of variables derived from flight tests or wind-tunnel tests, or a data pack provided by a manufacturer. This data should cover the flight envelope from 0 Kt to V_{ne} (the never exceed speed) and from ground level to the maximum operational altitude. The aerodynamic data includes the variables that contribute to the aerodynamic forces and moments, which can be functions of the angles of attack and sideslip, Mach number, flap, slat and speed-brake positions, undercarriage position and propulsive forces. For example, Figure 3.6 shows a plot for the rate of change of rolling moment with sideslip $C_{l\beta}$ for a light aircraft (Fink et al., 1970). Four plots are provided for four thrust settings and each plot contains three different flaps settings.

In the computation of aerodynamic derivatives, either the data is captured from plots and computed using spline fitting, applying pre-computed tables, as a function of one or two variables, or the variable is computed directly from an approximation. For example, the coefficient of drag C_D for the Boeing 747-100 is shown in the following fragment of code:

```
float Aero_AeroCd()
{
    float Cd;
    Cd = 0.061 - 0.12 * Model_AlphaWing +
         3.0 * Model_AlphaWing * Model_AlphaWing;
```

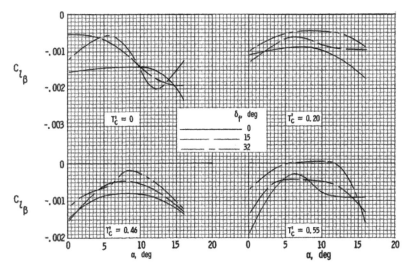

Figure 3.6 Aerodynamic Data for $C_{l\beta}$.

```
if (Model_Flaps < 0.667)
{
    Cd = Cd - (0.667 - Model_Flaps) * 0.05;
}
else
{
    Cd = Cd + (Model_Flaps - 0.667) * 0.288;
}
return Cd + (0.028 - Model_Flaps * 0.021) * Model_Gear;
}
```

where C_D is a function of the angle of attack of the wing, the flap setting and the undercarriage.

There is one further consideration with an aerodynamic model, that is, the time to compute the values of aerodynamic variables during each frame. Table lookup methods requiring interpolation and extrapolation can involve searches which may be proportional to n, the number of tabulated points, or possibly $\log(n)$, and considerable care is needed to ensure that the computations are completed within the simulation frame time, under all conditions.

3.4.1 Performance and Handling

The aerodynamic forces and moments contribute to both the aircraft performance, for example, cruise speed, rates of climb, landing speed, range and endurance, and also to the handling perceived by a pilot. In simple terms, the handling is the ease with which the pilot can manoeuvre the aircraft, which depends on the stability and responsiveness of the aircraft. Ideally, an aircraft should be stable so that if it is disturbed, it will return quickly to its previous stable state. At the same time, an aircraft should also respond quickly to pilot inputs, enabling the pilot to control the attitude of the aircraft. These two requirements are, to a large degree, contradictory, and for the aircraft designer a compromise is inevitably required between stability and response. Every aircraft is defined by a unique set of aerodynamic coefficients, which define the performance and handling of the aircraft.

For a specific set of aerodynamic coefficients of an aircraft, there are four important responses (in effect, signatures), which define a set of equations and enable the response and stability of the aircraft as well as the handling qualities to be determined:

1) The *short period phugoid* – the response to a disturbance in pitch, which should damp out in a few seconds.
2) The *long period phugoid* – also a response to a disturbance in pitch and although it may take several hundred seconds to damp out completely, it does not usually affect the ability of the pilot to control the aircraft, even though the effect is noticeable in most aircraft.
3) The *Dutch roll* – the response to a lateral disturbance causing the aircraft heading to oscillate, which may take several minutes to damp out. A lateral disturbance which causes sideslip can lead to the aircraft yawing, which in turn produces a rolling moment leading to a yawing moment and so on. Although the effect is undesirable (and mostly attenuated by the yaw damper of a transport aircraft), aircraft are designed to damp out Dutch roll, while at the same time giving the pilot sufficient authority in terms of rudder and aileron input to manoeuvre the aircraft, where responsiveness (e.g. during landing) is essential.
4) The *spiral mode* – the degree to which a rolling or turning manoeuvre can increase the bank angle of the aircraft as a result of the manoeuvre. In other words, having applied bank angle, the lateral dynamics of the aircraft are such that the rate of turn and bank angle may increase without further pilot input.

At a specific speed and altitude, the characteristics of the short and long period phugoids and the Dutch roll can be predicted from the transfer functions for longitudinal or lateral dynamics. Alternatively, the pitch response to elevator input or yaw response to rudder input can be plotted, as shown in Figure 3.7 for a Boeing 747-100 aircraft. Both techniques enable the natural frequency and damping of these modes to be identified.

In Figure 3.7(a), the response to an elevator doublet is clearly shown. The pitch rate is established in approximately 2 s. For a fighter aircraft, a much faster response would be expected. In Figure 3.7(b), the elevator pulse input causes a pitch up, causing the aircraft to climb. As the airspeed reduces, the pitching moment and lift also reduce to a point where the aircraft starts to descend. This motion repeats approximately every 50 s. In Figure 3.7(c), the aircraft responds to a rudder pulse input, resulting in an oscillation in heading (yaw) and sideslip with a period of approximately 18 s. With the yaw damper disengaged, this motion has mostly damped out in 50 s. There are two points to note from these examples. Firstly, the tests were applied with steady-state initial conditions, that is to say the aircraft was trimmed at the start of each test. Secondly, these results are specific to the entry speed, the aircraft mass, the aircraft configuration (particularly the flap setting and undercarriage position) and the applied input. These modes of motion are described in more detail in Chapter 10.

3.4.2 Coefficient of Lift

Lift results from the flow of air over the wing aerofoil and depends on the airspeed of the aircraft, the density of air and the geometry of the wing. Lift L is typically given by

$$L = \frac{1}{2}\rho V^2 s C_L \tag{3.53}$$

where ρ is air density, which varies with altitude, V is the true airspeed, s is the wing area and C_L is the coefficient of lift. C_L may also contain components of lift generated by the fuselage and tail. For light aircraft, C_L is a function of α and, for large transport aircraft and military aircraft, is also

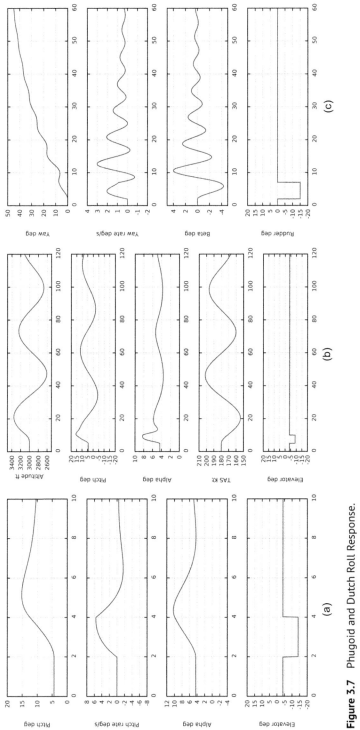

Figure 3.7 Phugoid and Dutch Roll Response.

a function of Mach number. In addition, C_L also varies with flap position and possibly the undercarriage position. Figure 3.8 shows the C_L curve for a light aircraft, based on data from full-scale wind-tunnel tests.

The plots are shown for three flap settings (0°, 15° and 32°). Note that this data is limited to α in the range from −5° to +20° and that, above approximately 8°, the curve exhibits significant nonlinearity. The stall angle occurs at the maximum value of C_L although an aircraft stall warning typically activates 2–3° below this value. The actual stall angle also varies with flap setting. Although this range is typical of light aircraft operations, values outside this range can be extrapolated, which is limited in accuracy to a few degrees. C_L is a particularly important parameter in flight simulation; it affects aircraft performance in terms of cruise speed and climb rates and the stall conditions. Note also that the stall is simply a function of the angle of attack, whereas the stall is often quoted as an airspeed in aircraft operating handbooks. The stall speed is the lowest speed to maintain level flight; nevertheless, an aircraft can still stall at much higher speeds. A stall warning sensor is based on the angle of attack, although few civil aircraft are fitted with an angle-of-attack meter and therefore stall conditions are referenced to stalling speeds.

3.4.3 Coefficient of Drag

Drag is the resistance to motion caused by the movement of the aircraft through the air. Both lift and drag increase with the angle of attack. These two variables interact in the sense that to fly fast,

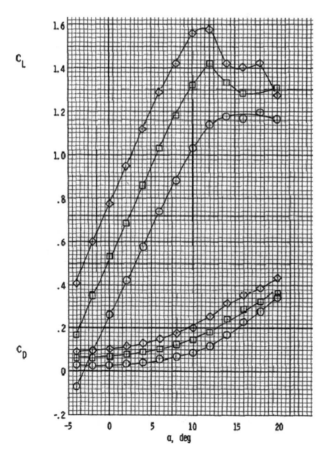

Figure 3.8 Coefficients of Lift and Drag for a Light Aircraft.

α must be small to reduce drag but must also generate sufficient lift. For any aircraft, there is a unique value of α for level flight at a specific airspeed. Although aircraft designers try to minimise aircraft drag, they are constrained by structural integrity so that, for all aircraft, there is a compromise between lift and drag and the lift over drag curve for an aircraft provides insight into maximum airspeed, maximum range and endurance. Aircraft drag force D is given by

$$D = \frac{1}{2}\rho V^2 s C_D \tag{3.54}$$

where C_D is the coefficient of drag. A typical plot of C_D versus α is shown in the lower traces of Figure 3.8. Two important points are evident from this graph. Firstly, C_D increases as a function of α^2 rather than linearly. Secondly, C_D also varies with the flap setting. As in the case of C_L, values of C_D can be estimated, or derived from wind-tunnel tests or are provided in a data pack. C_D depends not only on the wing drag, but also drag from the fuselage, flaps, slats, tail and undercarriage. C_D also varies significantly with Mach number.

C_D is normally defined as a combination of parasitic drag and induced drag. The parasitic drag coefficient is usually a constant, whereas the induced drag is a function of C_L^2, and C_D is usually defined as a *drag polar*, where

$$C_D = C_{DO} + C_{D1} C_L^2 \tag{3.55}$$

Unlike C_L which reduces post-stall, C_D continues to increase, so that airspeed reduces very quickly, which in turn causes the lift to reduce. Again, values outside the range given in Figure 3.8 can be determined by extrapolation and, to a large degree, drag is symmetric as the parasitic coefficient dominates at high positive and negative values of α. A number of other terms also contribute significantly to aircraft drag:

- Spoilers and speed brakes are activated specifically to decelerate an aircraft (particularly during descent) by increasing aircraft drag.
- The undercarriage creates considerable drag, particularly during an approach.
- Drag also varies with the angle of sideslip as the area of the fuselage in the slipstream changes.
- Wind-milling propellers also produce additional drag. For propeller-driven aircraft, some propellers can be feathered to reduce propeller drag in the case of an engine failure.

One consideration with the value of C_D is that although C_{DO} and C_{DL} are small values, the total aircraft drag is the same order of magnitude as the total engine thrust. Errors in determining these values are, effectively, amplified in the equations of motion. As with the coefficient of lift, C_D affects aircraft performance, particularly cruise and climb speeds. In addition, the increase in C_D from deployment of the flaps and undercarriage has a major effect on approach speeds and therefore the accuracy of C_D on aircraft performance is critical.

3.4.4 Coefficient of Side Force

Viewed from above, most aircraft are symmetric about the x axis. However, if the aircraft is subject to a cross-wind or the nose of the aircraft is displaced by movement of the rudder or asymmetric engine thrust, the angle of sideslip between the fuselage and the wind will result in a side force. The fuselage acts as an aerofoil, where the lateral angle of sideslip is equivalent to an angle of attack. Two terms contribute to the side force S, one caused by the angle of sideslip and one caused by rudder displacement, given by

$$S = \frac{1}{2}\rho V^2 s \left(C_{y\beta} \beta + C_{y\delta r} \delta r \right) \tag{3.56}$$

where δr is the rudder displacement. The stability derivative $C_{y\beta}$ is the change in side force result-ing from the angle of sideslip, mostly from the vertical tail with smaller contributions from the fuselage and wing. $C_{y\beta}$ affects the damping of lateral oscillations and is therefore negative. The stability derivative $C_{y\delta r}$ is the change in side force caused by rudder deflection and is usually positive.

The computation of the primary forces in the stability axes is given by the following code fragment:

```
Model_Lift      = 0.5 * Weather_Rho * Vc2 * Aero_s * (Model_Cl + Model_ClTail * De);
Model_Drag      = 0.5 * Weather_Rho * Vc2 * Aero_s * Model_Cd;
Model_SideForce = 0.5 * Weather_Rho * Vc2 * Aero_s *
                  (Model_CyDr * Dr + Model_CyBeta * Model_Beta);
```

where **Dr** is the rudder displacement (radians) and the contribution of lift by the tail is computed separately and added to the lift generated by the airframe. Note the convention used throughout these code examples; where an underscore is used, the prefix is the module name and the suffix is the variable name. For example, **Model_Beta** is the variable for **Beta** (sideslip) in the module **Model**.

3.4.5 Pitching Moment Coefficients

The pitching moment is the response to an elevator or stabiliser input. A pilot moves the elevator control to set the aircraft pitch attitude. In visual flight conditions, the pilot uses the distant horizon to judge the pitch attitude of the aircraft, whereas in instrument flight conditions, the pitch attitude is shown by the attitude indicator. The pitching moment M_{stab} in stability axes is given by

$$M_{stab} = \tfrac{1}{2}\rho V^2 s \bar{c}\left(C_{m0} + C_{m\alpha}\alpha + C_{m\delta e}\delta_e\right) + \tfrac{1}{4}\rho V s \bar{c}^2\left(C_{mQ}Q + C_{m\dot\alpha}\dot\alpha\right) \tag{3.57}$$

where ρ is air density, V is true airspeed, s is wing area, is wingspan, $\bar{c}$ is wing mean chord and δ_e is the elevator deflection (radians). The poles of the transfer function $\dfrac{\theta(s)}{\delta_e(s)}$ define the natural frequency and damping of the short and long period phugoid responses (McLean, 1990).

C_{m0} is the contribution to the pitching moment from the distribution of mass of the aircraft and depends on the position of the aircraft C/G; it also varies with flap setting and undercarriage posi-tion. $C_{m\alpha}$ is the main influence in the natural frequency of the short period mode, particularly the response to elevator input and gusts. $C_{m\alpha}$ is usually large and negative to provide sufficient author-ity for the pilot to control pitch (and therefore the angle of attack). However, increasing $C_{m\alpha}$ would require an increase in $C_{m\delta e}$, which is often referred to as elevator power or elevator effectiveness, as it determines the moment exerted by the lift produced by the elevator ($C_{l\delta e}$). $C_{m\delta e}$ also deter-mines the maximum forward position of the C/G which must be balanced by elevator input. $C_{m\delta e}$ should be sufficiently large to enable a pilot to take evasive action or to respond to a disturbance, such as turbulence.

C_{mq} is the pitching moment resulting from the pitch rate and, as the value is normally negative, is therefore a damping term, particularly for the short period mode and, to a lesser degree, the long period mode. C_{mq} has a significant influence on the overall handling qualities of an aircraft. $C_{m\dot\alpha}$

also affects the pitching moment as a result of the rate of change of angle of attack, and is usually negative, also contributing to the damping of the short period mode.

3.4.6 Rolling Moment Coefficients

The rolling moment results from the aileron and rudder inputs, producing an angle of bank. The rolling moment L_{stab} in stability axes is given by

$$L_{stab} = \tfrac{1}{2}\,\rho V^2 sb\left(C_{l\beta}\beta + C_{l\delta a}\delta_a + C_{l\delta r}\delta_r\right) + \tfrac{1}{4}\,\rho V sb^2\left(C_{lP}P + C_{lr}R\right) \tag{3.58}$$

where δ_a is aileron deflection and δ_r is rudder deflection. Care is needed with the value of b, which in some notations is the full wingspan and in other notations is the wing semi-span. The rolling acceleration is a function of the inputs, the angle of sideslip and the roll rate, giving a second-order transfer function for $\dfrac{\phi(s)}{\delta_a(s)}$. Normally, this mode, known as the Dutch roll mode, is a damped second-order system. On the one hand, the pilot needs to be able to control the bank angle (and therefore the turn rate) accurately. On the other hand, the Dutch roll mode can introduce undesirable oscillation, particularly in the yaw axis, which is reduced by rudder input. There is significant cross-coupling between aileron and rudder inputs, that is to say both aileron and rudder input can induce a change of bank angle.

The coefficient $C_{l\beta}$ is sometimes referred to as the *effective dihedral derivative* as it counteracts the rolling moment caused by sideslip and is the change in rolling moment caused by the aircraft side-slipping. $C_{l\beta}$ increases the damping of both the Dutch roll mode and the spiral mode. The stability derivative C_{lda} is also known as aileron effectiveness and is the rolling moment resulting from aileron deflection. Note that some aircraft have both inboard and outboard ailerons (with different values of C_{lda}) and, in addition, spoilers can also be used for roll control. With differential ailerons, care is needed to check whether the value of C_{lda} refers to a single aileron or to the pair of ailerons. The range of values depends on the type of aircraft, with relatively fast roll rates for agile military aircraft. C_{ldr} is the derivative contributing to rolling moment from rudder deflection, where a positive rudder input results in a positive rolling moment. For many aircraft, this value is small and is often neglected.

C_{lp} is the roll damping derivative due to roll rate and is negative to counter the roll rate. The main contribution is from the wing and increases with aspect ratio. The stability derivative C_{lr} is the rolling moment due to yaw rate. For spiral mode stability, C_{lr} should be a small positive value.

3.4.7 Yawing Moment Coefficients

The yawing moment is similar to the rolling moment in the sense that the roll and yaw components are cross-coupled. The yawing moment results from the input from rudder and ailerons, producing a yaw angle (or heading). As with the rolling moment, the transfer function $\dfrac{\psi(s)}{\delta_r(s)}$ is also a second-order response and influences the Dutch roll mode. The yawing moment N_{stab} in stability axes is given by

$$N_{stab} = \tfrac{1}{2}\,\rho V^2\,sb\left(C_{n\beta}\beta + C_{n\delta a}\delta_a + C_{n\delta r}\delta_r\right) + \tfrac{1}{4}\,\rho V sb^2\left(C_{nP}P + C_{nr}R\right) \tag{3.59}$$

The stability derivative $C_{n\beta}$ is often referred to as the weathercock derivative as it provides resistance to yawing motion. A positive sideslip will induce a positive yawing moment with the main component from the aircraft tail. $C_{n\beta}$ has a significant effect on the natural frequency of the Dutch roll mode and also on the spiral mode stability. $C_{n\beta}$ depends on the geometry of the tail and is usually positive and large to improve the overall handling qualities. $C_{n\delta a}$ is the derivative which produces a yawing moment from deflection of the ailerons. For negative values of $C_{n\delta a}$, if the ailerons are moved to bank the aircraft to the right, the aircraft will yaw to the left as a result of the drag from aileron deflection, and vice-versa, an effect known as adverse-yaw. Normally, $C_{n\delta a}$ is a very small positive value, particularly with differential ailerons. The derivative $C_{n\delta r}$, which is also known as rudder effectiveness, is the yawing moment produced by rudder deflection. $C_{n\delta r}$ is used to counteract (or produce) sideslip, particularly to reduce sideslip in turns, and makes an important contribution to handling qualities. For light aircraft, the effect of the propeller, particularly at high power settings, can have a major effect on rudder effectiveness and $C_{n\delta r}$ is sometimes given as a function of the coefficient of thrust C_T, given by

$$C_T = \frac{T}{\frac{1}{2} \rho V^2 s} \tag{3.60}$$

where T is engine thrust.

The stability derivative C_{np} is the change in yawing moment resulting from the roll rate. C_{np} makes an important contribution to Dutch roll damping, which increases with large negative values of C_{np}. The stability derivative C_{nr} is the yawing moment produced by the yaw rate of the aircraft and is the main contribution to damping of the Dutch roll mode, which depends mostly on the tail of the aircraft, and increases the spiral mode stability. The damping is increased with more negative values of C_{nr} but if this value is too large, the aircraft will be unresponsive to rudder input.

The computation of the pitching, rolling and yawing moments in stability axes is summarised as follows:

```
Model_Pmt = 0.5 * Weather_Rho * Vc2 * Aero_s * Aero_CBar *
            (Model_Cm0 + Model_CmAlpha * Model_AlphaWing + Model_CmDe * De) +
            0.25 * Weather_Rho * Model_Vc * Aero_s * Aero_CBar * Aero_CBar *
            (Model_CmQ * Model_Q + Model_CmAlphaDot * Model_AlphaDot);
Model_Rmt = 0.5 * Weather_Rho * Vc2 * Aero_s * Aero_b *
            (Model_ClBeta * Model_Beta + Model_ClDa * Da + Model_ClDr * Dr) + 0.25 *
            Weather_Rho * Model_Vc * Aero_s * Aero_b * Aero_b *
            (Model_ClP * Pstab + Model_ClR * Rstab);
Model_Ymt = 0.5 * Weather_Rho * Vc2 * Aero_s * Aero_b *
            (Model_CnBeta * Model_Beta + Model_CnDr * Dr + Model_CnDa * Da) +
            0.25 * Weather_Rho * Model_Vc * Aero_s * Aero_b * Aero_b *
            (Model_CnBetaDot * Model_BetaDot + Model_CnP * Pstab + Model_CnR * Rstab);
```

3.4.8 Mach Number

In many data packs, the main aerodynamic derivatives are also given as functions of Mach number. These effects are generally neglected for light aircraft but are significant for transport aircraft and military jet aircraft (Heffley and Jewell, 1972). Figure 3.9 shows an example of Mach number effects. In Figure 3.9(a), $C_{l\delta a}$ reduces for the inboard ailerons as a function of Mach number, which is constant up to $M = 0.75$. Figure 3.9(b) shows the aeroelastic effect on the inboard ailerons as a function of altitude and Mach number. Note that these values are given as normalised values rather than absolute values in the graphs.

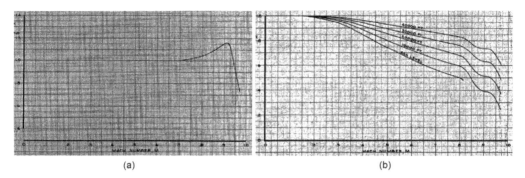

Figure 3.9 Effect of Mach Number on $C_{l\delta a}$ (best available copy).

An alternative method is to provide the terms C_{LU}, C_{DU} and C_{mU}, which enable C_L, C_D and C_m to be modified as a function of the airspeed U.

For the inboard ailerons of a Boeing 747-100, Figure 3.9(a) shows the change in $C_{l\delta a}$ with Mach number and Figure 3.9(b) shows the aeroelastic effect on $C_{l\delta a}$ as a function of altitude and Mach number. Note the nonlinearity of this aerodynamic derivative as a function of Mach number.

3.4.9 Observations

For the most part, aerodynamic derivatives are provided as non-dimensional derivatives, which implies that the contribution from terms such as, for example, $Cn\beta$, can be included in the yawing moment equation in the form $C_{n\beta} \cdot \beta$. However, in some texts, dimensional derivatives are provided. Care is needed that the terms used to compute the dimensional values correspond to the units and conventions used in the equations of motion. Non-dimensional coefficients can be derived from the dimensional coefficients, dividing by the scale factors given in Table 3.1.

It is particularly important to comply with axis conventions, which are right-handed orthogonal axes, as shown in Figure 3.3. With the provision of data from external sources, it is important to check that the same axis notations are used, otherwise some aerodynamic derivatives may have an incorrect sign. The force and moment equations given earlier include the commonly used aerodynamic derivatives, but in some simulations, additional aerodynamic derivatives may be provided and these are listed in Table 3.2.

The accuracy of aerodynamic derivatives depends very much on the source, which can include data from flight tests, wind-tunnel experiments, computational estimates and CFD programs. Generally, flight test data is the most accurate, particularly if parameter estimation methods

Table 3.1 Conversion of Aerodynamic Derivatives.

Dimensional term	Scale factor
$C_{l\beta}$, C_{lda}, C_{ldr}, $C_{n\beta}$, C_{nda}, C_{ndr}	$\frac{1}{2}\rho V^2 sb$
C_{lp}, C_{lr}, C_{np}, C_{nr}	$\frac{1}{4}\rho Vsb^2$
C_{m0}, $C_{m\alpha}$, C_{mde}	$\frac{1}{2}\rho V^2 s\overline{c}$
C_{mq}, $C_{m\dot{\alpha}}$	$\frac{1}{4}\rho Vs\overline{c}^2$

Table 3.2 Additional Aerodynamic Derivatives.

Derivative	Effect	Notes
$C_{D\alpha}$	Change of drag from α	Normally included in the drag polar as a function of C_l
$C_{L\dot{\alpha}}$	Change of lift from $\dot{\alpha}$	Normally ignored
$C_{D\dot{\alpha}}$	Change of drag from $\dot{\alpha}$	Assumed to be zero
C_{Lq}	Change of lift from pitch rate	Normally ignored if aerodynamic centre is near the C/G
C_{Dq}	Change of drag from pitch rate	Normally ignored for subsonic flight
$C_{L\delta e}$	Change in lift from elevator deflection	Normally ignored
$C_{D\delta e}$	Change in lift from elevator deflection	Assumed to be zero
C_{yp}	Change in side force from roll rate	Normally ignored
C_{yr}	Change in side force from pitch rate	Normally ignored
$C_{y\rho a}$	Change in side force from aileron deflection	Assumed to be zero

(Maine and Iliff, 1975) are used, but in cases where a derivative is a function of two or more variables, it is often difficult to alter one variable (e.g. an angle of attack) while maintaining other variables (e.g. airspeed) at a constant setting. Wind-tunnel data is limited in both the acquisition of data and running tests which cover the full-flight envelope. Moreover, scaling that involves airspeed, Reynolds numbers and gravity needs very careful consideration. The accuracy of methods estimated from aircraft dimensions may be as much as 50% in error. Data generated by CFD methods is generally limited to specific aerodynamic derivatives and aircraft components rather than the production of aerodynamic data for the complete aircraft.

The importance of the accuracy of aerodynamic data cannot be overstated. If the lift or drag coefficients are wrong, the aircraft cruise, or rate of climb or approach speeds are likely to be wrong. If the pitching, rolling or yawing moment derivatives are incorrect, the aircraft will pitch or bank at incorrect rates and the phugoid and Dutch roll responses in the simulator will be different from the aircraft. Bear in mind also that data that is correct at a low speed is likely to be in error at other speeds if compensation for Mach number is incorrect (or omitted). In the case where flight models are linearised, it is important to use the aerodynamic derivative values for a specific airspeed, altitude and configuration and to limit any tests to small excursions from these values. The temptation to 'fiddle' with aerodynamic derivatives should be avoided at all costs.

3.5 Equations of Motion

The equations of motion of any vehicle start from Newton's second law: acceleration is proportional to applied force or $f = ma$. For linear forces, if the applied force f and the mass m are known, the acceleration a can be computed. Similarly, for rotational motion, $M = Ja$, where M is the applied moment or torque, J is the moment of inertia and a is the angular acceleration (rad/s). Knowing acceleration, the velocity and position of the vehicle can be computed by applying mathematical integration. For an aircraft, the applied forces and moments are derived from the aerodynamic equations, the engine model, the undercarriage model and gravity. In principle, there is no difference in the equations of motion for an F-16 military aircraft, an Airbus A380, a Spitfire, a glider, an airship, a drone, a missile or a hovercraft. The derivation of forces and moments will certainly differ, but the computation of the resultant trajectory or flight path will follow the same set of equations.

There are numerous implementations of the equations of motion of an aircraft in the literature. Etkin outlines the classic 'flat-earth' equations for rigid body dynamics (Etkin and Reid, 1996). McFarland sets out in detail the equations of motion which extend to earth-centred, earth-fixed and earth-rotating coordinate frames (McFarland, 1975). Fogarty stresses the advantages of solving the equations in flight path axes rather than the conventional body axes (Fogarty and Howe, 1969). An extensive description of the various implementations is covered by Boiffier (1998).

Generally, for studies in flight dynamics, it is most convenient to derive the forces and moments in the body frame. However, for navigation and guidance, it is far more sensible to derive accelerations, velocities and position in the earth frame. Fortunately, the transformations between axes are well-known, straightforward to apply and are based on the following assumptions:

- The airframe is rigid – it does not distort in response to applied forces or moments.
- The aerodynamic forces F_x, F_y and F_z and moments M_x, M_y and M_z (roll, pitch and yaw) are computed in the body frame.
- The engine forces E_x, E_y and E_z and moments E_{mx}, E_{my} and E_{mz} are computed in the body frame.
- The undercarriage forces G_x, G_y and G_z and moments G_{mx}, G_{my} and G_{mz} are computed in the body frame.
- The gravitational acceleration is computed at the current location and altitude.
- The mass of the aircraft, which may vary during the flight, can be computed.
- The moments of inertia of the aircraft, which may vary during the flight, can be computed, typically as a function of aircraft mass.

There are effectively three dynamic regimes, one where the aircraft is airborne and the aerodynamic forces dominate, one where the aircraft is in contact with the ground and the undercarriage forces apply and one where the aircraft transitions between the ground and airborne regimes. In practice, if the forces are computed accurately in the body frame and the aerodynamic forces take account of ground effect and the change of air flow around the airframe during touchdown and take-off, these transitions should be continuous. In addition, during take-off, touchdown and taxiing, any incline of the runway or taxiway surface will add an extra gravitational component to the equations of motion. The integration of the equations of motion module with the aerodynamic, engine and undercarriage modules is shown in Figure 3.10.

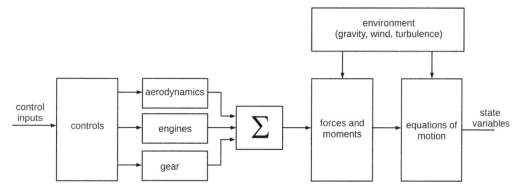

Figure 3.10 Integration of the Equations of Motion.

The flight controls, aerodynamics, engines and gear modules are specific to an aircraft, whereas computation of the forces and moments and the aircraft state variables is common to all aircraft models.

3.5.1 Forces

The aerodynamic forces and moments covered in Section 3.4 are computed in the stability axes and transformed to the body frame. The propulsion forces and moments are covered in Section 3.6. As the engines or propellers are attached to the aircraft, computation of the moments and forces in the body frame is straightforward. The derivation of undercarriage forces and moments, which are formulated as three forces and three moments in the body frame axes, is covered in Section 3.7. Gravity is derived from a gravity model, which takes account of the variation of gravity over the surface of the earth and the reduction of gravitational acceleration with increasing altitude. The gravity components are computed in the earth frame and transformed to the body frame.

The following equations are stated rather than derived as they are covered in most textbooks on aerodynamics and flight mechanics (Babister, 1980). Recalling that the aerodynamic forces are computed in stability axes rather than body axes and that the z component in the right-hand orthogonal body frame is negative up, then the three forces are resolved as follows:

$$F_x = L\sin\alpha - D\cos\alpha + Mg\sin\theta + E_x + G_x \tag{3.61}$$

$$F_y = S + Mg\sin\phi\,\cos\theta + E_y + G_y \tag{3.62}$$

$$F_z = -L\cos\alpha - D\sin\alpha + Mg\cos\theta\cos\phi + E_z + G_z \tag{3.63}$$

where L is lift force, D is drag force, S is side force, α is angle of attack, θ is pitch, ϕ is roll, M is mass, g is gravitational acceleration, and E and G are the engine and undercarriage forces, respectively. If the aircraft is in contact with the ground, the forces are modified as follows:

$$F_x = F_x - Mg\sin\gamma \tag{3.64}$$

where γ is the slope of the ground. If an undercarriage assembly is not in contact with the ground, then the forces generated by the undercarriage are zero. Of course, if the undercarriage is lowered, the additional aerodynamic forces should be included in the computation of lift and drag.

The accelerations and velocities in the body frame are computed from the body frame forces:

$$\dot{u} = \frac{F_x}{M} - qw + rv \tag{3.65}$$

$$\dot{v} = \frac{F_y}{M} - ru + pw \tag{3.66}$$

$$\dot{w} = \frac{F_z}{M} - pv + qu \tag{3.67}$$

$$u = \int \dot{u}\,dt \tag{3.68}$$

$$v = \int \dot{v}\,dt \tag{3.69}$$

$$w = \int \dot{w}\,dt \tag{3.70}$$

The code to compute the forces in the body frame is given in the following code fragment:

```
Model_XForce = Model_Lift * SinAlpha - Model_Drag * CosAlpha +
               Aero_Mass * Model_G * SinPitch + Engines_ThrustX + Gear_Fx;
if (Model_OnTheGround)
{
    Model_XForce -= Aero_Mass * Model_G * sin(DTED_Incline());
}

Model_YForce = Model_SideForce + Aero_Mass * Model_G * SinRoll * CosPitch +
               Engines_ThrustY + Gear_Fy;
Model_ZForce = -Model_Lift * CosAlpha - Model_Drag * SinAlpha +
               Aero_Mass * Model_G * CosPitch * CosRoll + Engines_ThrustZ + Gear_Fz;

Model_UDot = Model_XForce / Aero_Mass - Model_Q * Model_W + Model_R * Model_V;
Model_VDot = Model_YForce / Aero_Mass - Model_R * Model_U + Model_P * Model_W;
Model_WDot = Model_ZForce / Aero_Mass - Model_P * Model_V + Model_Q * Model_U;

U_body = Maths_Integrate(U_body, Model_UDot);
V_body = Maths_Integrate(V_body, Model_VDot);
W_body = Maths_Integrate(W_body, Model_WDot);
```

where the _body subscript denotes that these forces are the net forces in the body frame and the components of the engine thrust and undercarriage forces are computed in the Engine and Gear modules, respectively.

Generally, wind (including turbulence and wind shear) is provided as a set of velocities rather than forces, which are added to aircraft velocities in the body frame. Wind and turbulence components are usually defined in the local (north, east, down) frame and are transformed to the body frame as follows:

$$
\begin{bmatrix} w_u \\ w_v \\ w_w \end{bmatrix} = \begin{bmatrix} a_{11} & a_{21} & a_{31} \\ a_{12} & a_{22} & a_{32} \\ a_{13} & a_{23} & a_{33} \end{bmatrix} \begin{bmatrix} w_N \\ w_E \\ 0 \end{bmatrix}
\tag{3.71}
$$

where the matrix A is the appropriate DCM. Similarly, the turbulence terms are also defined in the local frame and transformed to the body frame,

$$
\begin{bmatrix} t_u \\ t_v \\ t_w \end{bmatrix} = \begin{bmatrix} a_{11} & a_{21} & a_{31} \\ a_{12} & a_{22} & a_{32} \\ a_{13} & a_{23} & a_{33} \end{bmatrix} \begin{bmatrix} t_N \\ t_E \\ t_d \end{bmatrix}
\tag{3.72}
$$

Note that the transformation matrix from the local frame to the body frame is the transpose of the matrix used for the transformation from the body frame to the local frame. The total velocities in the body frame are:

$$
u = u_b + w_u + t_u
\tag{3.73}
$$

$$
v = v_b + w_v + t_v
\tag{3.74}
$$

$$
w = w_b + w_w + t_w
\tag{3.75}
$$

The wind and turbulence components are added to the body velocities, as the angles of attack and sideslip are functions of the total body velocities u, v and w. By (incorrectly) adding the wind and turbulence components to the local velocities $[V_N, V_E, V_d]^T$, the wind would have no effect on the angles of attack and sideslip. There are also empirical considerations. For example, if the flight model does not support rearward flight, negative values of u should be clamped to zero.

The velocities can be transformed from the body frame to the local frame using the DCM defined in Section 3.2, as follows:

$$\begin{bmatrix} V_N \\ V_E \\ V_d \end{bmatrix} = \begin{bmatrix} a_{11} & a_{12} & a_{13} \\ a_{21} & a_{22} & a_{23} \\ a_{31} & a_{32} & a_{33} \end{bmatrix} \begin{bmatrix} u \\ v \\ w \end{bmatrix} \tag{3.76}$$

However, as will be seen later, additional accelerations occur in the local frame as a result of the earth's rotation, and if these terms are included, it is necessary to transform the accelerations from the body frame to the local frame:

$$\begin{bmatrix} \dot{V}_N \\ \dot{V}_E \\ \dot{V}_d \end{bmatrix} = \begin{bmatrix} F_N/m \\ F_E/m \\ F_d/m \end{bmatrix} = \begin{bmatrix} a_{11} & a_{12} & a_{13} \\ a_{21} & a_{22} & a_{23} \\ a_{31} & a_{32} & a_{33} \end{bmatrix} \begin{bmatrix} \dot{u} \\ \dot{v} \\ \dot{w} \end{bmatrix} \tag{3.77}$$

$$V_N = \int \dot{V}_N \, dt \tag{3.78}$$

$$V_E = \int \dot{V}_E \, dt \tag{3.79}$$

$$V_d = \int \dot{V}_d \, dt \tag{3.80}$$

From the local velocities, it is straightforward to compute the latitude and longitude rates:

$$\dot{\lambda} = \frac{V_N}{R} \tag{3.81}$$

$$\dot{\phi} = \frac{V_E}{R \cos \lambda} \tag{3.82}$$

$$\lambda = \int \dot{\lambda} \, dt \tag{3.83}$$

$$\phi = \int \dot{\phi} \, dt \tag{3.84}$$

3.5.2 Moments

The three moments in the body axes are derived from the aerodynamic moments L_{stab}, M_{stab} and N_{stab} in the stability axes, defined in Section 3.4, and the engine and undercarriage moments:

$$L = L_s \cos \alpha - N_s \sin \alpha + E_L + G_L \tag{3.85}$$

$$M = M_s + lift \left(cg - 0.25 \right) \bar{c} \cos \alpha + drag \left(cg - 0.25 \right) \bar{c} \sin \alpha + E_M + G_M \tag{3.86}$$

$$N = N_s \cos \alpha + L_s \sin \alpha + sideforce \left(cg - 0.25 \right) \bar{c} + E_N + G_N \tag{3.87}$$

where L_s, M_s and N_s are the moments computed in the stability axes, $\bar{c}$ is the wing mean chord, cg is the offset from the C/G datum, E_L, E_M and E_N are the engine moments and G_L, G_M and G_N are the undercarriage moments about the x, y, and z body axes, respectively. Note that the C/G position is nominally 0.25 of the wing mean chord $\bar{c}$. The C/G position, which affects aircraft handling, is normally set in a simulator session to replicate the effects of the distribution of passengers, fuel or cargo.

The three angular accelerations $\dot{p}$, $\dot{q}$ and $\dot{r}$ in the body axes about the x, y and z axes (roll, pitch and yaw axes) are given by the following equations:

$$\dot{p} = \frac{L + \left(I_{YY} - I_{ZZ} \right) qr + I_{XZ} \left(\dot{r}p + pq \right)}{I_{XX}} \tag{3.88}$$

$$\dot{q} = \frac{M + \left(I_{ZZ} - I_{XX} \right) rp + I_{XZ} \left(r^2 - p^2 \right)}{I_{YY}} \tag{3.89}$$

$$\dot{r} = \frac{R + \left(I_{XX} - I_{YY} \right) pq + I_{XZ} \left(\dot{p} - qr \right)}{I_{XX}} \tag{3.90}$$

giving the body rates:

$$p = \int \dot{p} \, dt \tag{3.91}$$

$$q = \int \dot{q} \, dt \tag{3.92}$$

$$r = \int \dot{r} \, dt \tag{3.93}$$

Note that the cross-coupled moments of inertia I_{XY} and I_{YZ} are assumed to be zero if the aircraft is symmetric about the x and z body axes, which applies to most aircraft. Generally, the moments of inertia are assumed to remain constant, but if data is available to define the moments of inertia as functions of mass, they are updated as the aircraft mass changes. The code to compute the moments in the body frame is as follows:

```
Pm = Model_Pmt + Model_Lift * (Aero_CgPosition - 0.25) * Aero_CBar * CosAlpha +
     Model_Drag * (Aero_CgPosition - 0.25) * Aero_CBar * SinAlpha +
     Engines_EnginePMT + Gear_My;
Rm = Model_Rmt * CosAlpha - Model_Ymt * SinAlpha + Engines_EngineRMT + Gear_Mx;
Ym = Model_Ymt * CosAlpha + Model_Rmt * SinAlpha +
     Model_SideForce * (Aero_CgPosition - 0.25) * Aero_CBar +
     Engines_EngineYMT + Gear_Mz;

Model_PDot = (Rm + (Aero_Iyy - Aero_Izz) * Model_Q * Model_R +
                  Aero_Ixz * (Model_RDot + Model_P * Model_Q)) / Aero_Ixx;
Model_QDot = (Pm + (Aero_Izz - Aero_Ixx) * Model_R * Model_P +
                  Aero_Ixz * (Model_R * Model_R - Model_P * Model_P)) / Aero_Iyy;
Model_RDot = (Ym + (Aero_Ixx - Aero_Iyy) * Model_P * Model_Q +
                  Aero_Ixz * (Model_PDot - Model_Q * Model_R)) / Aero_Izz;
Model_P = Maths_Integrate(Model_P, Model_PDot);
Model_Q = Maths_Integrate(Model_Q, Model_QDot);
Model_R = Maths_Integrate(Model_R, Model_RDot);
```

The transformations between axes and the computation of forces and moments are summarised in Figure 3.11.

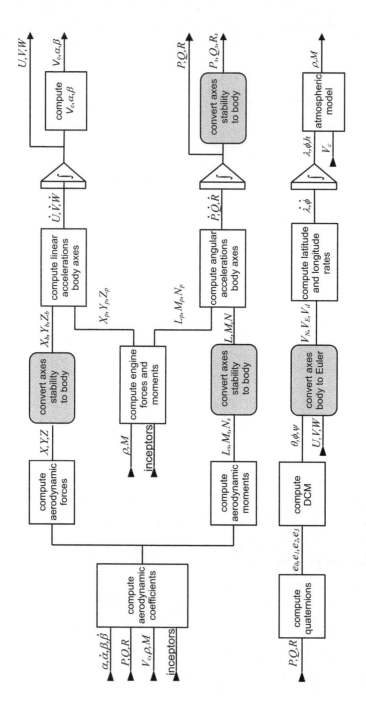

Figure 3.11 Implementation of the Equations of Motion.

In the equations involving angles and angular rates, where it is necessary to constrain values to the range from $-\pi$ to π, the following code is used to normalise angles:

```
float Maths_Normalise(float a)
{
    double x = (double) a;
    double r = fmod(x + M_PI, TWOPI);

    if (r < 0.0)
    {
        r += TWOPI;
    }
    return (float) (r - M_PI);
}
```

where the normalised angle is returned in radians in the range $-\pi$ to π. The function **Maths_Normalise** and other related trigonometric functions used in the simulation software are provided in the library *maths.c*.

3.5.3 Long-Range Navigation

The equations given in the preceding sections apply to a non-rotating earth. In many flight simulation applications, particularly flight of less than 100 nm or so, these equations are quite acceptable. Over longer distances, the equations of motion are modified to include the effect of the rotating earth frame. The first change is that the forces in the body frame, excluding the gravitation force, are resolved in the local frame:

$$F_N = F_x a_{11} + F_y a_{12} + F_z a_{13} \tag{3.94}$$

$$F_E = F_x a_{21} + F_y a_{22} + F_z a_{23} \tag{3.95}$$

$$F_d = F_x a_{31} + F_y a_{32} + F_z a_{33} \tag{3.96}$$

The distance of the aircraft from the centre of earth is

$$R = R_0 - P_z \tag{3.97}$$

where R_0 is the nominal earth radius and P_z is the aircraft altitude (negative up). The aircraft velocity in the local frame to include the Coriolis and centripetal terms is

$$\dot{V}_N = \frac{F_N}{m} - 2V_E \Omega \sin\lambda + \frac{V_N V_d - V_E^2 \tan\lambda}{R} \tag{3.98}$$

$$\dot{V}_E = \frac{F_E}{m} + 2V_N \Omega \sin\lambda + 2V_d \Omega \cos\lambda + \frac{V_E V_d - V_N V_E \tan\lambda}{R} \tag{3.99}$$

$$\dot{V}_d = \frac{F_d}{m} + g - 2V_E \Omega \cos\lambda + \frac{V_E^2 + V_N^2}{R} \tag{3.100}$$

$$V_N = \int \dot{V}_N \, dt \tag{3.101}$$

$$V_E = \int \dot{V}_E \, dt \tag{3.102}$$

$$V_d = \int \dot{V}_d \, dt \tag{3.103}$$

Note that gravitational acceleration g is added to $\dot{V}_d$, whereas previously components of g were added to the accelerations in the body frame. As the earth's rotation rate Ω is relatively small and R is a large value, the contributions of these additional terms are very small.

By resolving accelerations in the local frame, the wind and turbulence components are added in the local frame:

$$V_N = V_N + W_N + T_N \tag{3.104}$$

$$V_E = V_E + W_E + T_E \tag{3.105}$$

$$V_d = V_d + T_d \tag{3.106}$$

Note that the body frame velocities are now derived by the local to body transformation using the DCM in Section 3.2.3. Finally, the vehicle rates p, q and r are computed in the body frame and transformed to the local frame to take account of the earth's rotation rate:

$$\omega_p = \Omega \cos\lambda + \frac{V_E}{R} \tag{3.107}$$

$$\omega_q = -\frac{V_N}{R} \tag{3.108}$$

$$\omega_r = \Omega \sin\lambda + \frac{V_E \tan\lambda}{R} \tag{3.109}$$

$$p = p - \omega_p a_{11} + \omega_q a_{21} + \omega_r a_{31} \tag{3.110}$$

$$q = q - \omega_p a_{12} + \omega_q a_{22} + \omega_r a_{32} \tag{3.111}$$

$$r = r - \omega_p a_{13} + \omega_q a_{23} + \omega_r a_{33} \tag{3.112}$$

As before, the quaternions are updated from p, q and r, the DCM for the next frame is updated based on the quaternions, and the Euler pitch, roll and yaw angles are computed from the DCM.

Figure 3.12 shows a flight starting just north of Southampton, where the aircraft is at 30,000 ft, the ground speed is 520 Kt (indicated airspeed 320 Kt) and the heading is 360°M. The magnetic variation is zero and there is nil wind. In Figure 3.12(a) the effect of Coriolis acceleration is removed and the aircraft track is parallel to the 1.5 W meridian of longitude. In Figure 3.12(b), a drift of 34.5 nm is shown, resulting from rotation of the earth during the flight. As a cross-check, the eastward acceleration is given by $a = 2V\Omega\sin\lambda$, where $V = 267.5$ m/s, the earth's rotation rate $\Omega = 0.729\,115 \times 10^{-5}$ rad/s and the latitude = 51°N, giving a = 0.031 61 ms^{-2}. The distance travelled is $\frac{1}{2}at^2 = 69.15$ km or 37.3 nm.

3.6 Propulsion

Aircraft propulsion includes piston engines with propellers, turbojet engines, turboprops and turbofan engines. In addition, propellers can be classified as fixed-pitch and variable-pitch propellers. For most multi-engine aircraft, the engines are identical and one model can be used for all the engines. There are principally three methods to developing engine models:

1) Obtaining the engine data from the manufacturer in the form of a data pack.
2) Developing the model from the principles of thermodynamics.
3) Reverse engineering the model from aircraft performance data.

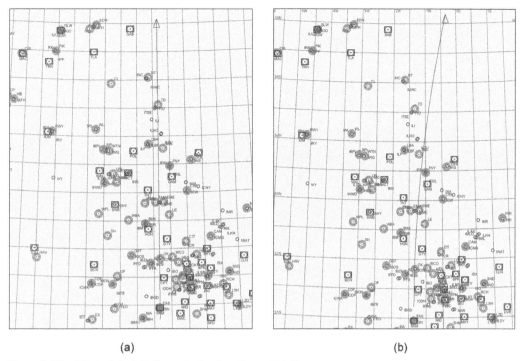

(a)   (b)

Figure 3.12 Effect of Coriolis Terms in the Equations of Motion.

The topic is an extensive subject and difficult to cover in depth in a section of a textbook. In the experience of the author, models fall into two categories: those developed from detailed data provided as a commercial data package and those produced from simplifications based on meeting the published performance data. An engine model includes the computation of thrust, RPM, fuel flows, temperatures and pressures (particularly for instrument displays). Although the time constants of engine dynamics are relatively long in comparison with the aerodynamic model, the equations must still be solved in real time. Generally, models of aircraft engines for flight simulation are based on dynamic models to compute the thrust forces. In full-flight simulators, the engine model may be as complicated as the aerodynamic model, particularly to include the engine control systems and the modelling of engine failures.

3.6.1 Piston Engines

Piston engines are rated by static horsepower (SHP), which is the maximum horsepower (HP) the engine is capable of producing. However, the power available depends on several variables. The throttle setting determines the fuel-flow rate into the engine, the mixture setting defines the fuel–air ratio, there are losses owing to friction, heat and sound and the propeller is a variable load on the engine. Engine performance also depends on air density, the ambient temperature of air entering the engine and the operating temperature of the engine. The engine normally provides power to drive the propeller but under some conditions, the propeller may have excess power and can drive the engine.

With variable-pitch propellers, the blade angle alters to maintain an engine RPM that is selected by a pitch lever. Settings for manifold pressure and RPM are given in the handbook for an aircraft, which provide an optimum RPM for the specified manifold pressure and airspeed; a governing

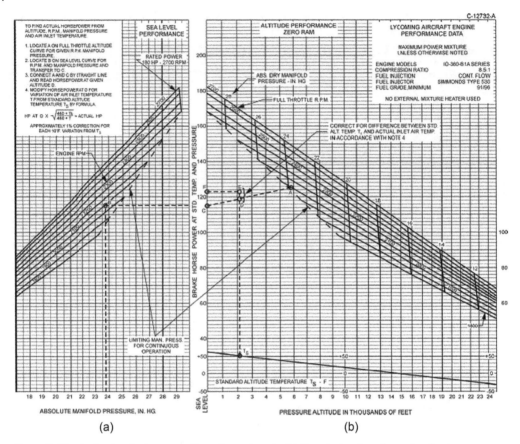

Figure 3.13 An Engine Performance Chart. Reproduced with Permission from Lycoming Engines, an Operating Division of Avco Corporation.

mechanism adjusts the blade angle to maintain the selected RPM, whereas for a fixed-pitch propeller, the blade is solid and the pitch angle is fixed. For a fixed-pitch propeller, RPM and EGT (exhaust gas temperature) gauges are normally provided and engine RPM is adjusted by the throttle. For a variable-pitch propeller, a manifold pressure gauge is also provided, which is used as a power indicator, as the RPM is normally governed to the RPM selected by the pitch lever.

Generally, engine performance data is in the public domain. For example, Figure 3.13 shows the performance charts for the Lycoming O-360-B1A engine. Figure 3.13(a) enables the engine power at sea level to be computed as a function of manifold pressure and engine RPM. Figure 3.13(b) provides adjustment for pressure altitude and ambient temperature. The notes on the data sheets explain the lookup method, which is easily mechanised. The main use of these graphs in aviation is to determine the fuel consumption for a given airspeed, altitude and engine setting. However, in simulator development, they provide a useful cross-check of the computation of engine power. A typical engine model is shown in Figure 3.14.

The small circle symbols denote inputs to the model and the arrows denote the outputs from the model. Typically, the movement of the levers can be implemented as a simple square law, where a large movement of the engine lever near to the maximum causes a relatively small throttle change. For each engine, the inputs are the throttle lever, pitch lever, mixture lever, ignition switches (magnetos) and a starter switch. The engine thrust is used in the equations of motion.

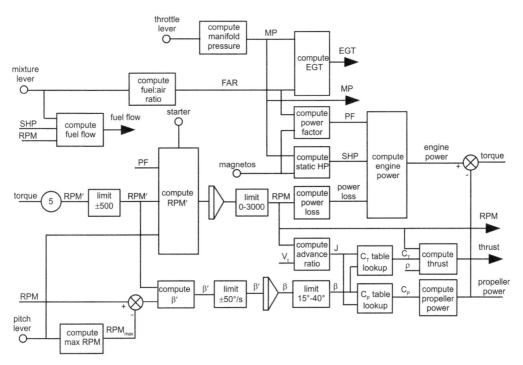

Figure 3.14 A Piston-Engine Model.

The inputs are read by the I/O system and returned in the range 0–1. The lever values used in the model are typically given by $y = 2x - x^2$, where x is the lever input position and y is the lever output, also in the range 0–1. Alternatively, the relationship of the throttle position to the fuel injection or carburettor input may be available from aircraft measurements. The governed RPM is set as follows:

```
RpmMax = 2700.0 * PitchLeverProfile(PitchLever);
```

where the pitch lever profile is shown in Figure 3.15.

The manifold pressure depends on the throttle lever, the engine RPM and the pressure altitude, which is computed in the weather model.

```
DeltaMP = (0.04635 * ThrottleLeverProfile(ThrottleLever) - 0.0469) * Engine->Rpm;
MP      = Weather_PressureAltitude + DeltaMP;
```

At tick-over, with the throttle closed and the engine idling at 600 RPM, ΔMP is -28.14 InHg. With the throttle fully open and the engine at 2700 RPM, ΔMP is 1.485 InHg. Note that, for most engines, data is provided in imperial units. The computations are therefore carried out using imperial units but are converted to SI units for use in the equations of motion.

The power available from a piston engine depends on the fuel–air ratio (FAR; Stinton, 1985), which is shown in Figure 3.16. Below a FAR of 0.05 (20:1) or above 0.125 (8:1), detonation cannot occur and the engine will stop. The maximum power is available at a ratio of 0.08 (12:1) and the

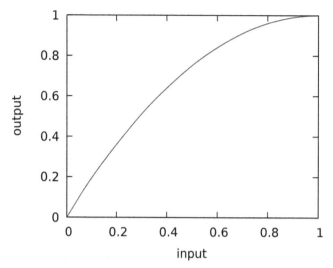

Figure 3.15 Engine Lever Profile.

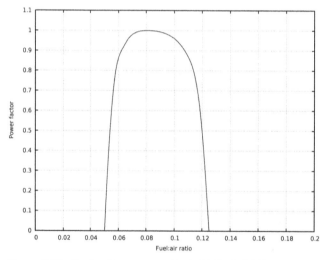

Figure 3.16 Engine Power versus Fuel–Air Ratio (FAR).

most economical setting (for range) is where the gradient of the curve is a maximum, near to 0.06 (16:1). The maximum mixture setting is used for take-off, climbing and landing (in case of a go-around) but in the cruise, the mixture can be leaned carefully for an economical setting. The power factor is computed as follows:

```
FAR = MixtureLeverProfile(MixtureLever) * Weather_DensityRatio * 0.1;
if (FAR <= 0.0625)
{
    Engine->Egt = MP * 1143.0 / 28.5 + 5714.0 * FAR;
}
else
{
    Engine->Egt = MP * 1857.0 / 28.5 - 5714.0 * FAR;
```

```
}

PowerFactor = SplineLib_Lookup1(PowerTab, 0, FAR);
if (PowerFactor < 0.0001 || Failed || Magneto == Engines_OffMag || FuelOut)
{
    PowerFactor = 0.0;
}
```

The fuel–air ratio depends on the mixture lever and air density and is used to compute the EGT which also depends on the manifold pressure. The power factor is computed from a spline fit to the graph shown in Figure 3.16. The engine will stop if the power factor is zero, both magnetos are off, there is no fuel or the engine has been failed by the instructor.

The engine power loss is proportional to the square of the RPM while the SHP depends on the manifold pressure and the engine RPM.

```
EnginePowerLoss = 0.0413 * Engine->Rpm * Engine->Rpm / 2700.0;
StaticHP = MP * (0.0039 * Engine->Rpm - 1.0);

if (Magneto != Engines_BothMags)
{
    StaticHP = StaticHP * 0.95;
}

EnginePower = StaticHP * PowerFactor - EnginePowerLoss;
```

For this engine, rated at 160 HP, the engine power loss is 111.51 HP at 2700 RPM and the gross SHP available is 271.605 HP giving a SHP of 160 HP. The actual power produced by the engine includes the power factor and the engine loss. The fuel flow is computed from the SHP, the RPM and the mixture lever position, based on values from the performance charts for the specific engine.

```
Engine->FuelFlow = (0.235 * StaticHP + 0.0125 * Engine->Rpm - 9.69) *
                   MixtureLever * 0.000225;
```

The propeller advance ratio J is given by

$$J = \frac{V}{nD} \tag{3.113}$$

where V is the true airspeed (the speed of the air passing through the propeller), n is the speed of rotation of the propeller and D is the diameter. For normal operations, J is in the range 0–2, but care is needed to avoid very large values of J at low RPM, for example, when the engine is started. Assuming the propeller diameter is 6 ft:

```
r = (Engine->Rpm < 300) ?300.0 : Engine->Rpm;
J = 16.9 * 1.9440 * Model_Vc / r;
```

The propeller thrust and power coefficients are given by

$$C_T = \frac{T}{\rho n^2 D^4} \tag{3.114}$$

$$C_P = \frac{P}{\rho n^3 D^5} \tag{3.115}$$

where T is thrust, P is power, ρ is air density, n is the propeller speed of rotation and D is the propeller diameter. Charts are produced by manufacturers of propellers, defining C_T and C_P as a function the advance ratio J and the blade angle β, which enable the propeller thrust and power to be computed. Propeller charts for a Piper PA-30, based on values of J from 0 to 1.2, are shown in Figure 3.17.

The blade angle β is shown for each curve and is fined towards 0° and coarsened towards 45°. In the fully feathered position, the blade angle is 90°. In the event of an engine failure, the pilot will feather the propeller of the failed engine to reduce the drag. A 2D spline lookup as a function of J and β is used to access C_T and C_P, using Neville extrapolation outside the range 0–1.2 for J.

```
Cp = CpValue(Engine->Beta, J);
Ct = CtValue(Engine->Beta, J);
```

The power needed to turn the propeller is given by

$$p = \frac{C_p n^3}{6,430,041\rho} \tag{3.116}$$

where the torque is computed from the difference in the power produced by the engine and the power absorbed by the propeller. Again, with the equation for torque, care is needed with division by the propeller RPM, to avoid generating excessive torque at very low RPM, for example, starting or stopping the engine.

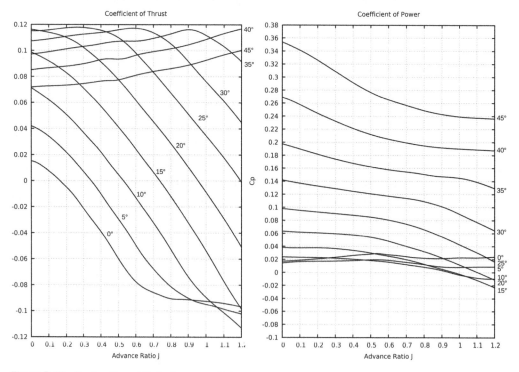

Figure 3.17 Engine C_T and C_P Performance Data.

```
PropPower = Cp * Engine->Rpm * Engine->Rpm * Engine->Rpm / 6430041.0 /
                Weather_DensityRatio;
r = (Engine->Rpm < 300.0) ? 300.0 : Engine->Rpm;
Torque = (EnginePower - PropPower) * 7120.91 / r;
```

As the moment of inertia of the propeller can be assumed to be constant, the angular acceleration of the propeller is proportional to torque and is limited to $\pm$ 500 rad/s^2. The propeller acceleration is adjusted by 100 rad/s^2 if the power factor is zero (i.e. the engine is stopping) or if the blade is being feathered by the pitch lever. In addition, if the starter motor is engaged below 650 RPM (the idle condition), the propeller acceleration is set to 500 rad/s^2.

```
RpmDot = Maths_Limit(Torque * 5.0, -500.0, 500.0);

if (PowerFactor <= 0.0001)
{
    RpmDot = RpmDot - 100.0;
    Engine->Egt = 0.0;
}

if (Engine->Rpm < 650.0 && Starter == IODefn_On)
{
    RpmDot = 500.0;
}
```

A governor is usually used to control the blade angle for a variable-pitch propeller and a simple controller is shown in Figure 3.18.

The blade angle and the rate of change of the blade angle are computed as follows:

```
BetaDot = (Engine->Rpm - RpmMax) / 10.0 + RpmDot / 20.0;
BetaDot = Maths_Limit(BetaDot, -50.0, 50.0);
Engine->Beta = Maths_Integrate(Engine->Beta, BetaDot);
Engine->Beta = Maths_Limit(Engine->Beta, 15.0, 40.0);
```

β and β' are limited to 15–40° and $\pm$50°/s, respectively. The engine RPM is obtained by integrating **RpmDot** and the manifold pressure is computed as a first-order lag with a small time constant and is limited to the range 0–40 Hg.

```
Engine->Rpm = Maths_Integrate(Engine->Rpm, RpmDot);
Engine->Rpm = Maths_Limit(Engine->Rpm, 0.0, 3000.0);
Engine->ManifoldPressure = Maths_Integrate(Engine->ManifoldPressure,
                                    MP - Engine->ManifoldPressure);
Engine->ManifoldPressure = Maths_Limit(Engine->ManifoldPressure, 0.0, 40.0);
```

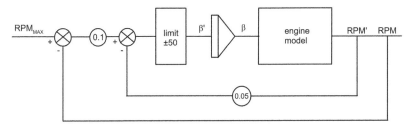

Figure 3.18 A Variable-Pitch Propeller Governor.

Finally, the engine thrust is computed from C_T, engine RPM and air density as follows:

```
Engine->Thrust = 0.0038121 * (Engine->Rpm * Engine->Rpm) * Ct /
                 Weather_DensityRatio;
```

Note that, in the case of a fixed-pitch propeller, the same code can be used, with the blade angle set to a constant value, typically 20°, omitting the code to compute the blade angle and the blade angle rate. One further complication is the ground effect, which can reduce thrust by as much as 30% when the aircraft is in contact with the ground and should be included when the aircraft is taxiing or during take-off, with the transition in thrust proportional to the altitude above the runway, rather than an abrupt change.

Dynamic testing requires applying throttle step changes and comparing the response times with aircraft data. Bear in mind that propeller charts are effectively simulator data and manufacturers expect to charge for this data. There are very few examples to predict propeller thrust or to produce propeller charts from propeller dimensions in the public domain. Worobel's reports (1971, 1972) include a FORTRAN program to predict propeller characteristics.

3.6.2 Turbofans

As with any engine, if an accurate aerodynamic model is available, it is possible to compute the thrust needed for level flight at specific airspeeds and altitudes. Applying a simple law between the engine lever position and thrust, it would be possible to produce a table of thrust as a function of engine lever position, airspeed and altitude. Of course, this is a gross oversimplification: a jet engine is a complex system of axial and radial gas flows at high temperature, in turbomachinery rotating at very high speed, and thrust is a function of several variables, mostly with nonlinear relationships. The position is further complicated as most civil and military engines are controlled by full authority digital engine control (FADEC) systems, which adjust the fuel flow for both safe and optimal operation of the engine. There are relatively few publications to predict jet-engine parameters to the degree of accuracy needed for a flight simulator. Consequently, most simulations of jet engines are dependent on the availability of a data package for the engine.

Jet-engine models are also complicated as they contain models for both forward and reverse thrust and the transition between these operating states. The engine model covered in this section is the Pratt and Whitney JT9D-7A engine covered in the NASA model of the Boeing 747-100 (Hanke, 1971). The engine model is limited to the engine lever position, the ambient temperature at altitude and the Mach number. These relationships are provided as plots of two variables and splines are used extensively for real-time interpolation of this data. One further consideration is that engine start and engine shutdown are dynamic and complex processes. An airline simulator will model the effect of APU connection, electrical loading, fuel selection, air condition bleeds and engine start (including problems associated with 'hot starts' and 'cold starts') and shutdown models, which are not included for reasons of brevity.

Typically, in a flight simulator, the engine lever position is detected by means of a potentiometer, giving a normalised output of 0–1 from minimum position to maximum position. However, engine levers may have a flight idle position, which is not necessarily zero. In addition, the engine lever can also be moved into a reverse thrust position, which is usually measured with a second potentiometer. Engine levers used with auto-throttle systems are actively controlled and, with all the associated mechanical linkage, backlash must also be taken into account. However, it is possible to measure the lever input for forward and backward movement to determine the backlash and to convert potentiometer measurements to lever angle.

The static engine pressure ratio (EPR) is computed as a function of the engine lever angle and the ambient engine intake temperature, as shown in Figure 3.19. Imperial units are used throughout this model with the resultant forces and moments converted to SI units.

The static EPR is computed from the lever angle and ambient temperature as follows:

```
float tempF = (Weather_Temperature(-Model_Pz) - 273.15) * 9.0f / 5.0f + 32.0f;
float lever_angle = 44.0 + ThrottlePosition * 81.0;
float e = LookupEPR(epr_table, lever_angle, tempF);
```

where spline interpolation is used to compute **e**. Note that the temperature is converted from K to °F. The inertia of the turbine is modelled as a first-order lag, with a five-second time constant, based on data provided in the report.

```
Engines_Engines[EngineNumber].Epr = Maths_Integrate(Engines_Engines[EngineNumber].Epr,
                    0.2 * (e - Engines_Engines[EngineNumber].Epr));
```

In addition, there is a small change in EPR from the deployment of the undercarriage at flaps 25 or 30 at flight idle in the approach and during touchdown. The reduction in EPR as a function of

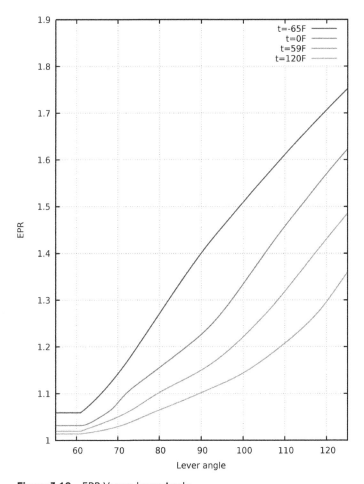

Figure 3.19 EPR Versus Lever Angle.

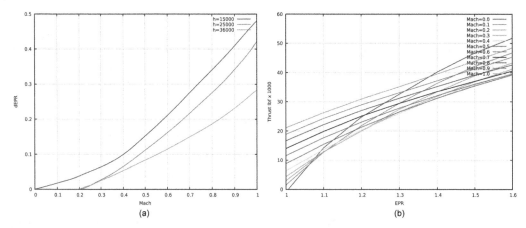

Figure 3.20 Change in EPR versus Mach Number and Net Thrust versus EPR.

Mach number is shown in Figure 3.20(a). For altitudes below 15,000 ft, the 15,000 ft curve is used, and for altitudes above 36,000 ft, the 36,000 ft curve is used; otherwise spline interpolation is used to compute the EPR value.

```
depr = SplineLib_Lookup2(dEPR_table, m, h);
e = e - depr;
```

The installed net thrust is then computed as a function of the engine EPR and the Mach number, based on the plots shown in Figure 3.20(b) for 11 values of Mach number.

Spline interpolation is used to compute the net thrust, which is converted from lbf to SI units. Note also that the thrust is limited to positive values in the forward thrust mode.

```
t = SplineLib_Lookup2(Net_thrust_table, Engines_Engines[EngineNumber].Epr, m) *
                  1000.0f * 4.448222;   /* lbf -> n */
if (t < 0.0f)
{
    t = 0.0;
}
Engines_Engines[EngineNumber].Thrust = t;
```

Based on the location and alignment of the engines with respect to the airframe, the combined forces and moments of the four engines are computed in the body frame as follows:

```
for (i = 0; i <= 3; i += 1)
{
    Engine(Engines_ThrottleLever[i], Engines_ReverseLever[i], i, Fail[i]);
}

Engines_EngineThrustX = Engines_Engines[0].Thrust + Engines_Engines[1].Thrust +
                        Engines_Engines[2].Thrust + Engines_Engines[3].Thrust;
Engines_EngineThrustY = 0.0349 * (Engines_Engines[0].Thrust +
                                  Engines_Engines[1].Thrust -
                                  Engines_Engines[2].Thrust -
                                  Engines_Engines[3].Thrust);
Engines_EngineThrustZ = -0.0436 * Engines_EngineThrustX;
```

```
Engines_EnginePMT = ZEO * (Engines_Engines[0].Thrust  +
                           Engines_Engines[3].Thrust) +
                    ZEI * (Engines_Engines[1].Thrust  +
                           Engines_Engines[2].Thrust);
Engines_EngineYMT = YEO * (Engines_Engines[0].Thrust -
                           Engines_Engines[3].Thrust) +
                    YEI * (Engines_Engines[1].Thrust -
                           Engines_Engines[2].Thrust);
Engines_EngineRMT = 0.0436 * Engines_EngineYMT;
```

The engines are numbered 0–3, where engine number 0 is the outer port engine number. As the engine alignment is fixed, the values 0.0349 and −0.0436 are the components of thrust in the Y and Z body axes, respectively. The values **YEI**, **YEO**, **ZEI** and **ZEO** are the offsets from the centreline in the Y and Z body axes, respectively (where the inner and outer engines are denoted by I and O, respectively).

Modelling of the reverse thrust includes a two-second delay to allow for the deployment of the reverse mode and a simpler derivation of thrust. A linear model of thrust as a function of EPR and Mach number is used to derive reverse thrust, which is converted from lbf to newtons, based on data provided in the report.

```
if (ReverseThrust)
{
    float t = -((Engines_Engines[EngineNumber].Epr - 1.0) * 29286.0 +
            m * 50000.0) * 4.4482;

    ThrottlePosition = ReversePosition;

    if (t > 0.0)
    {
        t = 0.0;
    }
    Engines_Engines[EngineNumber].Thrust = t;
}
```

3.7 Landing Gear

The undercarriage assembly of most modern aircraft comprises a nose wheel assembly and main gear assemblies under each wing. Although these assemblies may contain many wheels, from the simulation perspective, each assembly is normally treated as a single entity. The location of each undercarriage assembly is referenced to the nominal aircraft C/G, or some fixed datum. It is also common to treat each assembly as a spring/damper system where the spring and oleo (damper) are attached to the fuselage and to the wheels. In addition, the tyres of each wheel assembly can also be treated as springs, in terms of deformation under both vertical and lateral loads.

As with the aerodynamic and engine data, the data for the undercarriage assembly is normally provided by the aircraft manufacturer as part of the data pack. In cases where manufacturer data is not available, reasonable assumptions can be made from data provided in various publications on undercarriage and tyre dynamics. The spring element of the undercarriage provides sufficient resilience to absorb landing forces and to support the fuselage. The oleos are provided to dampen out the oscillations resulting from the spring suspension and to ensure that, in the case of a heavy landing, the undercarriage assembly does not fully compress the springs, which could result in structural damage.

For most aircraft, the nose wheel is steerable, with linkages to the rudder pedals for high-speed steering and a tiller for low-speed manoeuvring. The toe brakes are applied to the main undercarriage assemblies, but not to the nose wheel. Unless the undercarriage is fixed, it can be raised and lowered, with a known cycle time and, during this action, the drag and pitching moment (as the undercarriage is below the aircraft C/G) vary with the position of the undercarriage.

When in contact with the ground, tyres provide a vertical force which supports the undercarriage assembly. The tyres are also subject to lateral forces during steering. The runway friction, which depends on the surface material, also affects the friction of the tyres and side friction, which can lead to scuffing and scrubbing. All these forces need to be correctly modelled and generally they are computed in the frame of each undercarriage assembly and then transformed to forces and moments applied to the body frame.

Consider the undercarriage forces shown (not to scale) in Figure 3.21. The nose gear assembly exerts a force n_1 at a distance d_1 from the C/G and the two (identical) main gear assemblies each exert a force n_2 at a distance d_2 from the C/G. The moment arms are defined as follows:

$$2n_2 \cdot d_2 = n_1 \cdot d_1 \tag{3.117}$$

$$n_1 + 2n_2 = mg \tag{3.118}$$

where m is the aircraft mass and g is the gravitational constant, giving

$$n_2 = \frac{mg \cdot d_1}{2(d_1 + d_2)} \tag{3.119}$$

$$n_1 = mg - 2n_2 \tag{3.120}$$

For the Boeing747-100, $d_1 = 23.470$ m (77.0 ft), $d_2 = 3.658$ m (12 ft) and $m = 260,000$ kg, giving $n_1 = 212,550$ N and $n_2 = 1,169,025$ N. In other words, at rest, the nose gear assembly must provide 213 kN and each main gear must provide 1169 kN. In practice, the aircraft must cater for the worst-case rate of descent in a heavy landing. For the Boeing 747-400, the springs are compressed to approximately 80% of their travel at rest. However, the spring curve is nonlinear and the undercarriage is capable of supporting twice the nominal aircraft mass.

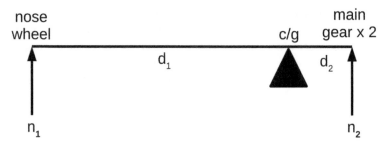

Figure 3.21 Static Undercarriage Loads.

The other important characteristic is the oleo damping, which must absorb high accelerations resulting from a heavy landing. The damping force is given by

$$f = 2Cv^2 \tag{3.121}$$

where f is the damping force, v is the velocity of the damper and C is a function of the oleo compression. Typically, C is nonlinear, providing most damping towards the maximum compression. The values for C are chosen so that the oscillations will be damped out within approximately 10 s. If the damping is too high, the oleo will be too stiff and there is a possibility of causing structural damage, but if the damping is too low, the aircraft response will appear springy and the oscillations of the fuselage could continue for an unreasonably long time. Typically transport aircraft descend at 500–600 ft/min, reducing the descent rate to less than 100 ft/min, at touchdown. The oleos should absorb this rate of descent, but if they are too stiff, there is a possibility that the aircraft will bounce back into the air, possibly leading to a go-around. In simulators used for pilot training, the modelling of the undercarriage and aircraft dynamics at touchdown is critical, otherwise the technique used to land the simulator will differ from the technique used in the aircraft, which is clearly unacceptable. The strut loading and strut damping for the Boeing 747-100 are shown in Figure 3.22.

Figure 3.22(a) was generated using a spline fit of the tabulated data, whereas Figure 3.22(b) was produced using linear segments, as implied by the nonlinear nature of the tabulated data in the report. Note that the strut loading is given in lbs rather than lbf. In other words, Figure 3.22(a) shows the mass that can be supported by the strut rather than the force produced. In addition to the oleo forces, the tyres also act as a spring for vertical applied forces (supporting the aircraft mass) and for lateral forces resulting from turning. Drag forces are also produced by the main gear braking system and by the coefficient of friction of the runway surface.

When the undercarriage is lowered, the oleos are at maximum displacement or zero deflection. On contact with the ground, one or more oleos will start to compress. It is assumed that the undercarriage assemblies are orthogonal to the body frame and that the undercarriage dynamics only apply when one or more of the undercarriage assemblies is in contact with the ground. Note also that the runway coordinates are usually defined in the visual system. It is necessary to compute the height of the wheels above the runway surface, either from information in the navigation database or by reference to pre-computed terrain elevation data.

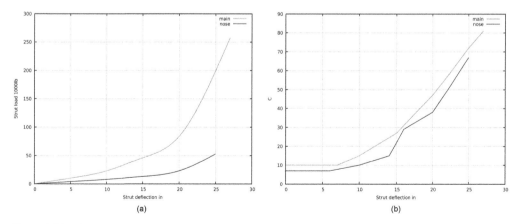

Figure 3.22 Strut Loading and Damping.

Let the three undercarriage assemblies be located at X_{Li}, Y_{Li} and Z_{Li} in the body frame axes, relative to the aircraft C/G, where X, Y and Z denote the respective body axes and $i = 1$ for the nose gear, $i = 2$ for the left main gear and $i = 3$ for the right main gear. Each gear assembly is approximated by the centre position of the assembly. The position of each gear assembly, relative to the ground, depends on the pitch and roll angles, assuming the runway coincides with the local frame. The strut compression ΔS_{Ti} and the compression rate $\Delta \dot{S}_{Ti}$ are given by

$$\Delta S_{Ti} = h + X_{Li} \sin\theta - Y_{Li} \sin\phi\cos\theta - Z_{Li}\cos\phi\cos\theta \tag{3.122}$$

$$\Delta \dot{S}_{Ti} = \dot{h} + X_{Li}\cos\theta\dot{\theta} + Y_{Li}\left(\sin\theta\sin\phi\dot{\theta} - \cos\phi\cos\theta\dot{\phi}\right) + Z_{Li}\left(\sin\theta\cos\phi\dot{\theta} + \sin\phi\cos\theta\dot{\phi}\right) \tag{3.123}$$

where h is the height of the undercarriage assembly above the runway.

The common trigonometric terms are computed as follows:

```
spitch = sin(Model_Pitch);
cpitch = cos(Model_Pitch);
sroll  = sin(Model_Roll);
croll  = cos(Model_Roll);
spsr   = spitch * sroll;
spcr   = spitch * croll;
cpsr   = cpitch * sroll;
cpcr   = cpitch * croll;
```

The code to compute the velocity of the oleos is given by the following:

```
Hat = DTED_PostHeight(Model_Latitude, Model_Longitude,
                      AeroLink_NavPkt.GroundLevel);
h = Maths_Feet(-Model_Pz + (float) Hat);
h_dot = Maths_Feet(-Model_Vd);

dst1 = (h + XL1 * spitch - YL1 * cpsr - ZL1 * cpcr) * 12.0;
dst2 = (h + XL2 * spitch - YL2 * cpsr - ZL2 * cpcr) * 12.0;
dst3 = (h + XL3 * spitch - YL3 * cpsr - ZL3 * cpcr) * 12.0;

dst1_dot = (h_dot + XL1 * cpitch * Model_Q +
            YL1 * (spsr * Model_Q - cpcr * Model_P) +
            ZL1 * (spcr * Model_Q + cpsr * Model_P)) * 12.0;
dst2_dot = (h_dot + XL2 * cpitch * Model_Q +
            YL2 * (spsr * Model_Q - cpcr * Model_P) +
            ZL2 * (spcr * Model_Q + cpsr * Model_P)) * 12.0;
dst3_dot = (h_dot + XL3 * cpitch * Model_Q +
            YL3 * (spsr * Model_Q - cpcr * Model_P) +
            ZL3 * (spcr * Model_Q + cpsr * Model_P)) * 12.0;
```

where **Hat** is the height above the terrain (relative to sea level) and **h** is the height of the aircraft C/G above the terrain. Note that the module **Model** uses SI units throughout, whereas the undercarriage dimensions are provided in feet and the oleo deflection used in the lookup table is given in inches. The approximations for $\dot{\theta}$ and $\dot{\phi}$ are the body rates q and p, respectively. Note also that ΔS_{Ti} is negative for compression.

The strut force V_{Fi} is computed from the graph in Figure 3.22(a) as a function of ΔS_{Ti} and the damping constant C is computed from the graph in Figure 3.22(b), which is given by

$$D_{fi} = 2C_i\Delta S_{ti}\left|\Delta S_{ti}\right| \tag{3.124}$$

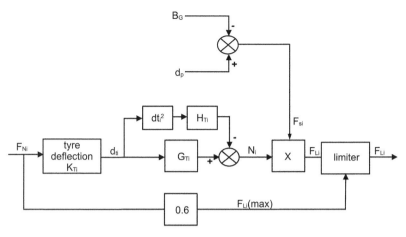

Figure 3.23 Wheel Side Force.

where D_{fi} is positive for strut compression. The net strut force F_{Gzi} is

$$F_{Gzi} = -\left(V_{Fi} + D_{fi}\right)^{\cdot}$$ (3.125)

The normal force of the tyre resulting from the oleo force F_{Gzi} is given by

$$F_{NGi} = \frac{F_{Gzi}}{\cos\theta\cos\phi}$$ (3.126)

Computation of the wheel side force is shown in Figure 3.23.

The tyre deflection is proportional to the normal force and a second-order law is used to compute the tyre side force N_i. Movement of the rudder pedals d_p deflects the nose wheel, producing an angle of the nose wheel relative to the fuselage. In addition, there is a side force B_G in each assembly caused by the difference between the heading of the aircraft and the direction of motion. The resultant side force is limited as a proportion of F_{Ni}.

The wheel drag force is computed from the coefficient of friction of the runway surface, the position of the brake pedals and rolling friction of all the aircraft components including breakout friction, as shown in Figure 3.24.

Typical values for runway friction vary from 0.02 on an icy surface, to 0.1 on a wet runway and 0.4 on a dry runway. The park brake is assumed to have the same coefficient of friction as the maximum application of toe brakes. A nominal value of 0.015 is used for rolling friction. The code for the nose gear (which is very similar to the main gears) is as follows:

```
if (dst1 < 0.0)                 /* nose gear */
{
    ug  = Model_U;
    vg  = Model_V + Model_R * Maths_Metres(XL1);
    ng  = ug * Model_A11 + vg * Model_A12;
    eg  = ug * Model_A21 + vg * Model_A22;
    gp  = (Model_U < 1.0) ? Model_Yaw : atan2(eg, ng);
    gd  = Model_Yaw;
    bg1 = gp - gd;
    bg1 = Maths_Normalise(bg1);

    c1      = lookup(-dst1, damping_nose, 7);
```

```
    vf1     = 2.0 * lookup(-dst1, spring_nose, 6);
    dforce1 = -2.0 * c1 * dst1_dot * fabs(dst1_dot);
    fgz1    = -(vf1 + dforce1);
    fng1    = fgz1 / cpcr;
    dt1     = fng1 * kt1;
    fs1     = (dt1 * gt1 - (dt1 * dt1 * ht1)) * (ds - Maths_Degrees(bg1));
    fs1     = Maths_Limit(fs1, -fabs(0.6 * fng1), fabs(0.6 * fng1));
    fmu1    = -((0.015 + bof) * fabs(fng1)); /* no braking on nose wheel */
    fx1     = fmu1 - fng1 * Model_Pitch - fs1 * Maths_Rads(ds);
    fy1     = fs1 + fng1 * Model_Roll;
    fz1     = fmu1 * Model_Pitch - fs1 * Model_Roll + fng1;
}
else
{
    fx1 = 0.0;
    fy1 = 0.0;
    fz1 = 0.0;
}
```

where **ug** and **vg** are the velocities in the body frame, **ng** and **eg** are the north and east velocities in the local frame, **gp** is the aircraft velocity vector and **gd** is the aircraft heading, which enables the sideslip angle **gp** to be computed. Note that **bg1** is not the aerodynamic sideslip (which is negligible at taxiing speed below 30 Kt) and below 1 m/s, **bg1** is zero. As this example is for the nose wheel, the braking component is omitted from the computation of **fmu1**. For the main gear, **fmu2** is computed as follows:

```
fb2 = Model_LeftBrake * 10.0 * 2.0 * 0.263 * w / g;
fb2 = Maths_Limit(fb2, 0.0, fbmax);
fmu2 = -((0.015 + bof) * fabs(fng2) + fb2);
```

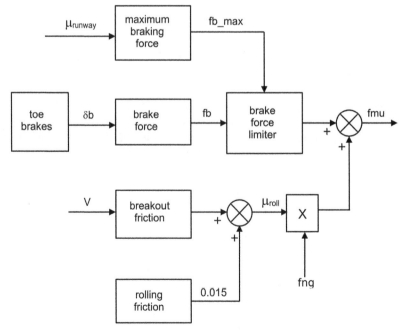

Figure 3.24 Wheel Drag Force.

The forces `fx1`, `fy1` and `fz1` are resolved in the body frame. These forces only apply if the oleo is under compression or, in other words, if `dst1` is negative; otherwise the forces are zero. The net forces from the three undercarriage assemblies are computed in SI units as follows:

```
Gear_Fx = (fx1 + fx2 + fx3) * 4.448222; /* convert to N */
Gear_Fy = (fy1 + fy2 + fy3) * 4.448222;
Gear_Fz = (fz1 + fz2 + fz3) * 4.448222;
```

Similarly, the moments in the body frame are computed in SI units as follows:

```
mx1 = fz1 * YL1 - fy1 * (-dst1 / 12.0);
mx2 = fz2 * YL2 - fy2 * (-dst2 / 12.0);
mx3 = fz3 * YL3 - fy3 * (-dst3 / 12.0);

my1 = -fz1 * XL1 + fx1 * (-dst1 / 12.0);
my2 = -fz2 * XL2 + fx2 * (-dst2 / 12.0);
my3 = -fz3 * XL3 + fx3 * (-dst3 / 12.0);

mz1 = fy1 * XL1 - fx1 * YL1;
mz2 = fy2 * XL2 - fx2 * YL2;
mz3 = fy3 * XL3 - fx3 * YL3;

Gear_Mx = (mx1 + mx2 + mx3) * 1.35581795; /* convert to Nm */
Gear_My = (my1 + my2 + my3) * 1.35581795;
Gear_Mz = (mz1 + mz2 + mz3) * 1.35581795;
```

The undercarriage forces **Gear_Fx**, **Gear_Fy** and **Gear_Fz** are added to the computation of the body frame forces and the moments **Gear_Mx**, **Gear_My** and **Gear_Mz** are added to the computation of the body frame moments in the equations of motion.

These equations provide the basis for general forms of aircraft undercarriage assemblies. As with the aerodynamic data and engine data, they depend on the provision of aircraft-specific data for the oleo characteristics, tyre characteristics, braking friction and so on. Without such data, it is difficult to estimate many of these terms and if incorrect values are used, the response of the undercarriage during taxiing and landing can be unrealistic. Nevertheless, the spring force needed to support the aircraft depends on the aircraft mass, and the oleo damping time constant is likely to be between 2 s and 10 s. Rolling friction and braking friction are universal and can be estimated from braking distances. The data on tyre pressures is published but tyre deformation data is limited to papers on vehicle dynamics. One final observation is that although offline testing is valuable, piloted tests for a range of taxiing tasks provide feedback on the quality of the undercarriage model, albeit a subjective validation.

Validation of the undercarriage depends on the availability of test data and invariably involves taxiing tests with pilots familiar with the aircraft type. Parameters to test include time and distance to brake from touchdown speed to taxiing speed, rudder effectiveness in steering at different speeds, tiller steering at low speed, effects of differential braking and ability to hold the brakes against specific engine thrust. From the developer's perspective, one very useful test is the drop test. Figure 3.25 shows the response of the undercarriage assembly for a simulated drop from 2 ft.

The oleo deflections are shown in inches, confirming the displacement under load against time in seconds, which shows that the spring response of the oleos is damped out after approximately 15 s.

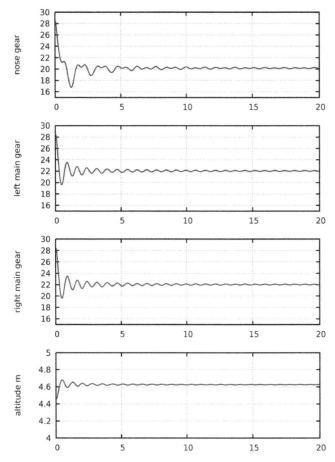

Figure 3.25 Oleo Response.

References

Anon. (1976), *U.S. Standard Atmosphere*, NASA-TM-X-74335.

Anon. (2004), *In-Flight Separation of Vertical Stabilizer, American Airlines Flight 587, Airbus Industrie A300-605R, N14053, Belle Harbor, New York, November 12, 2001*, Aircraft Accident Report NTSB/AAR-04/04, Washington D. C.

Babister, A. W. (1980), *Aircraft Dynamic Stability and Response*, Pergamon Press.

Beal, T. R. (1993), Digital Simulation of Atmospheric Turbulence for Dryden and Von Karman Models, *Journal of Guidance Control and Dynamics*, Vol. 16, No. 1, pp. 132–138.

Boiffier, J. L. (1998), *The Dynamics of Flight – The Equations*, John Wiley and Sons.

Etkin, B. and Reid, L. D. (1996), *Dynamics of Flight: Stability and Control*, John Wiley and Sons.

Fink, M. P., Delma, C., Freeman, Jr., and Greer, H. D. (1970), *Full-scale Wind-Tunnel Investigation of the Static Longitudinal and Lateral Characteristics of a Light Single-engine Airplane*, NASA TN-D-5700, Langley Research Centre.

Fogarty, L. E. and Howe, R. M. (1969), *Computer Mechanization of Six-degree of Freedom Flight Equations*, NASA CR-1344.

Hanke, C. R. (1971), *The Simulation of a Large Jet Transport Aircraft, Vol. I: Mathematical Model*, NASA CR-1756, Vol. II: Modelling Data, Boeing Report No. N73-10027.

Heffley, R. K. and Jewell, W. F. (1972), *Aircraft Handling Qualities Data*, NASA-CR-1244.

Maine, R. E. and Iliff, K. W. (1975), *A FORTRAN Program for Determining the Aircraft Stability and Control Derivatives from Flight Data,* NASA TN-D-7831.

McFarland, R. E. (1975), *A Standard Kinematic Model for Flight Simulation at NASA-Ames*, NASA-CR-2497.

McLean, D. (1990), *Automatic Flight Control Systems*, Prentice-Hall.

Moorhouse, D. J. and Woodcock, R. J. (1982), *Background Information and User Guide for MIL-F-8785C, Military Specification – Flying Qualities of Piloted Airplanes*, AFWAL-TR-81-3109, Wright-Patterson AFB.

Roberts, L. and Wan, T. (1985), *A Simplified Model of the Turbulent Microburst*, NASA CR-177108.

Spence, G. T., Le Moigne, A., Allerton, D. J. and Qin, N. (2007), Wake Vortex Model for Real-time Flight Simulation Based on Large Eddy Simulation, *AIAA Journal of Aircraft*, Vol. 44, No. 2, pp. 467–475.

Stinton, D. (1985), *The Design of the Aeroplane*, Sheridan House Inc.

Tomlinson, B. N. (1975), *Developments in the Simulation of Atmospheric Turbulence*, Royal Aircraft Establishment Technical Memorandum FS-46, Farnborough.

Worobel, R. (1972), *Computer Program User's Manual for Advanced General Aviation Propeller Study*, NASA CR-2066.

Worobel, R. and Mayo, M. G. (1971), *Advanced General Aviation Propeller Study*, NASA CR-114399, Hamilton Standard, Windsor Locks, CT.

4

Flight Control Systems

It wasn't that the X-1 would kill you, it was the systems in the X-1 that would kill you.

Chuck Yeager

4.1 Automatic Flight Control

A flight control system (FCS) in an aircraft is designed to improve the accuracy of navigation, to reduce pilot workload and to increase the overall safety of aircraft operations. A risk analysis is undertaken for all avionic equipment, including the FCS, where the likelihood of failure of the FCS is extremely low. If the FCS should fail, its impact either must be contained (often by means of duplicated redundant systems) or, alternatively, the flight crew must be able to identify a failed system and switch it out without endangering the aircraft. Consequently, an aircraft FCS is relatively simple in terms of design, to ensure the high degree of reliability required for the certification process.

Although an FCS in a flight simulator does not need to meet the same level of reliability, it must correctly replicate the behaviour of the equivalent FCS in all regimes of flight. Often, the data package for a flight simulator may contain the diagrams and details of the actual control systems used in the aircraft. However, if such data is not available, the simulator designer is required to implement these control functions to fully replicate the range of autonomous modes of the aircraft.

Students faced with the design of control systems are confronted with a plethora of control system design methods (Franklin et al., 1994), covering PID (proportional, integral and derivative) control, optimal control, adaptive control, predictive control, multi-variable control, H-infinity control and so on, plus numerous artificial intelligence (AI) methods including fuzzy logic, genetic algorithms and neural nets. It is not feasible to cover this wide range of design methods and the emphasis is to develop typical designs used in aircraft FCSs. These designs are not intended to be definitive; they provide a template, or starting point, for the development of control laws and FCSs and, consequently, many of these designs are based on standard PID control (Åström and Hagglund, 1995) and are open to criticism from control theory experts. The point is reiterated that these designs are intended as exemplars, illustrating the problems to be covered in simulating FCSs and enabling the developer to take these concepts further, in terms of effective control system design. For readers interested in the design of aircraft flight control laws, the topic is covered in depth in several definitive texts (Roskam, 2003; Stevens and Lewis, 2003; Etkin, 1996; Smetana, 1994).

Flight Simulation Software: Design, Development and Testing, First Edition. David Allerton.
© 2023 John Wiley & Sons Ltd. Published 2023 by John Wiley & Sons Ltd.
Companion Website: www.wiley.com/go/flightsimulationsoftware

In order to reduce pilot workload, automatic FCSs are used to control altitude, heading, airspeed and vertical speed. By providing control laws for these four modes, a pilot no longer operates the aircraft by means of the control column or engine levers but is able to select the altitude, heading, airspeed and rate of climb via a control panel. Automatic laws are also provided to intercept tracks to radio beacons. During an approach, automatic landing systems intercept and track the glide slope in the vertical plane and the localiser in the lateral plane, enabling flight crews to descend to 200 ft above the runway (or lower) without any manual intervention. Finally, the flight plan itself can be activated by the aircraft flight management system (FMS) to command the FCS to engage the guidance of the vertical and lateral segments of a flight plan.

In flight simulation, these control laws may be provided as diagrams in a data package or as transfer functions for the different modes. Alternatively, these functions are common to most transport aircraft and variations in the control laws arise from the mechanical dimensions and response of the primary actuators and the aircraft mass and inertias. However, the overall response of the automatic modes of modern transport aircraft are very similar and, without access to aircraft data, the simulator developer is required to implement the full range of automatic modes, such that the specific responses are in line with the operation of the aircraft in these modes.

4.2 Development of Flight Control Laws

In order to control altitude, heading, airspeed and vertical speed, controllers are developed to compute the inputs which are applied to the flight model and engine model to achieve the desired performance and response. In most cases, the controller is a transfer function which, in combination with the transfer function of the aircraft, provides the required response to selected altitude, heading, airspeed or vertical speed.

Given that the testing of complex software can take up to 50% of the project development time (Myers, 2004), modifying and testing software on a flight simulator requires setting up the simulator conditions and acquiring and analysing data from flight tests. Moreover, only one set of tests can be run at any time and the specialist nature of a flight simulator may require a pilot, who is experienced on the specific aircraft type, to undertake specific tests. An alternative approach is to transfer much of the development and testing to offline facilities.

4.2.1 The Case for Offline Development and Testing

Consider the diagrams shown in Figure 4.1.

In both diagrams, the main simulator modules are common and identical. The main difference is that in Figure 4.1(a) the simulator inputs are the pilot's controls and the outputs are the flight displays, including the IG images, whereas in Figure 4.1(b), the inputs are taken from a script defined in a text file and the outputs are written to further files, enabling the data to be plotted offline.

The significance of Figure 4.1 should not be overlooked. If the offline version is organised correctly, the main modules in both implementations should be identical. This means that a module or part of a module developed and tested in the offline version can be swapped into the online version directly, with no extra coding to accommodate changes to the software. Similarly, software running in the simulator can be used directly in the offline version. This observation is particularly relevant to university teaching, where only one or two students can use the simulator at any time,

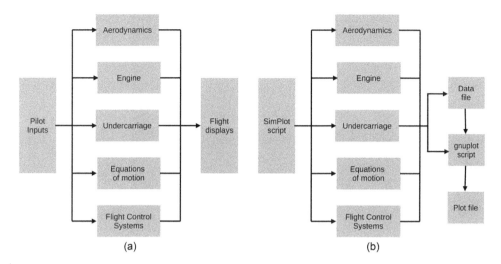

Figure 4.1 Flight Simulator Organisation.

but a large class can develop and test software in a classroom environment. Moreover, because there is no pilot in-the-loop, there is no real-time simulation requirement and offline simulation can run many times faster than real time.

There is one further advantage of offline testing that is common in both the validation and qualification of simulator software. If pilot inputs are captured in flight tests conducted in the simulator (or possibly from airborne tests) this data can also be used to provide sets of inputs for the offline simulation, to confirm that the offline software produces an identical set of results. In offline simulation the inputs are often 'perfect' doublets or steps, whereas in the aircraft a pilot is not able to apply an instantaneous input of exactly the correct magnitude. Actual pilot inputs, recorded during flight tests, can be used in offline testing for comparative analysis.

The use of offline testing is particularly important in the development of simulator software. It enables different versions of software to be tested under both online and offline environments with minimal disruption of the overall development. However, it is important to bear in mind that, if offline simulation is to be used in a development programme, then the interfaces and software environments need to be designed from the outset to accommodate this interchange of software. In many simulator programmes, offline testing is a precursor to online simulation, which subsequently provides the basis for airborne flight testing.

The term *script* is given to a set of commands to control the simulation, typically in the form of a text file. The commands are read from a file and executed. A *script* can be used to emulate pilot inputs, execute actual simulation code and record outputs. For example, a command might be to apply an elevator input or engage an automatic mode. Although these requirements are similar to Octave and MATLAB (Quarteroni and Saleri, 2003), even with the automatic coding tools provided for MATLAB and Simulink, it is far from straightforward to organise the code as a one-for-one replacement with simulator modules. There is one further consideration with this approach. If the offline and online code is identical, the real-time performance of the code can also be assessed offline with some degree of confidence that this performance will apply to code subsequently running in the simulator. However, if the offline code is based on MATLAB, there is a potential loss of performance as the code is interpreted or, alternatively, the code must be translated to a real-time programming language.

The software package developed for offline simulation used in this book is called SimPlot, which is a relatively small C program of just over 1000 lines. It contains a compiler to read the user commands and set up the appropriate conditions prior to a test and writes the outputs from the main modules to a text file, generating a gnuplot script file, which is used to produce graphical or plotted output. The choice of the gnuplot package is quite arbitrary. It is used because it is a popular plotting tool with numerous features for 2D and 3D plotting, is available for most platforms and is open source software. The majority of the plots obtained in this book were obtained using gnuplot. Nevertheless, it is straightforward to adapt the output generation to other plotting packages including Octave, MATLAB and Excel.

4.2.2   SimPlot

The SimPlot package performs the following functions:

- compiling the user commands specifying the inputs and outputs to be used in the simulation;
- initialising the simulation environment;
- running the simulation on a frame-by-frame basis;
- recording the specified outputs which are written to a data file;
- translating SimPlot plotting commands to generate an appropriate gnuplot script.

The script input file defines the inputs, the type or profile of inputs and the magnitude and duration of inputs. Typical inputs include a pulse, ramp, step, doublet, sinusoidal waveform or random noise. The outputs to be plotted are also defined in terms of the variables used in the simulation and the units and range of the variables. The script in Example 4.1 shows a small test program to measure aircraft performance during a ground roll from rest.

Example 4.1   A SimPlot Program to Test Ground Roll

```
set altitude 0 ft
set TAS 0 kts
set flaps 20
set gear 1.0
set engine_lever 0.2

plot TAS kts 0 180
plot thrust N 0 1000000
plot engine_lever 0 1.2

input engine_lever pulse 10 30 0.8

time 120 secs
```

The aircraft altitude is set to 0 ft, the true airspeed is set to 0 Kt, the flaps are set to 20°, the undercarriage is set to down (1.0) and the engine levers are set to 0.2 (flight idle). A pulse is applied to the engine levers and the engine lever input is increased by 0.8, that is, 100% throttle, starting at 10 s, for a duration of 30 s. Three variables are plotted, true airspeed (TAS) in the range 0–180 Kt, the total engine thrust in the range $0-10^6$ N and the engine lever in the range 0–1.2. The simulation is set to run for 120 s.

In this example, the data is generated as ASCII data in four columns, where the first column is the simulated time and the other three columns denote airspeed, engine thrust and engine lever position. A second file is also automatically generated to enable a plotting package to read the data

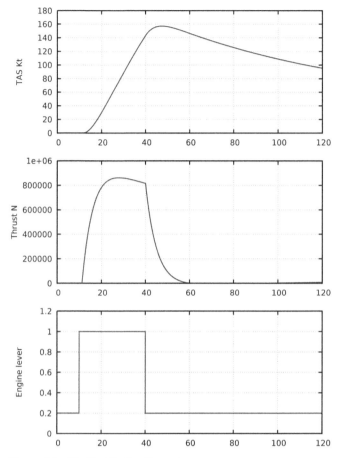

Figure 4.2 SimPlot Taxiing Example.

file and to generate a graphical plot suitable for insertion in a report or document. Numerous packages exist to plot data, which define the calls to set the plotting formats and type of output. An output plot produced by gnuplot for Example 4.1 is shown in Figure 4.2.

With offline development and testing of this form, the developer produces software for an algorithm or simulator module, writes a SimPlot script file to test specific aspects of the software and can check the results of the tests from the generated plots.

The commands defined in a SimPlot script cover the following classes of action:

- setting initial conditions;
- defining inceptor inputs and the type and duration of the inputs;
- defining variables to be recorded and their respective units;
- engaging and disengaging automatic flight modes;
- specifying the duration of a test.

The requirements of SimPlot are relatively basic. It should be easy to implement a test in terms of setting the conditions and recording the outputs: it should cover the full regime of flight testing appropriate to simulator validation and it should maintain compatibility with actual simulator code. Put another way, it should be possible to fully test simulator modules without requiring access to a flight simulator or a pilot.

4.2.3 Trimming

There is one important aspect of flight testing that is essential to replicate. Often a flight test is initiated from a trimmed state. Prior to an exercise, the pilot will trim the aircraft to ensure it is in a steady-state (non-accelerating) condition before applying the inceptor inputs. SimPlot replicates the pilot actions to trim an aircraft. In practice, most aircraft are symmetric about the roll and yaw axes, and trimming normally applies to the pitch axis, to set the elevator input and the throttle (or power) input.

For trimming in the longitudinal axis, it is necessary to determine the elevator and throttle settings, such that the aircraft is in a steady state, that is, the linear accelerations and the angular rates are zero. In other words, given a set of initial conditions, the pitch θ and the airspeed U are constant. The advantage for the simulator developer is that the equations of motion and the initial conditions are known and the rate terms can be ignored (they should be zero). From inspection of the equations of motion, the steady-state conditions occur if the body frame X and Z forces are zero. There is a slight complication in that the transients of a turbofan engine model and propeller dynamics can introduce a lag in the computation of engine thrust. However, as the equations defining the engine dynamics are also known, for the purposes of computing the trim conditions, the engine transients can be eliminated.

An iterative trimming method is used, selecting a trial value for the throttle and computing the elevator position by solving the equations of motion for one iteration, with all rate terms set to zero. As the throttle and elevator positions tend to their trimmed states, the forces in the X and Z axes in the body frame should also tend to zero. The method of convergence is to compute the resultant *Xforce* and *Zforce* terms; if *Xforce* > 0 there is excessive thrust and the throttle is reduced by a small amount and similarly, if *Xforce* < 0, there is insufficient thrust and the throttle is increased by a small amount. As *Xforce* approaches zero, the aircraft is trimmed in airspeed and the throttle position is the trimmed value. In fact, very small tolerances of 0.5 N were found to work satisfactorily for large transport aircraft. The algorithm is valid for any trimmed vertical speed as the airspeed and rate of climb are known, enabling the angle of attack to be computed relative to the flight path angle, giving a straightforward computation for the elevator.

The equations of motion are reasonably well-conditioned for most aircraft and the algorithm converges to the trimmed state within a few thousand iterations. Of course, it is possible to set initial conditions which will defy convergence to a trimmed state, for example, an airspeed well below the stall speed. In such cases, the number of iterations will exceed a predefined maximum, the solution will time out and the user can be notified that the trimming method has failed.

The algorithm, in pseudo-code, is summarised as follows:

```
inputs: the airspeed Uₜ, vertical speed Vₜ and altitude hₜ
set the initial values for elevator δe and throttle δp to mid-range
set the flap and undercarriage positions

compute the flight path angle  γ = tan⁻¹ (Vₜ/Uₜ)

set the initial angle of attack α = 0
set the iteration count tcount = 0

repeat
    compute air density ρ, density ratio ρ/ρ₀ and speed of sound a
    set the aircraft altitude and airspeed to the initial values hₜ and Uₜ
    set αw = α + wing incidence angle
    set θ = α + γ
```

```
        set w = Ut tan(α)
        set Vc = √(Ut² + w²)
        set Mach number M = Vc / a
        solve the engine equations to derive the current thrust
        solve the linear equations of motion to derive Xforce and Zforce
        δp = δp - Xforce / 10⁶
        limit δp to the range 0 to 1.0
        solve the angular equations of motion to derive δe
        tcount = tcount + 1
        if tcount > 10⁵
                the trimming has failed to converge (stop)
    until |Xforce| < 0.5 and |Zforce| < 0.5
    return α, δe and δp as trimmed initial values
```

where the subscript t denotes trim values. The value used to increment δp is based on the maximum values for *Xforce*. This selection is important; if the value is too large the convergence will be unacceptably slow and if it is too small, the throttle will oscillate between its minimum and maximum values. In practice, the method works well because the elevator has a small effect on airspeed and the throttle has a small effect on lift. The SimPlot program in Example 4.2 positions the aircraft at 3000 ft, with an indicated airspeed of 200 Kt and the gear down and the flaps set to 20°. The simulation runs for 120 s, plotting pitch, pitch rate, altitude, true air speed and elevator. Note the final line which activates the auto-trim function.

Example 4.2

```
set altitude 3000 ft
set IAS 180 kts
set flaps 20
set gear 0.0
plot pitch degs -15 15
plot pitch_rate deg/s -10 10
plot altitude ft 2500 3500
plot IAS kts 160 200
plot elevator degs -20 20
time 120 secs
autotrim
```

The output is shown in Figure 4.3(a), which shows that the aircraft maintains its initial state for 2 minutes with no fluctuations or drift. In this example, the solution converged after 4718 steps, with the following values:

```
trim: alpha= 5.25 deg, pitch= 5.25 deg, de=-4.82 deg, dp= 0.60 EPR= 1.12
TAS= 97.2 m/s (189 Kt) Mach=0.29 Alt=  914 m ( 3000 ft)
```

By default, the trim conditions are set for level flight, but if a vertical speed is set (positive for climbing, negative for descending), the trim conditions are computed for the specific rate of climb, as shown in Figure 4.3(b), where the following line was added to the SimPlot program:

```
set rate_of_climb 500 fpm
```

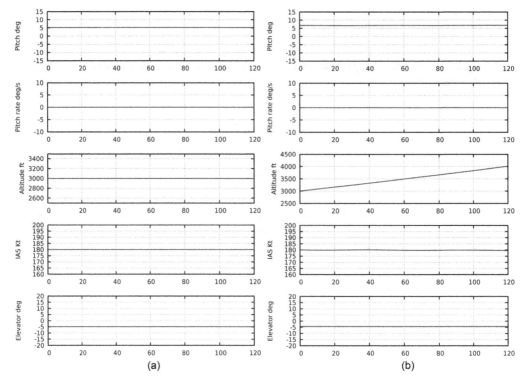

Figure 4.3 Trimming Results.

In this case, the solution converged after 4981 steps, with the following values:

```
trim: alpha= 5.20 deg, pitch= 6.71 deg, de=-4.32 deg, dp= 0.69 EPR= 1.18
TAS= 97.2 m/s (189 Kt) Mach=0.29 Alt=  914 m ( 3000 ft)
```

Note that in Figure 4.2(a), with level flight, the pitch angle and angle of attack are equal, whereas for climbing flight, the pitch angle is 6.71° and the angle of attack is 5.20°. The initial conditions for the throttle and elevator positions are computed at the start of the test and remain constant during the test.

4.3 PID Control

Many of the systems found in aircraft FCSs are in the class of controllers known as set-point systems. That is to say, the input is the desired value of a parameter and the control system attempts to move the aircraft actuators to achieve and maintain this value. A simple example of a set-point system in the home is a thermostat used to set the temperature of the house. If a change is made to the thermostat setting, the boiler will increase or decrease its output until the required temperature is detected by the thermostat and the boiler will provide sufficient output to maintain the set-point temperature, even if a door or window is subsequently left open or if the ambient temperature rises. A typical set-point system, implemented with a PID controller, is shown in Figure 4.4.

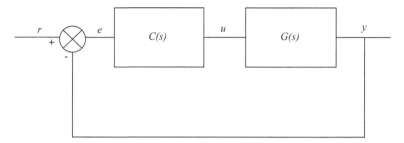

Figure 4.4 A PID Controller.

The error e is the difference between the actual system output y and the demanded output r. The controller output u is the input to the system dynamics $G(s)$ to achieve the desired output y. In practice, the control system may also have access to derivative and integral values of e.

The attraction of PID control is the simplicity of the controller $C(s)$, which is given by

$$C(s) = K_P + \frac{K_I}{s} + K_D s \tag{4.1}$$

where s is the Laplace operator. The design of the controller requires selection of K_P, K_I and K_D to achieve the desired overall system response, which is known as tuning the controller. Increasing K_P increases the loop gain. For systems where there is a steady-state error, K_I can reduce this error. K_D is usually introduced to improve the system response. However, these values are interdependent, for example, altering K_I may also affect the system gain or time response. PID controllers are often able to improve the response of first- and second-order systems and such systems are commonly found in flight dynamics. There is a further advantage of PID control in that it is defined as an equation in the Laplace operator s. Once the transfer function of the system and its controller is evaluated, techniques such as root locus or Nyquist, which are provided in modern control system design packages, can be used to determine the overall stability and response of the system. Indeed, several control system design packages are available specifically for PID applications. One other consideration with PID control is that all three terms are not necessarily required. For example, a system might not exhibit a steady-state error or an adequate response may be achieved with proportional control.

It is more convenient to rewrite equation 4.1 in the following form:

$$C(s) = K_P \left(1 + \frac{1}{T_I s} + T_D s \right) \tag{4.2}$$

where $T_I = \frac{K_I}{K_P}$ and $T_D = \frac{K_D}{K_P}$. Applying a first-order difference to approximate differentiation

$$\frac{dy}{dt} \Big|_n = \frac{y_n - y_{n-1}}{\Delta t} \tag{4.3}$$

Similarly, for integration, after n steps

$$\int y(t)dt = \sum_{k=1}^{n} y_k \Delta t \tag{4.4}$$

The discrete form of equation 4.2 is given by

$$m_n = K_p \left[e_n + \frac{1}{T_I} \sum_{k=1}^{n} e_k \Delta t + T_D \left(\frac{e_n - e_{n-1}}{\Delta t} \right) \right] \tag{4.5}$$

where m_n is the control applied at the nth step and Δt is the step length. This equation, which is executed at every time interval (or every frame in a real-time system), can be implemented in three lines of code as follows:

```
s = s + e;
m = Kp * e + Ki * s + Kd * (e - eold);
eold = e;
```

where **s** is the summation of errors over n steps (and should not be confused with the Laplace operator s) and **m** is the PID controller output. The variable **eold** is simply the value of **e** at the previous step. Note that the value of **s** must be initialised at the start of the simulation. Moreover, considerable care is needed if the simulation is reset at any time, to ensure that the values of **eold** and **s** are set to the values corresponding to the reset conditions. One concern is that the derivative term is a simple first-order difference, which can introduce noise in the controller. Bibbero (1977) proposes a fourth-order Adams–Bashforth method, where

$$\frac{dy}{dt} = \frac{1}{6T_s} (y_n + 3y_{n-1} - 3y_{n-2} - y_{n-3}) \tag{4.6}$$

where T_S is the time step and initial estimates are required for the first three steps.

PID control offers several advantages in developing control laws in a flight simulator:

- The equations for PID control can readily implemented in the C programming language.
- The rules for tuning PID controllers are straightforward.
- Control system design tools can be used to analyse the stability and response of the overall system.
- Testing and validation of the design are aided by the simplicity of the controller.
- A PID controller can, with care, be used with nonlinear systems.

However, PID controllers used in FCSs are not without problems. Firstly, differentiation can introduce noise. Secondly, the controller will be designed for a specific operating point and may not be stable at other points in the flight envelope. Thirdly, the desired PID controller output may exceed the acceptable range of the system inputs. For example, although the PID controller may demand 30° of elevator input, the elevator may be limited to 25°. In this case, the difference between the commanded input and the actual input may vary for a potentially long time, where the error term **e** is non-zero and the integral term **s** increases at each interval. When the system reaches its desired value, the accumulated value of **s** will continue to be (incorrectly) applied, a situation commonly referred to as *integrator wind-up*. The simplest solution is to inhibit the update of the integral term when the actuator is saturated. Unfortunately, this can introduce a problem: when the actuator comes out of the saturated condition, the value of **s** relates to the saturated state rather than the current state, resulting in a lag before the actuator responds correctly.

The solution is evident from the second line of the basic PID control law, which is modified as follows:

```
m = Kp * e + Ki * s + Kd * (e - eold) + M;
```

In the steady-state, **e** is zero and **M** equals the integral term. Although **M** may be zero in many systems, a minimum input value may still need to be applied at the minimum operating point. The equation is modified by adding **M**, which is effectively the means of setting the operating point for the controller.

Without including **M**, if the system is reset to a new operating point, this change may introduce a major disturbance rather than a smooth transition. For example, an autopilot may command a new altitude. What is required is known as a 'bumpless' transition. The solution is to compute the value of **M** under steady-state conditions, which is applied at the transition when the error term **e** is zero and **s** is set to zero and the output is set to **M**. Considerable care is needed in simulation, as the simulator may be repositioned or reset to a previous condition, and it is clearly undesirable to introduce unexpected perturbations under such conditions.

4.4 Automatic Modes

In manual flight, the pilot controls the throttle to manage airspeed, the elevator to control the pitch attitude, which establishes the rate of climb, and the ailerons to control bank angle, which determines the rate of turn. Responsibility for control of any or all of the airspeed, altitude, heading and vertical speed modes can also be given to the automatic control systems. If the flight plan is active, altitude, heading, airspeed and vertical speed computed by the FMS are passed to the FCS. Alternatively, the flight crew can select modes and enter altitude or heading or airspeed or vertical speed and override FMS selection. If the FMS is inactive or Flight Control Unit (FCU) modes are disengaged, control is returned to the flight crew for manual flight. At any time the flight crew can disengage the FCS.

These different levels of automatic modes provide a hierarchy of control systems, with control of body rates at the innermost level, height, heading, vertical speed and airspeed controlled at the next level, capture and tracking of signals from radio aids at a further outer layer and flight plan activation and execution at the outermost layer. At the lowest level, undesired lateral motion is filtered by a yaw damper and turn coordinator, the aircraft pitch attitude is controlled to set the flight path angle and the bank angle is controlled to produce the desired rate of turn. At the next level, airspeed is controlled by an auto-throttle so that climbing, descending and level flight can be established at a specific airspeed. Altitude is controlled by selecting a rate of climb, and heading is controlled by selecting a rate of turn. At the outermost level, the flight plan is executed by controlling airspeed, altitude, heading and vertical speed to follow the vertical and lateral profiles of the flight plan.

The examples given in this chapter are all based on a nonlinear model of the aircraft and engine dynamics of a Boeing 747-100 (Hanke, 1971).

4.4.1 Turn Coordinator

Aircraft are turned by applying ailerons to bank the aircraft, where the rate of turn is proportional to the angle of bank. If the net forces of lift and gravity are not acting perpendicular to the body frame, the flight crew and passengers are subject to undesirable lateral forces. The effect is similar to a car driving round a sharp bend, where the passengers feel themselves pushed against the car doors. The solution is relatively simple – the side force arises from the angle of sideslip, which can be minimised by application of the rudder during the turn. Indeed, this is the technique used in pilot training, where right rudder is applied in banking to the right and left rudder is applied in banking to the left. In manual flying the balance ball in a slip indicator is centred by applying rudder during the turn, whereas in automatic flight, rudder input is computed by the FCS to reduce the sideslip.

The controller can be implemented simply by applying rudder input to oppose the sideslip and rate of change of sideslip (McRuer et al., 1973), as follows:

$$\delta r = -(K_1 \beta + K_2 \dot{\beta}) \tag{4.7}$$

A turn coordinator is shown in Figure 4.5.

The equivalent code is given by the following:

```
const float k = 20.0;

void FCS_TurnCoordinator(float *Rudder)
{
    *Rudder = *Rudder - Model_Beta * k;
}
```

Figure 4.6 shows the effect of the turn coordinator in response to a pulse aileron input, where the aircraft is at 3000 ft, initially at 180 Kt with flaps 20 and gear up (a typical manoeuvring speed for holding). In this example, $k_1 = 30$ and $k_2 = 10$.

Although damping of the sideslip is shown clearly, there are two points to bear in mind with this example. Firstly, the controller is simplified in comparison with actual aircraft systems and, secondly, the rudder dynamics are not included. In practice, the time constant of the rudder actuator has a significant effect on the dynamic response and should be included in any detailed analysis.

4.4.2 Yaw Damper

When aileron input is applied to bank the aircraft, the resultant sideslip will produce a yawing moment which will subsequently induce a rolling moment, leading to oscillatory motion in the roll and yaw axes, which is known as *Dutch roll*. Generally, oscillatory yawing is undesirable. For example, during an approach, it is difficult to maintain heading if the aircraft oscillates in yaw in response

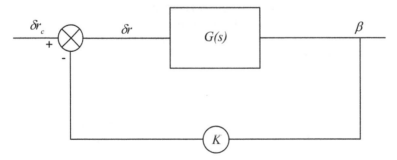

Figure 4.5 Turn Coordinator.

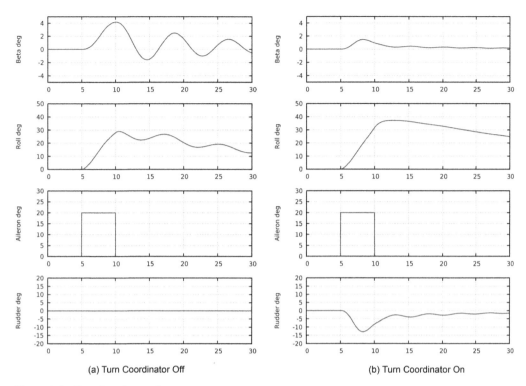

(a) Turn Coordinator Off   (b) Turn Coordinator On

Figure 4.6 Turn Coordinator Response.

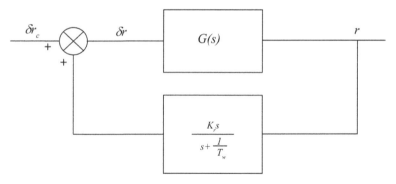

Figure 4.7 Yaw Damper.

to aileron or rudder input. Although applying rudder to counteract sideslip will dampen the Dutch roll, as in the case of a turn coordinator, it may also attenuate the turn rate. The function of a yaw damper is to supresses the Dutch roll response but without reducing the desired turn rate, which can be achieved using a washout filter in the feedback path with the following transfer function:

$$\frac{K_r s}{s + \dfrac{1}{T_w}} \tag{4.8}$$

where K_r is a constant and T_w is the time constant of the washout filter. A typical yaw damper configuration is shown in Figure 4.7.

Taking the general form of the transfer function $\frac{y(s)}{x(s)} = \frac{Ks}{s+a}$ for the feedback path, the equation

can be rearranged to give $y = kx - \frac{ay}{s}$, where $\frac{1}{s}$ is the integration operation and $a = \frac{1}{T_w}$. The

equivalent pseudo-code is as follows:

```
t = integrate(t, ay)
y = kx - t
```

where t is a temporary variable to hold the integral of **ay**. The important consideration is that the pilot still has sufficient authority to manoeuvre the aircraft, particularly for conflicting traffic or in response to an upset such as turbulence or engine failure.

The yaw damper code is as follows:

```
const float Kr=20.0
const float Tw=1.0

float FCS_YawDamper(float *Rudder)
{
    ydampint = Maths_Integrate(ydampint, ydamp);
    ydamp    = Model_R * Kr - ydampint / Tw;
    return Rudder + ydamp;
}
```

where **ydamp** is the filter output and **ydampint** is the intermediate integral term. Note that the rudder actuator transfer function is not included. The effect of the yaw damper is shown in Figure 4.8 for the aircraft response to a rudder doublet, that is, right rudder input (5 s) followed by left rudder input (5 s), which is the worst-case rudder disturbance. Notice the significant damping in yaw rate and the additional rudder input, with almost full rudder displacement provided by the yaw damper. In some aircraft, separate yaw dampers are provided for the lower and upper rudders.

A yaw damper is used on most transport aircraft to improve the overall ride qualities for passengers, particularly to reduce the effects of turbulence. Note the interaction of the yaw damper with the turn coordinator, where both functions are attempting to reduce sideslip. There is a potential for the two units to interfere with each other and, consequently, the turn coordinator and yaw damper are often combined in a single unit.

4.4.3 Pitch Rate Controller

By controlling the pitch rate, the pitch attitude can also be controlled. A pitch rate control law is shown in Figure 4.9, where the inner pitch rate controller provides a pitch rate proportional to the difference between the commanded pitch angle and the current pitch angle.

During banked manoeuvring, the vertical component is maintained by including the roll angle components (McLean, 1990), given by

$$e = q_c - (q\cos\phi - r\sin\phi) \tag{4.9}$$

where q_c is the commanded pitch rate. The code for the pitch rate controller is as follows:

```
float FCS_Pitch(float ThetaC)
{
    const float Pitch_Kp = -50.0;
    const float Pitch_Ki = -5.0;

    float de;
    float e;
    float s = Pitch_s;
    float Qc = ThetaC - Model_Pitch;

    Qc = Maths_Limit(Qc, -2.0 / ONERAD, 2.0 / ONERAD);
    e = Qc - (Model_Q * cos(Model_Roll) - Model_R * sin(Model_Roll));
    s += e;
    de = Pitch_Kp * e + Pitch_Ki * s;
    if (de > -1.0 && de < 1.0)
    {
        Pitch_s = s;
    }
    de = Maths_Limit(de, -1.0, 1.0);
    return de;
}
```

Values of -50 and -5 were used in the PI control law for the proportional and integral terms, respectively, and the elevator value is returned in the range ± 1.0. The pitch rate is limited to $2°/s$ on the basis that an excursion of $12°$ takes 5–6 s, giving the flight crew sufficient time to monitor the change of pitch. The response of the controller to a commanded pitch angle of $10°$ at 3000 ft, 250 Kt with the flaps and gear up is shown in Figure 4.10, together with the SimPlot script.

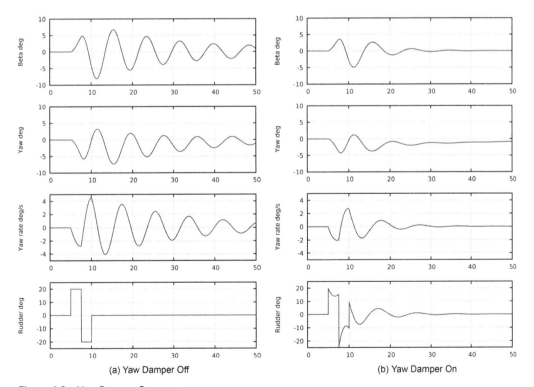

(a) Yaw Damper Off   (b) Yaw Damper On

Figure 4.8 Yaw Damper Response.

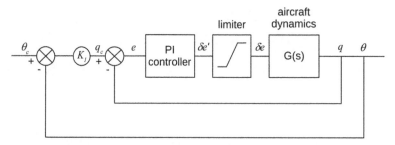

Figure 4.9 Pitch Rate Controller.

The test was conducted with the aircraft untrimmed, at an initial pitch angle of 0°. Note the pitch rate limit of 2°/s and the corresponding nonlinear response of the elevator.

A flight path angle controller can be implemented with a pitch angle controller, using the identity $\theta = \gamma + \alpha$. A similar algorithm is used for the flight path angle controller, with the code to compute **Gamma**, **Qc** and **e** modified as follows:

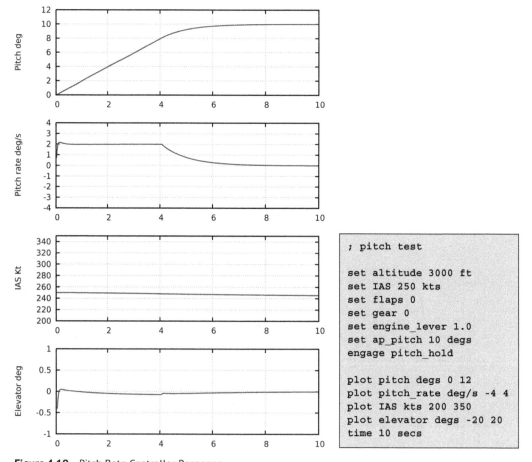

Figure 4.10 Pitch Rate Controller Response.

```
float Gamma = Model_Pitch - Model_Alpha;
float Qc    = (GammaC - Gamma) * 0.2;
float e     = Qc - (Model_Q * cos(Model_Roll) - Model_R * sin(Model_Roll)) - Model_UDot * 0.0015;
```

where **Gammac** and **Gamma** are the commanded and current flight path angles, respectively. The flight path angle controller is used in the altitude-hold and vertical-speed-hold controllers and in the auto-land controller. Note the additional feedback term from forward acceleration.

4.4.4 Auto-throttle

The operational handbook for an aircraft defines the aircraft performance in terms of airspeed. For example, after take-off, the initial climb with flaps will be at a predefined airspeed, but at different airspeeds as the flaps are progressively raised. Similarly, the descent segments will be flown at specific airspeeds and in particular the approach will be flown at reference speeds, depending on the aircraft mass and flap settings. In these situations, the pilot workload is considerably reduced if the airspeed is controlled by an auto-throttle, allowing the pilot to manage the vertical and lateral manoeuvring. Similarly, flight plan segments are likely to be based on economical or efficient speeds for aircraft operations and the FMS can select auto-throttle speeds for the different segments of the flight plan. A typical auto-throttle control system is shown in Figure 4.11.

In an aircraft, the airspeed U is derived from the air data computer and the acceleration dU/dt is provided by an accelerometer, whereas in simulation these variables are directly available in the equations of motion. The throttle position typically ranges from 0.2 (flight idle) to 1.0 (fully forward). In practice, the computation of the throttle position for large changes of speed can saturate, causing the integrator wind-up problem described in Section 4.3. The solution is to inhibit updating of the integral term when the computed throttle position is outside the operating range of the throttle lever (or levers). The code for an auto-throttle is given below, where $K_1 = 5.0$, $K_2 = 1.0$, $K_P = 1.0$ and $K_I = 0.02$. To avoid integrator wind-up, the updating of **SPD_s** is inhibited if the computed throttle position saturates.

```
float FCS_SPD(float Vref)
{
    const float SPD_Kp = 1.0;
    const float SPD_Ki = 0.02;
    const float K1     = 5.0;
    const float K2     = 1.0;

    float       IAS    = Model_U * sqrt(Weather_DensityRatio);
    float       e      = (Vref - IAS) * K2 - Model_UDot * K1;
    float       s      = SPD_s + e;
    float       tp     = SPD_Kp * e + SPD_Ki * s;

    if (tp > 0.2 && tp < 1.0)
    {
        SPD_s = s;
    }
    tp = Maths_Limit(tp, 0.2, 1.0);
    return tp;
}
```

The response of the auto-throttle accelerating from 240 to 270 Kt, in level flight, is shown in Figure 4.12. Note the slight decrease in pitch to maintain level flight with increasing airspeed.

The performance of the auto-throttle in the presence of turbulence, set to maintain 250 Kt at 5000 ft, is shown in Figure 4.13.

The design of an auto-throttle is particularly important in developing flight control laws for vertical and lateral modes, as airspeed is often constant for these modes. There are a few points to be considered with this algorithm. Firstly, it assumes instantaneous movement of the throttle,

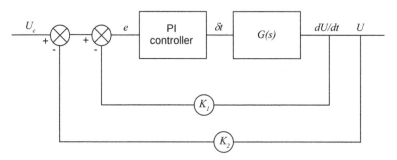

Figure 4.11 Auto-throttle.

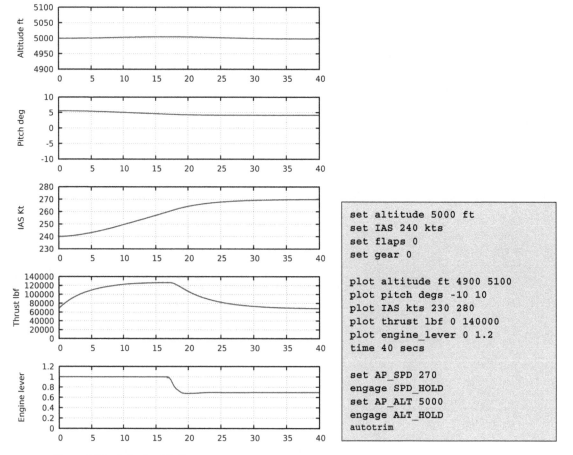

```
set altitude 5000 ft
set IAS 240 kts
set flaps 0
set gear 0

plot altitude ft 4900 5100
plot pitch degs -10 10
plot IAS kts 230 280
plot thrust lbf 0 140000
plot engine_lever 0 1.2
time 40 secs

set AP_SPD 270
engage SPD_HOLD
set AP_ALT 5000
engage ALT_HOLD
autotrim
```

Figure 4.12 Auto-throttle Response.

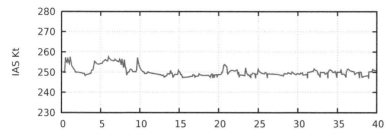

Figure 4.13 Auto-throttle Response in the Presence of Turbulence.

whereas in practice the throttles are driven by servomotors and their transfer functions should be included in a practical design. Secondly, the accuracy of U and dU/dt is far higher in simulation than measurements from aircraft sensors. Thirdly, the auto-throttle illustrates the problem of integrator wind-up in PI control and its solution. Arguably, this is a PID controller as dU/dt is also fed back to the error signal e.

4.4.5   Vertical Speed Hold

Climb and descent rates are determined by reference to the operational handbook for the aircraft, or as required in the flight plan and to comply with ATC requirements. Management of vertical speed can be implemented by control of the flight path angle as a function of airspeed. For normal operations, the climb and descent rates should not exceed 1000 ft/min, to reduce the 'ear-popping' effects noticed by passengers. Rate of climb values avoid reducing airspeed to values close to stalling speeds or exceeding airspeeds likely to cause structural damage to the aircraft.

Vertical speed control is achieved by commanding the flight path angle required for the demanded rate of climb and the TAS, as shown in the following code:

```
float FCS_VSPD(float Vref)
{
    float gs = sqrt(Model_Vn * Model_Vn + Model_Ve * Model_Ve);

    return FPA_Hold(Vref / gs);
}
```

Figure 4.14 shows a climb from trimmed level flight from 3000 ft at 250 Kt (flaps up and gear up) at a rate of climb of 1000 ft/min for 10 minutes.

The rate of climb is established within 13 s. Note the slight reduction in the flight path angle with increasing altitude to maintain a rate of climb of 1000 ft/min.

4.4.6   Altitude Hold

Above the transition altitude, air transport aircraft fly at flight levels spaced at 100 ft intervals, for example, FL234 corresponds to an altitude of 23,400 ft. Control of rate of climb is also useful for altitude control, as it can be used as a parameter to reach a desired altitude, where the rate of climb reduces as the altitude is reached. For example, more than 500 ft from the assigned altitude, the aircraft climbs or descends at a constant rate of climb, typically 1000 ft/min (fpm). If the aircraft is less than 500 ft above or below the assigned altitude, the rate of climb reduces linearly to zero at the assigned

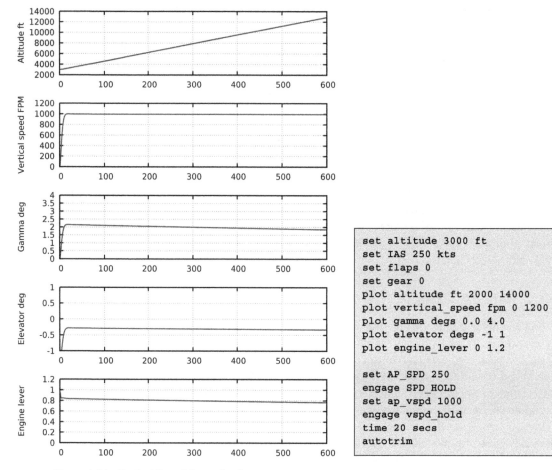

Figure 4.14 Vertical Speed Controller Response.

altitude. An altitude-hold control law is shown in the following code, where the reference altitude **Href** is given in metres and the rate of climb **vs** is limited to 5.08 m/s or 1000 fpm:

```
float FCS_ALT(float Href)   /* -ve up, metres */
{
    float vs = Maths_Limit(-Href + Model_Pz) * 0.03333, -5.08, 5.08);

    return FCS_VSPD(vs);
}
```

Figure 4.15 shows a climb from trimmed level flight at 3000 ft at 250 Kt to 8000 ft. Notice the levelling off from a vertical speed of 1000 fpm to zero. The controller commands the flight path angle control law as a function of the vertical distance between the aircraft altitude and the assigned altitude.

4.4.7 Heading Hold

In manual flying, aircraft heading is controlled by applying the ailerons to bank the aircraft, where the aircraft heading is displayed on an HSI or compass. It is important to appreciate the difference

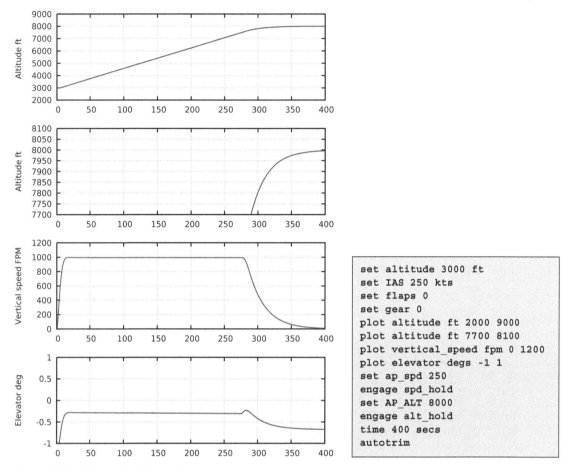

```
set altitude 3000 ft
set IAS 250 kts
set flaps 0
set gear 0
plot altitude ft 2000 9000
plot altitude ft 7700 8100
plot vertical_speed fpm 0 1200
plot elevator degs -1 1
set ap_spd 250
engage spd_hold
set AP_ALT 8000
engage alt_hold
time 400 secs
autotrim
```

Figure 4.15 Altitude-Hold Controller Response.

between heading and track: the aircraft heading is the instantaneous direction of the aircraft, whereas the track is the flight path followed by the aircraft over the ground. In the presence of wind, the track is the vector addition of the aircraft velocity vector and the wind vector. On modern aircraft, heading or track modes can be selected as the wind component can be derived from the navigation sensors.

In an automatic mode, the flight control law computes the aileron input to turn the aircraft until the desired heading is reached. However, for heading hold autopilots, the angle of bank is usually constrained to 20–25° in order to limit wing loading. In addition, the stall speed of the aircraft reduces with increasing bank angle. It is normally assumed that the aircraft heading is controlled by aileron input, with rudder inputs added separately by a turn coordinator. The turn rate is given by

$$\dot{\psi} = \frac{\phi g}{V} \tag{4.10}$$

where ϕ is the bank angle and V is the TAS. The turn rate is proportional to the bank angle and reduces with increasing airspeed. For transport aircraft, turns are mostly executed when the flight plan is active, or during holding patterns or in an approach. In these situations, the turn rate is typically less than 5°/s. In situations where a faster turn rate is required, for example in an evasive

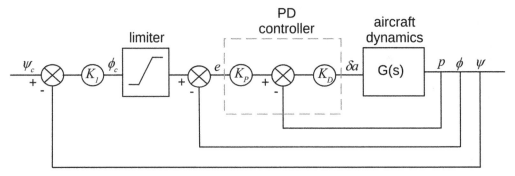

Figure 4.16 Heading Hold Controller.

action, the flight crew have the option to disengage the heading hold autopilot and bank the aircraft manually. A heading hold controller is shown in Figure 4.16.

The heading hold controller contains an inner bank angle PD controller, where the input is the commanded bank angle, with the PD control law given by

$$e = K_P(\phi_c - \phi) - K_D p \tag{4.11}$$

where p is equivalent to the roll rate $d\phi/dt$. There is no requirement for an integral term in the PID controller because the aileron value is zero when $\phi_c = \phi$ and $p = 0$. The commanded bank angle is limited to ±25° and by the gain K_1, which determines the heading value as the commanded bank angle reduces. For example, up to 10° of heading difference, the bank angle could be limited to 25°, reducing linearly to zero as the heading reaches the assigned heading. The code for the bank angle hold controller is given by the following code:

```
float FCS_BankAngle(float RollC)
{
    const float Bank_Kp = 3.0;
    const float Bank_Kd = 3.0;
    float      da;

    RollC = Maths_Limit(RollC, -DEG25, DEG25);
    da = (RollC - Model_Roll) * Bank_Kp - Model_P * Bank_Kd;
    da = Maths_Limit(da, -1.0, 1.0);
    return da;
}
```

where **DEG25** is 25° (radians) and the PD controller values **Kp** and **Kd** are 3.0. The function **BankAngleHold** returns the normalised aileron position, limited to the range ±1.0, to acquire or maintain the commanded bank angle **RollC**.

The heading hold controller is given by the following code:

```
float FCS_HDG(float HdgRef)
{
    float dHdg;
    float trc;

    HdgRef = HdgRef + (float) (AeroLink_NavPkt.MagneticVariation); /* degrees true */
    dHdg   = HdgRef - Model_Yaw;
    dHdg   = Maths_Normalise(dHdg);
```

```
    trc     = dHdg * 0.15;   /* (20 deg -> 3 deg/s */
    trc     = Maths_Limit(trc, -DEG3, DEG3);

    return FCS_BankAngle(trc * Model_U / Model_G);
}
```

The parameter **HdgRef** is the commanded heading in degrees magnetic, which is defined by the true heading plus the current magnetic variation. **dHdg** is the heading error which is normalised to the range ±180°. **trc** is the commanded turn rate, limited to 3°/s. The commanded bank angle is given by $\phi_c = \dfrac{\dot{\psi}_c u}{g}$. The response to a heading hold change from 30°M to 80°M at 3000 ft, 180 Kt and flaps 20, is shown in Figure 4.17.

The response was recorded with the turn coordinator and yaw damper enabled. The airspeed, altitude and flap settings are typical of values used in holding patterns. The bank angle stabilises at 25° with a turn rate of 2°/s during the main part of the turn. The completion of the turn and the

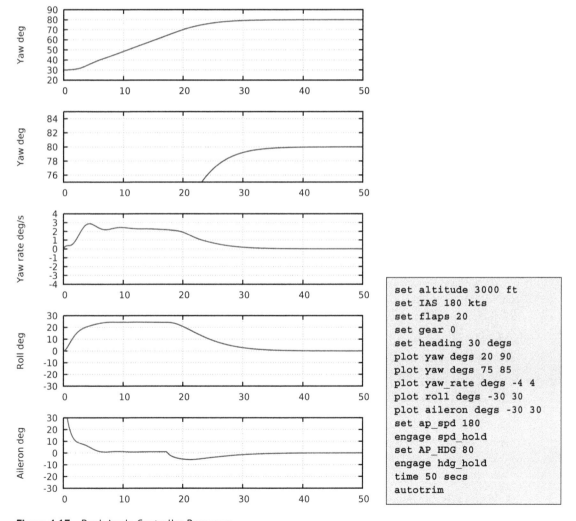

Figure 4.17 Bank Angle Controller Response.

reduction of the bank angle to zero is shown in the lower trace of the yaw angle. The turn through 50° is completed in 30 s. On the one hand, if the roll response is sluggish, it will take an unnecessarily long time to acquire a new heading. On the other hand, if the response is too fast, the roll input may not be able to reduce the turn rate sufficiently, overshooting the required heading.

4.4.8 Observations on Automatic Modes

The automatic control of airspeed, altitude and heading enables an aircraft to be flown autonomously, where the flight crew enter settings and monitor the displays. The controls outlined in this section provide the primary autopilot functions found on most modern transport aircraft. Often, with simulator documentation, there is a warning '*Not for use in flight*' and this applies to the examples given earlier. They are intended only to illustrate the basis of automatic control but they are not definitive. Experts in control theory will undoubtedly feel that better controllers could have been introduced. However, the aim of this textbook is not to focus on control theory, which would require far more description, but to cover the basics of each controller and provide working software, as a starting point for design, with the potential for improvement. One other point to note with these examples is that they have been tested with the nonlinear aerodynamic and engine models used in a real-time flight simulator. For analysis in most control design packages, both the aircraft dynamics and engine dynamics would need to be linearised. This comment is not to disregard the power of modern control system design packages but to observe that a design can be developed up to a point using a design package; however, eventually, it needs to be coded and tested in the real-time environment of an actual simulator. Students of flight simulation need to appreciate this distinction, otherwise they are likely to find that a controller developed in a control design package may have a very different performance when the controller is implemented in a real-time flight simulator.

4.5 Airbus Control Laws

In most aircraft developed since the 1930s, the movement of the elevators and ailerons is directly proportional to movement of the control column (or centre-stick) for both hydraulic actuation and direct mechanical linkage. However, aircraft designed by Airbus differ in two ways. Firstly, the pilot moves a side-stick rather than a control column and, secondly, the side-stick is connected to computer systems which acquire the side-stick position and move the control surfaces. This design decision, inherent in nearly all Airbus aircraft, has several implications:

- The feel or loading of the side-stick does not vary with airspeed, whereas conventional controls become much stiffer at high airspeed.
- The side-sticks of the pilot and co-pilot are not connected – the control logic decides how to combine the two inputs.
- The relationship between side-stick movement and the aircraft response is programmable – the control law is implemented as an algorithm, allowing different control laws to be applied in different regimes of flight.
- Potentially dangerous manoeuvres by a pilot, for example, inputs that could lead to stalling, can be over-ridden by the computer.

It is probably fair to say that Airbus have been rather guarded about the details of their control laws, releasing mostly general descriptions of their algorithms (Favre, 1994). Nevertheless, there is some information in the public domain to attempt to implement control laws that are close to the versions used in Airbus aircraft. The basic difference between pilot inputs in classical aircraft and Airbus aircraft is that the pilot controls pitch rate and roll rate in an Airbus aircraft. For example, in the pitch axis, a small displacement of the side-stick results in a small pitch rate whereas a large displacement produces a large pitch rate. In the pitch axis, there is a further distinction that the pilot also commands the normal force, known as N_Z, which is blended with the pitch rate.

The corollary of these conventions is that, if the side-stick is placed in the centre position, it commands zero pitch rate and zero roll rate and the aircraft should maintain its pitch and roll attitude, independent of changes in altitude, airspeed, bank angle or engine settings. A further implication is that there is no need for separate trimming. To hold the aircraft in its present state, the pilot moves the side-stick to the centre position. Of course, there are strong arguments for both the advantages and limitations of such a scheme and this textbook is not intended as a forum for these discussions.

The longitudinal control law is known as a C^* control law, which combines pitch rate demand with normal force (N_Z) demand. In other words, moving the side-stick commands a G force and a pitch rate. With the side-stick centred, the G force is 1.0 and the pitch rate is zero. The lateral control law is a rate controller controlling the aircraft rate of roll. In addition, the controllers also apply limits to manoeuvres in order to ensure that the aircraft operates within a safe region at all times in normal flight.

4.5.1 Pitch Normal Law

The normal force N_Z is given by

$$N_Z = \frac{\dot{w} + pv - qu}{g} \tag{4.12}$$

where $\dot{w}$ is the acceleration along the z axis in the body frame, p is the roll rate, q is the pitch rate, u is the velocity along the x axis, v is the velocity along the y axis and g is gravitational acceleration. The pitch normal law is shown in Figure 4.18.

The inputs to the control law are the side-stick position δe (normalised to the range ± 1), the pitch rate $d\theta/dt$ (given by q), the ratio $cos(\theta)/cos(\phi)$ and the normal force N_Z. The output is the commanded elevator angle. The upper part of the diagram shows the computation of N_Z and the lower part shows the combined Nz and pitch rate control law used to compute the elevator position. Two tests were flown at an altitude 3000 ft, an airspeed 180 Kt, with flaps 20° and undercarriage up, as shown in Figure 4.19. In Figure 4.19(a), the side-stick is pulled backwards after 3 s, producing a pitch rate of 1.5°/s. The pitch angle increases to 10° and then, after 12 s, the side-stick is pushed forwards, producing a pitch rate of −1.8°/s, returning the pitch attitude to 2.5° nose-up. The top trace shows the actual pilot side-stick input and the lower trace shows the elevator angle computed by the control law. The coding of the control law is straightforward, as follows:

```
float NzLaw(float Elevator)
{
    float c = cos(Model_Pitch) / cos(Model_Roll);
    float dPitch = Model_Q * cos(Model_Roll) - Model_R * sin(Model_Roll);
    float Nz = (Model_Lift * cos(Model_Alpha) +
                Model_Drag * sin(Model_Alpha)) / (Aero_Mass * Model_g);
    float Nzc = -Elevator + c - K1 * dPitch;
```

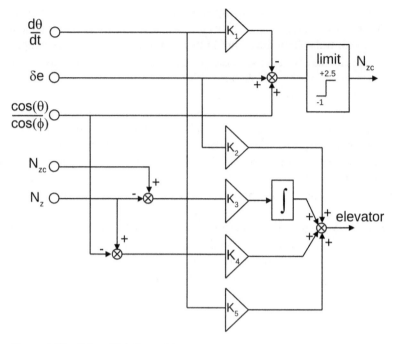

Figure 4.18   Airbus Pitch Normal Law.

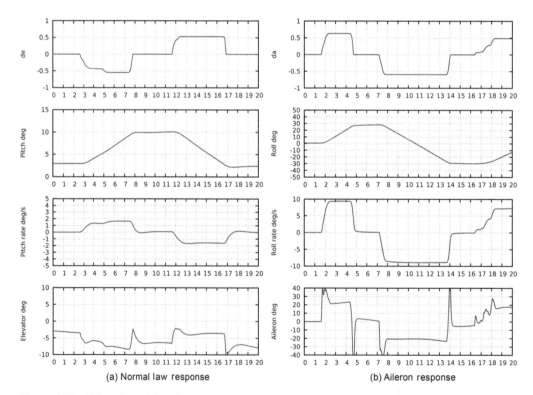

Figure 4.19   Airbus Control Law Responses.

```
        float dNz;

        Nzc = Maths_Limit(Nzc, -1.0, 2.5);
        dNz = Nzc - Nz;
        Nz_s = Maths_Integrate(Nz_s, dNz * K3);
        return Elevator * K2 + Nz_s + (Nz - c) * K4 + dPitch * K5;
    }
```

The following gains were used for the controller: $K_1 = 10.0$, $K_2 = 0.5$, $K_3 = -0.8$, $K_4 = 1.0$ and $K_5 = 10.0$. Note that, for the actual aircraft control laws, these values are modified according to the flight conditions, typically as functions of the TAS V_c and the aircraft configuration (flaps, spoilers and undercarriage).

4.5.2 Roll Rate Law

The roll rate control law provides a roll rate that is proportional to side-stick deflection and can be implemented as a PID controller as shown in Figure 4.20. The response to aileron input is shown in Figure 4.19(b), where right aileron is applied, producing a roll rate of 9°/s as the aircraft rolls to 29°. After 7.5 s, left aileron is applied producing a roll rate of −9°/s as the aircraft rolls to −30°. The top trace shows the actual pilot side-stick input and the lower trace shows the aileron input computed by the control law. After 17 s, the aircraft is restored to a wings-level attitude. Note that this response is also influenced by the response of the turn coordinator and the yaw damper. The aileron input δa is normalised to the range ±1. The value of K_{in} limits the maximum roll rate to 15°/s. The following gains were used for the PID controller: $K_p = 10.0$, $K_i = 0.5$ and $K_d = -5.0$. The code for the roll rate controller is as follows:

```
float RollRateHold(float RollRateC)
{
    float da;
    float olds = RollRate_s;
    float e = RollRateC * RollRate_Kin - Model_P;

    RollRate_s = RollRate_s + e;
    da = e * RollRate_Kp + RollRate_s * RollRate_Ki  + (e - RollRate_eold) * RollRate_Kd;
    RollRate_eold = e;

    if (da >= 1.0)
    {
        da          = 1.0;
        RollRate_s = olds;
    }
    else if (da <= -1.0)
    {
        da          = -1.0;
        RollRate_s = olds;
    }
    return da;
}
```

To avoid integrator wind-up, the integral term of the controller is not changed if the output saturates.

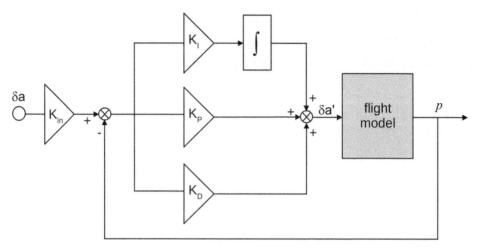

Figure 4.20 Airbus Roll Control Law.

These examples of Airbus controllers are considerably simplified. They illustrate a version of the laws for the basic modes. In the actual laws, many of the constants vary with flight conditions and there are alternative versions of the control laws. The software referenced in the appendix includes several of the protections that apply to Airbus aircraft, including limiting the angle of attack, pitch and roll. As with many of the examples in this book, they are working algorithms and are provided as a template to enable the reader to develop new algorithms or to provide a reference version for comparison with prototype versions.

4.6 Tracking

In modern aircraft, the position of the aircraft can be computed with reasonable accuracy and the positions of radio navigation aids are stored in an on-board navigation database, usually as part of the FMS. This information enables an aircraft to track directly towards (or from) a location or to intercept and fly along a track between two locations. In addition, with the management of airspeed and altitude, it is possible to follow a predefined 3D flight path. One further consideration is that navigation of this form will be affected by wind, which is likely to vary with location, altitude and time. Strictly, this last consideration is not a major problem in simulation. Firstly, if the aircraft position is known accurately, it is straightforward to compute drift resulting from wind. Secondly, in a simulator, the actual wind magnitude and direction will be known, potentially obviating the need to estimate the wind vector.

Although the availability of GPS has greatly improved navigation, the tracking in en-route navigation and approach guidance uses the localiser signals provided by VHF omni-directional range (VOR) and instrument landing system (ILS) radio aids, respectively, which are covered in Section 5.7. It is straightforward to compute the relative bearing of a navigation aid from the aircraft, as shown in Figure 4.21.

Assume that the aircraft at position A needs to intercept the track from B to C. The relative bearing of C from the aircraft is α and the relative bearing of B from C is β. In the case of VOR and ILS radio aids, β will be defined in a navigation database while the angular offset ε from BC is provided by a sensor. Alternatively, in the case of an FMS, the positions (latitude and longitude) of B and C may be specific waypoints rather than radio aids. The aircraft needs to fly in the direction from A to P. The choice of the point of interception P is important. If the aircraft flies in the direction towards AQ, it will intercept BC at right angles and will fly past BC, needing to recapture the track.

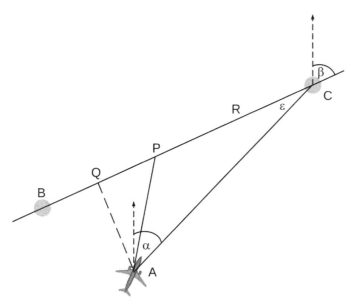

Figure 4.21 Interception of a Track.

Alternatively, if the aircraft flies in the direction towards AR, it may fail to intercept BC before reaching C.

Three possible methods can be used to compute the aircraft heading to intercept the track BC:

1) The point P can be computed from the geometry of the positions B and C. The choice of P can be fairly arbitrary, for example, three-quarters of the distance from C to Q. The heading h is given by the direction of the line AP.
2) If the relative bearing angles α and β are known, the commanded aircraft heading $h = \beta - g(\beta - \alpha)$, where g is a gain (typically 6).
3) If the offset error ε is known, the heading $h = \beta - g\varepsilon$, where g is a gain. This is similar to point 2, but ε is determined from the signal transmitted by the radio aid at C.

As the aircraft approaches the track, α will converge towards β until $h = \beta$. The method of interception is based on continuously re-computing and refining the heading h. As the aircraft closes towards BC, the point P (extrapolated from the aircraft position and heading) moves slightly closer to C and the aircraft heading increases (in this example) until the aircraft reaches BC, at which point the aircraft heading is β, with the aircraft positioned on the track BC, pointing directly towards C. By continuously updating the aircraft heading, the aircraft will make a curved path from the start of the interception to the completion. The curvature of the path depends on the choice of g. If it is too small, the curvature will be too gradual but if it is too large, the curvature will be too acute, with the possibility of overshooting BC. The interception is illustrated in Figure 4.22.

Initially, the aircraft at A flies towards P. At the position A_1, the aircraft points towards P_1 and flies along the track A_1P_1. At the position A_2, the aircraft points towards P_2 and flies along the track A_2P_2. At the position A_3, the aircraft points towards P_3 and flies along the track A_3P_3, and so on. By repeatedly computing the heading, the aircraft follows a curved track to intercept the desired track to C and also turns, so that its heading will be aligned with the track when it is intercepted. Even on the track to C, the process is still repeated and any deviation from the track results in recapture of the track. The acquisition of track offset from a VOR station is described in Section 5.7. The code to intercept a VOR track is given by the following function:

```
float FCS_LOCHold()
{
    float h;
    float dh;
    float error;

    h     = Maths_Rads((float) AeroLink_NavPkt.HSI_Crs);
    error = AeroLink_NavPkt.NAV1.LocaliserError;

    if ((error >= -DEG85) && (error <= DEG85))    /* TO */
    {
        /* do nothing */
    }
    else if (error <= -DEG95)    /* FROM */
    {
        error = -Maths_PI - error;
    }
    else if (error >= DEG95)     /* FROM */
    {
        error = Maths_PI - error;
    }
    else
    {
        error = 0.0;   /* cone of confusion */
    }

    dh = error * 6.0;
    dh = Maths_Limit(dh, -DEG60, DEG60);
    h = h + dh;
    h = Maths_Normalise(h);
    return FCS_HeadingHold(h);
}
```

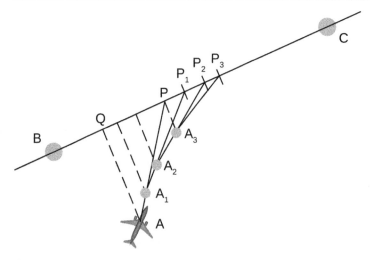

Figure 4.22 Gradual Interception of a Track.

where **h** is the VOR radial, which is the relative bearing of the track to the VOR station, and **error** is the angular offset of the aircraft from the VOR radial. If the aircraft is flying towards the VOR station and is inside a cone ±85° from the radial, or the aircraft is flying away from the VOR station and the error is less than −95° or greater than 95°, the track is captured as above, otherwise the VOR signal is not valid. The aircraft heading is given by **h=h+dh** where **dh=error*6.0**. A gain of 6 is a

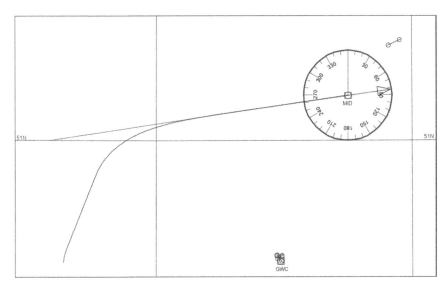

Figure 4.23 Capture of a VOR Radial.

compromise to capture the localiser at an angle of interception that is neither too shallow nor too steep. Note that that the capture angle is limited to ±60° and the heading commanded by the heading hold autopilot is normalised to the range ±180°. Figure 4.23 shows the capture of the 082 radial to Midhurst VOR (MID) from 25 nm south-west of the VOR.

Note that the intercept is a straight line until the error angle reduces to less than 14° and that the aircraft maintains the track after passing overhead the VOR.

4.7 Auto-land

An aircraft ILS, which is described in Section 5.7, receives two signals: a localiser error indicating the angular error from the (conceptual) line extending from the runway centreline and a glideslope error indicating the angular error from the runway glide path (typically 3°). The ILS localiser is more accurate than a VOR localiser (but has a shorter range) and its datum is the runway centre-line. The orientation of the runway, known as the QDM, is published in navigation charts. In an ILS approach, the localiser indicates the lateral angular offset. The ILS receiver indicates the vertical angular offset from the specific glideslope for the runway. Both receivers provide inputs to an ILS display to enable pilots to acquire and follow the localiser and glideslope.

The automated version of the ILS is known as an auto-land system (Hogge, 2004). For an airfield equipped with a CAT-IIIC ILS and an aircraft equipped with the appropriate ILS receiver, a fully trained flight crew are able to fly an approach to touchdown, even in zero visibility. In these conditions, the auto-land system is responsible for manoeuvring the aircraft and the role of the flight crew is to monitor the system and the flight displays and to take control of the aircraft for a go-around in the event of any failures.

An ILS localiser is illustrated in Figure 4.24 (not to scale). The receiver detects the localiser error ε and the control law determines the intercept track to acquire the localiser, which is similar to the VOR localiser tracking law. The aileron input computed by the *Heading Hold* function is given by

$$\delta a = HeadingHold\,(q + 10\varepsilon) \tag{4.13}$$

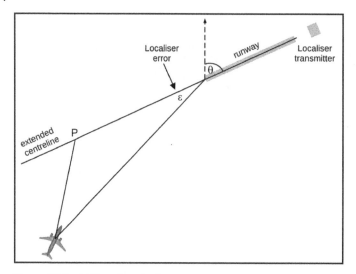

Figure 4.24 ILS Localiser Capture.

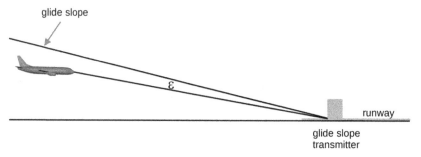

Figure 4.25 ILS Glideslope Capture.

where ε is the localiser error detected by the ILS receiver, q is the runway QDM and δa is the commanded aileron angle. The gain of 10 is larger than the gain used for the VOR localiser, as the ILS localiser measurement is more sensitive. Below 50 ft, close to touchdown, this gain is reduced to 1.0. After touchdown, the localiser tracking enables the aircraft to follow the runway centreline until the aircraft speed reduces to taxiing speed.

The ILS glide slope is shown in Figure 4.25. The tracking algorithm is similar to the localiser. The glide slope angle for the runway is defined in the navigation database and the glide slope error determines the commanded flight path angle to intercept the glide slope, given by

$$\delta e = -g_{ref} + 5\varepsilon \tag{4.14}$$

where ε is the glide slope error detected by the ILS receiver, g_{ref} is the specific runway glide slope angle and δe is the commanded elevator angle.

Below 50 ft, the throttle is set to flight idle and the control law commands a flight path angle of $-0.25°$ (slightly nose-up). Once in contact with the runway, the auto-land commands a flight path angle of half the pitch angle. During an approach, the glide path is captured from below the glide slope and the controller will command a flight path angle of zero to fly level until the glide slope is intercepted.

The code for the autopilot is as follows:

```
void FCS_Autoland(float *Elevator, float *Aileron)
{
    float Qdm;
    float h;
    float fpa;
    float gserr;

    if (AeroLink_NavPkt.ILS1.ILSBeacon)
    {
        Qdm = (float) (AeroLink_NavPkt.ILS1.RunwayQdm);
        Qdm = Maths_Normalise(Qdm);
    }
    else
    {
        return;
    }
    h = -Model_Pz + (float) (AeroLink_NavPkt.GroundLevel) + Aero_CGHeight;
    if (h < 15.0)
    {
        *Aileron = FCS_HeadingHold(Qdm + 1.0 * (float) (AeroLink_NavPkt.ILS1.LocaliserError));
        if (Model_OnTheGround)
        {
            *Elevator = PitchHold(Model_Pitch * 0.5);
        }
        else
        {
            *Elevator = FPAHold(-0.005);
        }
    }
    else
    {
        gserr = (float) (AeroLink_NavPkt.ILS1.GlideSlopeError);
        gserr = Maths_Limit(gserr, -DEG1, DEG1);
        fpa = -DEG3 + 5.0 * gserr;
        if (fpa > 0.0)
        {
            fpa = 0.0;
        }
        *Aileron   = FCS_HeadingHold(Qdm + 10.0 * (float) (AeroLink_NavPkt.ILS1.LocaliserError));
        *Elevator = FPAHold(fpa);
    }
}
```

The localiser and glide slope errors are computed in the navigation equations, described in Section 5.6. Note that, for reasons of brevity, this is an example of a very basic auto-land algorithm and omits any compensation for wind. The tracking is based on the functions developed in Section 4.6 for heading hold and flight path angle hold.

Figure 4.26 shows an auto-land into an airport from approximately 10 nm with the aircraft initially positioned at 2000 ft and offset from the centreline, in nil wind. Note the change in pitch attitude when the undercarriage is lowered and full flap is selected 6 nm from touchdown. The auto-throttle reference speed is set to 156 Kt. The horizontal axis shows the distance to touchdown in nautical miles.

The top trace shows the glide path, with the aircraft flying level below the glide path. The middle line is the 3° glide slope (not to scale) with the lower and upper limits of 2.3° and 3.7°, respectively.

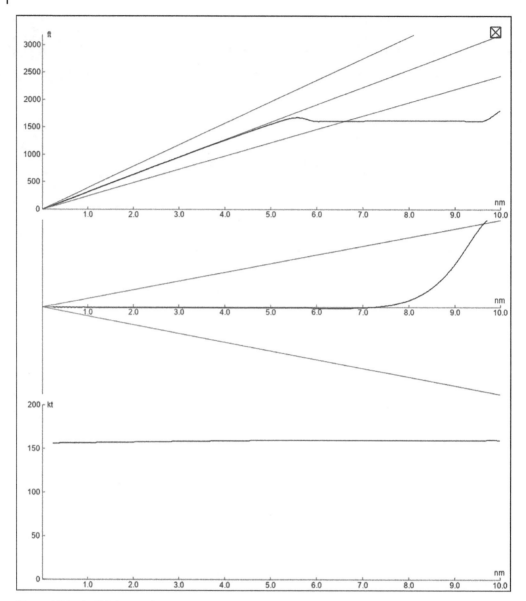

Figure 4.26 An ILS Approach.

The middle trace shows the localiser, with the aircraft intercepting the localiser from the right with the localiser limits ±2.5°. The lower trace shows the airspeed during the approach, with the auto-throttle maintaining a reference speed of 156 Kt.

4.8 Flight Director

The FCS computes the elevator and aileron positions in vertical and lateral guidance, respectively. If the flight director is enabled, the difference between the required pilot elevator and aileron inputs is displayed, typically in the form of crosshairs on the attitude indicator, indicating the control action. For example, if the vertical bar is offset by a small amount to the left, a small corrective input of left aileron is required, or if the horizontal bar is well below the centre position, a relatively large backward elevator input should be applied. This response is very similar to an auto-land controller in the sense that the display shows the corrective action for the elevator and aileron. However, there is an important distinction. In an auto-land system, the elevator and aileron move relatively quickly in response to the controller and the stability of the system is based on coefficients with relatively small time constants. In the case of a flight director, the pilot monitors the display and responds to the information by moving the elevator and aileron. This overall control loop has a relatively large time constant, which includes visual scanning, cognitive processing and muscle response of the pilot as well as the aircraft dynamics. Consequently, the commanded elevator and aileron positions need to be filtered to match the response of a human pilot (Weir et al., 1971). A rudimentary version of a flight director for an ILS approach is given in the following code:

```
void UpdateFlightDirector(float Elevator, float Aileron, int *fdh, int *fdv)
{
    float de;
    float da;
    float dh;
    float dv;

    FCS_Autoland(&de, &da);
    dh   = (Elevator - de) * 140.0;
    dh   = Maths_Limit(dh, -140.0, 140.);
    fdhx = Maths_Integrate(fdhx, 0.5 * (dh - fdhx));
    *fdh = intround(fdhx);

    dv   = (da - Aileron) * 140.0;
    dv   = Maths_Limit(dv, -140.0, 140.0);
    fdhy = Maths_Integrate(fdhy, 0.5 * (dv - fdhy));
    *fdv = intround(fdhy);
}
```

The elevator and aileron positions are computed by the algorithm used by the auto-land system described in Section 4.7. The gains of the flight director bars are scaled so that full deflection occurs for full deflection of the controls and, in this simple example, the movement of the bars is filtered by a first-order lag. The response of the display should be sufficiently fast to enable a pilot to respond to changes, without over-controlling, but avoiding a slow response where the display is unresponsive to corrections. For automatic modes selected by the FCU, the flight director provides vertical guidance for altitude hold and vertical speed hold and lateral guidance for heading hold.

4.9 Flight Management Systems

Automation on the flight deck of a modern civil aircraft enables the flight crew to select altitude, heading, airspeed and vertical speed via the FCU panel, engaging and disengaging automatic modes and setting specific values. There is no requirement for pilot input of the flying controls or throttles, which are managed by the FCSs. The FMS updates the flight plan and determines the altitude, heading and airspeed needed to execute the flight plan. The FMS commands the FCS to manage the altitude, heading and airspeed needed for the current segment of the flight plan. The flight crew can override any FMS setting by selecting altitude, heading, airspeed or vertical speed via the FCU and can also disable all automatic modes to resume manual flying of the aircraft.

Figure 4.27 Flight Control Unit (See Plate 3).

4.9.1 Flight Control Unit

A typical FCU panel, which is also known as a Master Control Panel (MCP) in Boeing aircraft, is shown in Figure 4.27. The FCU consists of push buttons which illuminate when active, selectors, switches and rotating knobs. For example, in the left-hand sub-panel, the LS button has been pushed to activate the ILS display on the primary flight display (PFD), but the FD has not been selected to activate the flight director. The knob enables barometric pressure to be displayed in inches of mercury or hectopascals and can be rotated to select a specific barometric pressure or pushed to revert to the standard setting (1013 hPa). The adjacent panel contains five buttons to select specific display modes on the navigation flight display (NFD), two selectors to define the display modes and range (in plan mode) and two switches to display either ADF or VOR selections. The main sub-panel to the right contains four knobs. The values set by these knobs are displayed in the panel above each knob. They allow the flight crew to select the auto-throttle airspeed, heading, altitude or rate of climb. If FCU selection is active, the knobs can be rotated to set new values, which will prompt an immediate response from the FCS. There are six push buttons: LOC engages VOR tracking, AP1, AP2 and A/ATHR engage the autopilots and auto-throttles respectively, EXPED will engage a faster descent and APPR will engage the auto-land mode. The three small push buttons are used for display modes: to display airspeed in Kt or as a Mach number, to display heading or track, or vertical speed or flight path angle, and to display altitude in feet or metres.

The operation of the FCU is straightforward. If a button is pressed or a selector is rotated, a display mode is changed or an automatic mode is engaged or disengaged. If a knob is turned, the value transmitted by the FCU is decoded and used as the value for the relevant autopilot mode. The host software detects changes, converts values transmitted by the FCU and updates autopilot values or modes. Note that one or more of the modes can be engaged; for example, airspeed and altitude can be controlled by the FCU, allowing heading changes to be made by the pilot with aileron input. In addition, some modes override other modes. For example, the FCU cannot control altitude or vertical speed if auto-land is selected, which disables altitude-hold and heading hold (if selected).

4.9.2 Flight Management Systems Simulation

In some simulators, an actual FMS is used and its inputs are stimulated, for example to provide current position, airspeed, altitude, heading, outside air temperature and engine information, possibly via an ARINC-429 interface. In other simulators, the FMS equipment and software are fully emulated. In addition to executing the flight plan, an FMS also provides a pilot interface in the form a multi-function control and display unit (MCDU), which contains a display panel with numerous buttons and keypad entry. The main MCDU inputs are to upload the airline flight plan and flight details and to monitor aircraft information during the flight, including fuel management. In addition, the flight crew are able to modify the flight plan via MCDU inputs.

The FMS will contain a database of worldwide radio navigation aids, flight plan waypoints, airspace regulations, airports, runways, airways, holding patterns, airport arrival and departure procedures and company routes. The FMS data includes the aircraft fuel load, the mass of the flight crew, passengers, luggage and cargo, which enables take-off parameters, climb rate, top of climb, top of descent and approach speeds to be computed. The FMS navigation database is accessed to implement holds, departures, arrivals and diversions and includes details of the engine performance to optimise fuel burn or to manage times at waypoints in order to comply with ATC. The complexity of an FMS is beyond the scope of this book but a few of the fundamental aspects of FMS computations are outlined to provide some insight into FMS operations. The FMS is capable of automatically tuning navigation aids to compute distance and bearing to radio beacons and will also have access to INS and GPS data to estimate aircraft position. Note that the FMS monitors the flight plan, determining position and time to change altitude, heading or airspeed. It does not perform any control functions, per se, but accesses the flight plan, aircraft performance database and navigation database (which includes standard navigation procedures) to command the FCS. The role of the flight crew during FMS operation is to monitor the flight plan closely.

The FMS computes the distance and bearing to a waypoint based on great circle tracks. The computation of distance and bearing is implemented in the navigation equations described in Section 5.6. Knowing the current aircraft position and the distance and bearing to the next waypoint, the FMS can intercept and track each waypoint segment. Indeed, for the majority of civil aircraft flights, the aircraft flies from one waypoint to the next waypoint. The navigation aids are automatically tuned and the FMS will select an airspeed and flight level for each segment, subject to ATC requirements.

There are four main variants of direct tracking as shown in Figure 4.28. If a waypoint is omitted, rather than modifying the flight plan, the flight crew can select to *go-direct* to a specific waypoint. In Figure 4.28(a), the aircraft flies from waypoint WP_1 to WP_4, omitting waypoints WP_2 and WP_3. In Figure 4.28(b), rather than flying directly to the waypoint before turning to the next waypoint, the FMS anticipates the turn to the next waypoint, making a far more efficient turn. In some situations, particularly in busy ATC environments, rather than tracking along the flight plan route, the FMS can select a parallel-offset track, allowing other aircraft to follow a similar track but ensuring safe separation, as shown in Figure 4.28(c). Finally, it is not uncommon for aircraft to be delayed before an approach is approved by ATC. In these circumstances, the aircraft executes a holding pattern, usually with reference to a navigation aid, typically an NDB or VOR beacon, as illustrated in Figure 4.28(d). Normally, the holding pattern takes 4 minutes per circuit, with the holding information taken from the on-board navigation database. Both 180° turns are flown at 3°/s, taking 1 minute and the two straight legs are also

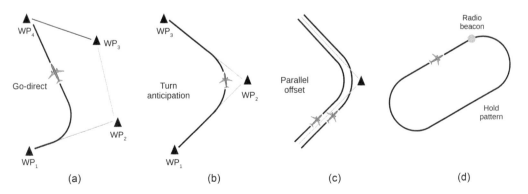

Figure 4.28 Flight Plan Tracking.

flown for 1 minute, returning to the starting point overhead the beacon. To execute the hold, the FMS extracts the location and direction of the holding pattern and simply commands a rate one turn (3°/s) to the right onto the reciprocal heading, then maintains this heading before anticipating the second turn, which captures the track back to the beacon. The tracking is identical to the VOR intercept covered in Section 4.6 but with turns flown at 3°/s. The FMS also takes account of the prevailing wind.

Turn anticipation is particularly important and requires the computation of the point along a flight plan segment where the turn onto the next segment should commence, as shown in Figure 4.29.

Assume the flight plan contains waypoints A, B and C. As the aircraft flies along the track AB, it will need to turn onto the track BC before reaching B. If the aircraft flies along the arc PQ, such that the turn rate is 3°/s, it should turn through the angle α so that its heading corresponds to the relative bearing of C from B. The position of the three waypoints will be known from the flight plan, enabling α to be computed. Assume also that the aircraft airspeed V is the same for both segments of the flight plan. Recalling that the bank angle ϕ is proportional to the rate of turn,

$$\tan\phi = \frac{\dot{\psi}V}{g} \tag{4.15}$$

or

$$\dot{\psi} = \frac{g\tan\phi}{V} \tag{4.16}$$

where $\dot{\psi}$ is the rate of turn. However, for a constant rate of turn, $V = R\dot{\psi}$, giving

$$R = \frac{V}{\dot{\psi}} \tag{4.17}$$

where R is the radius of the turn. Let $\beta = \angle ABO$, then

$$\beta = \frac{\pi - \alpha}{2} \tag{4.18}$$

$$\tan\beta = \frac{R}{d} \tag{4.19}$$

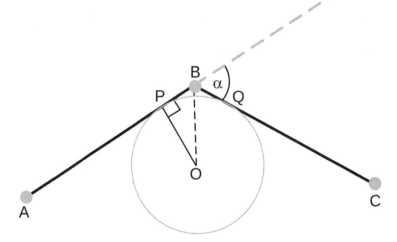

Figure 4.29 Turn Anticipation.

$$d = \frac{V^2}{g \tan \phi \tan \beta} \tag{4.20}$$

where $d = PB$, the distance from B to start the turn. In other words, assuming the maximum angle of bank is 25° under FMS guidance, d depends on the airspeed V and the change of direction α to the next waypoint segment. Strictly, this equation is simplified, because it takes a finite time to establish the maximum turn rate and, similarly, to reduce the turn rate to zero. In addition, the entry and exit speeds for the respective segments may be different, which may be overcome by flying the turn at the entry speed to the turn before changing to the airspeed of the next segment. In practice, the FMS algorithm will contain detailed information of the aircraft performance to better predict d.

Once the flight plan is loaded, it is updated each frame as follows:

```
if ((Nav_NumberOfWayPoints < 2) || (Nav_FlightPlan_Segment < 1))
{
  Nav_WayPoint.BeaconStatus = false;   /* deactivate flight plan */
  return;
}

if (Nav_FlightPlan_Segment <= (Nav_NumberOfWayPoints - 1))
{
  if (oldFlightPlan_Segment != Nav_FlightPlan_Segment)
  {
    Nav_WayPoint.BeaconStatus = true;   /* activate flight plan */
    if (Nav_FlightPlan_Segment == 1)
    {
      FCU_ALT = Nav_WayPoints[Nav_FlightPlan_Segment].WayPointAltitude;
      FCU_SPD = Nav_WayPoints[Nav_FlightPlan_Segment].WayPointSpeed;
    }
    FCU_ALT_Hold = true;
    FCU_HDG_Hold = false;
    FCU_SPD_Hold = true;
    FCU_APPR = false;

    oldFlightPlan_Segment = Nav_FlightPlan_Segment;

    cur_trk = NavLib_Bearing((double) Nav_WayPoints[Nav_FlightPlan_Segment].WayPointLatitude,
                             (double) Nav_WayPoints[Nav_FlightPlan_Segment].WayPointLongitude,
                             (double) Nav_WayPoints[Nav_FlightPlan_Segment+1].WayPointLatitude,
                             (double) Nav_WayPoints[Nav_FlightPlan_Segment+1].WayPointLongitude);
    Nav_WayPoint.RunwayQdm = (float) cur_trk - Nav_MagneticVariation;  /* magnetic hdg */
```

The conditions to cancel the flight plan are checked and the above code is activated if there is a new flight plan segment (initially, **Nav_FlightPlan_Segment** = 0). The altitude and indicated airspeed for the segment are extracted from the flight plan, the FCS altitude hold and speed hold are engaged and the bearing of the next waypoint to the current waypoint (°T) is computed. The turning distance at the next waypoint is computed from equation 4.20, as follows:

```
if (Nav_FlightPlan_Segment <= (Nav_NumberOfWayPoints - 2))
{
  next_trk = NavLib_Bearing((double) Nav_WayPoints[Nav_FlightPlan_Segment+1].WayPointLatitude,
                            (double) Nav_WayPoints[Nav_FlightPlan_Segment+1].WayPointLongitude,
                            (double) Nav_WayPoints[Nav_FlightPlan_Segment+2].WayPointLatitude,
                            (double) Nav_WayPoints[Nav_FlightPlan_Segment+2].WayPointLongitude);

  dtrk = next_trk - cur_trk;
  dtrk = Maths_Double_Normalise(dtrk);
```

```
        if (fabs(dtrk) < DEG2)
        {
            dturn = 1000.0;
        }
        else if (fabs(dtrk) > DEG170)
        {
            dturn = 10000.0;
        }
        else
        {
            vseg = GetTAS(Nav_WayPoints[Nav_FlightPlan_Segment].WayPointSpeed / 1.944,
                        Maths_Metres(Nav_WayPoints[Nav_FlightPlan_Segment].WayPointAltitude));
            beta = M_PI - dtrk;
            beta = Maths_Double_Normalise(beta);
            beta = fabs(beta);
            yrate = 9.81 * tan(DEG25) / vseg;
            if (yrate > DEG3)
            {
                maxbankangle = atan(DEG3 * vseg / 9.81);
            }
            else
            {
                maxbankangle = DEG25;
            }
            dturn = (vseg * vseg) / (9.81 * tan(maxbankangle) * tan(beta / 2.0)) + vseg * 5.0;
        }
    }
}
```

Care is needed with acute angles to avoid turning far too early and to ensure that the bank angle is limited to 25°. Finally, the track error and course correction are computed in the following code fragment:

```
bearing = NavLib_Bearing(NavLink_AeroPkt.Latitude, NavLink_AeroPkt.Longitude,
                         (double) Nav_WayPoints[Nav_FlightPlan_Segment+1].WayPointLatitude,
                         (double) Nav_WayPoints[Nav_FlightPlan_Segment+1].WayPointLongitude);
wpdist = NavLib_Distance((double) Nav_WayPoints[Nav_FlightPlan_Segment+1].WayPointLatitude,
                         (double) Nav_WayPoints[Nav_FlightPlan_Segment+1].WayPointLongitude,
                         NavLink_AeroPkt.Latitude, NavLink_AeroPkt.Longitude);

trackerr = bearing - cur_trk;
trackerr = Maths_Double_Normalise(trackerr);
Nav_WayPoint.LocaliserError = (float) trackerr;
```

Using a data structure for waypoints that is equivalent to VOR tracking allows the same tracking algorithm to be applied. The conditions to activate the next flight plan segment are determined, bearing in mind that a new segment may require a change of altitude and airspeed.

The following example of a flight plan is based on a simple route comprising five VOR transmitters in southern England:

```
WP01 51.053953, -0.625003, 5000, 350
WP02 51.017183,  0.116122, 5000, 370
WP03 51.304003,  0.597275, 4500, 350
WP04 51.330875,  0.034811, 4500, 360
WP05 51.305047, -0.447183, 4000, 320
WP06 51.053953, -0.625003, 5000, 350
```

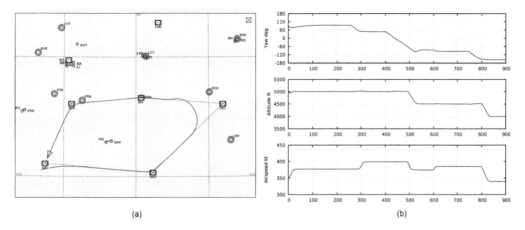

(a)   (b)

Figure 4.30   Execution of a Flight Plan.

The five waypoints are Midhurst (MID), Mayfield (MAY), Detling (DET), Biggin Hill (BIG) and Ockham (OCK). The flight plan contains the waypoint number, the latitude and longitude of the waypoint and the altitude (ft) and indicated airspeed (Kt) of each segment. The route is shown in Figure 4.30 together with the variation in altitude, TAS and heading.

In Figure 4.30(a), the flight plan is activated just south-west of MID and the aircraft intercepts the track from WP01 to WP02. At WP02, the turn to WP03 is anticipated. The turn at WP03 is also anticipated but the sharp turn towards WP04 also involves a change in altitude and airspeed. The computation of the start of the turn does not account for the change in airspeed and, consequently, the intercept overshoots slightly. After WP05, the aircraft returns to WP01 and the flight plan is cancelled just north of WP01. Figure 4.30(b) shows the heading, altitude and airspeed during the flight plan. The small initial variation is caused because the aircraft was not trimmed when the flight plan was activated.

References

Åström, K. J. and Hagglund, D. (1995), *PID Controllers: Theory, Design and Tuning*, ISA.

Bibbero, R. J. (1977), *Microprocessors in Instruments and Control*, John Wiley and Sons.

Etkin, B. and Reid, L. D. (1996), *Dynamics of Flight: Stability and Control*, John Wiley and Sons.

Favre, C. (1994), Fly-by-wire for Commercial Aircraft: The Airbus Experience, *International Journal of Control*, Vol. 59, No. 1, pp. 138–157.

Franklin, G. F., Powell, J. D. and Emani-Naemi, A. (1994), *Feedback Control of Dynamic Systems*, Addison-Wesley.

Hanke, C. R. (1971), *The Simulation of a Large Jet Transport Aircraft, Vol. I: Mathematical Model*, NASA CR-1756, Vol. II: Modelling Data, Boeing Report No. N73-10027.

Hogge, E. F. (2004), *B-737 Linear Autoland Simulink Model*, NASA/CR-2004-213021, Langley Research Centre, Virginia.

McLean, D. (1990), *Automatic Flight Control Systems*, Prentice Hall International.

McRuer, D., Irving, A. and Dunstan, G. (1973), *Aircraft Dynamics and Automatic Control*, Princeton, New Jersey.

Myers, G. J. (2004), *The Art of Software Testing*, John Wiley and Sons.

Quarteroni, A. and Saleri, F. (2003), *Scientific Computing with MATLAB and Octave*, Springer.

Roskam, J. (2003), *Airplane Flight Dynamics and Automatic Flight Controls*, Darcorporation.

Smetana, F. O. (1994), *Introduction to the Control of Dynamic Systems*, AIAA Education Series.

Stevens, B. L. and Lewis, F. L. (2003), *Aircraft Control and Simulation*, John Wiley and Sons.

Weir, D. H., Klein, R. H. and McRuer, D. T. (1971), *Principles for the Design of Advanced Flight Director Systems Based on the Theory of Manual Control Displays*, NASA/CR-1748.

5

Navigation Systems

I may not have gone where I intended to go, but I think I have ended up where I needed to be.

Douglas Adams

5.1 The Earth

The shape of the earth and the distribution of material within the earth affect gravitational acceleration over the surface of the earth. Although these changes are relatively small, for the modelling of inertial navigation and the simulation of vehicle dynamics, these second-order effects should be taken into account. Similarly, magnetic variation also varies over the surface of the earth and, moreover, varies slowly over time. Magnetic compasses have been used in aviation since the early days of flying and heading is given in degrees magnetic on charts and in navigation displays.

5.1.1 Gravity

The earth is an oblate sphere; it is an ellipsoid with a difference between the equatorial radius and the polar radius, as shown in Figure 5.1, where R is the radius of the earth and the following values are used to define the shape of the earth:

Semi-major axis R = 6,378,137.0 m
Semi-minor axis (polar axis) $r = R(1 - f)$ = 6,356,752.3142 m
Flattening of the ellipsoid $f = (R - r)/R$ = 1/298.257223563
Eccentricity of the ellipsoid $e = \left[f(2 - f)\right]^{0.5}$

Several equations are used to define gravitational acceleration, based on best-fit methods using harmonics. The equation adopted by the World Geodetic System in 1984 (WGS-84), and used in this book, is given by

$$g = \frac{9.7803267714\left(1.0 + 0.00193185138639 \sin^2 \lambda\right)}{\left(1.0 - 0.00669437999013 \sin^2 \lambda\right)^{0.5}} \ \mathrm{ms}^{-2} \tag{5.1}$$

where λ is latitude.

Flight Simulation Software: Design, Development and Testing, First Edition. David Allerton.
© 2023 John Wiley & Sons Ltd. Published 2023 by John Wiley & Sons Ltd.
Companion Website: www.wiley.com/go/flightsimulationsoftware

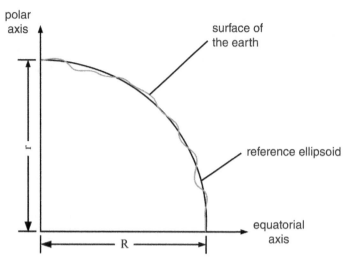

Figure 5.1 The Earth Reference Ellipsoid.

The code to derive gravitation acceleration as a function of latitude and altitude is as follows:

```
double Gravity(double latitude, double h)
{
    double s = sin(latitude);
    double s2 = s * s;
    double g = 9.78032667714 * ((1.0 + 0.001931851138639 * s2) / sqrt(1.0 - 0.0066943799013 * s2));

    return g * (Re * Re) / ((Re + h) * (Re + h));
}
```

where `Re` is the semi-major axis and `h` is the aircraft geometric altitude.

5.1.2 Magnetic Variation

There are two definitions of the north pole: the true north pole is on the spin axis of the earth at an internationally agreed location, whereas the magnetic north pole is the position on the earth pointed to by all magnetic compasses. The difference between true north and magnetic north varies over the surface of the earth as a result of the magnetic fields surrounding the earth and the mass of the earth's core, causing irregularities in the local magnetic field. In addition, these fields vary over time and charts are updated annually to include the variations. Magnetic variation can be as much as 10 degrees and the local magnetic variation should be included in navigation computations.

Although magnetic variation, which is also known as declination, can be set from a current chart in flight simulation, data for magnetic variation over the earth is available to compute the local magnetic variation. The National Oceanic and Atmospheric Administration (NOAA) publish data for worldwide magnetic variation (Chulliat et al., 2015). They also produce and distribute software which can be downloaded to enable organisations to estimate magnetic variation as a function of location and date. Although the software is complex and substantial, it is open software (provided as C code), which compiles and runs on most platforms and is supported with extensive documentation. The following code fragment computes the magnetic variation at a specific latitude, longitude and date:

```
CoordGeodetic.phi = degrees(Model_Latitude);
CoordGeodetic.lambda = degrees(Model_Longitude);
UserDate.DecimalYear = (double) (tm.tm_year + 1900) + (double) tm.tm_mon / 12.0;

MAG_GeodeticToSpherical(Ellip, CoordGeodetic, &CoordSpherical);
MAG_TimelyModifyMagneticModel(UserDate, MagneticModels[0], TimedMagneticModel);
MAG_Geomag(Ellip, CoordSpherical, CoordGeodetic, TimedMagneticModel, &GeoMagneticElements);
MAG_CalculateGridVariation(CoordGeodetic, &GeoMagneticElements);
MAG_WMMErrorCalc(GeoMagneticElements.H, &Errors);
```

where the local magnetic variation is given by `GeoMagneticElements.Decl`.

Latitude and longitude are given in degrees and, in this example, the date is given in years (to the nearest month). The actual coefficients are provided in a file '`WMM.COF`' and this data is updated by NOAA on a five-year cycle. A default ellipsoid and geoid is used in this example. The main call is `MAG_Geomag` which passes a data structure containing the geodetic coordinates and returns values in a data structure that includes the declination in the `Decl` field. Although the library is extensive, for most flight simulation applications, only the magnetic declination of the current region is needed. Generally, magnetic variation is given to the nearest degree in aircraft navigation. In modern transport aircraft, displays can be set to show either magnetic or true heading and the local magnetic variation will be included in Flight Management System (FMS) updates.

5.2 Sensor Modelling

In an aircraft, physical properties are measured by transducers which convert the measurements to electrical signals that are used in aircraft displays, flight control systems and engine systems. However, in a flight simulator, these variables are computed in the various modules, rather than measured, to significantly more accuracy than transducer measurements. For example, the airspeed in the aerodynamic model is probably accurate to within 1 mm/s, whereas in an aircraft the measurement is subject to several errors and is possibly accurate to 0.5 m/s. The transducer itself, which nowadays provides a digital output, will convert the measurement to 10 or 12 bits. Ten bits corresponds to approximately one part in a thousand; for an airspeed sensor with a range of 500 Kt, the measurement accuracy is limited to 0.5 Kt. In addition, the sensor will have calibration errors, caused by nonlinearities in the transducer, location of the sensor on the airframe or variation with temperature.

There is a further requirement with the modelling of sensors in a flight simulator and that is the modelling of sensor failures. In some cases, a failure may be the complete loss of a signal from a sensor. For example, the icing of an airspeed probe (or the failure of the probe heater) may give the effect of an unchanging or incorrect airspeed. But in other cases, sensor failures can result in spurious measurements with no obvious indication of the failure, for example, excessive precession of a gyro may still produce an output, even though the value is erroneous. Consequently, the modelling of a sensor in flight simulation focuses on replicating both the characteristics of the sensor and its failure modes. For example, a VHF Omnidirectional Range (VOR) receiver detects a line-of-sight VHF signal from a VOR transmitter. The signal may be lost because the path is blocked by a hill, the aircraft is too low to detect the signal or the transmitter is out of range. In these cases, the VOR receiver should fail in the same way that it would fail in an aircraft. Note the possible implications:

- The model of the VOR receiver needs to access the terrain database to detect visibility to the receiver.
- The relative altitude of the aircraft to the transmitter and the curvature of the earth determine visibility to the transmitter.
- The range of the transmitter from the aircraft, which is computed in the navigation equations, determines the signal strength.

In addition, the physics of the sensor is modelled to determine the actual failure mode, for example, how an instrument responds as a signal becomes weaker.

One other consideration for the developer is the choice between using actual aircraft equipment and fabricating equipment to emulate the functionality. In many cases, this choice depends on the difficultly of replicating the signals needed to drive the equipment. This limitation is particularly apposite where inputs are radio or radar signals, for example, a weather radar. In other cases, ensuring that all the modes, inputs and settings are replicated correctly may prove intractable. In recent years, some manufacturers have developed aircraft equipment which can also be used in a synthetic environment. One particular example is the Global Positioning System (GPS) receiver, where the equipment can be stimulated by position and altitude inputs generated by a computer and, in training exercises, enables a pilot to practice using all the operational modes of the equipment. Normally, the receiver would be connected via a cable from the GPS antenna but, in a training mode, the manufacturers provide access to intermediate stages of the receiver. Alternatively, vendors offer facsimile equipment for the industry, where the receiver is connected via a serial connection or Ethernet and the interface protocols are clearly defined. Even so, it is still essential to ensure full coverage of all the operating modes and failures and the developer needs to be confident that there is no difference between the aircraft equipment and the simulator equipment. Particularly in a training role, a sub-system with behaviour that is different from the aircraft has the potential to result in misuse of the equipment in the aircraft, with potentially fatal outcomes. For example, failure modes in auto-land equipment are categorised as safety-critical and, in certain modes, the flight crew have only a few seconds to initiate a go-around. In such cases, the conditions must be identical in both the aircraft and the simulator.

5.3 Navigation Principles

5.3.1 Position

In simple terms, navigation is based on the measurements of aircraft motion to provide guidance to the flight crew, to enable the flight to be managed safely and efficiently. This information enables the flight crew or flight guidance systems to control altitude, heading and airspeed and extends to following a flight plan, which defines the lateral and vertical flight path to be followed. In addition to errors caused by sensor malfunctions, navigation is also subject to magnetic variation over the surface of the earth, which is mostly known, and by the wind, which is generally unknown and varies with location, altitude and time. From the perspective of the simulator developer, flight crews should be able to follow the same navigation procedures and operate the navigation equipment in the same way as the aircraft. This requirement extends to worldwide navigation covering all the airports, navigation aids, routes and procedures throughout most countries.

Navigation sensors support three methods to determine position:

1) Direct measurement, for example providing altitude, heading and airspeed.
2) Dead reckoning, where the position is estimated from the vehicle velocity and direction from an initial known position, as a function of time.
3) Range and bearing measurements, typically from land-based radio transmitters and satellites.

In addition, navigation data is also available in the form of navigation databases and charts, which include the location and frequencies of radio beacons, airport approach and departure information, airways and en-route charts. For airlines, these navigation databases are updated monthly and considerable care is taken to ensure that the databases used in the simulators are up-to-date and identical to the databases used in the aircraft. A broad classification of navigation covers flight under visual flight rules (VFR) and flight under instrument flight rules (IFR). VFR cover both general aviation and much military aviation, whereas for commercial civil aviation, with the exception of taxiing, take-off and the final stages of an approach, most flights are conducted under IFR conditions. In addition, commercial civil aviation covers three forms of guidance:

1) Manual flying, where the flight crew apply direct inputs to the control column, rudder pedals and engine levers.
2) Automatic flight control, where responsibility for manoeuvring the aircraft is passed to the automatic flight control system (FCS) and the flight crew select altitude, heading and airspeed, typically via a flight control unit (FCU), and monitor the aircraft using the displays.
3) FMS guidance, where the aircraft flight management system determines the vertical and lateral flight path to execute the flight plan from the waypoints in the flight plan, and commands the FCS to select an altitude, heading, airspeed and vertical speed.

Note that the FMS is also required to take account of wind, which can be estimated from the difference between the desired flight path and the actual flight path.

Aircraft position is defined in terms of latitude, longitude and altitude, referenced to the earth frame shown in Figure 5.2. The parallels of latitude range from 0° at the equator to 90° at the poles. The meridians of longitude are aligned to the prime meridian at 0°, which passes through the village of Greenwich near London. The earth rotates through 360° of longitude in one day or 15° per hour, with lines of longitude ranging from 0° to ±180°. Distance is derived from latitude where one degree of latitude equals 60 nautical miles (nm) or 1 minute of latitude equals 1 nm. Note that 1° of longitude is different from 1° of latitude. The unit of one nautical mile is based on a nominal radius of the earth of 6,378,137 m, neglecting the slightly oblate shape of the earth, so that

$$1 \, \text{nm} = \frac{1}{60} \frac{1}{360} 2\pi R = 1852 \, \text{m} \left(6076 \, \text{ft} \right) \tag{5.2}$$

Note that a statute mile, which is 1609 m, is not based on the nautical mile and generally, in aircraft operations, nautical miles are used in preference to statute miles. There is also one further consideration with latitude and longitude; the units of these angles are usually radians. The accuracy of 32-bit floating-point arithmetic is approximately 2^{-23} or 1.19×10^{-7}. Therefore, 1 m corresponds to $\frac{1}{60} \frac{\pi}{180} \frac{1}{1852}$ radians, which is 1.57×10^{-7}. In other words, in equations involving

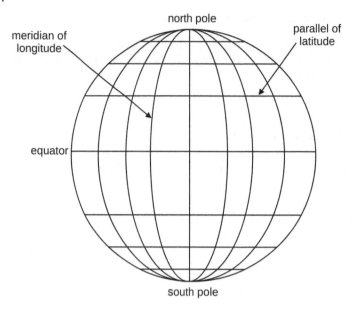

north pole

meridian of
longitude

parallel of
latitude

equator

south pole

Figure 5.2 Latitude and Longitude.

angles of latitude and longitude, numerical errors resulting from 32-bit floating-point operations could be of the order of 1 m. Although this may not seem to be significant in the computation of position, in the case of a visual system, for example, where an aircraft is taxiing slowly, if the aircraft position changes in 1-m increments, this jerky motion would be unacceptable. Consequently, double-precision arithmetic should be used in all navigation computations involving angles of latitude and longitude.

5.3.2 Airspeed

Aircraft velocity is measured in nautical miles per hour (Kt) and as a Mach number. In an aircraft, these values are computed in the air data computer from measurements of static air pressure, dynamic air pressure and outside air temperature (OAT). In an aircraft, the dynamic pressure is derived from a pitot-static probe. However, depending on the location of the probe, air entering this probe can vary with angle of attack and angle of sideslip, and the dynamic pressure also varies with air density. The measured airspeed is then corrected, whereas in a flight simulator, the true airspeed is known from the equations of motion and other variants of airspeed are computed from this value. The four commonly used aircraft velocities are:

1) True airspeed (TAS) – in an aircraft, all the corrections to measured airspeed have been applied. In flight simulation, the TAS $V_T = \sqrt{u^2 + v^2 + w^2}$, where u, v and w are the body frame velocities, which are used in the derivation of lift, drag and side force. V_T is the magnitude of the velocity vector of the centre of gravity of the aircraft, also known as the flight path vector.
2) Indicated airspeed (IAS) – in an aircraft, the airspeed indicator is a pressure gauge driven directly from dynamic pressure and varies with air density, whereas in a flight simulator, indicated airspeed is derived from the TAS, where $IAS = V_T \sqrt{\rho}$, where ρ is the density ratio computed by the weather module as a function of altitude. Performance values in pilot handling notes and flight plans are usually stated as IAS values. Take-off speeds, landing speeds, approach speeds, climb and cruise speeds are given as IAS values or Mach number, including FCU settings.

3) Calibrated airspeed (CAS) takes account of any errors resulting from the positioning of the airspeed sensors and errors in the conversion of dynamic pressure. CAS is used in general aviation to provide a more accurate airspeed for navigation but is compensated for by the air data computer in modern aircraft and, generally, is not used in flight simulation.
4) Ground speed is the velocity with respect to the ground and is a combination of the aircraft TAS and the prevailing wind. For example, an aircraft flying at 80 Kt into a 15 Kt headwind will have a ground speed of 65 Kt. In an aircraft, ground speed is measured by the inertial navigation system, GPS and Doppler radar. In flight simulation, V_N and V_E, the north and east velocity components, respectively, are computed in the equations of motion, and ground speed is given by $\sqrt{V_N^2 + V_E^2}$.

5.3.3 Altitude

In an aircraft, altitude is provided by the air data computer, converting static pressure to altitude, whereas in a flight simulator, altitude (P_Z) is derived in the equations of motion. Note that P_Z is negative upwards and is relative to a notional sea level. Height above terrain can also be acquired in aircraft equipped with a radar altimeter. In effect, an altimeter is a barometer, and if an aircraft is parked for several hours, as the local air pressure changes, the indicated altitude will also vary. The indicated altitude can be adjusted to set the altimeter to one of the following pressure settings:

- QFE – on an airfield, the altimeter barometric pressure knob can be turned until the altitude is zero. This barometric pressure setting can be used for local flying, as the altimeter indicates the altitude above the airfield. It can also be used when an aircraft rejoins the circuit, enabling the circuit to be flown at the altitude published for the airfield.
- QNH – for flight away from an airfield, within a region, all aircraft flying in that region are required to set their altimeters to the QNH setting provided by air traffic services, which is updated regularly. This pressure setting ensures that aircraft maintain altitudes referenced to sea level;
- Above the transition altitude, altitudes are defined by flight levels (e.g. FL235 is 23,500 ft) and all aircraft set their altimeters to the International Standard Atmosphere (ISA) pressure setting of 1013 HPa or 29.92 in Hg.

In flight simulation, the pressure setting is normally set by the instructor and flight crews will select QFE, QNH or standard (STD) altimeter settings, as they would in an aircraft. Following the correct procedures for altimeter setting is an essential part of aviation safety and compliance with air traffic control. Consequently, altimeters in a flight simulation should function exactly as aircraft equipment in terms of air pressure settings and flight crew settings.

5.3.4 Heading

In an aircraft, heading is provided by magnetic compasses and gyro compasses. The problem with magnetic compasses is that the magnetic variation changes (albeit slowly) over the earth and with time. In addition, any magnetic field, generated by aircraft equipment or from electrical storms, can introduce large errors in sensing aircraft heading. Furthermore, the magnetic compass is also subject to errors during accelerating manoeuvres. The problem with a gyro compass is that it drifts with time and needs to be regularly realigned to the magnetic compass. Fortunately, charts and databases of magnetic variation are published, for example, the International Geomagnetic

Reference Field Model published by NOAA (Thébault et al., 2015). Such information is included in FMS data uploaded with flight plans in commercial transport aircraft. Alternatively, charts of isogonal lines (lines of common magnetic variation) can be used in flight simulation to set the local magnetic variation.

Heading settings and flight plan values are normally provided in degrees magnetic although modern displays provide both magnetic and true heading. Within the navigation software, heading values are entered as magnetic headings, and all computations are performed using true headings, which are converted back to magnetic headings for displays. The conversion is simple:

$$\psi_m = \psi_t - \delta m \tag{5.3}$$

$$\psi_t = \psi_m + \delta m \tag{5.4}$$

where ψ_m is magnetic heading, ψ_t is true heading and δm is magnetic variation, which is positive for east variation and negative for west variation. Note that aircraft displays provide both heading and track. Heading is given by yaw in the flight model and is the instantaneous direction of the aircraft, whereas the track is the flight path over the ground. In navigation, an EFIS display will display the track commanded by the FMS and the aircraft heading to enable the flight crew to compensate for the wind. The difference between heading and TAS and track and ground speed (GS) is shown in Figure 5.3.

In a flight simulator, the wind speed and direction are likely to be modelled and the current wind values can be used directly in the navigation computations, whereas wind is estimated in aircraft navigation from position measurements along the track.

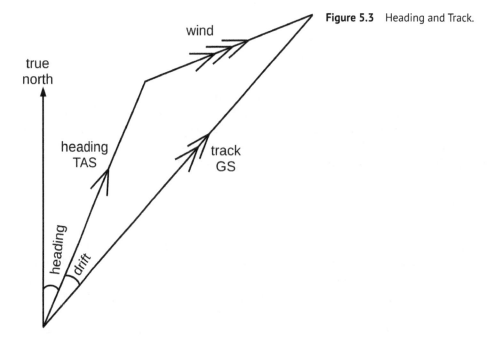

Figure 5.3   Heading and Track.

5.3.5 Distance and Bearing

In the navigation equations, aircraft position is given by latitude, longitude and altitude, and velocity is given by V_N, V_E and V_d. Similarly, the location of radio beacons and runways is defined in terms of latitude, longitude and altitude (above sea level).

The distance between two points is usually based on the distance on the surface of the earth along a great circle route between the points. Radio waves are assumed to follow great circle routes. Consequently, spherical geometry is applied to compute the distance and relative bearing between two points. The term 'relative bearing' implies a heading relative to north, as shown in Figure 5.4, where the bearing of B from A is 45° and the bearing of A from B is the reciprocal, 225°.

The distance d between two points (λ_1, ϕ_1) and (λ_2, ϕ_2), where λ and ϕ denote latitude and longitude, respectively, is given by

$$d = \sqrt{(\Delta\lambda)^2 + \cos\lambda_1 \cos\lambda_2 (\Delta\phi)^2} \tag{5.5}$$

where

$$\Delta\lambda = \lambda_2 - \lambda_1 \tag{5.6}$$

$$\Delta\phi = \phi_2 - \phi_1 \tag{5.7}$$

Figure 5.4 Relative Bearing.

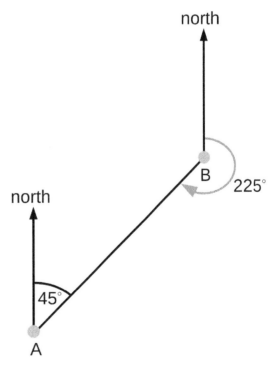

and d is in radians. The distance D, in metres, is given by

$$D = Rd \tag{5.8}$$

where R is the radius of the earth. The code in the NavLib library to compute distance is as follows:

```
double NavLib_Distance(double Lat1, double Long1, double Lat2, double Long2)
{
    double dLat;
    double dLong;

    dLat  = Lat2 - Lat1;
    dLong = Long2 - Long1;
    return EarthRadius * sqrt(dLat * dLat + cos(Lat1) * cos(Lat2) * dLong * dLong);
}
```

The relative bearing Ψ from (λ_1, ϕ_1) to (λ_2, ϕ_2) is given by

$$\psi = \tan^{-1}\left(\frac{\sin\lambda_2 - \sin\lambda_1 \cos d}{\cos\lambda_2 \ \sin\Delta\phi\cos\lambda_1}\right) \tag{5.9}$$

Care is needed with computation of the relative bearing. Firstly, $\tan^{-1}$ produces a result in the range $0–\pi/2$, whereas the bearing can be in the range from $-\pi$ to π. Secondly, if $\lambda_1 = 90°$, or $\lambda_2 = 90°$ or $\phi_1 = \phi_2$, the denominator of equation 5.9 is zero, resulting in numeric overflow. The solution is shown in the following code from the NavLib library to compute relative bearing:

```
double NavLib_Bearing(double Lat1, double Long1, double Lat2, double Long2)
{
    double x, y;
    double dLat;
    double dLong;
    double d;

    dLat  = Lat2 - Lat1;
    dLong = Long2 - Long1;
    d     = sqrt(dLat * dLat + cos(Lat1) * cos(Lat2) * dLong * dLong);
    x     = sin(Lat2) - sin(Lat1) * cos(d);
    y     = cos(Lat2) * sin(dLong) * cos(Lat1);
    return atan2(y, x);
}
```

The use of the maths library function **atan2** correctly returns an angle in the range $-\pi$ to π and, moreover, avoids division by zero.

There is one further consideration with the use of spherical coordinates. Over distances greater than a few hundred miles, the direct track (the shortest distance) between two points on the surface of the earth is curved. The track forms part of a great circle, with the plane of the circle passing through the centre of the earth. Moreover, there is only one great circle passing between two points. The implication for long-range navigation is shown in Figure 5.5.

The aircraft departs from A with a heading of approximately 30° but by the time it reaches point B, the heading has increased to approximately 80°.

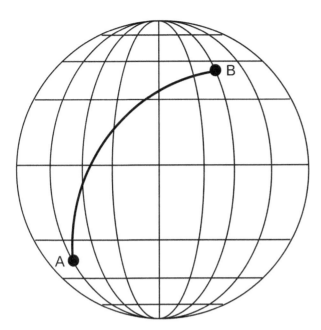

Figure 5.5 Great Circle Tracks.

5.4 Navigation Databases

Navigation databases contain the details of navigation aids, airfields, runways and flight plan information, including airways. Although this book focuses on open software (in the sense that source code and libraries can be downloaded for free), to the author's knowledge there are no extensive navigation databases available for free use in flight simulation. For airlines, the navigation databases used for aircraft and flight simulators are purchased from suppliers with monthly updates. Tremendous care is taken with the accuracy of information in these databases and the cost of acquisition, maintenance and distribution of the databases is significant. Jeppesen produces navigation databases which are used throughout aviation. The company Navigraph is licensed to distribute a subset of these databases to flight simulation users, which is regularly updated and Navigraph databases are used in this textbook. This is not intended as an endorsement of the product; it is left to the discretion of the reader to select alternative databases, where appropriate.

The Navigraph databases are provided in various formats and two databases are used with the software described in this book, one for navigation aids covering non-directional beacons (NDB), VOR, distance measuring equipment (DME) and instrument landing system (ILS) radio transmitters and a second covering airfields including runways and taxiways. The database referenced in this book is based on the Navigraph AIRAC cycle 2012, version 1, which was valid from 5 November 2020 to 3 December 2020. The variant of the database downloaded is X-Plane GNS430, which includes the files Airports.txt and Navaids.txt. In the cycle 2012 release, the database contains 13,652 navigation aids and 13,790 runways. A region for the UK, from 49°N to 61°N and 8°W to 2°E, contains 308 navigation aids and 756 runways.

The entries for the navigation aids at Southampton airport are shown in the following entry in the database:

```
EAS,SOUTHAMPTON EASTLEIGH,391.500,0,0,195,50.955028,-1.356119,0,EG,0
SAM,SOUTHAMPTON,113.350,1,1,195,50.955250,-1.345056,64,EG,0
```

EAS is an NDB operating on a frequency 391.50 KHz and **SAM** is a VOR with a DME operating on a frequency 113.35 MHz. The fields of this file are critical. The ident of the navigation aid is given, followed by the name. The other fields include the frequency of the aid, the latitude and longitude in degrees, the beacon range and elevation. The frequency of each navaid is given in MHz for ILS and VOR beacons and KHz for automatic direction finding (ADF) beacons.

Similarly, the entry for Southampton airport is shown in the entry below:

```
A,EGHI,SOUTHAMPTON,50.950278,-1.356667,44,6000,0,5600,0
R,02,20,5653,121,0,0.000,0,50.943572,-1.360528,31,3.00,50,1,0
R,20,200,5653,121,1,110.750,200,50.957186,-1.352944,44,3.10,51,1,0
```

The lines starting with the letter 'A' denote airfield information and the subsequent lines starting with the letter 'R' contain runway information. The first line includes the ICAO airport identifier, the latitude and longitude, the height above sea level (44 ft) and the lengths of the main runways (6000 ft and 5600 ft). The next line describes runway 02, with a QDM of 20°, the runway length (5653 ft), width (121 ft), a flag to indicate the existence and frequency of an ILS (if present), the latitude and longitude of the touchdown location, the height above sea level (31 ft), the glideslope angle (3.0°), the altitude at the runway threshold (50 ft), the type of surface and a status. The third line describes runway 20, which includes an ILS on a frequency of 110.75 MHz. The fields of the two database files are separated by commas. Each database is read by the navigation library at the start of a simulation and the entries for the region selected are formed in data structures used by the navigation software. A major part of the software to read the databases is validation of the data, to ensure that the data is within acceptable bounds and does not contain any erroneous entries.

Three global data structures are created, which enable the beacons and runways to be accessed by the navigation equations and the instructor station; and they also enable airfield data to be passed to the visual system. The data structure for the navaids contains the following fields:

```
typedef unsigned char NavLib_BeaconMode;
typedef unsigned int NavLib_BeaconType;

typedef struct
{
    float               BeaconLatitude;    /* rads */
    float               BeaconLongitude;   /* rads */
    unsigned int        Frequency;         /* freq * 100 */
    float               Qdm;               /* degs 1-360 iff ILS */
    unsigned int        Range;             /* not available */
    unsigned int        Elevation;         /* m */
    unsigned int        Runway;            /* 23R etc */
    char                Ident[6];
    NavLib_BeaconType   Navaid;
} NavLib_BeaconRecord;
```

The fields are also converted to the units used in the flight simulator. **BeaconLatitude** and **BeaconLongitude** are converted to radians. **Frequency** is converted to an integer, for example, 113.75 is stored as 11375. Where navaid frequencies are compared for equivalence, it is essential to use integer arithmetic rather than floating-point. The runway **Qdm** field is used in the navigation computations during an ILS approach. Unfortunately, range (which varies for NDB stations) is not provided in the database; it is computed as line-of-sight for VOR, DME and ILS stations. The **Elevation** of the navaid is its altitude above sea level. The **Runway** field is not used and the **Ident** field provides the characters used on charts, which are transmitted as Morse. Finally, the **Navaid** field identifies the type of navaid, including NDB, VOR, VOR/DME, ILS and ILS/DME. It is not possible to extract VORTAC or TACAN navigation transmitters from the information provided in the database.

The data structure for the runways contains the following information:

```
typedef struct
{
    float            RunwayLatitude;     /* rads */
    float            RunwayLongitude;    /* rads */
    float            RunwayEndLatitude;  /* rads */
    float            RunwayEndLongitude; /* rads */
    unsigned int     Frequency;          /* iff ILS */
    float            Qdm;                /* runway QDM true */
    unsigned int     Length;             /* m */
    unsigned int     Width;              /* not used */
    unsigned int     Elevation;          /* m */
    unsigned int     Runway;             /* 23R etc */
    char             Name[NavLib_NameSize];
} NavLib_RunwayRecord;
```

Note that latitudes and longitudes are provided for the ends of each runway, which enables the runway QDM to be computed more accurately than the value given in the database; the QDM is necessary for ILS and flight director computations as well as for auto-land guidance. The frequency only applies if there is an active ILS. The length, width and elevation (at the threshold) are provided, although the runway width is not used. As will be seen in Chapter 7, combining data from the navigation database with data used in the visual system can lead to inconsistencies, which is a potential problem in the interoperability of flight simulation software.

The navigation database provides essential information needed for the navigation equations. It can also provide useful data for the visual system, in the case where a generic runway is used; the runway in the visual system can be matched to the location, altitude and orientation of an actual runway in the navigation database. The drawback is that any lack of information in the database can lead to inconsistencies in the navigation software. For example, the location of an ILS transmitter and its altitude should match the runway data. In addition, the database may contain more than 10,000 beacons and runways and, consequently, searching for navaid frequencies or displaying navaid data on charts is potentially time-consuming.

A library is used to read the navigation databases as the formats are non-standard. At the lowest level, characters are read one at a time, with the following code:

```
int FileIO_Rdch()
{
    int ch;

    do
    {
        ch = fgetc(FileStream);
    } while (ch == CR);  /* ignore CR */

    LastCh = ch;
    if (ch == EOL)
    {
        LineNumber += 1;
    }

    return ch;
}
```

The system file function **fgetc** reads one character from the input stream. For compatibility between Linux and Windows systems, the carriage-return character (**CR**) is ignored. The variable **LineNumber** is updated if an end-of-line (**EOL**) character is found and the variable **LastCh** provides the basis for back-spacing after looking ahead to read the next character. The function **FileIO_Rdch** is defined as an **int** type, to enable the end of file character (**EOF**) to be detected (as a negative value).

This library includes functions to read strings, integers and floating-point numbers. For example, the function **FileIO_ReadInt** is as follows:

```
int FileIO_ReadInt()
{
    int n = 0;
    int ch;

    do
    {
        ch = FileIO_Rdch();
    } while (ch == ' ');

    while (ch >= '0' && ch <= '9')
    {
        n   = n * 10 + ch - (int) '0';
        ch = FileIO_Rdch();
    }
    return n;
}
```

Leading spaces are skipped and the value is computed by reading numeric ASCII characters ('0' to '9'), until a non-numeric character is encountered, which is treated as a terminator, and discarded. The important point about reading these databases is that the software should be tolerant of variations in formats (e.g. different lengths of names or arbitrary blank spaces) and be consistent in converting units in the database to units used in the navigation equations. Navigation databases are nowadays captured, stored and transferred electronically, with minimal scope for error, bearing in mind that the databases used in this book originated from data that is used by airlines.

5.5   Map Projections and Charts

Navigation over the earth is based on a spherical frame whereas a chart is a flat 2D surface. It follows that the mapping of a curved surface onto a flat surface will, inevitably, give rise to distortions. The requirement for charts is to devise a mapping from 3D coordinates in a spherical frame to coordinates in a 2D frame, which minimises these distortions. In simple terms, the view of the world printed on a chart or displayed on a screen should correspond to the view of the world seen from the flight deck or cockpit of an aircraft. In the case of general aviation, where a flight plan is likely to be drawn on a chart, the roads, railway lines, coastlines, shapes of towns and cities should enable a pilot to confirm position by identifying objects on the ground from information on the chart.

The mapping from (λ, ϕ) to (x, y), where λ and ϕ are latitude and longitude and x and y are Cartesian coordinates in the axes of the chart, requires three properties:

- *equivalence* – equal areas on the surface of the earth should have equal areas on the chart;
- *equidistance* – equal distances on the surface of the earth should have equal distances on the chart;
- *conformance* – angles should be preserved so that a shape on the surface of the earth has the same shape on the chart.

Ideally, the track between two points should be a straight line and the bearing between two points on the earth should be identical to the bearing between the two points on the chart.

If the angles are preserved, the mapping is termed *conformal* and includes the projection known as Lambert conformal mapping (Maling, 1992), first developed in 1772. Conceptually, Lambert projection can be thought of as a transparent cone placed over a sphere. Objects on the sphere are projected (or drawn) onto the cone, which is then unfolded onto a 2D surface. The arrangement is shown in diagrammatic form in Figure 5.6.

The point shown on the surface of the sphere is defined by its radius r from the apex of the cone and the angle θ subtended by the point, relative to a datum. The properties of the Lambert conformal mapping are particularly useful for charts used in aviation:

- Lines of latitude and longitude intersect at 90°, as they do on the earth sphere.
- Angles and bearings are preserved so that a shape on the surface of the earth has the same shape on the chart.
- Great circle lines are straight lines, which include lines of longitude and the equator.

The mapping from spherical coordinates to Cartesian coordinates uses intermediate polar coordinates:

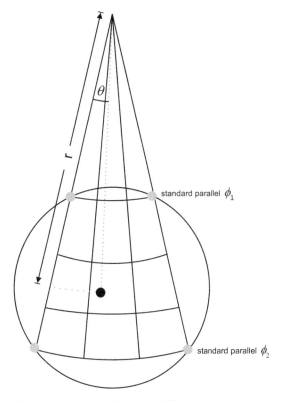

Figure 5.6   Lambert Conformal Mapping.

$$(\lambda, \phi) \rightarrow (r, \theta) \rightarrow (x, y) \qquad (5.10)$$

where

$$x = r \sin \theta \qquad (5.11)$$

$$y = r \cos \theta \qquad (5.12)$$

The angle θ is proportional to the longitude and is defined by a convergence factor n, which depends on the shape of the cone, which in turn is defined by the two lines of colatitude, ϕ_1 and ϕ_2, which are known as standard parallels and are shown in Figure 5.6. These values are chosen by the chart publisher to minimise the distortion over the chart and, for most charts, the standard parallels are indicated on the chart. Note that the convergence value n is a constant for a given chart. The transformation from (λ, ϕ) to (r, θ) is given by the following equations:

$$n = \frac{\log_e \sin x_1 - \log_e \sin x_2}{\log_e \tan\left(\dfrac{x_1}{2}\right) - \log_e \tan\left(\dfrac{x_2}{2}\right)} \qquad (5.13)$$

$$r = \frac{\sin x_1}{n} \left(\frac{\tan\left(\dfrac{x}{2}\right)}{\tan\left(\dfrac{x_1}{2}\right)} \right)^n \qquad (5.14)$$

$$\theta = n\lambda \qquad (5.15)$$

where

$$x = \frac{\pi}{2} - \phi \qquad (5.16)$$

$$x_1 = \frac{\pi}{2} - \phi_1 \qquad (5.17)$$

$$x_2 = \frac{\pi}{2} - \phi_2 \qquad (5.18)$$

Although these expressions seem complicated, the terms involving x_1 and x_2 are constant. Moreover, the inverse of the equations is straightforward to implement, for example, mapping from screen coordinates (x, y) to (λ, ϕ) to derive the latitude and longitude of the position. Lambert conformal projection is used for most aeronautical charts because the properties are well matched to aviation. Figure 5.7 shows two chart projections of a UK database, taken from an instructor station display.

Figure 5.7(a) shows a computer-generated chart of southern England. The curvature of the lines of latitude is apparent on this scale. Figure 5.7(b) shows the same chart zoomed out to cover the UK region from 8°W to 2°E and 49°N to 61°N. The standard parallels for these charts are based on the central latitude of the database ±6°. One final point to note is that charts are also displayed on aircraft flight displays and similar transformations and projections can be used.

Three functions are provided for mapping. **Map_SetMapCentre** sets the Lambert constants **Lambert_K** and **Lambert_N** for the current chart display, centred at (**Latitude, Longitude**).

```
void Map_SetMapCentre(float Latitude, float Longitude)
{
    float Chi;
    float lat1;
    float lat2;
    float r;
    float r1;
    float r2;

    Map_MapLatitude   = Latitude;
    Map_MapLongitude  = Longitude;
    Chi               = PIBY2 - fabs(Latitude);
    Lambert_N         = cos(Chi);
    Lambert_K         = tan(Chi) * pow(tan(Chi / 2.0), -Lambert_N);
    lat1              = Chi - DEG6;
    lat2              = Chi + DEG6;
    r                 = Lambert_K * pow(tan(Chi / 2.0), Lambert_N);
    r1                = Lambert_K * pow(tan(lat1 / 2.0), Lambert_N);
    r2                = Lambert_K * pow(tan(lat2 / 2.0), Lambert_N);
    msf               = D12 / fabs(r1 - r2);
    MapOffset         = r * msf;
}
```

The global variables **Lambert_K**, **Lambert_N**, **msf** (map scale factor) and **MapOffset** are computed for use in the mapping functions. The function **Map_GlobeToScreen** is the mapping from (**latitude, longitude**) to screen coordinates (**x, y**).

```
void Map_GlobeToScreen(float Latitude, float Longitude, float *x, float *y)
{
    float Chi;
    float Theta;
    float r;
    float t;

    Chi   = PIBY2 - fabs(Latitude);
    Theta = Lambert_N * (Longitude - Map_MapLongitude);
    r     = Lambert_K * pow(tan(Chi / 2.0), Lambert_N);
    *x    = (r * sin(Theta) * msf / Map_MapScaleFactor) + MapCentreX;
    t     = (r * cos(Theta) * msf - MapOffset) / Map_MapScaleFactor;
    if (Map_MapLatitude > 0.0)
    {
        *y = MapCentreY - t;
    }
    else
    {
        *y = MapCentreY + t;
    }
}
```

Similarly, the inverse function **Map_ScreenToGlobe** computes (**latitude, longitude**) from screen coordinates (**x, y**).

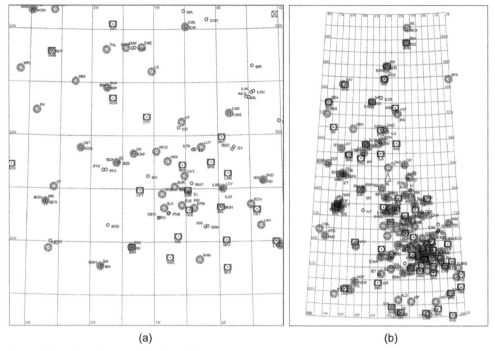

(a)   (b)

Figure 5.7   Chart Displays (see Plate 13).

```
void Map_ScreenToGlobe(float x, float y, float *Latitude, float *Longitude)
{
    float Chi;
    float Theta;
    float r;
    float dx;
    float dy;

    dx = (x - MapCentreX) * Map_MapScaleFactor;
    if (Map_MapLatitude > 0.0)
    {
        dy = (MapCentreY - y) * Map_MapScaleFactor + MapOffset;
    }
    else
    {
        dy = (y - MapCentreY) * Map_MapScaleFactor + MapOffset;
    }
    r         = sqrt(dx * dx + dy * dy) / msf;
    Theta     = asin((x - MapCentreX) * Map_MapScaleFactor / (r * msf));
    *Longitude = Theta / Lambert_N + Map_MapLongitude;
    Chi       = 2.0 * atan(pow(r / Lambert_K, 1.0 / Lambert_N));
    *Latitude = PIBY2 - Chi;
    if (Map_MapLatitude < 0.0)
    {
        *Latitude = -(*Latitude);
    }
}
```

Note that latitudes are positive in the northern hemisphere and negative in the southern hemisphere. Similarly, easterly longitudes are positive and westerly longitudes are negative. Both functions provide the correct mappings for all regions of the world.

Figure 5.8 Navigation Computations.

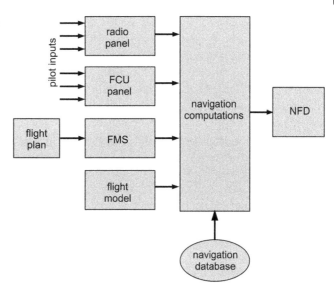

5.6 Navigation Computations

The navigation computations needed to display the essential navigation information on the navigation flight display (NFD) are shown in Figure 5.8.

The frequencies for communications and navigation beacons are entered via the flight plan or the radio panel by the flight crew. Similarly, the sources of displayed information and the display modes are entered via the FCU. The FMS provides data to display flight plan information including navaids, runways and waypoints. The flight model provides data including aircraft latitude, longitude, altitude, heading and wind. In addition, the navigation software has access to the location, altitude, frequencies and types of navigation aids in the navigation database. The outputs are used by the NFD software to display navigation information depending on the sources and display modes selected.

The data structures used by the navigation database software are described in Section 5.4. The data structure used by the navigation software, which applies to ILS, VOR and ADF computations, is as follows:

```
typedef struct
{
    unsigned int SelectedBeacon;
    float        LocaliserError;       /* radians */
    float        GlideSlopeError;      /* radians */
    float        SlantDistance;        /* m */
    float        BearingToStation;     /* radians */
    bool         BeaconStatus;         /* true if active */
    bool         IlsFrequency;         /* true if ILS */
    float        RunwayQdm;            /* QDM (T) radians */
} Nav_NavRecord;
```

where **SelectedBeacon** is a pointer to an entry in the navigation database. **LocaliserError** is the angular offset from the localiser and only applies to VOR and ILS transmitters.

GlideslopeError is the offset from the glideslope and only applies to ILS transmitters. **SlantDistance** is the line-of-sight distance from the aircraft to the transmitter. **BearingToStation** is the bearing from the aircraft to the station. **BeaconStatus** is true if the station is found in the database and is in range and the navigation channel is working correctly (i.e. has not been failed). For DME ground stations, only DME distance applies. **IlsFrequency** is true if the station is an ILS, to distinguish it from a VOR station. **RunwayQdm** is the QDM of the runway and only applies to an ILS approach.

The main navigation computation is performed by the function **NavCalculations**, which is used to update the ADF, VOR, DME and ILS systems each frame. The first parameter **Nav** is a pointer to the **Nav_NavRecord** data structure for the beacon. The second parameter is the course pointer value and the third parameter is the condition of the navaid, which, if false, indicates that it is not functioning or has been deliberately failed (e.g. by the instructor). A check is made that the beacon is a valid beacon in the navigation database. The value is set in the radio panel software when a new radio frequency is selected or the simulator is reset to a previous state, which includes any tuned navaids. The **Operational** parameter is used to simulate a receiver failure mode.

```
BeaconNumber = Nav->SelectedBeacon;
OK = BeaconNumber > 0 && Operational;
```

The beacon is accessed from the navigation database to determine the type of beacon (ILS, VOR or ADF) and, in the case of a VOR or ILS station (VHF frequencies are line-of-sight), the height of the aircraft above the station is computed either from the station elevation in the navigation database or the altitude of the current runway (**Nav_GroundLevel**).

```
if (OK)
{
    bcn = &NavLib_Beacons[BeaconNumber];

    IlsBeacon = NavLib_ILS & bcn->Navaid;
    VorBeacon = NavLib_VOR & bcn->Navaid;
    AdfBeacon = NavLib_NDB & bcn->Navaid;
    if (IlsBeacon | VorBeacon)  /* VHF line-of-sight */
    {
        z = -(float) (NavLib_Beacons[BeaconNumber].Elevation) - (float) NavLink_AeroPkt.Pz;
    }
    else
    {
        z = Nav_GroundLevel - (float) NavLink_AeroPkt.Pz;
    }
    if (z < 0.0)
    {
        z = 0.0;
    }
}
```

Computation of the range for VHF stations is given by the distance to the horizon, based on the curvature of the earth.

$$R \approx 1.22\left(\sqrt{H_T} + \sqrt{h}\right) \tag{5.19}$$

where R is the range in statute miles, H_T is the height of the transmitter above sea level in feet and h is the aircraft altitude in feet. In SI units, the range is given by

$$R = 3568\left(\sqrt{H_T} + \sqrt{h}\right) \tag{5.20}$$

The maximum range of the transmitter is computed as follows:

```
MaxRange = 3568.0 * (sqrt(z) + sqrt(-NavLink_AeroPkt.Pz));
if (MaxRange < 3000.0)
{
    MaxRange = 3000.0;
}
else if (MaxRange > 160000.0)
{
    MaxRange = 160000.0;
}
if (AdfBeacon)
{
    MaxRange = (float) bcn->Range * 1852.0;
}
```

The minimum range is limited to 3 km and the maximum range is limited to 160 km, based on the typical power values for VHF transmitters. The range for NDB transmitters depends on the power of the transmitter. Unfortunately, the range of all NDBs in the Navigraph database is set to 195 nm and is therefore ignored. The ground and slant distance is computed to check that the transmitter is in range:

```
dground = (float) NavLib_Distance(NavLink_AeroPkt.Latitude, NavLink_AeroPkt.Longitude,
                         (double) bcn->BeaconLatitude, (double) bcn->BeaconLongitude);
dslant = sqrt(dground * dground + z * z);
if (dslant <= MaxRange)
{
    Nav->BearingToStation = (float) NavLib_Bearing(NavLink_AeroPkt.Latitude,
                                    NavLink_AeroPkt.Longitude,
                                    (double) bcn->BeaconLatitude,
                                    (double) bcn->BeaconLongitude);
```

The bearing to the station is computed from the latitude and longitude of the aircraft and the transmitter, respectively. If an ILS transmitter is detected, the follow code computes the glideslope and localiser errors:

```
if (IlsBeacon)
{
    Nav->RunwayQdm = NavLib_Beacons[BeaconNumber].Qdm;   /* true */
    Maths_Normalise(&Nav->RunwayQdm);
    qTrue = Nav->RunwayQdm;
    Nav->LocaliserError = Nav->BearingToStation - qTrue;
    t1 = dground * sin(Nav->LocaliserError);
    t2 = dground * cos(Nav->LocaliserError);
    t2 += 416.0;   /* G/S transmitter typically 400m past start of the runway */
    dground = sqrt(t1 * t1 + t2 * t2);
    dslant = sqrt(dground * dground + z * z);
    Nav->GlideSlopeError = 0.052359877 - atan2(z, dground);
    t2 -= 416.0;   /* reset to start of runway */
```

```
    if (Nav_CurrentRunway > 0)
    {
        t2 = t2 + NavLib_Runways[Nav_CurrentRunway].Length;
    }
    else
    {
        t2 = t2 + 3000.0;
    }
    Nav->LocaliserError  = atan2(t1, t2);
}
```

Note that headings and bearings on charts and in displays are given in degrees magnetic, whereas all the navigation equations use degrees true, which is an important convention as the conversion from magnetic angles to true angles and vice-versa is only needed for the input and output of navigation data, enabling the computations to be independent of magnetic variation. As the range to the runway does not justify the use of spherical coordinates, the positions **t1** and **t2** are computed from the ground distance to the runway threshold, which is derived from the runway coordinates given in the navigation database, enabling the slant distance to the glideslope transmitter and localiser transmitter to be computed. The localiser error is computed as $\tan^{-1}(t2/t1)$, using the **atan2** function to avoid division by zero.

If the beacon is not an ILS, the bearing and distance values are required for VOR and NDB stations, particularly to compute the localiser error of a VOR. The code fragment for the VOR and NDB stations is as follows:

```
else
{
    Nav->RunwayQdm = 0.0;
    Nav->LocaliserError = Nav->BearingToStation - Obs - Nav_MagneticVariation;
}
Nav->LocaliserError = Maths_Normalise(Nav->LocaliserError);
Nav->SlantDistance = dslant;
if (bcn->Navaid == NavLib_DME)
{
    DmeOnly = true;
}
```

In the case of a VOR station, the OBS value (converted to degrees true) is subtracted from the relative bearing to give an offset from the selected radial. The slant distance is added to the data structure and a check is made to distinguish between VOR/DME, ILS/DME and DME-only stations. Note the use of **Maths_Normalise** in these computations to constrain angles to the range $\pm\pi$, where angles may exceed this range in the computations of bearings.

The other requirement from the navigation software is to establish the navigation sources needed for the displays. For EFIS displays which include an NFD, the function **Nav_Update** sets the sources for the NFD. Each frame, the CRS setting and the distance, bearings and status for the ILS, VOR and ADF channels are updated.

```
Nav_HSI_Crs = Radio_Radios[0].CrsKnob;
NavCalculations(&Nav_ILS1, Maths_Rads((float) Nav_HSI_Crs), true);
NavCalculations(&Nav_VOR1, Maths_Rads((float) Nav_HSI_Crs), true);
NavCalculations(&Nav_VOR2, Maths_Rads((float) Nav_HSI_Crs), true);
NavCalculations(&Nav_ADF1, 0.0, true);
NavCalculations(&Nav_ADF2, 0.0, true);
```

The main compass and expanded compass displays show either the ILS or the VOR selections. In ILS mode, the localiser indicator, glideslope indicator, displayed frequency and DME are set for the selected ILS station. Otherwise the display uses VOR1 as its source, as follows:

```
if (FCU_ModeSelector == NavDefn_ModeILS)
{
    Nav_HSI_Localiser = Nav_ILS1.LocaliserError;
    Nav_HSI_GlideSlope = Nav_ILS1.GlideSlopeError;
    Nav_HSI_ILSMode = Nav_ILS1.IlsFrequency;
    Nav_HSI_Status = Nav_ILS1.BeaconStatus;
    if (Nav_ILS1.BeaconStatus)
    {
        Nav_DmeDistance = Nav_ILS1.SlantDistance;
    }
    else
    {
        Nav_DmeDistance = -10000.0;
    }
}
else
{
    Nav_HSI_Localiser = Nav_VOR1.LocaliserError;
    Nav_HSI_GlideSlope = Nav_VOR1.GlideSlopeError;
    Nav_HSI_ILSMode = Nav_VOR1.IlsFrequency;
    Nav_HSI_Status = Nav_VOR1.BeaconStatus;
    if (Nav_VOR1.BeaconStatus)
    {
        Nav_DmeDistance = Nav_VOR1.SlantDistance;
    }
    else
    {
        Nav_DmeDistance = -10000.0;
    }
}
```

The radio magnetic indicator (RMI) modes are taken from FCU selections of either VOR or ADF sources, as follows:

```
if (FCU_NavSwitch1 == NavDefn_NavADF)
{
    Nav_Rmi_Dir1 = Nav_ADF1.BearingToStation - NavLink_AeroPkt.Yaw;
}
else if (FCU_NavSwitch1 == NavDefn_NavVOR)
{
    Nav_Rmi_Dir1 = Nav_VOR1.BearingToStation - NavLink_AeroPkt.Yaw;
}
else
{
    Nav_Rmi_Dir1 = Maths_PIBY2;
}
if (FCU_NavSwitch2 == NavDefn_NavVOR)
{
    Nav_Rmi_Dir2 = Nav_VOR2.BearingToStation - NavLink_AeroPkt.Yaw;
}
else if (FCU_NavSwitch2 == NavDefn_NavADF)
{
    Nav_Rmi_Dir2 = Nav_ADF2.BearingToStation - NavLink_AeroPkt.Yaw;
```

```
}
else
{
    Nav_Rmi_Dir2 = Maths_PIBY2;
}
```

Finally, the runway database is checked every frame to see if there is a closer runway than the currently selected runway. This search is necessary if the visual system uses a generic airfield, which is replicated throughout the navigation region.

```
Nav_CurrentRunway = NavLib_NearestRunway(NavLink_AeroPkt.Latitude, NavLink_AeroPkt.Longitude,
                                         NavLib_Beacons[Nav_ILS1.SelectedBeacon].Frequency);

if (Nav_CurrentRunway > 0)
{
    Nav_GroundLevel = -(float) (NavLib_Runways[Nav_CurrentRunway].Elevation);
}
else
{
    Nav_GroundLevel = 0.0;
}
```

The pointer to the current runway is updated (if necessary) and the overall ground level is set to the altitude of the selected airfield.

5.7 Radio Navigation Aids

In VFR conditions, although a flight plan will be checked by visual reference to match features on the chart to roads, towns, rivers and railway lines, etc., radio navigation beacons provide a useful cross-check. In IFR conditions, the flight plan is followed entirely from flight instruments and navigation is by reference to radio navigation beacons.

In IFR conditions, radio beacons provide relative bearing to a beacon, distance from a beacon and an angular offset from an approach path in both lateral and vertical planes. If the relative bearings and distances to beacons are known, then a position fix can be obtained from two relative bearings, two distances or a relative bearing and a distance (also known as rho-theta navigation). Although the availability of GPS has revolutionised both the automotive and the aeronautical industries, providing accurate position fixes over the complete earth and at all times, radio beacons are still used for en-route guidance and approaches and it is necessary to model airborne receivers and their displays in a flight simulator.

5.7.1 Automatic Direction Finding (ADF)

Non-directional beacons have been in use since the 1950s and evolved from technology developed during the Second World War. Users of an amplitude modulated (AM) radio are aware that the signal quality is very dependent on the orientation of the radio antenna. ADF is based on the same principle; a loop antenna detects the strongest signal from a ground transmitter when the antenna is parallel to the incoming signal. Originally, these antennas were physically rotated, but nowadays the components in two antennas are resolved electronically to give the relative bearing of the strongest signal. The advantage of ADF is that both the transmitter and the receiver are relatively simple and, at the low operating frequencies in the range 190–1750 KHz, the signal is mostly a ground wave with some reflection off the ionosphere. Therefore, ADF is not dependent on line-of-sight to the transmitter and, with sufficient power in the transmitter, ranges of up to 200 miles are achievable. The drawback, and the reason ADF is not approved for primary navigation, is that it is subject to errors from electrical fields, reflections from the ionosphere, electrical storms in the atmosphere and interference from other AM sources. Nevertheless, ADF transmitters are still widespread in Europe and North America and, over short ranges of less than 15 nm, low-power (25 W) NDB transmitters, located near to airfields and known as locator beacons, are used in approach and holding procedures, for which their use is approved.

One other source of error results from the orientation of the antenna during turning manoeuvres, which is known as 'dip' error. Typically, the dip error is simulated by adding half the bank angle (up to ±10°) to the relative bearing. Errors up to 5° are within ICAO limits and generally an ADF is only used in approach procedures and holding patterns as the accuracy for en-route navigation is relatively poor.

The basic ADF instrument comprises a rotating compass card and a pointer. The compass card is aligned with the magnetic heading and the needle points towards the selected beacon and is independent of aircraft heading. In earlier ADF instruments, known as a radio bearing indicator (RBI), the card could be rotated using a rotary knob, to enable the pilot to select the current heading. In modern aircraft, the ADF display is known as an RMI, usually with two needles for two selected frequencies, allowing the selection of NDB or VOR sources. Figure 5.9 illustrates a typical ADF display and the situation.

In Figure 5.9(a), the aircraft is heading 140°M and the relative bearing to the NDB is 190°M with the needle pointing to 190°M. Figure 5.9(b) shows the view above the aircraft, with the aircraft heading 140°M and the NDB station 50° to the right, with the ADF needle pointing towards the station. Figure 5.9(c) shows the chart view, with a magnetic variation of 2°W. The aircraft symbol is shown as a triangle, heading 140°M. The track from the aircraft to the Cranfield NDB (CIT) is 21.3 nm from the aircraft on a track of 190°M.

The relative bearing is computed from the current position of the aircraft and the location of the selected ADF beacon. The ADF pointer is simply the relative bearing minus the aircraft heading, as follows:

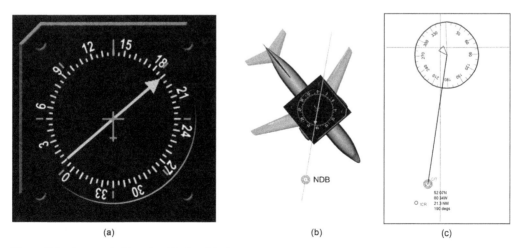

(a)   (b)   (c)

Figure 5.9 Automatic Direction Finding Display.

```
Nav->BearingToStation = (float) NavLib_Bearing(NavLink_AeroPkt.Latitude, NavLink_AeroPkt.Longitude,
                        (double) bcn->BeaconLatitude, (double) bcn->BeaconLongitude);
Nav->RmiPointer = Nav->BearingToStation - (float) NavLink_AeroPkt.Yaw;
```

In the display of the ADF, the background is rendered as a texture with an empty dial, and the compass card is drawn as a texture rotated by the aircraft heading. The needle is also rendered as a texture, rotated by the relative bearing. The main part of the code is as follows:

```
void Adf_Adf(int x0, int y0, float Hdg, float rb1, GLuint texobj1, GLuint texobj2)
{
    float rb;

    rb1 = Maths_Normalise(rb1);
    rb = -Maths_Degrees(rb1);

    if (rb > (oldrb + 0.5))
    {
        rb = oldrb + 0.5;
    }
    if (rb < (oldrb - 0.5))
    {
        rb = oldrb - 0.5;
    }

    oldrb = rb;

    Glib_SetTexture(texobj1);
    Glib_DrawTexture(x0 - 128, y0 - 128, 256, 256, 0.0, 0.0, 1.0, 1.0, 1.0);
    Glib_SetTexture(texobj2);
    Glib_DrawTextureRotated(x0, y0, 256, 256, 0.0, 0.0, 1.0, 1.0, Maths_Degrees(Hdg), 1.0);
```

The relative bearing is normalised to the range $\pm 180°$ and is constrained to rotate at a maximum rate of $25°/s$. The instrument background and the compass card textures are 256×256.

5.7.2   VHF Omni-directional Range

A VOR receiver is similar to an ADF receiver, in the sense that it computes the relative bearing to a VOR transmitter. However, the transmission method and signal detection are based on the detection of the phase difference between two signals, modulated at 30 Hz, where one transmitter is rotating and directional and the other is omnidirectional. The signal is demodulated to provide a bearing to the VOR station. As the transmitter operates in the VHF band (112–118 MHz), it is not affected by the errors associated with ADF transmission, although interference from ground reflections near the transmitter and reflections from airport buildings can affect the errors, which are generally less than 2° (ICAO limit).

Rather than displaying the relative bearing, the VOR display enables a pilot to select a specific radial to or from a VOR transmitter and the display shows the angular error relative to that radial, limited to $\pm 10°$. The display also indicates whether the aircraft is positioned to or from the transmitter, with respect to the selected radial. By selecting a VOR station, a pilot can determine the bearing to the VOR station by turning the compass dial until the needle is centred. Similarly, if the aircraft drifts to the left of right of the radial, this situation is indicated by the VOR display showing the angular error from the current radial, enabling the pilot to change heading to recapture the desired track towards a VOR station.

Figure 5.10 shows a typical VOR display. In Figure 5.10(a), the OBS knob has been turned to select the 072 radial. The VOR localiser bar shows that the aircraft is 6° to the left of the localiser to the VOR. The inner circle corresponds to 2° of offset and each mark represents a further 2° from the selected localiser. In Figure 5.10(b), the aircraft shown by the arrowhead is approximately 20 nm west of the Midhurst VOR (MID), heading 120°M and is positioned 6° left of the 072 radial to the VOR transmitter (which is also the 252 radial *from* the VOR transmitter). Often a VOR display is combined with an ILS display and, in this example, the red LS flag denotes that the ILS receiver is not active. VOR stations are commonplace in Europe and North America and a position fix can be obtained from the intersection of two VOR radials. The closest VOR stations are automatically tuned in FMS computations to obtain position fixes for tracking. Note that an error of 2° corresponds to a maximum position error of 2 nm in 60 nm.

There is an added complication with VOR in that a display would give similar indications for the reciprocal radials, where an aircraft that is left of the 070 radial is also right of the 250 radial. To avoid this ambiguity, the *TO* and *FROM* indications are displayed. For example, if an aircraft is south-west of a VOR, the 045 radial would be shown as *TO* and the 315 radial would be shown as *FROM*.

Navigation displays on modern EFIS systems display relative bearings to both VOR and ADF stations, as shown in Figure 5.11. In Figure 5.11(a), the green pointer shows NAV1 tuned to Midhurst VOR (MID) on a frequency 114.0 MHz. The transmitter is on a bearing of 250°M, directly behind the aircraft. The transmitter tuned to NAV1 is shown in the bottom left-hand corner. The blue pointer shows NAV2 tuned to the EPSOM NDB (EPM) on a frequency 316.0 KHz. The transmitter is on a bearing 355°M to the left of the aircraft heading. The transmitter tuned to NAV2 is shown in the bottom right-hand corner. Figure 5.11(b) shows the map display, with the aircraft (shown as an arrowhead) on an approach into London Gatwick at 5.2 nm. The two black lines from the aircraft to the respective VOR and ADF transmitters show the respective relative bearings. The magenta bar in the NFD shows the localiser offset for either the VOR or ILS, in this case the ILS localiser for London Gatwick (IGG). The transmitter tuned to the ILS is shown in the top right-hand corner. For the EFIS NFD, in ILS and VOR mode, the white front and back pointers, the white to/from triangle and the magenta CDI bar

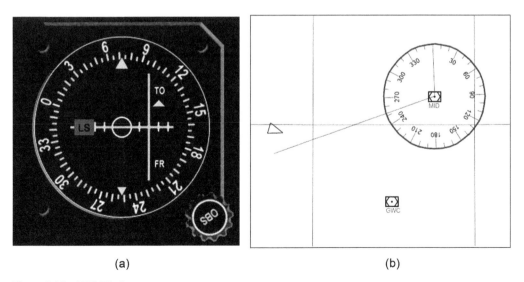

(a)   (b)

Figure 5.10 VOR Display.

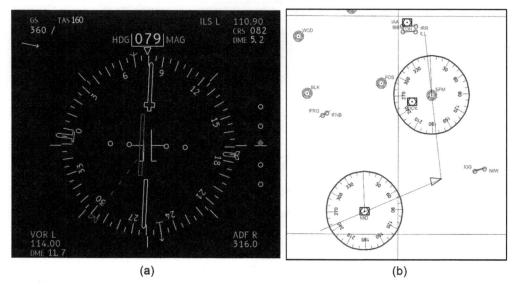

Figure 5.11 EFIS NFD (ILS Mode).

rotate with the CRS, 082°M in Figure 5.11(a). The fragment of code to compute the offset is as follows:

```
if (LocaliserError >= -DEG85 && LocaliserError <= DEG85)
{
    HsiFlag = ToFlag;
}
else if (LocaliserError < -DEG95)
{
    LocaliserError = -Maths_PI - LocaliserError;
    HsiFlag = FromFlag;
}
else if (LocaliserError > DEG95)
{
    LocaliserError = Maths_PI - LocaliserError;
    HsiFlag = FromFlag;
}
if (ILSMode)
{
    ScaleFactor = 1970.975;
    HsiFlag     = OffFlag;
}
else
{
    ScaleFactor = 492.743;
}
Cdi = LocaliserError * ScaleFactor;
Cdi = Maths_Limit(Cdi, -108.0, 108.0);
```

where `LocaliserError` is the angular error from the radial to the VOR station. The *TO/FROM* condition is determined from the localiser error, noting that no valid signal occurs when the aircraft is ±10° from the perpendicular to the selected radial. The scale factor for VOR and ILS modes is the mapping from radians to pixels, where the ILS sensitivity is four times the VOR value. `Cdi` is the course deviation index or the displayed localiser error. The code to render the rotated pointers, the *TO/FROM* triangle and the course deviation index is as follows:

```
Glib_PushMatrix();
Glib_Rotate(Maths_Degrees(-Crs));
Glib_Line_Loop(0, 0, 12, CrsFrontPtr);

for (i = -86; i <= 86; i += 43)
{
    if (i != 0)
    {
        Circle5(i, 0);
    }
}
Glib_Line_Loop(0, 0, 4, CrsBackPtr);

Glib_Colour(Glib_MAGENTA);
Glib_Line_Loop(0, 0, 3, CdiBar);

Glib_Colour(Glib_WHITE);
if (HsiFlag == ToFlag)
{
    Glib_Line_Loop(0, 0, 2, ToFromFlagPtr);
}
else if (HsiFlag == FromFlag)
{
    Glib_Rotate(Maths_Degrees(M_PI));
    Glib_Line_Loop(0, 0, 2, ToFromFlagPtr);
}
Glib_PopMatrix();
```

The display is rotated by `-Crs` degrees, the front and back pointers are each drawn as a line-loop, the four circles are rendered by `Circle5` and the CDI bar is drawn in magenta, also as a line-loop. The white triangle displays the *TO/FROM* indications and is rotated by 180° for the *FROM* indication.

5.7.3 Distance Measuring Equipment

Although a position fix can be obtained from two relative bearings to VOR stations, it can also be obtained from the combination of a relative bearing and a distance, known as *rho-theta* navigation, where *rho* denotes range and *theta* denotes bearing. Consequently, many VOR transmitters are co-located with DME transmitters and their frequencies are co-paired, so that only one frequency needs to be tuned. It is important to appreciate that DME operates in the VHF band and therefore the distance measured is line-of-sight to the station, providing slant distance rather than the distance over the ground. For example, flying at 6000 ft directly overhead a DME transmitter, the equipment will display the distance as 1.0 nm. The range of a DME is limited to 150 nm and 75 nm below 20,000 ft.

The DME in an aircraft measures distance from a DME transmitter by timing the interval between transmitting a pulse and detecting a reply from the transmitter. In practice, the distance measurement errors are 100–300 m. In flight simulation, the computation of slant distance should include similar errors. Typically, DME displays show the distance to the DME station as a numeric value to the nearest 0.1 nm and some DME displays also indicate the time to the DME station computed from the DME distance and the current airspeed. A fragment of the code to compute DME distance in the navigation equations follows:

```
dground = (float) NavLib_Distance(NavLink_AeroPkt.Latitude, NavLink_AeroPkt.Longitude,
                        (double) bcn->BeaconLatitude, (double) bcn->BeaconLongitude);
dslant = sqrt(dground * dground + z * z);
```

where **z** is the height of the aircraft above the transmitter, **dground** is the ground distance to the transmitter and **dslant** is the slant distance to the transmitter.

5.7.4 Instrument Landing System

Although visual approaches are used in the majority of light aircraft operations, for transport aircraft and most commercial aircraft, the flight crew are qualified to fly an instrument approach using the ILS display. In poor visibility, an ILS approach can be flown to an altitude of 200 ft above the ground, at which point, if the runway visibility is sufficient, the landing can be completed as a visual approach. Otherwise the landing must be aborted and the aircraft can either go-around to attempt another landing or divert to another airfield. Typically, an ILS approach has a 3° glideslope, which is a gradient of approximately 1 in 20. In other words, at an altitude of 200 ft, the aircraft will be 4000 ft from the touchdown point on the runway or two-thirds of a nautical mile.

The ILS is a navigation aid used solely during the landing phase. Often, an ILS is co-paired with a DME to provide the flight crew with both altitude and distance information, as a cross-check during an approach. An ILS is based on two transmitters. The localiser transmitter is located beyond the far end of the runway and transmits a signal that provides an angular offset from the runway centre line. The glide path transmitter is placed slightly to the left of the runway, adjacent to the touchdown point, and transmits a signal that enables the receiver to measure the angular offset from the ideal glide path. The actual glide path angle is fixed for a specific runway.

Although ILS also uses the VHF band, the method of transmission and modulation is very different from VOR. The localiser has an error of 0.1–0.5°, providing a maximum localiser angle of ±2.5° (rather than the 10° of VOR) and a range of 25 nm. The glide path error is 0.1–0.2° with a range of 10 nm. In addition to the localiser and glide path transmitters, three locator beacons transmit a narrow vertical beam which is detected as the aircraft passes overhead. On the flight deck, the locator beacon signals provide a flashing indicator and an aural tone, which is a cross-check of altitude and distance during an approach. The middle marker is typically located half a mile from touchdown and is indicated by a yellow flashing lamp, and the outer marker is typically located 5–6 miles from touchdown and is indicated by a blue flashing lamp.

The ILS is shown in diagrammatic form in Figure 5.12, with the localiser shown as a vertical plane from the localiser transmitter and the glide path shown as a sloping plane from the glide path transmitter. In the aircraft, the ILS receiver detects the difference in signal strength in two lobes by demodulating the transmitted signal for both the localiser and glide path transmitters.

The glideslope is illustrated in Figure 5.13.

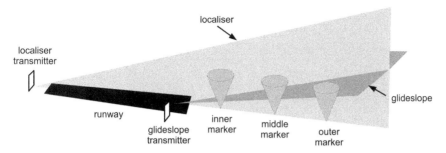

Figure 5.12 The ILS System.

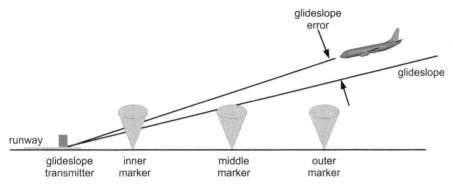

Figure 5.13 ILS Glideslope.

The location of the glideslope transmitter may be available from the navigation database or the location of the touchdown runway markings in the visual system. In flight simulation, the glideslope error is calculated from the aircraft altitude, the runway altitude, the slant distance to the glideslope transmitter and the runway glideslope angle.

The localiser is illustrated in Figure 5.14. The runway QDM is available in the navigation database. The location of the localiser may also be provided in the navigation database, otherwise the location can be estimated from the runway length and QDM. The localiser error is the relative bearing from the localiser transmitter to the aircraft minus the runway QDM. Note that the localiser is located past the end of the runway but safely below the clearance height for departing aircraft.

Figure 5.15 shows a conventional ILS display and an EFIS display. In Figure 5.15(a), the aircraft is positioned 1° to the left of the localiser and ½° below the glide path. The pilot has set the course to the runway QDM 082°M and, depending on the distance to touchdown, is likely to turn onto a heading of 087, to capture the localiser and then maintain a heading of 082°M. The pilot will probably reduce the rate of descent by a small amount to capture the glide path and then maintain the rate of descent depending on the glide path angle and the airspeed.

In the PFD shown in Figure 5.15(b), the aircraft is on an approach passing 1820 ft and the glide path is shown by the white diamond symbol to the right of the attitude indicator with the glide path scale shown by the four white circles. The aircraft is just below the glide path. The localiser is shown by the white diamond symbol at the bottom of the attitude indicator, with the localiser scale shown by the four white circles, with the aircraft approximately ½° to the right of the localiser.

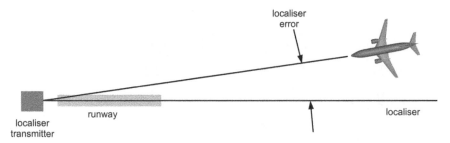

Figure 5.14 ILS Localiser.

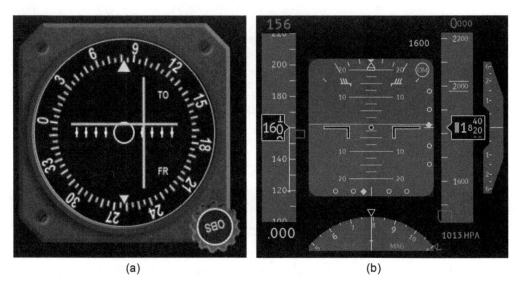

(a)   (b)

Figure 5.15 ILS Displays.

Note that the outer marker symbol is shown in the top right-hand corner of the attitude indicator, which flashes as the aircraft passes over the marker beacon. There are several points to bear in mind with ILS displays:

- The indications are independent of heading, airspeed or rate of descent. In other words, they are simply a 'snapshot' of the current position. For example, the aircraft could be exactly on, but flying perpendicular to, the localiser.
- The rate of closure varies with the distance from the ILS transmitters. To capture the localiser, a pilot needs to take account of the distance, airspeed, heading and turn rate.
- False nulls can occur above the glide path, which are caused by reflections from the ground immediately in front of the transmitter. Consequently, capture of the glide path is always from below the glide path.
- If either the localiser or glide path signals are lost or inconsistent, warning flags are activated on the display; otherwise the two ILS needles might falsely centre under such conditions, giving incorrect and potentially dangerous indications.

The ILS localiser, which is very similar to the VOR localiser, is shown as a CDI bar. However, for an EFIS display, the glideslope is shown as a diamond symbol which moves vertically relative to a datum, which represents the nominal glide path, as shown in Figure 5.15(b). The code for the glideslope display is as follows:

```
void Compass_GlideSlope(intCompassX, intCompassY, float GlideSlopeError, bool
ILSMode, bool Selected)
{
    int h;

    Glib_LoadIdentity();
    Glib_Translate((float) CompassX, (float) CompassY);

    h = intround(GlideSlopeError * 7039.2);
    if (h > OldGlideslope_h)
    {
        h = OldGlideslope_h + 1;
    }
    else if (h < OldGlideslope_h)
    {
        h = OldGlideslope_h - 1;
    }
    if (h < -108)
    {
        h = -108;
    }
    else if (h > 108)
    {
        h = 108;
    }
    if (Selected)
    {
        Glib_Colour(Glib_MAGENTA);
        Glib_Triangle(259-8, h, 259, h+8, 259, h-8);
        Glib_Triangle(259, h+8, 259+8, h, 259, h-8);
    }
    OldGlideslope_h = h;
}
```

The display is drawn relative to the centre of the compass display. The glideslope error is converted to the symbol displacement in pixels, with the movement limited to 50 pixels per second, to avoid abrupt changes. The diamond symbol is rendered as two triangles.

5.8 Traffic Collision Avoidance Systems

Traffic collision avoidance systems (TCAS) detect the trajectory of other aircraft and provide guidance to flight crews to avoid potential conflicts, displaying the situation on the vertical speed indicator, showing advisory rates of climb/descent and the relative positions of intruding aircraft. Although the aircraft interrogations are based on Mode-S radar, in flight simulation, detection of conflicting traffic, which is usually introduced as synthetic agents, is computed from the position of the aircraft and the traffic, where the response is identical to an airborne system, and the traffic is displayed on the navigation display. A TCAS conflict is shown in Figure 5.16.

The simulator aircraft A at (x_1, y_1) has a velocity vector v_1 and the intruding aircraft B at (x_2, y_2) has a velocity vector v_2. The requirement is to detect if the separation between the two aircraft will become closer than a predefined minimum distance if they continue along their current tracks. By subtracting the vector v_1 at A and from the vector v_2, aircraft A can then be considered to be at rest and the relative velocity of B to A is given by the vector S, giving a miss distance M. The coordinates and velocity vectors would be known in a simulation, whereas

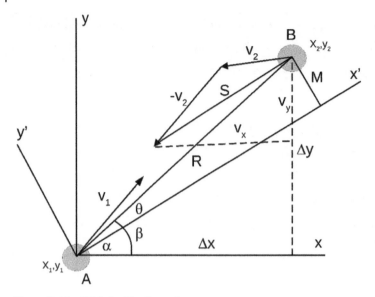

Figure 5.16 TCAS Conflict Detection.

actual TCAS equipment is based on airborne measurements of the range R and the range rate dR/dt. From Figure 5.16

$$\Delta x = x_2 - x_1 \tag{5.21}$$

$$\Delta y = y_2 - y_1 \tag{5.22}$$

$$\beta = \tan^{-1}\left(\frac{\Delta y}{\Delta x}\right) \tag{5.23}$$

$$v_x = v_{x2} - v_{x1} \tag{5.24}$$

$$v_y = v_{y2} - v_{y1} \tag{5.25}$$

$$\alpha = \tan^{-1}\left(\frac{v_y}{v_x}\right) \tag{5.26}$$

$$\theta = \beta - \alpha \tag{5.27}$$

$$m = R\tan\theta \tag{5.28}$$

$$\vec{S} = \vec{v}_2 - \vec{v}_1 \tag{5.29}$$

$$\dot{R} = -S\cos\theta \tag{5.30}$$

$$\tau = \frac{-R}{\dot{R}} \tag{5.31}$$

The value τ is given in seconds and is, in effect, the time to impact. If τ falls below a critical value, a resolution advisory is displayed, giving flight crews sufficient time to manoeuvre to avoid the conflict.

5.9 Inertial Navigation Systems

The inertial navigation system (INS) has been used in military and commercial aircraft since the 1960s. An INS comprises three gyros mounted orthogonally and three accelerometers, also mounted orthogonally. Before the 1990s, the gyros were predominantly based on mechanical devices and the INS was a mechanically stabilised platform. However, with the advances in miniaturisation in microelectronic devices and the advent of the ring laser gyro, the size of INS equipment has reduced and both accuracy and reliability are considerably improved. INS devices include a processor that can solve the equations digitally and at high speed. In military applications, the INS is covert, avoiding reliance on ground transmitters for position location. In civil aviation, the INS is a safety-critical component, used in both navigation and auto-land, and typically, at least three INS units are used to provide sufficient redundancy. More recently, INS systems have been integrated with GPS systems as the error characteristics complement each other (Titterton and Weston, 2004).

In flight simulation there are two approaches to the modelling an INS. As the error characteristics are known, errors can be added to the true position (computed by the flight model) to emulate the characteristics of an INS. The alternative method, and the one adopted in this book, is to model the physical properties of an INS so that the error properties are included in the model. There are four fundamental problems with the modelling of an INS:

- It is not possible to use actual aircraft equipment as the accelerations in a simulator are very different from the accelerations in an aircraft.
- Any bias or friction in an INS will generate additional accelerations, which lead to increasing errors and it is, therefore, essential to understand the error properties of a specific INS.
- An INS is not necessarily located at the centre of motion of an aircraft and care is needed with transformations from body axes to world axes.
- As the accelerations and rates can be very small, accuracy of the computations is particularly important.

Three sets of axes are used in modelling an INS:

- the *e-frame*, an earth-centred, earth-fixed, non-rotating frame, with its origin at the centre of the earth;
- the *n-frame*, a local or navigation frame, which is continually realigned to point north, east and down with its origin directly below the aircraft on the surface of the earth;
- the *b-frame*, the body frame covered in Chapter 3. The origin is at the centre of gravity of the aircraft with the axes pointing forwards, along the starboard wing and orthogonal to the underside of the aircraft.

Three axis transformations are used:

- accelerations measured in the *b-frame* to the *n-frame* to derive V_N, V_E, V_d, λ, ϕ and h;
- the earth rotation rate in the *e-frame* to the *n-frame*;
- rates in the *n-frame* to the *b-frame*, to derive the quaternion rates.

Note that textbooks on INS design use different notations to describe transformations. The convention adopted in this book is that a rotation ω_{ab}^c is a rotation from frame a to frame b, defined in frame c. The structure of an INS is shown in Figure 5.17, where the transformations are shown in grey.

The accelerometers sense accelerations and the gyros sense angular body rates in the body axes. The body accelerations are transformed to the navigation frame and the local gravity value is computed as a function of latitude, longitude and altitude, also in the navigation frame. The earth's rotation rate is transformed to the local navigation frame and V_N, V_E and latitude enable the aircraft angular rates in the earth frame to be computed in the navigation frame. The rates are transformed to the body frame and combined with the aircraft body rates P, Q and R. Using the same transformations as the flight model, the quaternion rates are integrated to form the quaternions, which enable measurements of the aircraft roll, pitch and yaw to be computed. The accelerations in the navigation frame are integrated to derive V_N, V_E and V_d, which enable latitude, longitude and altitude to be computed. The INS inputs are the three accelerometer measurements and the three rate gyro measurements in the body frame. The INS outputs are pitch, roll, yaw, latitude, longitude, altitude and the north, east and down velocities.

The earth's rotation rate Ω in the earth frame is given by

$$\omega_{ie}^e = \begin{bmatrix} 0 \\ 0 \\ \Omega \end{bmatrix} \tag{5.32}$$

The earth rotates once per 23 hours and 56 minutes giving $\Omega = 7.292\,115 \times 10^{-5}$ rad/s. In the navigation frame, the earth's rotation is given by

$$\omega_{ie}^n = \begin{bmatrix} \Omega \cos\lambda \\ 0 \\ -\Omega \sin\lambda \end{bmatrix} \tag{5.33}$$

where i is the inertial frame and e denotes the earth frame. The aircraft rotation, known as the transport rate, is given in the navigation frame by

$$\omega_{en}^n = \begin{bmatrix} \dfrac{V_E}{R+h} \\ \dfrac{-V_N}{R+h} \\ \dfrac{-V_E \tan\lambda}{R+h} \end{bmatrix} \tag{5.34}$$

where R is the earth's radius (6,378,137 m), h is the aircraft altitude above the surface of the earth, λ is the aircraft latitude and V_N and V_E are the aircraft's north and east velocities, respectively. These two rotation vectors are combined to give the transformation from the inertial frame to the earth frame in the navigation frame

$$\omega_{in}^n = \omega_{ie}^n + \omega_{en}^n \tag{5.35}$$

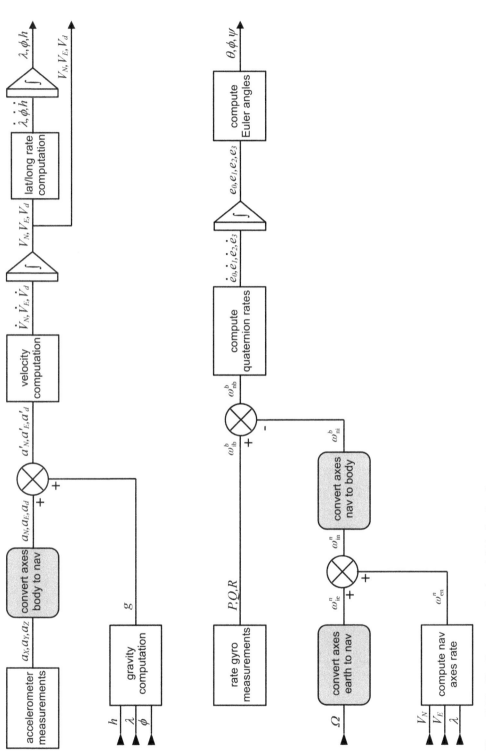

Figure 5.17 The Structure of an Inertial Navigation System.

The transformation to the body frame is given by

$$\omega_{in}^b = C_n^b \omega_{in}^n \tag{5.36}$$

This vector is subtracted from the body rates (sensed by the rate gyros) to give the rotation vector in the body frame, which is used in the computation of the quaternions.

$$\omega_{nb}^b = \omega_{ib}^b - \omega_{ni}^b \tag{5.37}$$

The accelerometer outputs are measured in the body frame and transformed to the navigation frame using the body to Euler frame DCM. Note that an accelerometer is unlikely to the positioned at the aircraft C/G, and a further transformation is needed in the body frame to allow for the location of the accelerometers.

As the earth's gravitational acceleration is used in the equations, any errors in an approximation will accumulate in the INS equations. Two models of the gravity component, as a function of altitude and latitude, are commonly used. The WGS-84 model of the normal gravity (g_0) on the surface of the reference ellipsoid is given by

$$g_0 = \gamma_e \frac{1 + k \sin^2 \lambda}{\sqrt{1 - e^2 \sin^2 \lambda}} \tag{5.38}$$

where
γ_e is the gravity at the equator = 9.780 325 3359 ms^{-2}
k is the gravity formula constant = 0.001 931 852 652 41
e is the first eccentricity of the earth, where $e^2 = 0.006\,694\,379\,990\,14$
λ is the geocentric latitude
An alternative form is the Airy formula (Biezad, 1999).

$$g_0 = g_e \left(1 + 0.005\,2884 \sin^2 \lambda - 0.000\,059 \sin^2 2\lambda\right) \tag{5.39}$$

where g_e is the gravitational acceleration at the equator and λ is the geocentric latitude. The variation of gravity with altitude is given by

$$g = g_0 \frac{R^2}{(R+h)^2} \tag{5.40}$$

The gravity component is added to the accelerations measured in the navigation frame to give

$$\begin{bmatrix} a_N \\ a_E \\ a_d \end{bmatrix} = c_b^n \begin{bmatrix} a_x \\ a_y \\ a_z \end{bmatrix} + \begin{bmatrix} 0 \\ 0 \\ g \end{bmatrix} \tag{5.41}$$

The accelerations from the earth's rotation and the transport rate can then be added to the accelerations in the navigation frame

$$\dot{V}_N = a_N - 2\,V_E\Omega\,\sin\lambda + \frac{V_NV_d - V_E^2\,\tan\lambda}{R+h} \tag{5.42}$$

$$\dot{V}_E = a_E + 2\,V_N\Omega\,\sin\lambda + 2\,V_d\Omega\,\cos\lambda + \frac{V_NV_E\,\tan\lambda + V_EV_d}{R+h} \tag{5.43}$$

$$\dot{V}_d = a_d - 2\,V_E\Omega\,\cos\lambda + \frac{V_E^2 + V_N^2}{R+h} \tag{5.44}$$

The accelerations in the navigation frame are integrated to give

$$V_N = \int \dot{V}_N\,dt \tag{5.45}$$

$$V_E = \int \dot{V}_E\,dt \tag{5.46}$$

$$V_d = \int \dot{V}_d\,dt \tag{5.47}$$

The INS latitude, longitude and altitude outputs are computed as follows:

$$\dot{\lambda} = \frac{V_N}{R+h} \tag{5.48}$$

$$\dot{\phi} = \frac{V_E}{\cos\lambda\left(R+h\right)} \tag{5.49}$$

$$\lambda = \int \dot{\lambda}\,dt \tag{5.50}$$

$$\phi = \int \dot{\phi}\,dt \tag{5.51}$$

$$h = \int \dot{h}\,dt \tag{5.52}$$

These terms include $\tan\lambda$ and are therefore not valid for navigation near to the poles and wander-axes methods are used to implement the equations (Siouris, 1993).

As these equations correspond to the equations of motion covered in Chapter 3, they represent the equations of a perfect INS. However, in practice, an INS will contain errors which accumulate over time, which include:

- initial alignment of the accelerometers and gyros – the position of the aircraft may not be known accurately during initialisation or the aircraft may be moving when the INS is realigned;
- bias in the accelerometers – small internal mechanical forces, particularly friction, are included in the sensed accelerations;

- gyro drift rate – the rate measurements are used in several transformations and any errors will affect the computations;
- computational errors – the conversion from analogue measurements to digital values can introduce rounding errors and computational accuracy is limited by the word length of the INS computer;
- gravity vector – variations in gravity occur over the surface of the earth and these are approximated in equation 5.27.

For a perfect INS, $x = \iint \ddot{x}\,dt\,dt$, where $\ddot{x}$ is referred to the specific force. A bias error B will be included in the double integration of the measured specific force, giving the resultant distance error

$$\Delta d = \iint B\,dt\,dt = \frac{Bt^2}{2} \tag{5.53}$$

Similarly, a drift rate W in a gyro will introduce a tilt error $\Delta\theta = Wt$, such that

$$\Delta d = \iint gWt\,dt\,dt = \frac{gWt^3}{6} \tag{5.54}$$

In other words, the distance error increases with t^2 for an accelerometer bias error and with t^3 for a gyro drift rate error. While B and W are reduced to small values in the manufacture of an INS, they cannot be eliminated. Accurate initialisation is provided at most airport stands, otherwise the error propagation can only be reduced by realigning the INS in flight, for example, overflying a landmark at a known latitude and longitude. Nowadays, in aircraft equipped with satellite navigation systems (GNSS), the INS is realigned to the GNSS position when the INS error exceeds the maximum GNSS error. The error terms for the various sources of error are not derived here but are summarised in Table 5.1 and can be validated in a simulation model where ω_s is the Schuler frequency, R_0 is the earth's radius and t is the time since the previous alignment. Note that the errors are bounded by the Schuler tuning with the exception of the gyro bias, where the error increases

Table 5.1 Summary of INS Errors.

Error	Effect
Initial position δx_0	δx_0
Initial velocity δV_0	$\delta V_0 \dfrac{\sin\omega_s t}{\omega_s}$
Initial attitude $\delta\theta_0$	$\delta\theta_0 R_0\left(1 - \cos\omega_s t\right)$
Accelerometer bias δf_{xb}	$\delta f_{xb}\left(\dfrac{1 - \cos\omega_s t}{\omega_s^2}\right)$
Gyro bias $\delta\omega_{xb}$	$\delta\omega_{xb} R_0\left(t - \dfrac{\sin\omega_s t}{\omega_s}\right)$

with time. A detailed INS model is required in studies involving sensor fusion and sensor platforms with multiple inputs and to provide an INS model with specific error characteristics in the analysis of military mission systems using integrated sensors. The main code of an INS model is as follows:

```
const double B = 0.001;  /* accelerometer bias 0.001 m/s^2 */
const double W = Radians((double) (0.002 / 60.0 / 60.0); /* gyro bias 0.002 deg/s

g = Gravity(latitude, h);

Pitch = asin(-A31);
Roll = atan2(A32, A33);
Yaw = atan2(A21, A11);

Ax = 0.0 + B;  /* accelerometer bias added to Ax */
Ay = 0.0;
Az = -g;

Pgyro = EarthRate * cos(latitude) + W; /* gyro bias added to Pgyro */
Qgyro = 0.0;
Rgyro = -EarthRate * sin(latitude);

Wn1 = EarthRate * cos(latitude) + Ve / (Re + h);
Wn2 = -Vn / (Re + h);
Wn3 = -EarthRate * sin(latitude) - (Ve * tan(latitude)) / (Re + h);
nav2body(&Wb1, &Wb2, &Wb3, Wn1, Wn2, Wn3);

P = Pgyro - Wb1;
Q = Qgyro - Wb2;
R = Rgyro - Wb3;

Quaternions(P, Q, R);
SetDCM();

body2nav(&An, &Ae, &Ad, Ax, Ay, Az);

VnDot = An -
        2.0 * Ve * EarthRate * sin(latitude) +
        (Vn * Vd - Ve * Ve * tan(latitude)) / (Re + h);
VeDot = Ae +
        2.0 * Vn * EarthRate * sin(latitude) +
        2.0 * Vd * EarthRate * cos(latitude) +
        (Vn * Ve * tan(latitude) + Ve * Vd) / (Re + h);
VdDot = Ad -
        2.0 * Ve * EarthRate * cos(latitude) -
        (Ve * Ve + Vn * Vn) / (Re + h) + g;
Vn = Integrate(Vn, VnDot);
Ve = Integrate(Ve, VeDot);
Vd = 0.0; /* INS cannot fall */;

latitudeDot = Vn / (Re + h);
longitudeDot = Ve / ((Re + h) * cos(latitude));
hDot = -Vd;

latitude = Integrate(latitude, latitudeDot);
longitude = Integrate(longitude, longitudeDot);
h = Integrate(h, hDot);
```

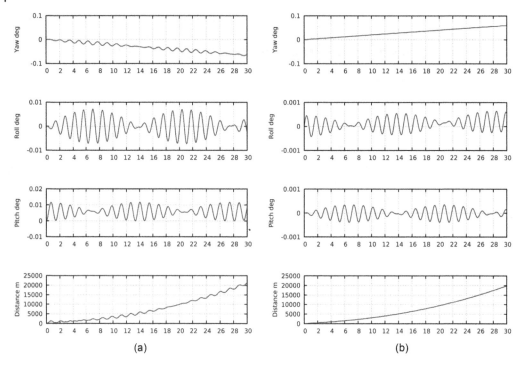

Figure 5.18 Simulation of INS Errors.

An INS model of this form allows the developer to introduce specific errors or to simulate errors to match an INS specification.

One method of validating the INS model is to position the INS at rest and monitor the errors over a simulated period of several hours. Figure 5.18 shows the simulation of an INS located at 45°N, 0°E, at rest, with an initial heading of 360°N and pitch and roll angles of zero. A simulation of 30 hours was run offline, with an update rate of 100 Hz and the output plotted every second. Figure 5.18(a) shows an accelerometer bias of 0.001 ms^{-2} added to **Ax** (the accelerometer x-axis) and Figure 5.18(b) shows a gyro bias of 0.002°/s added to **Pgyro** (the gyro roll axis).

The period and bound of the Schuler loop and the growth of distance error with time are clearly shown. Without realignment of the INS, the drift is approximately 20 km in 30 hours, or 0.7 km/hour, for the error values used in these examples. There are several important issues raised by this model:

- It is essential to use double length floating-point precision throughout, particularly as the angles and angular rates (and their products) are relatively small.
- The accuracy of the integration algorithm should correspond to the system characteristics.
- In the examples, **Vd** is reset to 0.0 in the equations as the INS is stationary.

This relatively small module provides a reasonable model of the common sources of error in an INS and provides a useful tool in the analysis and simulation of inertial sensors.

5.10 Satellite Navigation

The availability of satellite navigation since the 1980s has brought about radical change to transportation, particularly in motor vehicles and aircraft. Up to that time, the accuracy of navigation

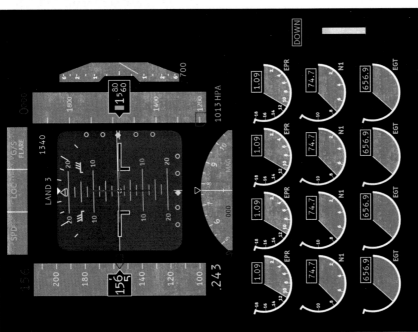

Plate 1 EFIS Display.

Plate 2 Light Aircraft Instrument Panel.

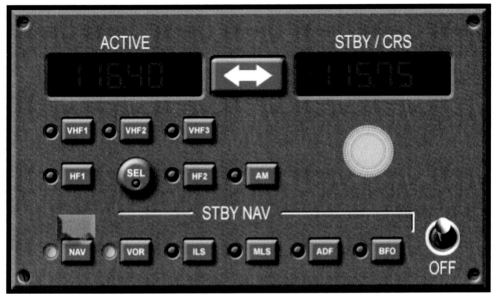

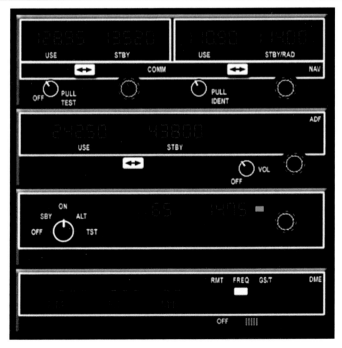

Plate 3 Soft Panel Displays.

Plate 4 A Visual Database Produced Using Blender.

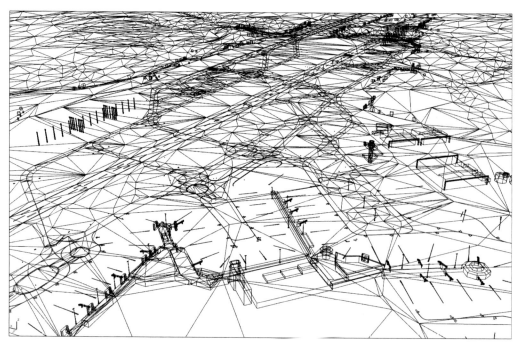

Plate 5 Wire-Frame Version of a Visual Database.

Plate 6 Approach to Bristol Lulsgate Airport in 5 km Visibility.

Plate 7 Manchester International Approach 2 nm.

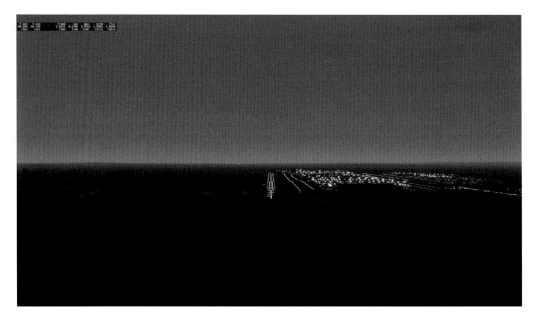

Plate 8 London Heathrow Approach Dusk 1 nm.

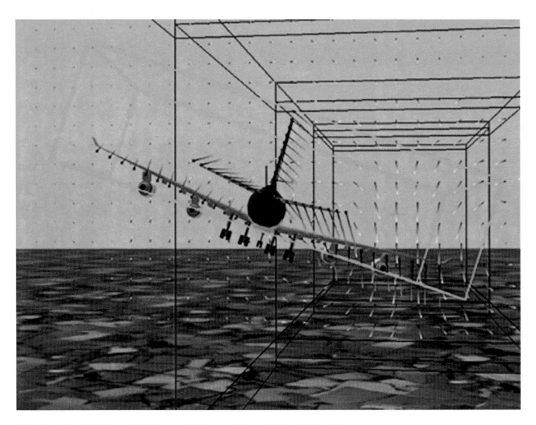

Plate 9 Visualisation of Air Flow and Forces during a Wake Vortex Encounter.

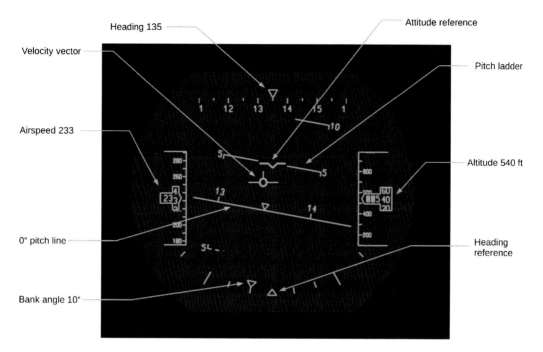

Plate 10 A HUD Format.

Plate 11 A HUD Overlay.

OpenGL   GTK

Plate 12   Flight Data Plot Pitch Rate.

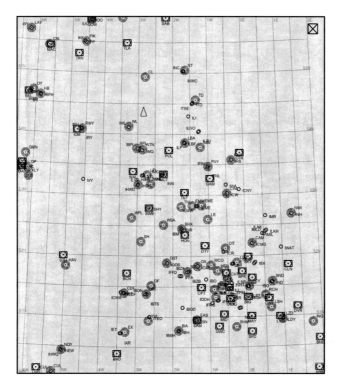

Plate 13   IOS Chart Display.

had been inversely proportional to range, with accuracies for worldwide navigation, typically Loran and Omega, measured in several hundreds of metres. GPS has afforded continuous worldwide coverage to accuracies better than 20 m for civilian users.

One of the problems with GPS is that the constellation of satellites changes continuously and therefore the accuracy varies with the constellation and time. One option to model the errors would be to take recordings of the position of a stationary GPS receiver over several days. The variation in accuracy would be evident from the measurements, which could be added to the exact position of the aircraft in a simulation. An alternative approach, and the one adopted in this book, is to model the physical properties of GPS in order to derive the appropriate error characteristics.

The constellation of satellites currently comprises 32 satellites. At any time, roughly half of the satellites are on the reverse side of the earth and many of the visible satellites may be at a low elevation where the signal is inaccurate. An aircraft GPS receiver needs to detect the signals from four of the visible satellites with the minimum measurement error, to estimate the range to the satellites and compute the aircraft position.

GPS is a complex system and it is left to the developer just how much of the system should be simulated. The details of the satellites and their orbits are available in the public domain and almanacs defining the satellite orbits can be downloaded. In theory, if the position of four satellites is known in an earth-centred coordinate frame, the user position can be computed from the equations defining these positions. The computation of position is based on three assumptions:

1) The satellites all transmit their messages at exactly the same time. In practice, the satellites contain very accurate caesium clocks, which are monitored and regularly realigned by ground stations.
2) Although the relative accuracy of clocks used in GPS receivers is poor, this error term can be eliminated in the computations.
3) As the radio signals pass through the ionosphere, the electric fields can cause refraction so that the signal path from the satellite to the receiver is curved, depending on the state of the ionosphere at the time and the elevation of the satellite relative to the GPS receiver. The curvature extends the signal path length and results in a small delay at the GPS receiver.

A simulation of GPS therefore contains the following elements:

- the equations to compute the range from each satellite;
- modelling of the delays caused by the ionosphere and the troposphere;
- solution of the range equations to estimate the position of the receiver.

Note that a model of this complexity neglects the spread-spectrum methods used to acquire the signal and other errors associated with satellite navigation and is limited to the non-military version of GPS. However, in solving the same set of equations as a GPS receiver, it is possible to model the statistical properties from four selected satellites, such that the noise in the model correctly replicates the main characteristics of a GPS receiver. There is one further characteristic of GPS signals: as a satellite goes below the horizon or another satellite rises above the horizon, the selection of satellites accessed by the receiver will change, leading to a possible 'jump' in range measurements and therefore an abrupt change in position.

The constellation of satellites is organised in six orbital planes inclined at 55° to the equatorial plane with four or more satellites in each orbit. Almanac data can be downloaded from several public-domain sources in various formats including https://celestrak.com/GPS/almanac/Yuma. For example, the data for satellite number 25 for the week beginning 7 September 2003 is as follows:

```
******** Week  211 almanac for PRN-25 ********
ID:                          25
Health:                      000
Eccentricity:                0.1059770584E-001
Time of Applicability(s):    61440.0000
Orbital Inclination(rad):    0.9425317251
Rate of Right Ascen(r/s):   -0.8011762293E-008
SQRT(A)  (m 1/2):            5153.704102
Right Ascen at Week(rad):    0.2315520405E+001
Argument of Perigee(rad):   -1.679881985
Mean Anom(rad):              0.2146618485E+001
Af0(s):                      0.5149841309E-004
Af1(s/s):                    0.0000000000E+000
week:                        211
```

The *ID* field gives the satellite number (1–32 for GPS), the *Health* field shows the operational status of the satellite (000 denotes a 'healthy' satellite) and the *Week* field identifies the date of the almanac. The main fields of interest are the *Right Ascen at Week* and *Argument of Perigee* fields, which are both defined in radians. The right ascension field Ω_0 defines the longitude where the orbit intersects the equatorial plane and the argument of perigee field θ_0 defines the angle of the satellite at the time of the almanac (midnight) along its orbit from the intersection with the equatorial plane. Table 5.2 shows the values for Ω_0 and θ_0 in degrees for the 28 operational satellites for the specific almanac.

GPS is based on an ECEF coordinate frame, where the position (S_x, S_y, S_z) of each satellite is given by

$$\theta = \theta_0 + t\frac{360}{48,082} \tag{5.55}$$

$$\Omega = \Omega_0 - t\frac{360}{86,164} \tag{5.56}$$

$$S_x = R_s\left(\cos\theta\,\cos\Omega - \sin\theta\,\sin\Omega\,\cos55°\right) \tag{5.57}$$

$$S_y = R_s\left(\cos\theta\,\sin\Omega + \sin\theta\,\cos\Omega\,\cos55°\right) \tag{5.58}$$

$$S_z = R_s\left(\sin\theta\,\sin55°\right) \tag{5.59}$$

where θ is the phase of the satellite in its orbit plane, Ω is the rising longitude of the orbital plane taking account of the earth's rotation and t is the time since the start of the almanac. These equations define the positions of the satellites based on data for a specific version of the almanac.

Satellites below an elevation of 5° are ignored as the refraction is excessive. The position of each satellite with respect to a tangential plane of the receiver is given by

$$dx = S_x - R_x \tag{5.60}$$

$$dy = S_y - R_y \tag{5.61}$$

Table 5.2 Almanac Data.

Satellite	Ω_0	θ_0
1	78.53042	−98.5241
2	−168.161	−101.192
3	−106.695	27.86253
4	−41.3816	−9.50852
5	−166.787	42.31026
6	−103.764	−117.944
7	−105.395	−107.201
8	139.1253	132.5993
9	135.4144	55.93217
10	17.75434	13.90936
11	−47.9001	−10.531
13	77.48031	42.85565
14	77.0286	−67.7352
15	−38.4826	122.6411
16	−163.057	−95.8007
17	−36.1177	−164.456
18	19.85077	−174.989
20	16.86114	101.2392
21	−39.8758	166.1483
23	20.65981	−94.5828
24	−40.0065	−84.7436
25	132.6695	−96.2501
26	77.6695	29.92697
27	134.2225	−129.629
28	−162.426	−138.709
29	75.84648	−90.288
30	−164.527	72.96847
31	−105.447	53.83715

$$dz = S_z - R_z \tag{5.62}$$

$$p_N = -dx\sin\lambda_r\cos\phi_r - dy\sin\lambda_r\sin\phi_r + dz\cos\lambda_r \tag{5.63}$$

$$p_E = -dx\sin\phi_r + dy\cos\phi_r \tag{5.64}$$

$$p_d = dx\cos\lambda_r\cos\phi_r + dy\cos\lambda_r\sin\phi_r + dz\sin\lambda_r \tag{5.65}$$

where the receiver position in ECEF coordinates is (R_x, R_y, R_z), λ_r and ϕ_r are the receiver latitude and longitude, respectively, and the position of the satellite is given in NED coordinates.

Typically, there are at least 10 satellites above 5°, with the position of four satellites needed to solve the navigation equation. The selection of four satellites from 10 is given by the binary numbers in the range $0–1023_{10}$ containing four bits. The DCM can be computed for each potential set of four satellites, to search for the optimal position dilution of precision (PDOP), where

$$A = \begin{vmatrix} \alpha_{11} & \alpha_{12} & \alpha_{13} & 1 \\ \alpha_{21} & \alpha_{22} & \alpha_{23} & 1 \\ \alpha_{31} & \alpha_{32} & \alpha_{33} & 1 \\ \alpha_{41} & \alpha_{42} & \alpha_{43} & 1 \end{vmatrix} \tag{5.66}$$

where

$$\alpha_{i1} = \left(\hat{R}_x - S_{ix} \right) / \hat{R}_i \tag{5.67}$$

$$\alpha_{i2} = \left(\hat{R}_y - S_{iy} \right) / \hat{R}_i \tag{5.68}$$

$$\alpha_{i3} = \left(\hat{R}_z - S_{iz} \right) / \hat{R}_i \tag{5.69}$$

and $\left(\hat{R}_x, \hat{R}_y, \hat{R}_z \right)$ is the estimated receiver position in the ECEF frame and $\left(S_{ix}, S_{iy}, S_{iz} \right)$ is the position of the ith satellite ($i = 1$ to 4). For four satellites, the geometric dilution of precision (GDOP) is given by

$$\text{GDOP} = \sqrt{trace\left[\left(A^T A \right)^{-1} \right]} \tag{5.70}$$

where

$$trace(A) = \sum_{i=1}^{4} A_{ii} \tag{5.71}$$

For 10 satellites, there are 210 distinct bit patterns from the 1024 possible bit patterns, containing 4 bits (logic 1). This set of satellites is searched for the minimum PDOP and these four satellites are used in computation of the navigation equation. The four range equations to be solved are given by

$$R_i = \sqrt{\left(R_x - S_{ix} \right)^2 + \left(R_y - S_{iy} \right)^2 + \left(R_z - S_{iz} \right)^2} + cb \tag{5.72}$$

where R_i is the pseudo-range from each satellite measured by the receiver, which includes the errors in the measurement, (S_{ix}, S_{iy}, S_{iz}) is the ECEF position of the ith satellite, c is the speed of light and b is the receiver clock bias. Notice that there are four equations in four unknown variables, which enables the receiver position R_x, R_y, R_z to be computed and allows the clock bias b to be discarded. Although the equations are nonlinear, they can be linearised and solved by an iterative Gauss–Jordan method (Westlake, 1968), which is straightforward to implement and generally converges with reasonable stability. The linearised solution is

$$x = A^{-1}r \qquad (5.73)$$

where A is the matrix defined in equation 5.66, x is a vector of the relative position errors $[\Delta x, \Delta y, \Delta z, c\Delta t]^T$ and r is a vector of the relative pseudo-range errors $[\Delta R_1, \Delta R_2, \Delta R_3, \Delta R_4]^T$, where

$$x = \hat{x} + \Delta x \qquad (5.74)$$

$$y = \hat{y} + \Delta y \qquad (5.75)$$

$$z = \hat{z} + \Delta z \qquad (5.76)$$

$$t = \hat{t} + \Delta t \qquad (5.77)$$

$$R_i = \hat{R}_i + \Delta R_i \qquad (5.78)$$

The solution of these equations generates improved estimates for (x, y, z, t) which are used in the next iteration of the solution. The algorithm repeats until the estimated values converge to an acceptable accuracy, typically in less than 10 iterations.

In the simulation model, the position of the satellites and the position of the receiver are known and the exact range can be computed accurately. However, to simulate the pseudo-range measurement, the errors inherent in GPS are added to the exact range. These errors can include delay through the ionosphere and the troposphere caused by refraction, multi-path reflections, the clock bias and noise in the receiver computations, for example, rounding and truncation errors. In fact, GPS receivers include error models of the ionosphere and troposphere, to reduce the errors in the pseudo-range measurements. The approximation for delay through the ionosphere depends on the elevation of the satellite (with respect to the receiver), the latitude of the receiver and the time of day; the estimate is based on Klobuchar's method (Klobuchar, 1987), given by

$$\Delta t = F\left(5 \times 10^{-9} + \sum_{i=0}^{3} \alpha_i \phi_m^i \left(1 - \frac{x^2}{2} + \frac{x^4}{24}\right)\right) \qquad (5.79)$$

where F is given by

$$F = 1 + 16\left(0.53 - E\right)^3 \qquad (5.80)$$

where E is the elevation of the satellite, ϕ_m is the geomagnetic latitude of the receiver and x is given by

$$x = \frac{2\pi\left(t - 50,400\right)}{\sum_{i=0}^{3}\beta_i} \qquad (5.81)$$

where t is the time (seconds) since the last GPS epoch and α and β are passed in the ephemeris data transmitted by each satellite and are available in ephemeris downloads. For night-time values of t, a fixed delay of 5 ns is used. The delay through the ionosphere is based on the model used in the NOAA GPS toolbox (https://geodesy.noaa.gov/gps-toolbox).

The delay through the troposphere depends on the temperature, pressure and water vapour in the atmosphere. The model developed by Smith and Weintraub (1953) can be used to estimate the delay:

$$N = 77.6 \times \frac{P}{T} + 3.73 \times 10^5 \times \frac{e}{T^2} \tag{5.82}$$

where P is the total atmospheric pressure (millibars), T is the absolute temperature (K) and e is the partial water vapour pressure. P and T are computed for a specific aircraft altitude in the weather module.

The software model updates at 1 Hz, which is a typical rate for a GPS receiver. Double-precision (64-bit) floating-point arithmetic is used throughout the model to achieve the desired accuracy. Unfortunately, *cglm* only supports single-precision (32-bit) floating-point arithmetic and dedicated 64-bit functions were developed for the 4×4 matrix operations, including inversion, transpose, multiplication and computation of determinants. ECEF vectors are defined in the following structure:

```
typedef struct
{
    double x;
    double y;
    double z;
} vertex3;
```

The following functions were written but, for reasons of brevity, are only summarised. The full code is provided in the companion code and referenced in the Appendix.

Function	Computation
`void ecef2geodetic(double *latitude,` `double *longitude,` `double *h,` `double x, double y,` `double z);`	Conversion of ECEF coordinates (`x,y,z`) to geodetic coordinates (`latitude,longitude,altitude`)
`void geodetic2ecef(double *x, double *y,` `double *z, double` `latitude, double` `longitude, double h);`	Conversion of geodetic coordinates (`latitude, longitude,altitude`) to ECEF coordinates (`x,y,z`)
`double Rads(double x);`	Degrees to radians
`double Degrees(double x);`	Radians to degrees
`double Normalise(double x);`	`x` normalised to the range 0–360°
`double range(vertex3 a, vertex3 b);`	Distance from **a** to **b** in ECEF

(Continued)

Function	Computation

`double getGDOPs(unsigned int bestsats[],`
`            vertex3 satellites[],`
`            vertex3 p,`
`            unsigned int visiblesats[],`
`            unsigned int nsats);`
	Returns the GDOP, where **nsats** is the number of satellites in view, **visiblesats** is the array of satellites in view, **p** is the receiver position, **satellites** is the array of satellites and **bestsats** is the 4 satellites selected.

`void invert(double a[5][5], double m[5][5]);`	**a** is the inversion of the matrix **m**
`double determinant(double m[5][5]);`	Returns the determinant of the matrix **m**
`double elevation(vertex3 sat, vertex3 p);`	Returns the elevation of the satellite **sat** with respect to the receiver position **p**
`double azimuth(vertex3 sat, vertex3 p);`	Returns the azimuth of the satellite **sat** with respect to the receiver position **p**
`double distance(double Lat1, double Long1,` `              double Lat2, double Long2);`	Returns the geodetic distance from (**Lat1,Long1**) to (**Lat2,Long2**)
`void solve(double a[5][5], double b[5],` `           double x[5])`	Solves $x = A^{-1}b$, where **a** is a 4×4 matrix and **b** and **x** are vectors of dimension 4.

`double ionosphere(double latitude,`
`              double longitude,`
`              double elev,`
`              double azimuth,`
`              double tow);`
	Returns the delay through the ionosphere at position (**latitude, longitude**), for the satellite elevation and azimuth, where **tow** is the time of the week

The function `nav_equation(double t, vertex3 satellites[], unsigned int sats[], vertex3 *aircraft)` solves the navigation equation, where **t** is the time of day, **satellites** is the array of satellites in the constellation, **sats** is an array of the four selected satellites and the estimate of the receiver position is returned in **aircraft**. The main body of the function is as follows:

```
geodetic2ecef(&shef.x, &shef.y, &shef.z,
              Rads(sheffield_lat), Rads(sheffield_long), sheffield_elev);

p.x = 0.0;  /* initial estimates of x, y, z, t */
p.y = 0.0;
p.z = 0.0;
dt  = 1.0;

do
{
    oldp.x = p.x;  /* remember p */
    oldp.y = p.y;
    oldp.z = p.z;

    for (i=1; i<=4; i+=1)
    {
        ss = sats[i];
        elev = elevation(satellites[s], p);
        az = azimuth(satellites[s], p);
```

```
        ionerr = ionosphere(sheffield_lat, sheffield_long, elev, az, t);
        troperr = 0.1 / (sin(Rads(elev)) + 0.012);
        cbias = speed_of_light * 0.001;  /* assume clock bias 1 ms */

        r[i] = range(satellites[s], shef) + ionerr + troperr + cbias;
        re = range(satellites[s], p);

        a[i][1] = (p.x - satellites[s].x) / re;
        a[i][2] = (p.y - satellites[s].y) / re;
        a[i][3] = (p.z - satellites[s].z) / re;
        a[i][4] = 1.0;
        b[i]    = r[i] - re;
    }

    solve(a, b, x);

    p.x = p.x + x[1];
    p.y = p.y + x[2];
    p.z = p.z + x[3];
    dt  = dt + x[4];
} while (fabs((p.x - oldp.x) * (p.x - oldp.x) +
              (p.y - oldp.y) * (p.y - oldp.y) +
              (p.z - oldp.z) * (p.z - oldp.z)) > 1.0);

aircraft->x = p.x;
aircraft->y = p.y;
aircraft->z = p.z;
```

where the receiver is positioned at (**sheffield_lat, sheffield_long, Sheffield_elev**), defined as (53°22′53.55″N, 01°28′41.39″W, 290 m). A fixed clock bias of 1 ms is used. The main part of the test code to simulate 24 hours of GPS signals is as follows:

```
t = 0.0;   /* start from GPS epoch */

for (i=0; i<86400; i+=1)   /* 24 hours = 86,400s */
{
    nsats = 0;
    for (s=1; s<=MaxSats; s+=1)
    {
        theta = Rads(Normalise(theta0[s] + t * 360.0 / 43082.0));
        omega = Rads(Normalise(omega0[s] - t * 360.0 / 86164.0));
        satellites [s].x = Rs*(cos(theta)*cos(omega) - sin(theta)*sin(omega)*cos(Rads(55.0)));
        satellites [s].y = Rs*(cos(theta)*sin(omega) + sin(theta)*cos(omega)*cos(Rads(55.0)));
        satellites [s].z = Rs*sin(theta)*sin(Rads(55.0));

        el = elevation(satellites[s], aircraft);
        if (el >= 5.0)
        {
            Nsats += 1;
            visiblesats[nsats] = s;
        }
    }
    md = getGDOPs(satsfound, satellites, aircraft, visiblesats, nsats);
    if (md > 12.0)
    {
        printf("gdop problem: i=%d gdop=%f\n", i, md);
    }
    nav_equation(t, satellites, satsfound, &aircraft);
    ecef2geodetic(&user_lat, &user_long, &user_alt, aircraft.x, aircraft.y, aircraft.z);
```

```
printf("%f %f\n", t / 3600.0,
        sqrt((shef.x - aircraft.x) * (shef.x - aircraft.x) +
             (shef.y - aircraft.y) * (shef.y - aircraft.y) +
             (shef.z - aircraft.z) * (shef.z - aircraft.z)));

t += 1.0;
if (t > (double) (sec_per_day * 7))
{
    t = 0.0;
}
}
```

theta and **omega** are updated at 1 Hz in order to position the satellites, the elevation of each satellite determines the number of visible satellites and the navigation equation is solved, enabling the receiver position error to be recorded.

The GPS model was run over a simulated 24-hour period, updating at 1 Hz, with the receiver positioned at a location in the University of Sheffield, England. Figure 5.19(a) shows the number of visible satellites (above 5° elevation) and Figure 5.19(b) shows the variation in PDOP over 24 hours.

The position error is shown in Figure 5.20. The almanac for week 211 in September 2003 shows 28 operational satellites. The number of satellites in view varies from five to 14 and the PDOP varies mostly from two to five but exceeds 8 m for a short time. Towards the end of the 24-hour period, only five or six satellites are in view with a concomitant increase in the PDOP. The published values give the following conditions for the dilution of precision values:

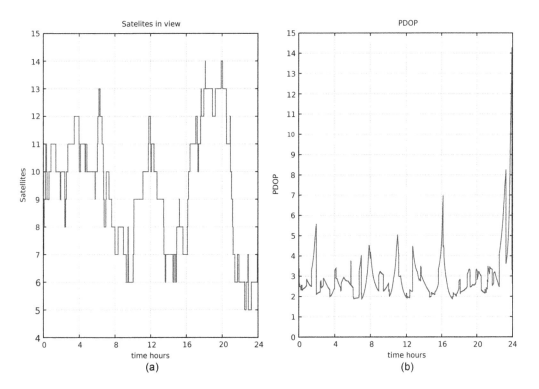

Figure 5.19 GPS Simulation.

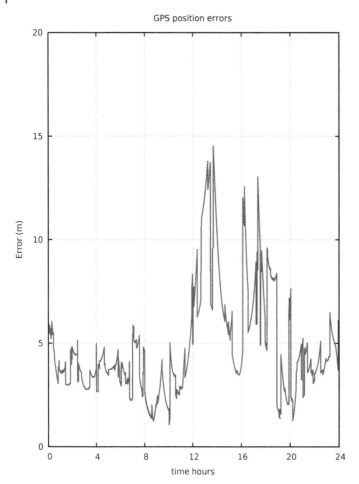

Figure 5.20 GPS Errors.

DOP	Value	Conditions
1	Ideal	Highest possible confidence – used for applications requiring the highest precision
1–2	Excellent	Sufficiently accurate for most applications
2–5	Good	Minimum level for accurate decision, sufficient for reliable en-route navigation
5–10	Moderate	Measurements can be used but some satellites may be obscured
10–20	Fair	A low level of confidence, providing only a rough estimate of position
> 20	Poor	Measurement accuracy can exceed 300 m and should be discarded

For the majority of the period, the position accuracy is 2–5 m but with worst-case values of 14 m. Note also the abrupt changes which coincide with changes in the number of visible satellites. With the increasing number of GPS applications, a model of a GPS receiver with appropriate statistical characteristics is a very useful tool in real-time aircraft simulation.

References

Biezad, D. J. (1999), *Integrated Navigation and Guidance Systems*, American Institute of Aeronautics & Astronautics.

Chulliat, A., Macmillan, S., Alken, P., Beggan, C., Nair, M., Hamilton, B., Woods, A., Ridley, V., Maus, S. and Thomson, A. (2015). *The US/UK World Magnetic Model for 2015-2020: Technical Report*, National Geophysical Data Center, NOAA. doi:10.7289/V5TB14V7.

Klobuchar, J. A. (1987), Ionospheric Time-delay Algorithm for Single Frequency GSP Users, *IEEE Trans.*, Vol. AES-23, No. 3, pp. 325–331.

Maling, D. H. (1992), *Coordinate Systems and Map Projections*, Elsevier Science and Technology, Butterworth_Heinmann.

Siouris, G. M. (1993), *Aerospace Avionics Systems*, Academic Press.

Smith, E. K. and Weintraub, S. (1953), The Constants in the Equation for Atmospheric Refraction Index at Radio Frequencies, *Proc. IRE*, No. 41, pp. 1035–1037.

Thébault, E. et al. (2015), International Geomagnetic Reference Field: The 12th Generation, *Earth, Planets and Space 2015*, Vol. 67, p. 79.

Titterton, D. and Weston, J. (2004), *Strapdown Inertial Technology*, Institution of Engineering and Technology.

Westlake, J. R. (1968), *A Handbook of Numerical Matrix Inversion and Solution of Linear Equations*, John Wiley & Sons.

6

Aircraft Displays

An algorithm must be seen to be believed.

Donald Knuth

6.1 OpenGL

6.1.1 The Development of OpenGL

By the early 1990s, several manufacturers had produced minicomputer systems with processing speeds to support computer-aided design applications and visualisation. However, there were no common graphics standards for software or hardware, resulting in considerable duplication of effort to provide high-performance graphics for workstations. Silicon Graphics Inc. (SGI) had developed OpenGL for their workstations and in 1992 released the first version of OpenGL. One attraction of OpenGL was that it was platform-independent. Device drivers were provided by the various vendors for their workstations, so that programmers required no knowledge of the graphics hardware.

At the heart of OpenGL was the concept of matrix transformations of the vertices of lines and polygons and the provision of a viewpoint which could be positioned and orientated with several forms of projection. The style, width and colour of lines could be defined with polygons drawn in outline form or in-filled in colour or rendered as a texture. In addition, the provision of anti-aliasing significantly improved the quality of rendered images. 2D graphics was treated as a subset of 3D graphics. All these features of OpenGL were embedded in the drivers for specific computers, allowing the developer to focus on the graphical rendering of objects and scenes. The success of OpenGL and its widespread uptake were partly a result of the well-thought-out design of OpenGL as an application programming interface (API) but also because of the provision of a standard and supporting documentation. As OpenGL was platform-independent, the portability of software applications increased considerably in this era.

This development of OpenGL applications extended to the late 1990s with the arrival of graphics cards for PCs and workstations, with vendors providing drivers and generally complying with the standard. A significant advance of the graphics cards was the provision of on-board memory to store the image, as opposed to sharing the memory with the host processor. In effect, OpenGL procedure calls were passed to the graphics cards as tasks and the special-purpose graphics chips performed the graphics operations, management of the frame-store and generation of the video

Flight Simulation Software: Design, Development and Testing, First Edition. David Allerton.
© 2023 John Wiley & Sons Ltd. Published 2023 by John Wiley & Sons Ltd.
Companion Website: www.wiley.com/go/flightsimulationsoftware

output signal. It was arguably the demand for workstation graphics for PCs in the 1990s that fuelled the growth and advances in graphics technology and possibly the growth in computer games that established OpenGL as a graphics API.

There was a major shift in the use of OpenGL around 2010. With advances in microelectronics, the amount of processing and memory on a single graphics card had advanced to the point where a card could contain 100 or more computing cores. The bottleneck of passing graphics data between the host and the graphics engine was reduced because graphics objects and their textures, which had previously consumed a large amount of host memory, could now be stored in the graphics engine memory. Because graphics rendering is inherently parallel, the benefits of processing objects in parallel on several hundred cores increased the processing performance to a remarkable degree. With the availability of multiple GPUs on a graphics card, the use of a C-like shader language was provided to enable developers to write code for the GPUs on the host computer and download compiled programs to the GPUs.

In constructing detailed visual databases, packages such as OpenFlight (Anon, 2018) and 3DS-Max enabled users to construct and visualise complex scenes, which are reduced to triangles and can be read directly into OpenGL. Considerable realism can be added to models where texture is, in effect, painted onto surfaces; the large amount of memory needed to support textures is provided by modern graphics cards. One deficiency with OpenGL is that text is not supported directly, although several packages provide character fonts in the form of textures, enabling text to be rendered at high speed.

Currently, there is concern that advances in graphics hardware are inhibited by the limitations of OpenGL which was developed at the time of rendering on a single processor. To a degree, the shader languages, supported by OpenGL, enable programmers to develop code to run on GPUs. But there is a growing consensus that APIs such as Vulkan (Sellers and Kessenich, 2016) are much closer to the GPU hardware and afford advances in image rendering that are not achievable with OpenGL.

In 2008, version 3.0 of the OpenGL specification was released. Many of the features of previous versions were deprecated and this version focused on access to the GPUs. Versions of OpenGL prior to version 3.0 are referred to as legacy OpenGL. However, a considerable amount of OpenGL legacy software exists and many of the underlying concepts of OpenGL are more closely related to graphics programming than to GPU programming. Vendors of graphics cards vary in the degree of backward compatibility offered with OpenGL, but it is fair to say that OpenGL from version 4.0 onwards is likely to be retained as the OpenGL standard in the future, unless superseded by Vulkan or other competitors.

6.1.2 Legacy OpenGL

Reference to OpenGL in this section implies legacy versions of OpenGL. OpenGL covers four aspects of graphics programming:

- rendering of a variety of types of geometric shapes;
- geometric transformations, particularly the translation, scaling and rotation of objects, provided by a matrix that is applied to geometric operations;
- provision of an implicit stack that enables the rendering state to be saved and restored;
- system-wide functions to control the characteristics of displayed graphics.

Note that the setting up of the graphics windows and management of the frame-store are not provided by OpenGL and are normally implemented by an API such as Graphics Library Framework (GLFW).

The geometric primitive types are defined in the *OpenGL Programming Guide* (Woo et al., 1996) and include dots, lines, triangles, quadrilaterals and polygons. Matrix operations on objects are implemented via OpenGL function calls. A transformation matrix is maintained by OpenGL, which is set initially to the identity matrix. The functions `glTranslate` and `glRotate` enable subsequently rendered objects to be translated and rotated with respect to the OpenGL axes. In many applications, graphics transformations are applied to sets of objects and the provision of a stack and operations to explicitly push and pop the transformation matrix simplify graphics operations on graphics structures.

The primitives provided by OpenGL treat 2D operations as a subset of 3D operations and support both integer and floating-point operations. For example, the following code renders a 2D triangle, which is rotated by 30° about the z axis (the axis pointing outward from the 2D screen) and translated by 500 pixels and 600 pixels in the x and y directions, respectively:

```
glRotatef(30.0, 0.0, 0.0, 1.0);
glTranslatef(500.0, 600.0, 0.0);
glBegin(GL_TRIANGLE);
  glVertex2f(100.0, 100.0);
  glVertex2f(200.0, 200.0);
  glVertex2f(300.0, 100.0);
glEnd();
```

A further set of OpenGL functions are provided to define graphics states. For example, `glLineWidth(w)` sets the width of lines to `w` pixels. Similarly, the colour of rendered objects can be set by `glColor3f(r,g,b)` where `3f` implies there are three floating-point arguments and `r`, `g` and `b` are the red, green and blue colour components, respectively, with each component defined in the range 0.0–1.0. Other system-wide functions provide different line types, (e.g. dashed and dotted lines), specify a clipping region and implement various forms of anti-aliasing.

Although legacy OpenGL is well suited to the rendering of aircraft displays, the limitation of the single pipeline and the relatively low bandwidth between the host computer and the graphics card very much constrain the performance of the GPUs on a graphics card. The move to maximise the performance of the GPUs since version 3.0 has changed both the basic functions of OpenGL and the method of passing graphics objects to the GPU for geometric processing.

6.1.3 OpenGL Version 4

This section refers to versions of OpenGL since version 3.0 and, in particular, up-to-date versions since version 4.0. Bear in mind that OpenGL is a specification in the form of an API rather than a library. Since 2006, the Kronos Group (a not-for-profit organisation supported by the computer industry) has taken on responsibility for the specification of OpenGL and has ensured that improvements to OpenGL keep pace with advances in graphics technologies. However, one important consideration for OpenGL programmers is that OpenGL versions since 4.0 will only run on compatible graphics cards.

It is not the intention to provide a user guide to writing modern OpenGL; this topic is addressed by numerous textbooks, particularly the recent editions of the *OpenGL Programming Guide* (Kessenich et al., 2017). This section focuses on the use of OpenGL to render 2D aircraft displays and, consequently, omits aspects of OpenGL which are outside the scope of this application.

In legacy OpenGL, to draw an object, the transformation values, the rendering primitive and the coordinates of the object to be rendered are passed to the GPU. This arrangement rather defeats the benefit of modern GPUs, as the GPU must wait for the host to provide this information before rendering the object. The approach with modern OpenGL is to transfer details of the objects to be rendered to the GPU and to program the GPUs to access data, held in the GPU memory, which includes transformations and control of the colour of individual pixels.

The OpenGL Programming Guide (Kessenich et al., 2017) summarises the major operations in OpenGL as follows:

- Specify the data for constructing shapes from OpenGL's geometric primitives.
- Execute various *shaders* to perform calculations on the input primitives to determine their position, color and and other rendering attributes.
- Convert the mathematical description of the input primitives into their *fragments* associated with locations on the screen. This process is called *rasterisation*. (A fragment in OpenGL is what becomes a pixel, if it makes it all the way to the final rendered image.)
- Finally, execute a *fragment shader* for each of the fragments generated by rasterisation, which will determine the fragment's final color and position.
- Possibly perform additional per-fragment operations such as determining if the object that the fragment was generated from is visible or blending the fragment's color with the current color in that screen location.

The client, the user's application, defines the structures of the data to be rendered and passes this information to the GPU. Within the GPU, shaders perform the geometric operations, such as translation, scaling or rotation, and other shaders perform the computations needed to determine the colour of each pixel. The software for each shader can be written in code very similar to C, cross-compiled and loaded into the GPU. These changes are a radical improvement on the legacy OpenGL pipeline. The graphics engine can be considered as a very high-performance system with hundreds of cores operating in parallel. Computer graphics is just one application of this architecture although the capabilities of the fragment shaders are optimised to display computer-generated images (both 2D and 3D). For 2D graphics, only vertex and fragment shaders are normally used.

From the software developer's perspective, the paradigm of OpenGL is that the programmer defines vertex buffer objects (VBOs) and vertex array objects (VAOs) as a means of defining objects to be passed from the host memory to the GPU memory and then sends a call to OpenGL to tell the GPU which objects to process and how to process them. This is a fundamental change from legacy OpenGL and, particularly for programmers used to legacy OpenGL, requires a complete redesign of graphics applications and their implementation. There is very little overlap or compatibility between these two versions of OpenGL. There is one further consideration; although the bandwidth between the host memory and GPU memory is high, it is far more efficient to transfer a few large areas of memory to the GPU rather than many small areas.

Having transferred the graphics data to the GPU, the host computer can instruct the GPU which graphics objects to render, with the code in the vertex shaders and fragment shaders executed by the GPUs to render the object. For the OpenGL programmer, the graphics description must be clearly defined, enabling the graphics objects to be copied to the GPU memory, but it is also necessary to write the code to be executed by the shaders. This change introduces considerable scope for errors because the OpenGL code in the host computer, the vertex shader and the fragment shader must all be correct. From experience, this programming environment is unforgiving and even minor errors can result in a blank screen.

OpenGL maintains the primitive types of legacy OpenGL, except for the polygon and quadrilateral types. Each type is based on vertices, as shown in Figure 6.1.

The use of VBOs, VAOs and shaders is illustrated in the following code example to draw a green rectangle with the bottom left corner at (100, 100) and the top right corner at (300, 200). More practical examples of aircraft displays are presented in subsequent sections. For reasons of brevity, the following simplifications are made:

- As only one rectangle is rendered, the array offsets are defined as constants.
- The program runs with a display resolution of 1920 × 1080 – these values are defined as constants in both the program and the shaders and must be redefined if the display resolution is altered.
- The shader programs are written in GLSL, the OpenGL shading language (GLSL), to compile, link and load the code for both the vertex shader and the fragment shader.
- The program runs under the GLFW library, which includes an error call-back and a key call-back to detect an ESC key press to terminate the program.
- An up-to-date version of the graphics library *glew* is referenced in the program and must be linked with the compiled program.
- The drawing of the rectangle is given in a separate C function, to improve the clarity of the software.

The rectangle is formed from two triangles, as shown in Figure 6.2.

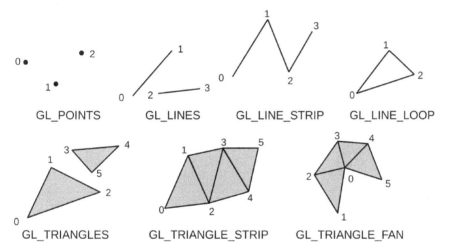

Figure 6.1   OpenGL Geometric Primitive Types.

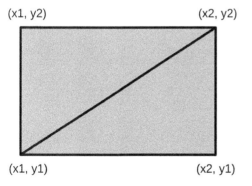

Figure 6.2   Reduction of a Rectangle to Triangles.

The preliminary definitions are as follows:

```
#include <stdio.h>
#include <stdlib.h>

#include <GL/glew.h>
#include <GLFW/glfw3.h>

#include "shader.h"

#define SCREENWIDTH   1920
#define SCREENHEIGHT 1080

#define TRIANGLES 3
```

The standard header files are included together with the header files for *glew* and GLFW, which are referenced in the program. In addition, the header file **shader.h** enables the vertex and fragment shaders to be compiled, linked and loaded. The specific screen dimensions are defined. The constant **TRIANGLES** corresponds to one of the primitive types that have just been defined. Although other types are used for aircraft displays, this example is limited to triangles. Global data needed for the program is declared as follows:

```
GLuint program;

GLuint vaoHandles[6];
GLuint vboHandles[2];

GLuint positionBufferHandle;
GLuint colorBufferHandle;

GLFWwindow* window;

GLfloat positionData[100];
GLfloat colorData[100];
```

The variable **program** is used to identify the current GPU core needed for the program. Only two VBOs are defined and, although only one VAO is used, six are declared to include other possible primitive types. **positionBufferHandle** and **colorBufferHandle** point to the vertex positions and the vertex colours, respectively. Note that these are not address pointers but values used to identify VBOs. **GLFWwindow** defines the window needed by GLFW. **positionData** and **colorData** define storage for the VBOs.

The VBOs and VAOs are initialised with the following code:

```
glGenBuffers(2, vboHandles);
positionBufferHandle = vboHandles[0];
colorBufferHandle = vboHandles[1];

glGenVertexArrays(6, vaoHandles);
glBindVertexArray(vaoHandles[TRIANGLES]);

glEnableVertexAttribArray(0);  // Vertex position
glEnableVertexAttribArray(1);  // Vertex color
```

```
glBindBuffer(GL_ARRAY_BUFFER, positionBufferHandle);
glVertexAttribPointer(0, 2, GL_FLOAT, GL_FALSE, 0, (void*) 0);

glBindBuffer(GL_ARRAY_BUFFER, colorBufferHandle);
glVertexAttribPointer(1, 3, GL_FLOAT, GL_FALSE, 0, (void*) 0);
```

glGenBuffers creates handles to access two buffers, where **positionBufferHandle** and **colorBufferHandle** are the handles to the two VBOs. **glGenVertexArrays** creates six handles to VAOs, even though only one will be used in this example. **glBindVertexArray** selects the specific VAO associated with the triangles to be rendered. As there is only one path between objects declared in the host computer memory and GPU memory, binding implies selecting or identifying the object to be referenced by the GPU. By default, VBOs are not active and are specifically activated by **glEnableVertexAtributeArray** which refers to the attribute (data organisation) of the VBOs (0 for position and 1 for colour, in this example). The numbering scheme for the VBOs must be the same as the VBO numbering used in the vertex shader. **glBind-Buffer** selects a specific VBO and **glVertexAttribute** defines the storage used for each VBO. Two floats are defined for the x and y coordinates of the vertices and three floats are defined for the red, green and blue components of the vertex colours. The remaining arguments specify that the values are floating-point types, the stride is zero (implying that the data values are stored contiguously) and the data starts at the first byte.

The vertex and fragment shaders are compiled, linked and loaded in a single call.

```
program = LoadShaders("test.v.glsl", "test.f.glsl");

if (program == 0)
{
    printf("Shader compilation failed\n");
    return 0;
}
```

Although only one call is made to load both shaders, any errors detected by **LoadShaders** are flagged, causing the program to halt. On termination of the program, **glDeleteProgram** instructs the GPU to remove any loaded programs and free their resources.

```
void free_resources()
{
    glDeleteProgram(program);
}
```

The positions of the six vertices of the two triangles of the rectangle are organised as follows in local storage:

X1	Y1	X2	Y2	X1	Y2	X1	Y1	X2	Y1	X2	Y2

Similarly, the local storage for the colours of the six vertices is organised as follows:

r1	g1	b1	r2	g2	b2	r3	g3	b3	r4	g4	b4	r5	g5	b5	r6	g6	b6

The function to draw an in-filled rectangle, with the bottom left-hand corner at $(x1, y1)$ and the top right-hand corner at $(x2, y2)$, formed from two triangles, is given below. Note that all the vertices are bright green, so that the green component is 1.0 and the red and blue components are 0.0.

```
void Rectangle(float x1, float y1, float x2, float y2)
{
    float r = 0.0;
    float g = 1.0; /* colour set to green */
    float b = 0.0;

    /* first triangle (x1, y1), (x2, y2), (x1, y2) */
    positionData[0]  = x1;positionData[1]  = y1;
    positionData[2]  = x2;positionData[3]  = y2;
    positionData[4]  = x1;positionData[5]  = y2;

    /* second triangle (x1, y1), (x2, y1), (x2, y2) */
    positionData[6]  = x1;positionData[7]  = y1;
    positionData[8]  = x2;positionData[9]  = y1;
    positionData[10] = x2;positionData[11] = y2;

    colorData[0]  = r;    colorData[1]  = g;    colorData[2]  = b;
    colorData[3]  = r;    colorData[4]  = g;    colorData[5]  = b;
    colorData[6]  = r;    colorData[7]  = g;    colorData[8]  = b;
    colorData[9]  = r;    colorData[10] = g;    colorData[11] = b;
    colorData[12] = r;    colorData[13] = g;    colorData[14] = b;
    colorData[15] = r;    colorData[16] = g;    colorData[17] = b;
}
```

During each frame, the rectangle is rendered by the function Display, as follows:

```
void Display()
{
    glClearColor(1.0, 1.0, 1.0, 1.0);
    glClear(GL_COLOR_BUFFER_BIT);

    glUseProgram(program);

    Rectangle(100.0, 100.0, 300.0, 200.0);

    glBindBuffer(GL_ARRAY_BUFFER, positionBufferHandle);
    glBufferData(GL_ARRAY_BUFFER, 12 * sizeof(float), positionData, GL_DYNAMIC_DRAW);

    glBindBuffer(GL_ARRAY_BUFFER, colorBufferHandle);
    glBufferData(GL_ARRAY_BUFFER, 18 * sizeof(float), colorData, GL_DYNAMIC_DRAW);

    glBindVertexArray(vaoHandles[TRIANGLES]);
    glDrawArrays(GL_TRIANGLES, 0, 6);
}
```

The first two lines set the background colour to white and clear the frame-store. **glUseProgram** informs the GPU to use the loaded shader and the function **Rectangle** writes a copy of the VBOs to local memory (see earlier). **glBindBuffer** is called to bind each VBO for GPU transfers. **glBufferData** is called, firstly for the vertex positions, which comprise 12 floating-point values stored at **positionData** (for the six vertices), and then for the vertex colours, which comprise 18 floating-point values stored at **colorData**. The last two lines of this fragment render the two triangles; **glBindVertexArray** binds the VAO used for triangles and **glDrawArrays** notifies the GPU that the data is passed as triangles with six vertices, that is, two triangles.

The main function is as follows:

```
int main(int argc, char* argv[])
{
    glfwSetErrorCallback(error_callback);

    if (!glfwInit())
        exit(EXIT_FAILURE);

    window = glfwCreateWindow(SCREENWIDTH, SCREENHEIGHT, "Lines", 0, NULL);

    if (!window)
    {
        glfwTerminate();
        exit(EXIT_FAILURE);
    }

    glfwMakeContextCurrent(window);
    glfwRestoreWindow(window);
    glfwSwapInterval(1);

    glfwSetKeyCallback(window, key_callback);

    GLenum glew_status = glewInit();
    if (GLEW_OK != glew_status)
    {
        fprintf(stderr, "Error: %s\n", glewGetErrorString(glew_status));
        return 1;
    }

    if (!GLEW_VERSION_2_0)
    {
        fprintf(stderr, "No support for OpenGL 2.0 found\n");
        return 1;
    }

    init_resources();

    while (!glfwWindowShouldClose(window))
    {
        Display();

        glfwSwapBuffers(window);
        glfwPollEvents();
    }

    free_resources();

    glfwDestroyWindow(window);
    glfwTerminate();

    return 0;
}
```

A window is created and call-backs are set up for errors and keyboard input. There is a further check with **glewInit** that *glew* is correctly initialised and is at least version 2.0. The resources needed for the VBO and VAO (as described earlier) are initialised. The main loop is common to most programs using GLFW. When the run-time loop terminates, the resources are freed and the GLFW window is closed. A separate update function could also be called in the main loop to update any non-graphics variables. For example, the position of the rectangle could be altered in each frame to follow the mouse movement but is omitted because the rectangle is static in this example. The GLSL code for the vertex shader is as follows:

```
#version 430 core

#define width  1920.0f
#define height 1080.0f

layout (location = 0) in vec2 VertexPosition;    // Position in attribute location 0
layout (location = 1) in vec3 VertexColor;       // Color in attribute location 1

out vec3 Color;                                  // Output a color to the fragment shader

void main()
{
    gl_Position.x = VertexPosition.x * 2.0f / width - 1.0f; // screen coords -> +/- 1.0
    gl_Position.y = VertexPosition.y * 2.0f / height - 1.0f;
    gl_Position.z = 0.0f;
    gl_Position.w = 1.0f;

    Color = VertexColor;
}
```

The first line informs the compiler that the graphics card hardware must support at least version 4.30. The screen height and width are defined, which must correspond with the screen coordinates used in the host software. The layout directive informs the compiler of the offsets in the VBO for the vertex position information and the vertex colour information, where **vec2** implies two elements and **vec3** implies three elements. The code in the function **main** computes the x and y coordinates, transforming from screen coordinates in the range $(0,0)$ to (**width-1**, **height-1**) to the normalised range from $(-1, -1)$ to $(+1, +1)$. The vector **gl_Position** has four components: the screen x and y coordinates, the z coordinate (which is zero for 2D graphics) and a fourth element denoting transparency, which is 1.0. The output from the vertex shader is passed to the fragment shader as a vector of three elements, that is, the red, green and blue colour components of each pixel. Note that no translation or rotation is needed in this example and it is more straightforward to define coordinates in pixel units in the C code in the host computer and to transform to normalised coordinates in the shader. Alternatively, the C code could use normalised coordinates, avoiding the need for transformation in the shader. The fragment shader contains the following code:

```
#version 430 core

in vec3 Color;          // Colour value from the vertex shader (smoothed)
out vec4 FragColor;     // Colour for the fragment

void main()
{
    FragColor = vec4(Color, 1.0f);    // Add alpha value of 1.0
}
```

Again, a check is included that the hardware version is at least 4.30. The input variable **Color** is the first three elements of the output variable from the vertex shader and the output **FragColor** is a vector of the three colour components computed by the vertex shader, with the fourth element (the alpha transparency) set to 1.0. Note that the fragment shader is trivial (but essential) in this example, but could contain much more complex arithmetic to determine colour mappings.

Although this example seems more complex than an equivalent legacy OpenGL version, requiring careful coding of the vertex shader, the fragment shader and the program in the host computer to organise the graphics data structures and access the GPU, the number of transfers to the GPU per frame is significantly reduced. Moreover, the graphics task is allocated to the GPU cores, operating in parallel, giving a significant improvement in the rendering speed. Once the conventions to access the GPU are understood, programming the GPU for other graphics primitives is straightforward.

6.2 *glib* – A 2D Graphics Library for Flight Simulation

In flight simulation, 2D graphics is used to emulate aircraft displays, to provide a user interface for the instructor station and to support the display of flight data and charts. In addition to the drawing of lines and polygons, a graphics library should also support management of the display and the rendering of high-quality character fonts and textures. An additional requirement is to meet the real-time constraints of the simulation frame rate, implying that performance of the library is critical. Not all these requirements are found in general-purpose 2D graphics packages, which are mostly used as graphic design tools rather than the provision of an API.

One option is to use OpenGL as the API for rendering. However, OpenGL is relatively low level and, in its basic form, does not support character generation of the widely available fonts. Rather than using a 2D graphics API such as OpenVG or Direct2D, a library *glib* was developed to provide the primitives specific to the requirements of real-time flight simulation. The advantage is that the library is small, written in C and optimised for real-time 2D graphics. The drawback is that it excludes features found in general-purpose packages that are not applicable to flight simulation.

The requirements to support real-time 2D graphics in flight simulation include:

- Management of the display – setting the screen resolution, selecting colours, enabling anti-aliasing and clipping and control of double buffering.
- Rendering of basic shapes, including dots, vectors and triangles.
- Rendering of character strings of various formats and sizes.
- Rendering of textures defined in standard graphics formats.
- Provision of 2D transformations for translation, rotation and scaling as matrix operations.
- Interaction with input devices, particularly a mouse and keyboard.
- Efficient access to the GPU to optimise the rendering rate.
- Compatibility with Windows and Linux platforms.
- Portability by basing the library on OpenGL.

The management of the frame-store, the display parameters and user input is provided by GLFW as part of the IDE rather than OpenGL. The provision of 2D and 3D transformations in legacy OpenGL was moved in version 4.0 from the OpenGL code to the shader code, requiring the user to implement these transformations by coding both the user application and the shaders. The requirement of meeting the real-time criteria is largely satisfied by the performance of modern GPUs with multiple cores, if it can be demonstrated that the rendering of the aircraft displays and the instructor station displays can be implemented within the simulator frame rate.

A particular attraction of OpenGL is that textures can be rendered at very high speed. Fortunately, character fonts can be stored as textures, enabling character generation to be implemented by rendering textures. However, the level of interfacing should reflect the objects used in simulation rather than forcing the user to work directly with low-level OpenGL primitives. This is essentially the purpose of all libraries: to hide unnecessary detail from the user, allowing the user to focus on design rather than details of an implementation.

6.2.1 GPU Software Interface

The use of VBOs and VAOs in Section 6.1 outlined the mechanism to define sets of graphics data which are passed to the GPU for rendering. Strictly, the host computer performs very little actual graphics – the rendering is executed by the GPU, including any transformations and colouring of graphic objects. The responsibility of the programmer is to organise the graphics data in the host computer, activate the transfers and write the code to be executed by the shaders.

The interface between the host computer and the GPU, used to transfer data to the GPU, introduces small delays and, if the objects are transferred to the GPU one at a time, the accumulation of these delays would be unacceptable. The solution is to write the graphics data to the host computer memory organised as a cache. The graphics data is written directly to the cache in main memory with no appreciable overhead. When the cache is sufficiently full, or needs emptying, the content of the cache is transferred to the GPU in a single operation. The actual rendering of these objects is then executed by the GPU cores operating in parallel. The responsibility of the host computer is to store the graphics data to be rendered in a queue and activate the GPU to access and render the graphics data.

If items written into the cache have a common state, which is known to the GPU, they can be accumulated in the cache. However, if a new item to be written into the cache has a different state from the items stored in the cache, the cache must be emptied and the caching restarts with a new state. For example, if a set of vectors is written into the cache with a common angle of rotation but a different rotation is applied to a new vector, the cache is emptied because the characteristics of data in the cache and data subsequently written to the cache will be different and the GPU has no way of detecting this change. The state of the cache and the GPU must be synchronised so that objects in the cache are rendered with the correct state. In other words, the caching of objects only applies where the state is common to all the objects in the cache. From the user's perspective, vectors are simply drawn by a library call. The writing of a vector to the cache and the management of the cache and the GPU state are handled by the library and are transparent to the user. The specific conditions for flushing the cache are:

- setting the transformation matrix to the identity matrix;
- resetting the transformation matrix, for example, popping it from a stack;
- translation, rotation or scaling operations that alter the transformation matrix;
- change of the current texture;
- change of the current font (assuming the fonts are implemented as textures);
- change of line width (assuming an OpenGL call is used);
- change of anti-alias settings;
- change of a clipping region.

As far as reasonably possible, the amount of flushing of the cache in an application should be minimised. Figure 6.3 shows the maximum amount of cached vectors, triangles, textures and strings for a desktop display of a Boeing 747-400, over 3000 frames (60 s), under worst-case conditions, where the displays are changing rapidly. The primary flight display (PFD) and engine-indicating and crew-alerting system (EICAS) are displayed on the left-hand side of the screen and the navigation flight display (NFD) and IOS can be toggled and displayed on the right-hand side of the screen. The state of the display is indicated in the text at the top of Figure 6.3.

The NFD contains very few vectors, indicating that the maximum caching of PFD vectors is approximately 100. When the IOS is displayed, the vectors change with zooming in and out of the map display increasing to 500 vectors. The number of triangles for this 2D display does not exceed 40. The number of textures varies from 120 to150 when the PFD and IOS are displayed as many of the chart symbols are textures. Otherwise only 10 textures are cached for the PFD

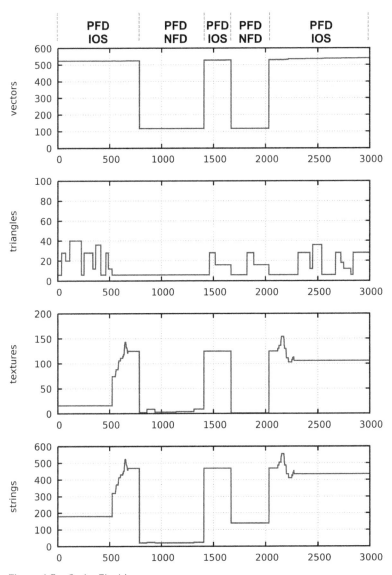

Figure 6.3 Cache Flushing.

and NFD. Some 500–600 strings are cached for the IOS, particularly the map display. The rendering of vectors and symbols for the instructor station chart display does not require any translations or rotations, nor are there any changes of line width or clipping, reducing the need to flush the cache, whereas for the PFD many of the vectors and triangles are rotated, requiring repeated flushing of the cache and several of the sub-displays use clipping functions. These tests should be viewed together with the frame time measurements to confirm the frame loading. With a basic graphics card, both displays occupy less than 1 ms of the 20 ms frame using the *glib* library.

There is one final point to consider with caching. Although vectors are written into the cache in the order they are drawn, and will be rendered in that order, care is needed with the layering (or prioritisation) of graphics. For example, if a texture and a vector are both cached, the vector may be drawn over the texture, or vice-versa, depending on the sequence of drawing operations. For aircraft displays, there is minimal layering and objects are rendered from the cache in the following order: dots, vectors, triangles, textures and strings. For example, a string will always be rendered over vectors or textures. In the case of the instructor station user interface, the text on an icon such as a button will be guaranteed to be drawn last. For most displays, text or vectors are normally drawn into empty spaces created within a texture; it would be unusual for vectors to be rendered directly over textures. In cases where the default layering does not apply, the user has the option to force flushing of the cache to ensure the rendering order. This implicit layering is a further example of the application-specific nature of the graphics library.

6.2.2 Dots, Vectors and Triangles

Although dots are a special case of a vector, they are recognised as distinct entities by OpenGL. Arguably, a vector is a special case of a triangle but OpenGL provides both forms. The distinction between the different types is the number of vertices passed to the GPU. A dot is represented by a single vertex (x, y), a vector by two vertices (x_1, y_1, x_2, y_2) and a triangle by three vertices $(x_1, y_1, x_2, y_2, x_3, y_3)$. When vertices are transferred to the GPU, the first argument of the function `glDrawArrays` defines the type as `GL_POINTS`, `GL_LINES` or `GL_TRIANGLES`.

Vectors are the lines rendered in an image and are ubiquitous in 2D graphics. Vectors are written to the cache in the host computer and passed as a block of vectors to the GPU (Wolf, 2013), either because there are no further vectors to be drawn or because the OpenGL state has changed and no longer applies to these stored vectors, for example, if the transformation matrix is changed. Clearly, it would be possible to pass the transformation matrix with each vector, but this would introduce a considerable overhead, and is particularly inefficient where the translation or rotation is common to a large group of vectors. Similarly, a set of vectors with a common colour could be drawn as a single block of vectors of a specific colour. In such cases, there is a design choice in developing the library, whether to associate a group of vectors with common attributes or whether to pass the coordinates *and* the attributes with each vector. The important point to appreciate is that the format selected in the client computer must be replicated in the GPU, that is, the GPU needs to know the format of graphics data passed to it. The use of *uniform* variables with shaders allows a variable to be shared between the client computer and the GPU and to be updated by the client as required. Of course, this adds an extra graphics operation in addition to passing the vector vertices. The alternative, of passing both the vectors and its attributes, increases the amount of memory needed by both the client computer and the GPU.

The translation, rotation and colour of a vector are not the only characteristics which can vary; the line width, line type (e.g. continuous, stippled or dotted), anti-aliasing and clipping conditions could also be defined on a per-vector basis. For the developer, depending on the application, it is necessary to choose which characteristics are generic and which are specific to each object. In OpenGL, generic primitives can be set which apply to groups of objects in a scene until the setting is changed, and include transformations, anti-aliasing and clipping. For aircraft displays, the following criteria were applied in *glib* for efficient transmission of vectors to the GPU:

- vertices – stored in the cache of vertex coordinates;
- colour – stored in the cache with vertex data for each vector;
- line width – set by either **glLineWidth** or rendering lines as filled rectangles;
- transformation – the matrix is passed as a uniform variable;
- anti-aliasing – use of **glEnable(GL_LINE_SMOOTH)** and **glDisable(GL_LINE_SMOOTH)** to define the aliasing state of vectors;
- clipping – use of **glEnable(GL_SCISSOR_TEST)** and **glDisable(GL_SCISSOR_TEST)** to enable and disable clipping, respectively, and **glscissor** to set the clipping window coordinates;
- line types – not applicable as the displays mostly contain continuous lines.

Initially, space is reserved in the cache for dots, vectors and triangles. This memory can be allocated statically, to cater for the worst-case number of objects. It is important that the bounds of the individual caches are checked for overflow as items are inserted into the cache. Dots comprise one vertex (two floats) and one colour (three floats); vectors comprise two vertices (four floats) and two colours (six floats); and triangles comprise three vertices (six floats) and three colours (nine floats) with the cache memory allocated as follows:

```
GLfloat DotPosition[MAXDOTS * 2];
GLfloat DotColor[MAXDOTS * 3];
GLfloat VectorPosition[MAXVECTORS * 4];
GLfloat VectorColor[MAXVECTORS * 6];
GLfloat TrianglePosition[MAXTRIANGLES * 6];
GLfloat TriangleColor[MAXTRIANGLES * 9];
```

The code to initialise the vertex buffers for vectors (the code for dots and triangles is similar, but is omitted) is as follows:

```
glGenVertexArrays(4, &vaoHandles[0]);

glGenBuffers(2, vboHandles);
positionBufferHandle = vboHandles[0];
colorBufferHandle = vboHandles[1];
    .
    .
    .
glBindVertexArray(vaoHandles[VECTORS]);

glBindBuffer(GL_ARRAY_BUFFER, positionBufferHandle);
glBufferData(GL_ARRAY_BUFFER, MAXVECTORS * 4 * sizeof(float), VectorPosition, GL_STATIC_DRAW);
glVertexAttribPointer(0, 2, GL_FLOAT, GL_FALSE, 0, (void*) 0);
glEnableVertexAttribArray(0);   /* Vertex position */

glBindBuffer(GL_ARRAY_BUFFER, colorBufferHandle);
glBufferData(GL_ARRAY_BUFFER, MAXVECTORS * 6 * sizeof(float), VectorColor, GL_STATIC_DRAW);
glVertexAttribPointer(1, 3, GL_FLOAT, GL_FALSE, 0, (void*) 0);
glEnableVertexAttribArray(1);   /* Vertex color */
```

A vertex array **vaoHandles** is generated for dots, vectors and triangles and two vertex buffers are generated for the vertices and colours. The vertex array and the vertex buffer are bound and allocated for vertex positions, defining the size and address of the buffer. The vertex is at location zero and contains two floats. Vertex array zero is then enabled and the process is repeated for the colour component, which is at location one and contains three floats. The function **Glib_Draw** writes vectors to the cache for subsequent rendering as follows:

```
void Glib_Draw(int x1, int y1, int x2, int y2)
{
    float r = ColourTable[CurrentColour][0];
    float g = ColourTable[CurrentColour][1];
    float b = ColourTable[CurrentColour][2];

    VectorPosition[nvectors * 4 + 0] = (float) x1;
    VectorPosition[nvectors * 4 + 1] = (float) y1;
    VectorPosition[nvectors * 4 + 2] = (float) x2;
    VectorPosition[nvectors * 4 + 3] = (float) y2;

    VectorColor[nvectors * 6 + 0] = r;
    VectorColor[nvectors * 6 + 1] = g;
    VectorColor[nvectors * 6 + 2] = b;
    VectorColor[nvectors * 6 + 3] = r;
    VectorColor[nvectors * 6 + 4] = g;
    VectorColor[nvectors * 6 + 5] = b;

    nvectors += 1;

    if (nvectors >= MAXVECTORS)
    {
        printf("Vectors overflow (%d)\n", MAXVECTORS);
        exit(-1);
    }
}
```

Rather than testing for overflow, the cache could be flushed when the overflow condition is detected. However, the check on overflow provides assurance that the size of the cache is adequate. For aircraft displays, the colour of the two vertices of a vector will be identical. The vertices are copied to the array **VectorPosition** and the colours are written to the array **VectorColor**. A check is also made for overflow of the arrays. The emptying of the vector cache and transferring the vectors to the GPU are given in the following fragment of code:

```
if (nvectors > 0)
{
    glUseProgram(VectorProgram);
    glBindVertexArray(vaoHandles[VECTORS]);
    glUniformMatrix4fv(vtloc, 1, GL_FALSE, (float *) Transform);
    glBindBuffer(GL_ARRAY_BUFFER, positionBufferHandle);
    glBufferData(GL_ARRAY_BUFFER, nvectors * 4 * sizeof(float), VectorPosition, GL_STATIC_DRAW);
    glBindBuffer(GL_ARRAY_BUFFER, colorBufferHandle);
    glBufferData(GL_ARRAY_BUFFER, nvectors * 6 * sizeof(float), VectorColor, GL_STATIC_DRAW);
    glDrawArrays(GL_LINES, 0, nvectors * 2);
    nvectors = 0;
}
```

The shader program **VectorProgram** is activated. The vertex array is bound, the transformation matrix is passed as a uniform array, found at the offset **vtloc**, the position vertex buffer is bound and its contents are passed; the sequence is repeated for the vertex colours. The vector is actually passed by **glDrawArrays** with the qualifier **GL_LINES** and the number of vectors times two, as there are two vertices per vector. The cache is emptied of vectors by setting **nvectors** to zero. The code for dots and triangles is very similar, with **glBufferData** defining the structure of the data to be passed.

The GPU vertex shader used for vectors contains the following code:

```
#version 430 core

#define width   1920.0f
#define height 1080.0f

layout (location = 0) in vec2 VertexPosition;   // Position in attribute location 0
layout (location = 1) in vec3 VertexColor;      // Color in attribute location 1

out vec3 Color;                                 // Output a colour to the fragment shader

mat4 ProjectMatrix =
{
    {2.0f/width,         0.0f,    0.0f,    0.0f}, // column-row ordering
    {      0.0f, 2.0f/height,    0.0f,    0.0f},
    {      0.0f,         0.0f,   -1.0f,    0.0f},
    {     -1.0f,        -1.0f,    0.0f,    1.0f}
};

uniform mat4 TransformationMatrix;   // translation and rotation

void main()
{
    gl_Position = ProjectMatrix * TransformationMatrix * vec4(VertexPosition, 0.0, 1.0);
    Color = VertexColor;
}
```

The version number is provided to ensure compatibility. The display resolution is defined and must match the host display resolution. The location of the vertex position input and colour input are defined (to match the application definitions) with an output variable **Color** passed to the fragment shader. The projection matrix **ProjectMatrix** provides conversion from pixel coordinates (provided by the host computer) to OpenGL coordinates used by the shader in the range ±1. The uniform matrix defines the 4×4 transformation matrix passed by the host computer. The position of each vertex is then transformed, multiplying the vertex by the transformation matrix and the projection matrix. Note that the z vertex is set to 0.0 for 2D graphics. The input vector **VertexColor** is copied to the vector **Color** and passed as an output to the fragment shader, which is as follows:

```
#version 430 core

in vec3 Color;          // Color value from the vertex shader (smoothed)
out vec4 FragColor;     // Color for the fragment

void main()
{
    FragColor = vec4(Color, 1.0f);   // Add alpha value of 1.0
}
```

The input is the vector **Color** and the output is the vector **FragColor** simply formed from **Color** with the addition of an alpha component set to 1.0.

6.2.3 Textures

Using 2D graphics, it would be possible to draw a radio panel as a collection of vectors to represent the knobs and switches on the panel. For example, a toggle switch could be constructed from a set of triangles, possibly requiring 100 triangles to form an acceptable image of a toggle switch. An alternative approach is to capture a digital image of a toggle switch which can be stored as a 2D texture and rendered as a single object rather than numerous triangles. Similarly, the bezel and markings of a flight instrument can either be drawn as vectors or rendered as textures extracted from a photograph.

The use of textures affords two advantages in 2D graphics. Firstly, the images are clearly more realistic and, secondly, the number of graphics operations to render an object is considerably reduced. The main drawback of this approach is the amount of memory needed to store a textured image. However, there have been two advances with GPU technologies, where the advantages of applying textures far outweigh vector generation methods. Improvements in the capacity and speed of memory technologies enable very large amounts of texture to be stored in GPU memory. In addition, the processing provided by GPUs has been optimised for the geometric operations needed for high-speed rendering of textures in both 2D and 3D applications.

A texture has a planar 2D coordinate frame and the image is formed from many colour dots, which are referred to as texels rather than pixels. One major problem with digitised images is that translations, rotations and scaling of textures can lead to distortion or jaggedness in images during rendering. However, algorithms developed to filter digital images and reduce these artefacts are implemented in hardware by most GPUs. Consequently, part of the design process for aircraft displays is to decide which elements should be rendered as vectors and triangles and which elements are best rendered as textures.

Most textures are captured as photographic images although a number of graphical editing tools enable complex images to be constructed from selections of texture to create realistic knobs, switches, selectors, panels, levers and so on for 2D aircraft displays. Numerous graphics formats are available to store digital images, including *bmp*, *tiff*, *jpeg* and *png* formats. For each format, the image is encoded, according to the specific standard, particularly the memory organisation of the image and the colour values of each texel.

Textures are mostly provided as binary files, which are converted to a bitmap image, which is stored in memory and can be accessed by OpenGL to render the texture and apply 2D and 3D transformations, including translation, rotation and scaling. Several packages are available to load textures, and the *freeimage* package, outlined in Section 6.3.4, which is open source and available for most platforms, enables images in most common formats to be converted to a bitmap image. For the most part, images used in the flight simulation projects covered in this book are *png* files, which can also be viewed with standard tools. The *glib* function `Glib_LoadTexture`, uses *freetype* functions to load a texture file.

The initialisation of the textures used by *glib* is similar to initialisation of the vectors and is as follows:

```
void Init_Textures()
{
    glGenVertexArrays(1, &TextureVAO);
    glBindVertexArray(TextureVAO);

    glGenBuffers(1, &TextureVBO);

    glBindBuffer(GL_ARRAY_BUFFER, TextureVBO);
```

```
    glBufferData(GL_ARRAY_BUFFER, MAXTEXTURES * 24 * sizeof(float), TextureVertices,
                GL_DYNAMIC_DRAW);

    glVertexAttribPointer(0, 2, GL_FLOAT, GL_FALSE, 4 * sizeof(float),
                          (void *) (0 * sizeof(float)));
    glEnableVertexAttribArray(0);

    glVertexAttribPointer(1, 2, GL_FLOAT, GL_FALSE, 4 * sizeof(float),
                          (void *) (2 * sizeof(float)));
    glEnableVertexAttribArray(1);

    TextureProgram = LoadShaders("texture-v.glsl", "texture-f.glsl");
    if (TextureProgram == 0)
    {
        printf("Shader compilation failed\n");
        exit(-1);
    }

    ttloc = glGetUniformLocation(TextureProgram, "TransformationMatrix");
    if (ttloc < 0)
    {
        printf("Texture shader: cannot find TransformationMatrix\n");
        exit(-1);
    }

    uniform_Texture_Sampler = glGetUniformLocation(TextureProgram, "texture_sampler");
    if (uniform_Texture_Sampler < 0)
    {
        printf("Texture shader: cannot find uniform_Texture_Sampler\n");
        exit(-1);
    }
}
```

A vertex array object **TextureVAO** is generated and bound. A single vertex buffer object **TextureVBO** is generated and bound. **glBufferData** defines the storage to be accessed by the GPU for the textures. Each texture is formed as two triangles, giving six vertices per texture, where each vertex is given by its (x, y) screen coordinate and its texture coordinate (u, v), giving a total of 24 floating-point values, which are stored in the array **TextureVertices**. Both VBOs (0 and 1) are enabled. The vertex and fragment shader programs **texture-v.glsl** and **texture-f.glsl** are compiled and loaded. The variable **TextureProgram** refers to the shader program dedicated to the rendering of textures. **glGetUniformLocation** enables the host computer to locate the shader variables **TransformationMatrix** and **texture_sampler**, via the handles stored in **ttloc** and **uniform_texture_sampler**, respectively, and checks that the GPU shader variables are defined.

The code to select an active texture is straightforward; the array **TextureTable** contains a pointer to access the texture.

```
void Glib_SetTexture(unsigned int t)
{
    if (TextureTable[t] == 0)
    {
        printf("Missing texture %d\n", t);
        exit(-1);
    }
```

```
    if (t != oldtexture)
    {
        Glib_Flush();
        CurrentTexture = TextureTable[t];
        oldtexture = t;
    }
}
```

A check is made that the texture file **t** has been loaded and, if so, the global variable **CurrentTexture** is set to point to the specific texture memory. If the current texture changes, existing objects in the cache are flushed.

The rendering of textures in *glib* is very similar to the rendering of vectors. The coordinates are written to the specific data structure in the host memory. Two versions are provided: firstly, where the texture is aligned orthogonally, for example, in an instrument panel, and secondly, where the texture is rotated about its centroid, for example a selector or knob. The *glib* function to render a texture is

```
Glib_DrawTexture(int x0, int y0, int sx0, int sy0, float tx1, float ty1,
                 float tx2, float ty2, float Alpha)
```

where the texture is drawn at (**x0,y0**), with sides of length **sx0** and **sy0**. The texture coordinates are the bottom left-hand corner at (**tx1,ty1**) and the top right-hand corner at (**tx2,ty2**). The alpha value provides transparency, allowing underlying objects to be seen through a texture. The checks for overflow of the array holding the textures are similar to the checks applied to vectors. The texture coordinates are copied to host memory, where the texture rectangle is represented as two triangles:

```
TextureVertices[ntextures * 24 + 0]  = x1;   /* bottom left */
TextureVertices[ntextures * 24 + 1]  = y1;
TextureVertices[ntextures * 24 + 2]  = tx1;
TextureVertices[ntextures * 24 + 3]  = ty1;
TextureVertices[ntextures * 24 + 4]  = x2;   /* bottom right */
TextureVertices[ntextures * 24 + 5]  = y1;
TextureVertices[ntextures * 24 + 6]  = tx2;
TextureVertices[ntextures * 24 + 7]  = ty1;
TextureVertices[ntextures * 24 + 8]  = x2;   /* top right */
TextureVertices[ntextures * 24 + 9]  = y2;
TextureVertices[ntextures * 24 + 10] = tx2;
TextureVertices[ntextures * 24 + 11] = ty2;

TextureVertices[ntextures * 24 + 12] = x2;   /* top right */
TextureVertices[ntextures * 24 + 13] = y2;
TextureVertices[ntextures * 24 + 14] = tx2;
TextureVertices[ntextures * 24 + 15] = ty2;
TextureVertices[ntextures * 24 + 16] = x1;   /* top left */
TextureVertices[ntextures * 24 + 17] = y2;
TextureVertices[ntextures * 24 + 18] = tx1;
TextureVertices[ntextures * 24 + 19] = ty2;
TextureVertices[ntextures * 24 + 20] = x1;   /* bottom left */
TextureVertices[ntextures * 24 + 21] = y1;
TextureVertices[ntextures * 24 + 22] = tx1;
TextureVertices[ntextures * 24 + 23] = ty1;
```

Notice, as before, that these values are stored in the host memory cache, which is flushed for the similar conditions that apply to vectors.

When the texture cache is flushed, the following fragment of code, in the function **flush**, copies the set of cached rectangular textures to the GPU:

```
case TEXTURES:
    if (ntextures > 0)
    {
        glUseProgram(TextureProgram);
        glBindVertexArray(TextureVAO);
        glUniformMatrix4fv(ttloc, 1, GL_TRUE, &Transform[0]);
        glBindTexture(GL_TEXTURE_2D, CurrentTexture);
        glUniform1i(uniform_Texture_Sampler, 0);
        glBindBuffer(GL_ARRAY_BUFFER, TextureVBO);
        glBufferData(GL_ARRAY_BUFFER, ntextures * 24 * sizeof(float),
                    TextureVertices, GL_DYNAMIC_DRAW);
        glDrawArrays(GL_TRIANGLES, 0, ntextures * 6);
        ntextures = 0;
    }
    break;
```

The GPU program dedicated to textures is activated and the VAO of the textures is bound. The current transformation matrix is copied to the GPU. The current texture is bound and the texture sampling is set. Finally, the VBO is bound, the texture data for all the textures in the cache is copied to the GPU and the rendering is activated as triangles, with six vertices per texture.

The vertex shader for textures is as follows:

```
#version 430 core

#define width  1920.0f
#define height 1080.0f

layout (location = 0) in vec2 pos;
layout (location = 1) in vec2 texture_coord;

out vec2 texture_coord_from_vshader;

vec4 position = vec4(pos, 0.0f, 1.0f);

mat4 ProjectMatrix =
{
    {2.0f/width,        0.0f,    0.0f,    0.0f}, // column-row ordering
    {      0.0f, 2.0f/height,    0.0f,    0.0f},
    {      0.0f,        0.0f,   -1.0f,    0.0f},
    {     -1.0f,       -1.0f,    0.0f,    1.0f}
};

uniform mat4 TransformationMatrix;   // translation and rotation

void main()
{
    gl_Position = ProjectMatrix * TransformationMatrix * position;
    texture_coord_from_vshader = texture_coord;
}
```

The version number provides a check on the hardware implementation and the screen size is defined to enable the coordinates to be normalised. In addition to defining the vertex position, the coordinates of the texture are also defined. In this case, the output to the fragment shader is the incoming texture coordinate. The position vector is extended to **vec4** as before and used in the transformation of the **ProjectMatrix** (for normalisation) and **TransformationMatrix** for translation and rotation. Note that both the vertex position and texture coordinates are passed to the fragment shader, which is as follows:

```
#version 430 core

in vec2 texture_coord_from_vshader;
out vec4 out_color;

uniform sampler2D texture_sampler;

void main()
{
    out_color = texture(texture_sampler, texture_coord_from_vshader);
}
```

The 2D sampler is defined as a uniform because it is shared between shaders. The pixel colour is generated by the texture function, which applies the texture sampler to the texture coordinate. The output pixel colour is derived by the function **texture**, which uses the uniform texture sampler and the 2D coordinates supplied by the vertex shader.

Two approaches can be used for 2D textures. If the textures are relatively small, they can be placed in a single large texture and then accessed via their specific texture coordinates and size. This method is used for the knobs and switches of a flight control unit (FCU), and is described in Section 6.7. For larger texture files, each texture can be loaded separately and accessed individually. This method is applicable to aircraft displays comprising separate instruments, where each instrument is captured as a separate texture.

The library *glib* provides the following functions for the rendering of textures:

```
extern void Glib_LoadTexture(char filename[], unsigned int texturenumber);
extern void Glib_SetTexture(unsigned int t);
extern void Glib_DrawTexture(int x0, int y0, int sx0, int sy0,
                    float tx1, float ty1, float tx2, float ty2, float Alpha);
extern void Glib_DrawTextureRotated(int x0, int y0, int sx0, int sy0, float tx1, float ty1,
                    float tx2, float ty2, float R, float Alpha);
```

Glib_LoadTexture loads a texture from the file **filename** and associates a unique number with the texture. **Glib_SetTexture** makes the texture **t** the currently active texture. **Glib_DrawTexture** renders the part of the texture defined by the texture coordinates (**tx1, ty1**) and (**tx2, tx2**) at the screen coordinates (**x0, y0**) with sides of length **sx0** and **sy0**, respectively. **Glib_DrawTextureRotated** is similar to **Glib_DrawTexture**, except that the texture is rotated about the point (**x0, y0**) by the angle **R**.

6.2.4 Fonts

Aircraft displays require textual characters to display numbers, or units or captions. Unfortunately, neither OpenGL nor GLFW provides character generation and this task is left to the programmer. Alphabetic and numeric characters can be generated in one of three ways: as bitmaps, vectors or textures. A character drawn as a bitmap is shown in Figure 6.4.

20
19
18
17
16
15
14
13
12
11
10
9
8
7
6
5
4
3
2
1
0
0 1 2 3 4 5 6 7 8 9 10 11 12

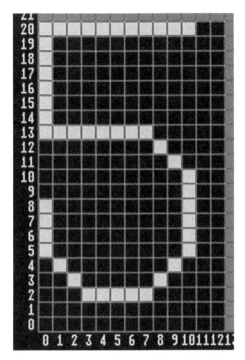

Figure 6.4 Bitmap Character Generation.

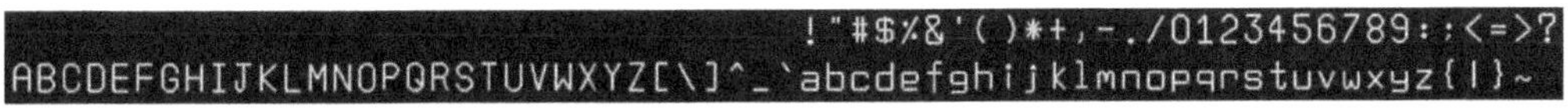

!"#$%&'()*+,-./0123456789:;<=>?
ABCDEFGHIJKLMNOPQRSTUVWXYZ[\]^_`abcdefghijklmnopqrstuvwxyz{|}~

Figure 6.5 Vector Character Generation.

In this example, the character '5' is drawn on a grid of 24 rows and 12 columns. The attraction of this method is that characters can be stored in memory, where the row length corresponds to the byte ordering of memory, so that storing characters and reading characters from memory is straightforward and, consequently, the rendering rate is relatively fast. Of course, with character sets of 128 or possibly 256 characters in a font, considerable effort is needed to capture the complete font set. However, the method has several limitations. Firstly, the font size is fixed and characters may not necessarily map to byte boundaries. Secondly, the method is not amenable to scaling or rotation, owing to the implicit storage on an integer grid. Thirdly, and most importantly, sloping lines are jagged prior to rendering and these effects are not removed by anti-aliasing.

An alternative method is to connect the segments of a bitmap character, which are rendered as vectors, as shown in Figure 6.5. Clearly the characters are smoother than bit-mapped characters and there is no observable aliasing as the lines are rendered as vectors. However, the character 'S' shown in Figure 6.6 contains 13 segments, implying 13 vector operations to render the character.

While this method is amenable to rotation, a character set based on vectors lacks many of the characteristics of modern fonts found in publishing and in software text formatting packages. The characters vary in width, not all characters are aligned on the same vertical line, increasing the curvature implies adding more segments to be rendered and the line width is constant throughout the character. In particular, any in-filling of lines is achieved by the selection of line width.

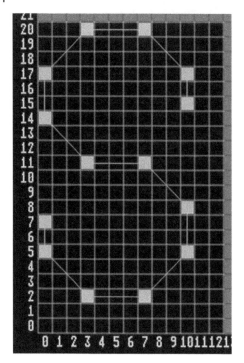

Figure 6.6 Vector Character Generation.

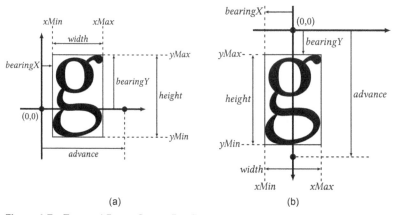

Figure 6.7 Textured Fonts. *Source*: FreeType organisation https://www.freetype.org.

The third and most popular method is to use the character sets provided on modern computers and construct textures derived from these character sets. The typography of true-type font characters is defined in the documentation issued by the *FreeType* organisation and is shown in Figure 6.7.

The character thickness can vary, different radii of curvature are provided within the character, and the height, width and origin of each character are defined, so that the spacing of characters is no longer fixed but is appropriate to the font spacing. This approach provides access to fonts commonly available in word-processing packages and specialised fonts.

Figure 6.8 Examples of True-type Fonts.

The *FreeType* organisation provides a range of tools to support the creation and management of fonts, particularly the font families known as true-type fonts that are available for most platforms. A particular attraction of using the font tools provided by *FreeType* is that the software is open. For simulation purposes, the *FreeType* library enables a font to be loaded at run-time, typically constructed as a texture atlas, so that individual characters can be extracted with minimal overhead and rendered as textures. Depending on the organisation of the storage of the font, *FreeType* provides graphics operations including scaling, rotation and anti-aliasing at the time the font is created in memory. It should be noted that *FreeType* is a software font engine designed for applications involving high-quality text, which are not necessarily real-time. Nevertheless, this approach to font rendering exploits the texture rendering provided by GPUs, enabling high-quality fonts to be rendered in real-time. Although the time to convert a font to a texture is substantial, this conversion only occurs once, at the start of an application.

Generally, for aircraft displays, rather than scaling fonts at run-time, copies of a font can be generated as fonts of specific sizes. The amount of storage needed for each font is not significant and certainly the quality of a font is far better if scaled by *FreeType* at the time the font is loaded, rather than applying a scaling transformation at run-time. Several true-type fonts are shown in Figure 6.8.

In this example, the lower two rows are an Arial regular font with font heights of 12 and 24 pixels, respectively. The next four rows are the B612 regular font with font heights of 11, 16, 20 and 32 pixels, respectively. The top row is an LCD font with a font height of 32 pixels. Notice the quality of the fonts and that it is possible to devise any symbol for a font character, either by using an available font or by employing the various font editing tools. *FreeType* allows a font file to be loaded and converted to an *atlas*, which is a texture holding all the characters of a font of a specific height. The code to load a font is as follows:

```
void Glib_LoadFont(char filename[], unsigned int fontnumber, unsigned int height)
{
    struct atlas *a;
    FT_Face      face;
    FT_Library   ft;

    if (FT_Init_FreeType(&ft))
    {
        fprintf(stderr, "Could not initialise freetype library\n");
        exit(-1);
    }

    if (FT_New_Face(ft, filename, 0, &face))
    {
        fprintf(stderr, "Could not open font file %s\n", filename);
        exit(-1);
    }
```

```
    if (fontnumber < 1 || fontnumber > MAXFONTFILES)
    {
        printf("Font %d out of range\n", fontnumber);
        exit(-1);
    }

    if (FontTable[fontnumber] != NULL)
    {
        printf("Font %d clash\n", fontnumber);
        exit(-1);
    }

    a = (struct atlas *) malloc(sizeof(struct atlas));
    if (a == NULL)
    {
        fprintf(stderr, "No space for font atlas %s\n", filename);
        exit(-1);
    }

    CreateAtlas(a, face, height);
    FontTable[fontnumber] = a;
}
```

The function is called with a font file, for example, `c:/windows/fonts/arial.ttf`, a font number which is associated with the font and the height (or size) of the font in pixels. The first part of the function checks that the library can be initialised and opened, that the font number is valid and a font with the same number does not already exist. Space is then allocated for the atlas, which is created, and the font number is stored in the array **FontTable** for subsequent access.

The following structure is provided for 128 characters based on the specification of the *FreeType* library:

```
struct atlas
{
    GLuint tex; // texture object

    unsigned int w; // width of texture in pixels
    unsigned int h; // height of texture in pixels

    struct
    {
        float ax; // advance.x
        float ay; // advance.y

        float bw; // bitmap.width;
        float bh; // bitmap.height;

        float bl; // bitmap_left;
        float bt; // bitmap_top;

        float tx; // x offset of glyph in texture coordinates
        float ty; // y offset of glyph in texture coordinates
    } c[128];        // character information
};
```

The fields are self-explanatory and relate to the definitions in Figure 6.7. The first part of the code for **CreateAtlas** generates the textures for 128 characters, as follows:

```
void CreateAtlas(struct atlas *a, FT_Face face, int height)
{
    unsigned int roww = 0;
    unsigned int rowh = 0;
    int          ox = 0;
    int          oy = 0;
    int          i;

    FT_Set_Pixel_Sizes(face, 0, height);
    FT_GlyphSlot g = face->glyph;

    a->w = 0;
    a->h = 0;

    memset(a->c, 0, sizeof(a->c));

    /* Find minimum size for a texture holding all visible ASCII characters */
    for (i = 32; i < 128; i+=1)
    {
        if (FT_Load_Char(face, i, FT_LOAD_RENDER))
        {
            fprintf(stderr, "Loading character %c failed!\n", i);
            continue;
        }
        if (roww + g->bitmap.width + 1 >= MAXWIDTH)
        {
            a->w = MAX(a->w, roww);
            a->h += rowh;
            roww = 0;
            rowh = 0;
        }
        roww += g->bitmap.width + 1;
        rowh = MAX(rowh, g->bitmap.rows);
    }

    a->w = MAX(a->w, roww);
    a->h += rowh;
    .
    .
    .
```

CreateAtlas is called with a pointer to the atlas, a pointer to the face and the font height. **FT_Set_Pixel_Sizes** sets the pixel sizes required for the font, where the zero denotes that the width is calculated dynamically as a function of height. **g** is the glyph for the font and the 128 characters are set to zero by **memset**. Only the visible ASCII characters from 32 to 127 are generated in the for-loop. The ith character is loaded by **FT_Load_Char** to find the width and height needed for the atlas.

The code for the linkage between the atlas and the GPU is almost identical to the code to initialise textures, described in the previous section. Finally, the following code sets the various pointers and offsets in the atlas for all 96 characters:

```
    rowh = 0;

    for (i = 32; i < 128; i+=1)
    {
        if (FT_Load_Char(face, i, FT_LOAD_RENDER))
        {
            fprintf(stderr, "Loading character %c failed!\n", i);
            continue;
        }

        if (ox + g->bitmap.width + 1 >= MAXWIDTH)
        {
            oy += rowh;
            rowh = 0;
            ox = 0;
        }

        glTexSubImage2D(GL_TEXTURE_2D, 0, ox, oy, g->bitmap.width,
                        g->bitmap.rows, GL_ALPHA, GL_UNSIGNED_BYTE, g->bitmap.buffer);
        a->c[i].ax = g->advance.x >> 6;
        a->c[i].ay = g->advance.y >> 6;

        a->c[i].bw = g->bitmap.width;
        a->c[i].bh = g->bitmap.rows;

        a->c[i].bl = g->bitmap_left;
        a->c[i].bt = g->bitmap_top;

        a->c[i].tx = ox / (float)a->w;
        a->c[i].ty = oy / (float)a->h;

        rowh = MAX(rowh, g->bitmap.rows);
        ox += g->bitmap.width + 1;
    }
```

A running count of the character offsets is maintained and, for each character, the relative pixel positions **ax**, **ay**, the width and height, the left and top positions and the texture width and height are updated. The essential information for the GPU is passed via **glTexSubImage2D** which specifies the offsets **ox** and **oy** in the texture and the width and height (in rows). The texture is alpha-blended (**GL_ALPHA**) so that the background of the character is not visible. The memory address of the glyph is passed in **g-> bitmap.buffer**.

In the library *glib*, the maximum number of fonts is defined as a constant **MAXFONTFILES** which can be increased to support more fonts. The code to render a string is given in the function **Glib_Chars**. The writing of the texture to the cache is very similar to the code for rendering textures. However, it is necessary to extract the texture information for each character from the atlas and this is achieved with the following fragment of code:

```
    for (i=0; i<strlen(text); i+=1)
    {
        int ch = (int) text[i];
        int j;

        /* Calculate the vertex and texture coordinates */
        float x2 = x + a->c[ch].bl;
        float y2 = -y - a->c[ch].bt;
        float w = a->c[ch].bw;
```

```
        float h = a->c[ch].bh;

        /* Advance the cursor to the start of the next character */
        x += a->c[ch].ax;
        y += a->c[ch].ay;

        /* Skip glyphs that have no pixels */
        if (!w || !h)
        {
            continue;
        }

        Strings[nstrings * 42 + 0]  = x2;
        Strings[nstrings * 42 + 1]  = -y2;
        Strings[nstrings * 42 + 2]  = a->c[ch].tx;
        Strings[nstrings * 42 + 3]  = a->c[ch].ty;
        Strings[nstrings * 42 + 7]  = x2 + w;
        Strings[nstrings * 42 + 8]  = -y2;
        Strings[nstrings * 42 + 9]  = a->c[ch].tx + a->c[ch].bw / a->w;
        Strings[nstrings * 42 + 10] = a->c[ch].ty;
        Strings[nstrings * 42 + 14] = x2;
        Strings[nstrings * 42 + 15] = -y2 - h;
        Strings[nstrings * 42 + 16] = a->c[ch].tx;
        Strings[nstrings * 42 + 17] = a->c[ch].ty + a->c[ch].bh / a->h;
        Strings[nstrings * 42 + 21] = x2 + w;
        Strings[nstrings * 42 + 22] = -y2;
        Strings[nstrings * 42 + 23] = a->c[ch].tx + a->c[ch].bw / a->w;
        Strings[nstrings * 42 + 24] = a->c[ch].ty;
        Strings[nstrings * 42 + 28] = x2;
        Strings[nstrings * 42 + 29] = -y2 - h;
        Strings[nstrings * 42 + 30] = a->c[ch].tx;
        Strings[nstrings * 42 + 31] = a->c[ch].ty + a->c[ch].bh / a->h;
        Strings[nstrings * 42 + 35] = x2 + w;
        Strings[nstrings * 42 + 36] = -y2 - h;
        Strings[nstrings * 42 + 37] = a->c[ch].tx + a->c[ch].bw / a->w;
        Strings[nstrings * 42 + 38] = a->c[ch].ty + a->c[ch].bh / a->h;

        for (j=4; j<=39; j+=7)
        {
            Strings[nstrings * 42 + j + 0] = r;
            Strings[nstrings * 42 + j + 1] = g;
            Strings[nstrings * 42 + j + 2] = b;
        }
    }
```

where the array `text` is the string of characters to be rendered. As each character is stored as a set of ordered textures, the ASCII value of each character in the string is used as an index to locate the relative pixel coordinates (`x2`, `y2`) and the texture coordinates (`tx`, `ty`) plus the red, green and blue components for each of the six vertices of the two triangles representing the textured rectangle of the character. As with the vectors and textures, these values are written to the cache, in this case, as a texture. When it is necessary to flush the cache, the stored character textures are rendered with the following code fragment:

```
case STRINGS:
    if (nstrings > 0)
    {
        glUseProgram(StringProgram);
```

```
          glBindVertexArray(StringVAO);
          glUniformMatrix4fv(stloc, 1, GL_TRUE, &Transform[0]);
          glBindTexture(GL_TEXTURE_2D, CurrentFont->tex);
          glUniform1i(uniform_tex, 0);
          glBindBuffer(GL_ARRAY_BUFFER, StringVBO);
          glBufferData(GL_ARRAY_BUFFER, nstrings * 42 * sizeof(float), Strings, GL_DYNAMIC_DRAW);
          glDrawArrays(GL_TRIANGLES, 0, nstrings * 6);
          nstrings = 0;
      }
    break;
```

The GPU program **StringProgram** is activated, the VAO for strings is bound and the uniform transformation matrix **Transform** located at **stloc** is transferred to the GPU. The texture of the character is bound to give the reference to the font texture and the texture mode. The VBO is then bound and the texture data for the set of string textures is passed as pairs of triangles.

The shaders for strings are similar to the shaders used for textures. The vertex shader is as follows:

```
#version 430 core

#define width  1920.0f
#define height 1080.0f

layout (location = 0) in vec2 pos;
layout (location = 1) in vec2 texcoord;
layout (location = 2) in vec3 stringcolor;

out vec2 texpos;
out vec4 color;

vec4 position = vec4(pos, 0.0f, 1.0f);

mat4 ProjectMatrix =
{
    {2.0f/width,       0.0f,    0.0f,    0.0f}, // column-row ordering
    {     0.0f, 2.0f/height,    0.0f,    0.0f},
    {     0.0f,       0.0f,   -1.0f,    0.0f},
    {    -1.0f,      -1.0f,    0.0f,    1.0f}
};

uniform mat4 TransformationMatrix;  // translation and rotation

void main(void)
{
    gl_Position = ProjectMatrix * TransformationMatrix * position;
    texpos = texcoord;
    color = vec4 (stringcolor, 1.0f);
}
```

In the vertex shader, the colour of characters of the string (**stringcolor**) is an additional input and is passed to the fragment shader, which is given by the following code:

```
#version 430 core

in vec2 texpos;
uniform sampler2D tex;
```

```
in vec4 color;

out vec4 gl_FragColor;

void main(void)
{
    gl_FragColor = vec4(1.0f, 1.0f, 1.0f, texture2D(tex, texpos).a) * color;
}
```

The resultant pixel colour `gl_FragColor` is computed from the texture value and the string colour.

This method of representing fonts as an atlas of textures is very powerful. The fonts can be based on published fonts and there is no reasonable limit to the number of active fonts used by an application. The rendering of the fonts as textures by the GPU is extremely fast and the translation and rotation of fonts (e.g. the characters around the circumference of a compass card), only requires setting the relevant values in the transformation matrix. The colour of the character set is also passed with the texture coordinates. For the developer, the library *glib* provides the following functions for the rendering of strings:

```
void Glib_LoadFont(char filename[], unsigned int fontnumber, unsigned int height);
void Glib_SetFont(unsigned int font);
void Glib_Char(char, int x, int y);
void Glib_Chars(char str[], int x, int y);
int  Glib_StringSize(char str[]);
```

`Glib_LoadFont` is called at the start of a program to load the required fonts, associates a user-defined number with each font and defines the font height in pixels. `Glib_SetFont` enables a user to switch between fonts, to select a font as the currently active font. `Glib_Chars` renders a string of characters at the screen coordinate (**x**, **y**). Similarly, `Glib_Char` renders a single character. As the width of characters varies, `Glib_StringSize` returns the width of the string **str** in pixels, which is particularly useful in aligning a string inside a box or centring a string. In addition, `Glib_Colour` can be called to set the colour of rendered strings.

6.2.5 Matrix Transformations

In OpenGL version 4.0, the user is responsible for the matrix transformations applied to dots, vectors, triangles and textures. The actual transformations are implemented by the GPUs and, consequently, the transformation matrix is shared between the host computer and the GPUs. A copy of the 4×4 transformation matrix is declared and maintained in the host computer and transferred to the GPU, by a `glUniformMatrix4v` call. The *cglm* library, described in Section 6.3.2, is used to implement the matrix operations. The code to reset the transformation matrix to the identity matrix is as follows:

```
void Glib_LoadIdentity()
{
    Glib_Flush();
    glm_mat4_identity(Transform);
}
```

`Glib_LoadIdentidy` is called at the start of each frame and can be used to reset the transformation matrix at any time, which is declared in *glib* as follows:

```
static mat4          Transform;
```

As this function resets the transformation matrix, the cache is flushed. Note that 4×4 matrices are used in the shader programs, even for 2D graphics operations. The other important point is that matrices are stored as linear arrays, in column–row order. The 2D translation function is given by

```
void Glib_Translate(float x, float y)
{
    vec4 t = { x, y, 0.0, 0.0 };

    Glib_Flush();
    glm_translate(Transform, t);
}
```

Similarly, the 2D rotation function is given by the following code:

```
void Glib_Rotate(float a)
{
    Glib_Flush();
    glm_rotate(Transform, a / ONERAD, (vec4) { 0.0, 0.0, 1.0, 0.0 } );
}
```

where **a** is the angle of rotation in degrees, **ONERAD** is defined as $180.0/\pi$ and the 2D rotation is a 3D rotation about the *Z* axis.

Two further matrix functions are provided to save and restore the state of the transformation. These functions are equivalent to the *glPushMatrix* and *glPopMatrix* functions provided in legacy OpenGL, enabling the transformation state to be saved on, and restored from, a dedicated stack. The functions are as follows:

```
void Glib_PushMatrix()
{
    if (stackp >= MAXSTACK)
    {
        printf("PushMatrix overflow\n");
        exit(-1);
    }

    glm_mat4_copy(Transform, stack[stackp]);
    stackp += 1;
}
```

```
void Glib_PopMatrix()
{
    if (stackp <= 0)
    {
        printf("PopMatrix underflow\n");
        exit(-1);
    }
```

```
    Glib_Flush();
    stackp -= 1;
    glm_mat4_copy(stack[stackp], Transform);
}
```

Note that the cache is not flushed in PushMatrix as the transformation matrix is copied to the stack, whereas the cache is flushed in **Glib_PopMatrix** because the transformation matrix is restored from the stack. If stack underflow or overflow is detected, the fault is logged and the program is terminated as this event is a serious failure.

Depending on the simulator, a display can be orientated in either portrait or landscape mode. Assuming the software in *glib* is written for landscape mode with the origin $(0, 0)$ in the bottom left-hand corner, then by adding a translation immediately after setting the identity matrix, the complete display can be rotated through $90°$, which is far simpler than providing two versions of *glib*. The following code in glib modifies the orientation of the complete display for portrait mode:

```
void Glib_LoadIdentity()
{
    Glib_Flush();
    glm_mat4_identity(Transform);
    #if (PORTRAIT)
        Glib_Translate(w, 0.0);
        Glib_Rotate(90.0);
    #endif
}
```

where **w** is the height of the display in portrait mode and the Boolean constant **PORTRAIT** defines the portrait or landscape mode. Similarly, scaling can be applied to the complete display to adjust for variations in the display dimensions.

6.2.6 Summary of glib Functions

The complete list of functions developed for the *glib* library, to display characters, strings, lines, dots, triangles and textures, and the 2D matrix transformations are given in Table 6.1.

These functions are specific to the graphics operations needed for 2D rendering of aircraft displays and are optimised to make effective use of GPUs. In addition, six functions are provided to support the rendering of 3D objects in flight displays, particularly for the attitude indicator and the magnetic compass used in the simulation of aircraft instruments.

6.3 Graphics Libraries

Several public-domain libraries are used to support the development of real-time graphics for flight simulation. There is not sufficient space to describe these libraries in detail, but their main functions are summarised to provide some insight into their capability and conventions. These libraries are available as open-source code and are free to download from the various organisations.

Table 6.1 Summary of *glib* Functions.

Function	Action
`void Glib_Char(char ch, int x, int y);`	The character **ch** is drawn at $(\mathbf{x},\mathbf{y})$
`void Glib_Chars(char str[], int x, int y);`	The string **str** is drawn at $(\mathbf{x},\mathbf{y})$
`int Glib_StringSize(char str[]);`	Returns the size of the string **str** in pixels
`void Glib_Draw(int x1, int y1, int x2, int y2);`	A line is drawn from $(\mathbf{x1},\mathbf{y1})$ to $(\mathbf{x2},\mathbf{y2})$
`void Glib_Dot(int x, int y);`	A dot is drawn at $(\mathbf{x},\mathbf{y})$
`void Glib_Rectangle(int x1, int y1,` `                 int xside, int yside);`	A filled rectangle **xside** × **yside** is drawn with the bottom left-hand corner at $(\mathbf{x1},\mathbf{y1})$

`void Glib_Triangle(int x1, int y1,`
`                 int x2, int y2,`
`                 int x3, int y3);`
	A filled triangle is drawn with vertices at $(\mathbf{x1},\mathbf{y1})$, $(\mathbf{x2},\mathbf{y2})$ and $(\mathbf{x3},\mathbf{y3})$

`void Glib_DrawTexture(int x0, int y0,`
`                 int sx0, int sy0,`
`                 float tx1, float ty1,`
`                 float tx2, float ty2,`
`                 float Alpha);`
	A texture **sx0** x **sy0** is drawn at $(\mathbf{x0},\mathbf{y0})$. The bottom left corner and top right corner are $(\mathbf{tx1},\mathbf{ty1})$ and $(\mathbf{tx2},\mathbf{ty2})$. **Alpha** is transparency

`void Glib_DrawTextureRotated(int x0, int y0,`
`                 int sx0, int sy0,`
`                 float tx1, float ty1,`
`                 float tx2, float ty2,`
`                 float R, float Alpha);`
	A texture **sx0** × **sy0** is drawn at $(\mathbf{x0},\mathbf{y0})$. The bottom left corner and top right corner are $(\mathbf{tx1},\mathbf{ty1})$ and $(\mathbf{tx2},\mathbf{ty2})$. The texture is rotated by the angle **R** (degrees). **Alpha** is transparency

`void Glib_Line_Strip(int x, int y,` `                 unsigned int n, int v[]);`	An OpenGL line strip **v** of **n** vertices is drawn at $(\mathbf{x},\mathbf{y})$
`void Glib_Line_Loop(int x, int y,` `                 unsigned int n, int v[]);`	An OpenGL line loop **v** of **n** vertices is drawn at $(\mathbf{x},\mathbf{y})$
`void Glib_DrawLines(int x, int y, int n,` `                 int v[]);`	OpenGL lines in the array **v** of **n** vertices are drawn at $(\mathbf{x},\mathbf{y})$
`void Glib_Colour(unsigned int col);`	The current colour is set to **col**
`void Glib_LineWidth(float w);`	The current line width is set to **w**
`void Glib_SetFont(unsigned int font, int spacing);`	The current font is set to **font** (**spacing** is not used)
`void Glib_AntiAliasing(bool Mode);`	Anti-aliasing is turned on or off
`void Glib_ClipWindow(int x, int y, int xs, int ys);`	A clipping window **xs** × **ys** is positioned with the bottom left-hand corner at $(\mathbf{x},\mathbf{y})$ and enabled
`void Glib_RemoveClipWindow(void);`	The current clipping window is disabled
`void Glib_Flush();`	Any pending graphics are flushed
`void Glib_Init();`	Initialisation of the library, called before any graphics calls
`void Glib_LoadIdentity();`	The transformation matrix is set to the identity matrix

Table 6.1 (Continued)

Function	Action
`void Glib_Rotate(float a);`	Rotation by **a** degrees
`void Glib_Translate(float x, float y);`	Translation by **x** and **y** pixels
`void Glib_Close(void);`	Closing the library on exit

`void Glib_LoadFont(char filename[],`
`                    unsigned int fontnumber,`
`                    unsigned int fontheight);`
	The font file **filename** is loaded and assigned the number **fontnumber** and height **fontheight** in pixels

`void Glib_LoadTexture(char filename[],`
`                       unsigned int`
`                       texturenumber);`
	The texture file **filename** is loaded and assigned the texture number **texturenumber**

`void Glib_SetTexture(unsigned int t);`	The current texture is set to **t**
`bool Glib_Errors();`	Any errors are written to *stdout*
`void Glib_Info();`	Information about the graphic library is written to *stdout*
`void Glib_PushMatrix();`	The transformation matrix is pushed onto the *glib* stack
`void Glib_PopMatrix();`	The transformation matrix is popped from the *glib* stack

Table 6.2 Additional *glib* Functions.

`void Glib_SetNormal();`	The normal matrix is computed for use in lighting applications

`void Glib_DrawSegment(`
`   GLfloat px, GLfloat py, GLfloat pz,`
`   GLfloat nx, GLfloat ny, GLfloat nz,`
`   GLfloat u, GLfloat v);`
	A segment of a sphere given by position coordinates (px, py, pz) and normal coordinates (nx, ny, nz) is rendered with texture coordinates (u, v)

`bool Glib_RotatePitch3D(float p);`	A rotation about the z axis by an angle **p**°
`bool Glib_RotateRoll3D(float r);`	A rotation about the x axis by an angle **r**°
`bool Glib_RotateYaw3D(float q);`	A rotation about the y axis by an angle **q**°

`void Glib_Lighting(vec3 pos, vec3`
`                    intensity,`
`                    vec3 Ka, vec3`
`                    Kd, vec3 Ks,`
`                    float shine);`
	The light source is set to the position, intensity and the ambient, specular, diffuse and shininess parameters

6.3.1 GLFW

For a long time, the OpenGL Utility Toolkit (GLUT) (Kilgard, 1996) provided an API for OpenGL as a platform-independent library. Although GLUT has been superseded by *freeglut* (Anon, 2013), a compatible package, *freeglut* retains several of the limitations of GLUT. However, GLUT has not been maintained in recent years and GLFW provides a clearer perspective of the real-time loop for graphics applications. Early editions of the OpenGL Programming Guide gave examples in GLUT, but more recent editions have been updated for GLFW. All examples in this book are based on GLFW for Windows and Linux implementations.

The general structure of a GLFW program is given by the following pseudo-code:

```
set up GLFW call backs
initialise GLFW
create a GLFW window
set up the GLFW state
initialise OpenGL
while (GLFW is not terminated)
{
    update non-graphics software
    update OpenGL graphics for the current frame
    swap the frame buffer
    check for mouse or keyboard input
}
close OpenGL
close the window
close GLFW
```

GLFW provides a set of call-backs, particularly to trap errors and user input. For example, to access the mouse cursor position in a graphics application, user code is called directly by GLFW if movement of the mouse is detected. GLFW allows the user to define a dedicated OpenGL window, including the window resolution and how the frame-store is set up and refreshed, including single and double buffering of the frame-store. The example in Section 6.1.3 illustrates initialisation of the GLFW environment for a real-time application, managing the real-time rendering and closing down the GLFW environment.

6.3.2   cglm

The *cglm* library is a mathematics library, developed at MIT, for matrix operations and is particularly suited to 2D and 3D graphics and can be downloaded from https://github.com/recp/cglm. The library is similar to the *glm* library but is written in C. The functions of the library shown in Table 6.3 are used in the *glib* graphics library for 2D transformations, which support the shader graphics operations.

Note that *cglm* is restricted to single-precision (32-bit) floating-point arithmetic. Two versions of *cglm* are provided, a compiled version that can be linked with a user application and a version using in-line macro substitution. The latter version is used with the *glib* library. The *cglm* library is stable and extensive and, subject to these constraints, users developing matrix operations would be well advised to consider the facilities afforded by *cglm*.

6.3.3   PngLib

The public-domain library *libpng*, which provides numerous functions for applications using *png* graphics files, can be downloaded from http://www.libpng.org/pub/png/libpng.html. The code in *pnglib*.c uses the OpenGL functions **glPixelStorei** and **glReadPixels** to access the image, which is then packed and organised to comply with the PNG ISO Standard (ISO/IEC 15948:2003 (E)), which is available at https://www.w3.org/TR/2003/REC-PNG-20031110 (a W3C

Table 6.3 *cglm* Library Functions.

Function	Action
void glm_mat4_copy(mat4 *mat*, mat4 *dest*)	The matrix **mat** is copied to the matrix **dest**
void glm_mat4_identity(mat4 *mat*)	The matrix **mat** is set to the identity matrix
void glm_translate(mat4 *m*, vec3 *v*)	The transformation matrix **m** is set to include the translations defined in the vector **v**
void glm_rotate(mat4 *m*, float *angle*, vec3 *axis*)	The transformation matrix **m** is set to include the rotations defined in the vector **v**
void glm_mat4_print(mat4 *mat*, FILE * *fstream*)	The matrix **mat** is written to the file stream **fstream**
void glm_mat4_inv(mat4 *mat*, mat4 *dest*)	The matrix **mat** is inverted and copied to the matrix **dest**
void glm_mat4_transpose(mat4 *mat*)	The matrix **mat** is set its transpose
void glm_mat4_pick3(mat4 *m*, mat3 *dest*)	The upper left matrix of **m** is copied to the matrix **dest**
void glm_rotate(mat4 *m*, float *a*, vec3 *v*)	The rotations defined by the vector **v** and the angle **a** are applied to the matrix **m**

recommendation). The image is written as a binary file, which can be imported into other applications. The library *glib* provides the following function to capture a region of the screen and generate a corresponding *png* file:

```
int PngLib_SavePngFile(char FileName[], int x, int y, int width, int height)
```

where **Filename** is the name of the *png* file to be created, (**x**, **y**) are the screen coordinates of the lower left-hand corner of the rectangle to be captured, and **width** and **height** define the size of the screen region in pixels. The file is written by the *libpng* function **png_write_image(png,rows)**, where **png** is the structure set up to hold the bitmap and **rows** is an array of rows of pixels formed by *libpng*. The source code of *pnglib* is provided in the companion software, with a link in the appendix. The main application is to take a copy of flight displays or the instructor station display as a *png* file for use in documentation and reports.

6.3.4 FreeImage

FreeImage is a library to manage the reading of graphics images in various formats and can be downloaded from https://freeimage.sourceforge.io. The *FreeImage* library functions are only used within *glib*. The user needs no knowledge of *FreeImage* and is provided with functions to load an image as a texture, to select a specific texture and to render the current texture. The advantage of using the *FreeImage* library is that it can load textures in many different formats and that only a few basic calls are needed to load and access a texture. Images generated by photography or produced with graphics editing tools are converted to an internal bitmap format that is compatible with the texture rendering functions provided in OpenGL. In particular, *FreeImage* makes rigorous checks of the graphics format and enables any number of textures to be used in an application.

The *glib* function `Glib_LoadTexture` loads a texture file:

```
void Glib_LoadTexture(char tname[], unsigned int texturenumber)
{
    unsigned int width;
    unsigned int height;
    FIBITMAP *bitmap;
    BYTE *data;
    FREE_IMAGE_FORMAT fif;

    glGenTextures(1, &TextureTable[texturenumber]);
    glBindTexture(GL_TEXTURE_2D, TextureTable[texturenumber]);

    // Get the format of the image file
    fif = FreeImage_GetFileType(tname, 0);

    // Load the data in bitmap if possible
    if (fif != FIF_UNKNOWN && FreeImage_FIFSupportsReading(fif))
    {
        bitmap = FreeImage_Load(fif, tname, 0);
    }
    else
    {
        printf("Unable to load the image file %s\n", tname);
        exit(-1);
    }

    width = FreeImage_GetWidth(bitmap);
    height = FreeImage_GetHeight(bitmap);
    data = (BYTE*) FreeImage_GetBits(bitmap);

    glPixelStorei(GL_UNPACK_ALIGNMENT, 1);
    glTexParameteri(GL_TEXTURE_2D, GL_TEXTURE_WRAP_S, GL_CLAMP);
    glTexParameteri(GL_TEXTURE_2D, GL_TEXTURE_WRAP_T, GL_CLAMP);
    glTexParameteri(GL_TEXTURE_2D, GL_TEXTURE_MIN_FILTER, GL_LINEAR);
    glTexParameteri(GL_TEXTURE_2D, GL_TEXTURE_MAG_FILTER, GL_LINEAR);

    glTexImage2D(GL_TEXTURE_2D, 0, 4, width, height, 0, GL_BGRA, GL_UNSIGNED_BYTE, (GLvoid*) data);
    glGenerateMipmap(GL_TEXTURE_2D);

    FreeImage_Unload(bitmap);
}
```

The name of the image file **tname** is passed to **LoadTextureImage** and **texturenumber** provides a user-defined number to associate with the texture, which is stored in the array **TextureTable**. The function **FreeImage_GetFileType** checks that it can access the specific graphics format and, if it is supported, the function **FreeImage_Load** loads the file into memory, returning a pointer to the bitmap holding the image. Otherwise an error is logged and the program terminates. The width and height of the file are set by **FreeImage_GetWidth** and **FreeImage_GetHeight**, respectively, and a pointer to the data area of the image is returned to the pointer **data** by **FreeImage_GetBits**. **glPixelStorei** is used to define how the image is stored, particularly the word orientation and byte alignment. The value of 1 implies that bytes are consecutive, in the order they are read. Note that texture coordinates are referred to as *s*, *t*, *p* and *q*. **glTexParameteri** specifies that the texture is a 2D texture, to be clamped at the *s* and *t* edges, and that a linear filter is applied. **glTexImage2D** specifies the 2D texture image, a level-of-detail of 0 (base level), four colour components, the width and height of the texture, a border of 0, the colour format, the type of storage and a pointer to the image data in memory. Finally, **glGenerateMipmap** generates the mipmap, in this case for a 2D texture. At this point, the bitmap is no longer needed and is unloaded.

6.3.5 FreeType

FreeType is a library for rendering fonts used for the aircraft displays and the instructor station and can be downloaded from https://www.freetype.org. *FreeType*, which is written in C, produces high-quality fonts based on glyph images. The main use of the library is the production of character atlases, which, once constructed, can be used as textures to render text strings. The advantages of using *FreeType* are the fast rendering speed and that it can render the true-type fonts provided for Windows and Linux platforms. The characters are rendered as textures and exploit the fast texture rendering of GPUs, particularly the translation and rotation of textures.

The *FreeType* library functions are only used within *glib*. The user needs no knowledge of *FreeType* and is provided with two functions: to load a font and to select a specific font as the current font for the rendering of strings. An internal structure is used to define an atlas for each font. `Glib_LoadFont` is called to load a true-type font file, where the font height is defined in pixels. The user-defined number allocated to each loaded font enables fonts to be switched to render strings in different font faces or sizes. The two functions provided in *glib* are as follows:

```
void Glib_LoadFont(char filename[], unsigned int fontnumber, unsigned int height)
void Glib_SetFont(unsigned int fontnumber)
```

The details of `Glib_LoadFont`, which generates a font atlas using the *FreeType* faces and provides straightforward access to the textures formed for each font character, are described in Section 6.1.

6.3.6 Compiling, Linking and Loading Shader Programs

The C program shader.c is based on sample code provided by the Kronos Group and contains the function `LoadShaders` to compile, link and load vertex and fragment shader programs written in the OpenGL Shading Language (glsl). `glCreateShader` creates a data structure for a shader program, which is read into an array of characters, as raw text. The code is compiled by `glCompileShader` and linked by `glLinkProgram`, which apply extensive checks to the shader code. If the compilation and linking are successful, the value returned by `LoadShaders` identifies a specific shader program, which can be referenced by `glUseProgram` to activate vertex and fragment shaders, to access uniform variables in shader programs and in passing data to the GPU using `glDrawArrays`. In *glib*, the initialisation of the vectors, triangles, textures and fonts includes the compilation and linking by `LoadShaders`. The code in `shader.c` separates management of the shader programs from the run-time access to the shaders in *glib*. Different vertex and fragment shaders are used for triangles, textures, text and 3D objects in *glib*, which are loaded, linked and compiled during the initialisation of these graphic types.

6.3.7 SVG

There are several ways to produce textures needed for aircraft displays. Packages such as Photoshop, Blender and Inkscape enable shapes, colours, textures and images to be captured and edited using graphics tools on workstations. Packages such as Inkscape enable textured images to be captured as vector graphics and exported in standard graphics formats, such as *png* files. The advantages of these packages is the freedom to design images and the user interface, which provides a direct view of the image.

One other option is to render images at run-time, but this poses two problems. Firstly, vectors and triangles are rendered as pixels, possibly introducing aliasing problems and, secondly, rendering a static texture image is a single operation, whereas rendering the image as vectors, triangles and texture may require numerous graphics operations and is wasteful of processing. An alternative is to render images off-line as textures, using basic drawing primitives to produce vector graphics rather than pixel images.

SVG (Scalable Vector Graphics) is an XML-based markup language to describe 2D vector graphics and is used widely in web applications. A graphics library was developed, based loosely on the functions in the library *glib*, used for aircraft displays, described in Section 6.2. Rather than generating OpenGL graphics, the library functions generate XML primitives, which can be viewed with packages such as Inkscape and converted to a high-resolution *png* texture. Because SVG is text-based, it is straightforward to modify the code to generate textures using a conventional text editor.

As an illustration, consider a simple image which contains a transparent background with a green rectangle, overlaid with a blue circle containing white text. The following program generates an XML file which can be viewed in a browser or in Inkscape:

```
int main(int argc, char *argv[])
{
    int    j;
    float x = 200.0;
    float y = 200.0;

    background(1000, 1000);

    colour(GREEN);
    rect(200, 200, 400, 300);
    colour(BLUE);
    circle(350, 350, 100);
    colour(WHITE);
    setfont("arial.ttf", 16);
    text(300, 320, "Test string");

    printf("</svg>\n");
}
```

For example, the code for the function **rect** is the equivalent of the following macro substitution:

```
void rect(float x, float y, float xs, float ys)
{
    y = YMAX - y - ys;

    printf("<rect fill=\"%s\" stroke-width=\"%d\" stroke=\"%s\" x=\"%f\" y=\"%f\"
           width=\"%f\" height=\"%f\"/>\n",
       Colours[CurrentColour], CurrentLineWidth, Colours[CurrentColour], x, y, xs, ys);
}
```

The corresponding XML file generated by the C program is as follows:

```
<svg version="1.1"
  width="1000" height="1000"
  xmlns="http://www.w3.org/2000/svg">
<rect fill="green" stroke-width="1" stroke="green" x="200.000000" y="500.000000"
    width="400.000000" height="300.000000"/>
<circle fill="blue" stroke-width="1" stroke="blue" cx="350.000000" cy="650.000000"
    r="100.000000"/>
<style> font: arial.ttf </style>
<text x="300.000000" y="680.000000" font-family="B612" style="fill: white; stroke: white;
    font-size: 16px;">Test string</text>
</svg>
```

This SVG file is exported as a *png* file by Inkscape, with a resolution of 1000 × 1000 pixels at 600 dpi, as shown in Figure 6.9. Note the smooth edge to the circle and the clear font characters in the Arial font, point size 16.

In the case of the Boeing 747-400 EFIS displays, several objects are generated on a transparent background (shown in light grey to highlight the components) 1000 × 1000 pixels, to provide both static and dynamic components of the PFD, as shown in Figure 6.10.

The static part of the attitude indicator includes black fillets to render the curved corners against a black background. The pitch lines are translated and rotated in response to the pitch and roll attitude of the aircraft. It would be very difficult, and time-consuming for the EICAS gauge lines and text to be

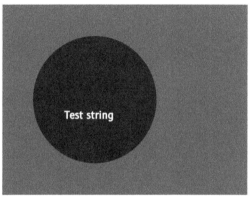

Figure 6.9 An Image Generated Using SVG.

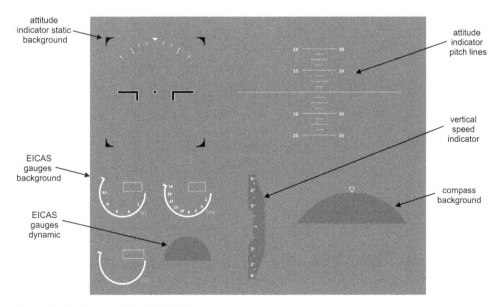

Figure 6.10 Boeing 747-400 EFIS Components.

rendered so sharply using conventional graphics and similarly for the compass background. The small grey semicircle is rotated for each gauge to indicate the particular engine settings. The positions of the components in the texture map are defined in the C program, making it straight-forward to compute the relative component texture coordinates to render the components as textures.

6.4 Design Considerations

6.4.1 Absolute and Relative Rendering

Consider an aircraft EFIS display, shown in schematic form in Figure 6.11, depicting an airspeed indicator, an attitude indicator, an altimeter and a compass display.

Typically, the screen origin is at (0, 0) and the top right-hand corner is at (*width*-1, *height*-1), where *width* is the screen width and *height* is the screen height in pixels, respectively. It would be possible to reference all objects in the display to the screen origin. There are two problems with this convention. Firstly, every graphic reference to an object would need to add the origin of the object as an offset. Secondly, any rotation must be translated to the screen origin, rotated, then translated back to the local origin. An alternative is to use local coordinates. Once a translation has been made to the local origin, all graphics operations are relative to that origin, avoiding the need to compute any offsets and simplifying rotations. If an object within a display is subsequently relo-cated, this can be achieved with a single translation. With relative drawing, all the elements of an object must be referenced to a local coordinate frame, for example, the centre of the current object.

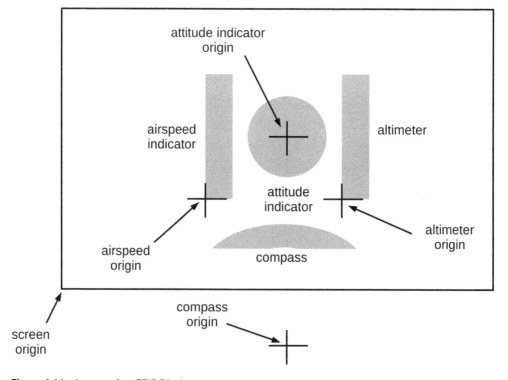

Figure 6.11   Layout of an EFIS Display.

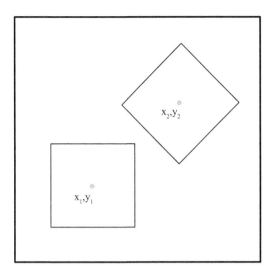

Figure 6.12   Coordinate Frames.

In legacy OpenGL, the *PushMatrix* and *PopMatrix* operations ensured that the system transformation matrix could be reset on completion of rendering an object to its state at the start of a rendering sequence. With modern OpenGL, the user is responsible for loading, reloading and setting values in the transformation matrix, which is passed to the GPU. However, the use of these primitives is particularly powerful in applications where displays are rendered in a local coordinate frame; they enable the rendering to be computed in the new coordinate frame before restoring to the previous coordinate frame.

As an example of using a local coordinate frame, Figure 6.12 shows a display of two squares of side p, one centred at (x_1, y_1) and the other centred at (x_2, y_2) and rotated through 45°.

In absolute coordinates, assuming the transform matrix is set to the identity matrix, the code would take the following form:

```
draw(x1-p/2, y1-p/2, x1-p/2, y1+p/2)
draw(x1-p/2, y1+p/2, x1+p/2, y1+p/2)
draw(x1+p/2, y1+p/2, x1+p/2, y1-p/2)
draw(x1+p/2, y1-p/2, x1-p/2, y1-p/2)
draw(x1-cos(45)*p/2, y1-sin(45)*p/2, x1-cos(45)*p/2, y1+sin(45)*p/2)
draw(x1-cos(45)*p/2, y1+sin(45)*p/2, x1+cos(45)*p/2, y1+sin(45)*p/2)
draw(x1+cos(45)*p/2, y1+sin(45)*p/2, x1+cos(45)*p/2, y1-sin(45)*p/2)
draw(x1+cos(45)*p/2, y1-sin(45)*p/2, x1-cos(45)*p/2, y1-sin(45)*p/2)
```

Note the computations required by the client side computer to pass the coordinates to the GPU. A similar version using relative coordinates is shown as follows:

```
translate(x1, y1)
drawsquare(-p/2, -p/2, p/2, p/2)
translate(x2-x1, y2-y1)
rotate(45)
drawsquare(-p/2, -p/2, p/2, p/2)

void drawsquare(int x)
```

```
{
    draw(-x,  -x,  -x,   x)
    draw(-x,   x,   x,   x)
    draw( x,   x,   x,  -x)
    draw( x,  -x,  -x,  -x)
}
```

where **p** is the length of the sides of the square. The computations inherent in translation and rotation would be performed by the GPU. Relocation of the display elements is simply modification of the translation parameters.

As PFD and NFD displays contain elements requiring translation and rotation, it is arguably better to use local coordinates, particularly to avoid the need to translate objects to the origin prior to rotation and apply the inverse translation after rotation. In the case of the instructor station, which is normally rendered in landscape mode, there is minimal translation or rotation and it is more convenient to use absolute coordinates. Note that setting the transformation matrix to the identity matrix is effectively setting absolute coordinates as the origin is set to $(0, 0)$ and the rotation about the z axis (for 2D graphics) is zero. However, care is needed with the physical location of the screen origin. In some systems it is at the bottom left-hand corner, while in other systems, it may be at the centre of the screen.

6.4.2   Memory Organisation

Graphics items such as vectors and textures are stored in the memory of the host computer and transferred to the memory of the GPU. However, not all computers address memory in the same way. Consider a 32-bit floating-point number stored as four bytes on two 32-bit computers, with different endian addressing, as shown in Figure 6.13.

In computers with little endian addressing, the least significant byte is the zeroth byte, whereas in big endian addressing, the zeroth byte is the most significant byte. The ordering of bytes within words also applies in networks, where blocks of data are transferred between computers; sequential copying of the data can result in a change of byte ordering at the destination computer and corruption of the data. In OpenGL, a buffer is copied from host memory to the GPU memory in the byte order 0, 1, 2, 3 ... and care is needed to ensure the data is transferred correctly.

The other consideration of memory addressing in OpenGL is the storage of matrices, where the ordering of the matrix elements can be row-major or column-major. Consider a 4×4 matrix defined as follows:

$$\begin{bmatrix} a_{11} & a_{12} & a_{13} & a_{14} \\ a_{21} & a_{22} & a_{23} & a_{24} \\ a_{31} & a_{32} & a_{33} & a_{34} \\ a_{41} & a_{42} & a_{43} & a_{44} \end{bmatrix}$$

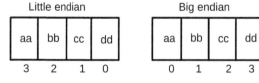

Figure 6.13   Endian Numbering.

The matrix is passed to the GPU as a linear array of 16 values. In row-major order, the data sequence is $a_{11}, a_{12}, a_{13}, a_{14}, a_{21}, a_{22}, a_{23}, a_{24}, a_{31}, a_{32}, a_{33}, a_{34}, a_{41}, a_{42}, a_{43}, a_{44}$, whereas in column-major order the sequence is $a_{11}, a_{21}, a_{31}, a_{41}, a_{12}, a_{22}, a_{32}, a_{42}, a_{13}, a_{23}, a_{33}, a_{43}, a_{14}, a_{24}, a_{34}, a_{44}$. Typically, shared data, for example a transformation matrix, is passed to the GPU using the `glUniformMatrix4fv` function, where the third parameter is `GL_TRUE` if the order is column-major and `GL_FALSE` if the order is row-major. Note that matrices defined in *cglm* use column-major ordering.

6.5 EFIS Displays

Electronic displays were introduced in the Boeing 757 and 767 aircraft and the Airbus A310 as early as 1978. These displays were based on CRT technology but the full 'glass cockpit', based on LCD technology with several flat-screen displays, arrived with the Boeing 747-400 and Airbus A340 aircraft. The early CRT and LCD displays were 8-inch-square displays (Spitzer, 1987). The Airbus A380 has eight 6 × 8 inch displays and the Boeing 787 has five 9 × 12 inch displays. Electronic Flight Instrument System (EFIS) displays include the PFD, the NFD and EICAS displays (Dawson, 1992). In modern commercial aircraft, these displays are LCD panels, with the PFD and NFD displays located directly in front of both pilots and the EICAS display in the centre of the console (Pallett, 1992). Figure 6.14 shows a typical EFIS display from a research flight simulator.

The PFD, which is based loosely on a Boeing 747-400 (Koniche, 1988), is positioned in the top left of the display. The EICAS display is directly below the PFD and is based on Boeing 767 engine displays. The NFD, which is also based on a Boeing 747-400, is positioned in the top right of the display. The FCU, which is modelled on an Airbus A340 FCU, is shown below the NFD. These particular displays run on a single screen with a display resolution of 1920 × 1080 pixels and are representative of the displays found in modern transport aircraft.

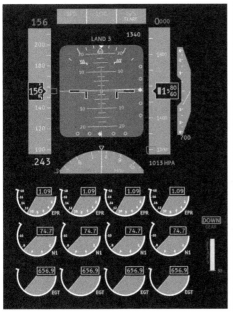

Figure 6.14 Emulation of EFIS Displays (see Plate 1).

The PFD comprises five separate displays. The airspeed is shown as a vertical strip to the left, indicating an airspeed of 156 Kt, a Mach speed of 0.243 and an autothrottle reference speed of 156 Kt. Similarly, the strip to the right indicates altitude, indicating an altitude of 1570 ft, a barometric pressure of 1013 mb and a reference altitude of 0 ft. The vertical speed indicator is to the right of the altitude strip, indicating a rate of descent of 800 ft/min. In the centre of the PFD, the attitude indicator displays pitch and roll and incorporates a glideslope indicator, a localiser indicator, a bank and sideslip indicator and magenta flight director bars. Below the attitude indicator, the compass shows the aircraft heading, currently 078°M, and a heading bug set to 045°M.

The EICAS displays are grouped together for the four engines, displaying EPR, RPM and EGT in three rows. The grey regions provide an indication of engine state with the actual engine parameters shown as a digital value. The NFD depends on the mode selection but displays aircraft heading and is often overlaid with a chart display of the area or an approach. In the mode shown, the ILS glide path is indicated as a magenta diamond symbol to the right of the display, with the localiser shown as a magenta bar. The display also shows selected navigation aid frequencies, wind speed and DME distance (4.3 nm from touchdown). In addition, as part of the validation process for this particular display, the scale in the bottom right shows a frame time of 6.3 ms occupied by the graphics rendering. There are several aspects of the design that need further consideration:

- The attitude indicator contains regions of blue and brown and reference pitch markers which rotate in pitch and roll and an angle of bank indicator, which rotates with bank angle.
- The digits indicated for airspeed and altitude rotate vertically, similar to tumbling digits on mechanical instruments.
- At least two different sizes of fonts are used.
- There is layered priority of displayed information.

6.5.1 Primary Flight Display

The displays shown in Figure 6.14 are produced using OpenGL for an off-the-shelf LED display and are reverse-engineered from photographs and drawings of the actual aircraft equipment. The mapping from physical measurements to pixel dimensions is critical, as it defines the size and layout of the displays as well as the size of fonts used for the displays. With the availability of a public-domain version of Airbus true-type fonts, textured fonts were used for this display. Textures could also be used for the sliding scales and the compass card, where the airspeed and altitude strips would need to cover the range 0–500 Kt and 0–50,000 ft, although only 120 Kt and 800 ft would be displayed at any time, respectively. Moreover, a texture for the rotating compass segment would require a very large texture. Consequently, the scales are rendered as vectors with textured characters.

One other aspect to consider with the PFD display is the use of clipping provided in OpenGL. The OpenGL function `glScissor(x,y,xs,ys)` ensures that the GPU only renders graphics in the screen region bounded by a rectangle of width `xs` and height `ys` with the left-hand bottom corner at (x,y). In other words, all graphics outside this window are discarded until either clipping is disabled by `glDisable(GL_SCISSOR_TEST)` or the clipping window is redefined. Clipping occurs in three regions of the PFD. For the airspeed and altitude digital displays, the characters rotate vertically, so that the only the parts of individual characters inside the white rectangular boxes are visible. By clipping the character strings, rows above and below the current character string can be written as conventional strings with the clipping removing the text as the digits rotate into or out of the box containing the digits. Similarly, for the compass segment, it would be particularly complicated to compute the parts of the compass card to be displayed. However, knowing that the segment extends by approximately ±40° from the current heading, the white marks and digits can be rotated and rendered but clipped to the left, right and lower bounding edges of the compass segment, also avoiding the need to rotate the grey segment.

To cover all the aircraft displays in detail would quadruple the length of this book. Consequently, the description is limited to the algorithms where the display of information is non-obvious. The airspeed indicator and the altimeter are very similar; they contain a sliding vertical strip, centred on the current value which is also displayed in a small window. For the airspeed indicator, the strip is marked in 10 Kt intervals with the airspeed digits every 20 Kt. As the marks repeat every 10 Kt, the offset of the strip marking is given by the airspeed remainder 10. The code fragment is as follows:

```
Offset = (Kts - (float) (KtsDigits / 10 * 10)) * 3.7;
s1 = MinSpeed;
s2 = MaxSpeed;
y  = -TapeHalfHeight;

do
{
    if (s1 >= 30)
    {
        Glib_Draw(AsiX + TapeWidth - 15, AsiY - Offset + y, AsiX + TapeWidth - 1, AsiY - Offset + y);
        if (s1 % 20 == 0)
        {
            Wrn3(s1, AsiX + 23, AsiY - Offset + y - 6);
        }
    }
    y  = y + TickSpacing;
    s1 = s1 + 10;
} while (!(s1 >= s2));
```

where **MinSpeed** is the current airspeed rounded to the nearest 10 Kt and **MaxSpeed** is given by **MinSpeed** plus 140. The offset corresponds to 3.7 pixels per Kt. For the airspeed indicator, the three digits are of equal size and roll over, going from 9 to 0 (increasing) or from 0 to 9 (decreasing) in the least significant digits. As the least significant digit changes from 9.0 back to 0.0 in increments of 0.1, the adjacent digit rolls, for example, from 3 to 4, in 10 increments, as the airspeed increases from 39 to 40 Kt. Similarly, the most significant digit rolls when the middle digit changes from 9 to 0 or 0 to 9. There is a slight complication with the altimeter; the two least significant digits change in 20 ft increments and, for clarity, the thousands digit is larger than the other digits. The code fragment for the centre rolling digit is as follows:

```
n = KtsDigits % 10;
if (n >= 9)
{
    Dig2Offset = (int) ((Kts - (float) (KtsDigits / 10 * 10 + 9)) * 40.0);
}
else
{
    Dig2Offset = 0;
}

if (KtsDigits >= 10)
{
    Glib_Char(KtsDigits % 100 / 10 + '0',        AsiX + 17, AsiY - 12 - Dig2Offset);
    Glib_Char((KtsDigits + 1) % 100 / 10 + '0', AsiX + 17, AsiY + 28 - Dig2Offset);
    Glib_Char((KtsDigits + 9) % 100 / 10 + '0', AsiX + 17, AsiY - 52 - Dig2Offset);
```

```
}
n = KtsDigits % 100;
if (n >= 99)
{
    Dig1Offset = (int) ((Kts - (float) (KtsDigits / 10 * 10 + 9)) * 40.0);
}
else
{
    Dig1Offset = 0;
}
```

where **n** is the least significant digit of airspeed. If **n** is 9, the digits offset is computed on the basis that 10 ft is equivalent to 40 pixels. If the airspeed is greater than 9 Kt, the middle digit is converted to its ASCII equivalent value and the current airspeed digit is drawn in the airspeed window, the digit plus one is drawn above the centre value and the digit less one is drawn below the centre value. In other cases, when the 9 digit is not transitioning, the offset is zero. The other parts of the airspeed indicator display are straightforward, with digits for airspeed above the display, digits for the Mach number below the display and an arrow indicating the predicted airspeed (over a 10 s interval). Also note the speed bug shown in magenta. The tick spacing between the 10 Kt markings is exactly 37 pixels. However, with an integer pixel grid, an FCU selected airspeed of 155 Kt, for example, corresponds to an offset of 5×3.7 or 18.5 pixels. Care is needed with rounding of such values, to avoid the position of the speed bug jittering by one or two pixels.

The compass display is similar to the airspeed and altitude displays, except that it has a dynamic rotary scale rather than a vertical sliding scale. The actual centre of the compass is 411 pixels directly below the attitude indicator. A clipping window is set so that only the uppermost segment is rendered, with the left and right clipping edges in line with the edges of the attitude indicator. The grey segment is drawn as a pre-computed texture, spanning approximately 70°, which defines the minimum and maximum angular offsets from the current heading. As the compass marks repeat every 5° with digit spacing of 10°, the offset, in terms of rotation, increases from the minimum displayed heading digits. The fragment of code to generate the compass markings and the compass digits is as follows:

```
offset = Maths_Rads((float) MinHdg) - Hdg;
Maths_Normalise(&offset);
h1 = MinHdg;
h2 = MaxHdg;

Glib_LoadIdentity();
Glib_Translate((float) CompassX, (float) CompassY);  /* rotation of compass card */

do
{
    Glib_PushMatrix();

    t = offset * ScaleFactor;
    Glib_Rotate(Maths_Degrees(-t));
    if (h1 % 10 == 0)
    {

        Glib_Draw(0, radius - 10, 0, radius);
        if (h1 % 30 == 0)
        {
            Glib_SetFont(Glib_EFONT16, 12);
            if (h1 > 90)
            {
                Glib_Char(h1 / 100 + '0', -10, radius - 10 - 20);
                Glib_Char(h1 / 10 % 10 + '0', 2, radius - 10 - 20);
```

```
                }
                else
                {
                    Glib_Char(h1 / 10 % 10 + '0', -5, radius - 10 - 20);
                }
            }
            else
            {
                Glib_SetFont(Glib_EFONT12, 8);
                if (h1 > 90)
                {
                    Glib_Char(h1 / 100 + '0', -8, radius - 10 - 16);
                    Glib_Char(h1 / 10 % 10 + '0', 2, radius - 10 - 16);
                }
                else
                {
                    Glib_Char(h1 / 10 % 10 + '0', -4, radius - 10 - 16);
                }
            }
        }
        else
        {
            Glib_Draw(0, radius - 5, 0, radius);
        }
        offset = offset + DEG5;
        if (h1 >= 360)
        {
            h1 = 5;
        }
        else
        {
            h1 = h1 + 5;
        }
        Glib_PopMatrix();
    } while (!(h1 == h2));
```

The angular offset is computed as the difference between the current heading and the minimum displayed heading. The display is translated to the compass coordinates and within the for-loop, the transformation matrix is pushed at the start of the loop and popped at the end of the loop ready for the next iteration. The actual rotation **t** is computed and includes a scale factor to match the compass dimensions to pixels. The remainder of the code simply rotates the 5° and 10° marks and converts the digits to an ASCII string every 10°. Potentially, the compass is a relatively complex display of both fixed and rotating vectors, characters and symbols (rendered as vectors). By using basic OpenGL primitives with translation and rotation, the main body of the code is reduced to approximately 50 lines.

The attitude indicator (AI) displays the pitch and roll angles provided by the inertial reference systems in an aircraft but derived from the equations of motion in a flight simulator. Maintaining accurate attitude is essential for efficient flying of commercial aircraft and the operational handbook for an aircraft provides power and attitude settings for specific aircraft performance. The AI is marked with lines of pitch and has bank markings to enable a pilot to select and maintain angles of pitch and bank accurately. The non-moving parts of the AI, which include the bank angle markings, the black reference bars and circle and the corner fillets are rendered as separate textures.

In flight simulation, the display of an AI is straightforward. At rest, with zero pitch and zero bank angle, the display comprises a rectangle with a blue upper half and a brown lower half, a white line across the rectangle and pitch markings. The AI is also known as an artificial horizon because it is stabilised in space

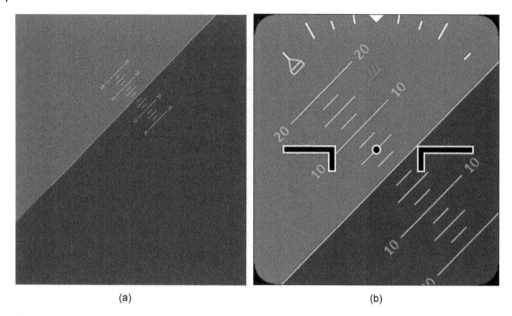

(a)   (b)

Figure 6.15 Rendering the AI.

and follows the actual horizon, with the blue and brown regions depicting sky and ground, respectively. The rectangle is rotated by the angles of pitch and roll within the square frame of the AI display. However, there is a problem; rotation of the rectangle could lead to a gap at high angles of bank or pitch. The solution is two-fold: firstly, the rectangle is drawn much larger than the AI display, to ensure that it covers the display region for all angles and, secondly, the rectangle is clipped to the display square. The rendering of the AI is shown in Figure 6.15. In Figure 6.15(a), without any clipping, the rectangle covers the full screen, which ensures there are no gaps in the display over the full range of pitch and roll angles. In Figure 6.15(b), the AI has been clipped to a clipping window the size of the AI display.

Note that the translation and rotation also apply to the pitch lines, the angle of bank indicator and the sideslip indicator.

6.5.2 Navigation Flight Display

The NFD is based on compass representations which are derived from the compass display used for the PFD. The NFD display modes are shown in Figure 6.16.

The displays show the ILS, VOR, NAV and ARC modes during an approach into London Gatwick airport with the aircraft 11.7 nm from touchdown. The ILS, VOR and NAV modes show the complete compass while the ARC mode shows an expanded segment of the compass. The track pointer, the localiser bar and the heading bug also rotate with the displayed compass card and are included in the rotation of the compass. The display also includes two radio magnetic indicator (RMI) pointers, blue and green for selected ADF and VOR navaids, respectively, each with front and back pointers to indicate direction. Rendering of the 360° compass card is given by the following code, where the display has been translated to the centre of the compass. The compass is marked at 5° and 10° intervals with the primary angles spaced every 30°.

```
static void CompassCard(float Hdg)
{
    int          IntHdg;
    unsigned int h;

    if (Hdg < 0.0)
    {
        IntHdg = -intround(Maths_Degrees(-Hdg));
    }
    else
    {
        IntHdg = intround(Maths_Degrees(Hdg));
    }
    IntHdg = Norm360(IntHdg);
    Glib_Colour(Glib_WHITE);
    Glib_LineWidth(2.0);
    for (h = 0; h <= 355; h += 5)
    {
        Glib_PushMatrix();
        Glib_Rotate(-((float) h - Maths_Degrees(Hdg)));
        if (h % 10 == 0)
        {
            Glib_Draw(0, 176, 0, 200);
            if (h % 30 == 0)
            {
                Glib_SetFont(Glib_EFONT16, 12);
                if (h > 90)
                {
                    Glib_Char(h / 100 + '0', -10, 156);
                    Glib_Char(h / 10 % 10 + '0', 2, 156);
                }
                else
                {
                    Glib_Char(h / 10 % 10 + '0', -5, 156);
                }
            }
        }
        else
        {
            Glib_Draw(0, 189, 0, 200);
        }
        Glib_PopMatrix();
    }
    WrHdg(IntHdg, -30, 225);
}
```

For each rotation, in the for-loop from 0° to 355°, the transformation matrix is pushed, the marks and characters are rendered and the transformation matrix is then popped to restore the previous state. Finally, the aircraft heading is displayed as three digits at the top of the compass. Almost identical code is used for the expanded form of compass, which is also similar to the PFD compass code, where the minimum and maximum displayed compass values are computed to minimise the amount of rendering for the individual compass arcs, marks and characters.

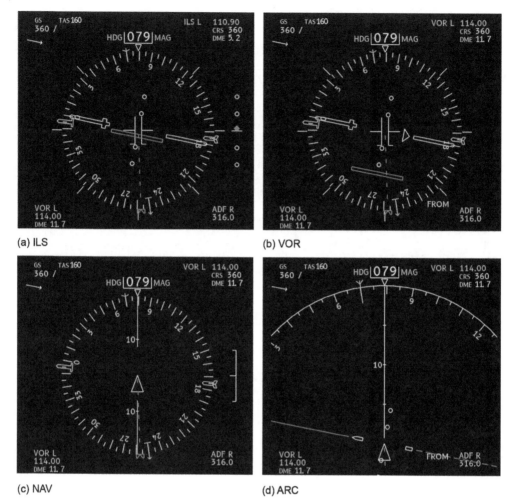

(a) ILS   (b) VOR

(c) NAV   (d) ARC

Figure 6.16 NFD Display Modes (See Plate 1).

6.5.3 EICAS Display

The EICAS on modern aircraft displays the main engine parameters and sub-systems indications for flaps and undercarriage. It may also display oil pressures, oil temperatures and oil and fuel quantities. With multi-engine aircraft, it is essential that the engine numbering is self-evident from the orientation of the displays. Typically, EICAS displays are clustered by information, for example, all RPM gauges are aligned horizontally and are separated so that the information relating to each engine is grouped vertically.

As variation between engines is generally small in most regimes of flight, the information is presented in a format where any variation between engines is clearly evident and quickly identified. For example, it is more important to monitor the EICAS displays and be aware that all RPM gauges are indicating about three-quarters and are static, rather than reading an N1 value of 71.7%. For the majority of simulator applications, the engine gauges are developed to replicate the aircraft instruments, either to provide a graphical representation of an analogue instrument or using 2D graphics to emulate an EICAS display. An EICAS display based on the Boeing 767 EICAS displays is shown in Figure 6.17.

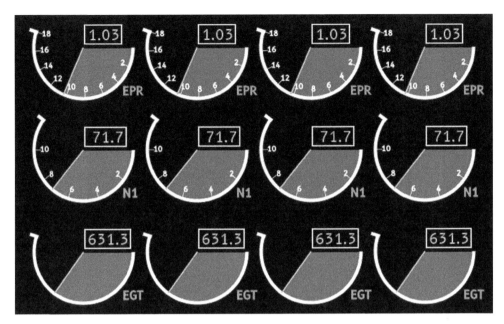

Figure 6.17 Typical EICAS Displays.

The leftmost column indicates the EPR, N1 and EGT values for engine 1, the port outer engine, and the three rows are each organised to display a specific parameter for engines 1–2–3–4. The grey shaded areas provide a quick comparison between engines, the white line provides a coarse indication of a parameter and the digital readout provides an exact value. The code for the EPR gauge is as follows:

```
void EICAS_EprGauge(int EprX, int EprY, unsigned int EngineNumber)
{
    int    x, y;
    float  e;
    char   str[20];

    Glib_LoadIdentity();
    Glib_Translate((float) EprX, (float) EprY);

    e = Maths_Rads(AeroLink_EngPkt.Engines[EngineNumber].Epr * 110.0);
    if (e < 0.0)
    {
        e = 0.0;
    }
    else if (e > DEG198)
    {
        e = DEG198;
    }
    if (e > DEG180)
    {
        Glib_DrawTextureRotated(0, 0, 140, 140, 0.3-0.07, 0.1-0.07, 0.3+0.07, 0.1+0.07,
                                Maths_Degrees(-e), 1.0);
        Glib_DrawTextureRotated(0, 0, 140, 140, 0.3-0.07, 0.1-0.07, 0.3+0.07, 0.1+0.07, 180.0, 1.0);
    }
    else
    {
        Glib_ClipWindow(EprX - 70, EprY -70, 140, 69);
```

```
            Glib_DrawTextureRotated(0, 0, 140, 140, 0.3-0.07, 0.1-0.07, 0.3+0.07, 0.1+0.07,
                            Maths_Degrees(-e), 1.0);
        Glib_RemoveClipWindow();
    }
    /* draw arc */
    Glib_DrawTexture(-78, -73, 163.0, 105.0, PFDTextures[0].tex_umin, PFDTextures[0].tex_vmin,
                            PFDTextures[0].tex_umax, PFDTextures[0].tex_vmax, 1.0);
    Glib_Colour(Glib_WHITE);
    Glib_LineWidth(2.0);
    x = intround(70.0 * cos(e));
    y = intround(-70.0 * sin(e));
    Glib_Draw(0, 0, x, y);

    e = AeroLink_EngPkt.Engines[EngineNumber].Epr;
    if (e < 0.0)
    {
        e = 0.0;
    }
    Glib_Colour(Glib_WHITE);
    Glib_SetFont(Glib_EFONT16, 12);
    sprintf(str, "%4.2f", e);
    Glib_Chars(str, 7, 10);
}
```

Rendering the gauges is a potentially complicated graphics task. The white arcs are drawn with the pre-computed textures shown in Figure 6.10, which ensures that the curved lines are crisp. However, in-filling the grey regions to two straight edges and a curved edge is far from straightforward. The grey semi-circular texture, also shown in Figure 6.10, is rotated. If the rotation of the semicircle is greater than 180°, the texture is rendered twice, firstly rotated by **e** and secondly by 180°, in order to cover the exposed segment, otherwise the texture is rotated and clipped below a horizontal line through the centre of the gauge. The actual engine EPR is formed as a string and rendered by `Glib_Chars`. Also note that, for these gauges, the digits in the actual aircraft displays do not tumble.

6.6 Flight Instruments

The flight instruments in light aircraft and in civil and military aircraft prior to 1990 were, in effect, transducers. The particular flight variable was converted to a pressure or electric current in order to drive the needle (or needles) of an instrument or gauge. Although it would be possible to replicate flight instruments using vector graphics, in practice most conventional flight instruments contain a static background and dynamic elements, including pointers, rolling digits, rotating cards and warning flags. These characteristics are more amenable to the rendering of textures rather than vectors. The background of an instrument can be captured from photographs or by using photo-editing tools. Similarly, a pointer can also be captured as a texture. A typical instrument display for a light aircraft is shown in Figure 6.18.

Rendering of instrument displays of this form is relatively straightforward. The background static texture is rendered at the location of the flight instrument and a pointer is rendered as a second texture, rotated about the centre of the instrument. An example of a typical instrument, the airspeed indicator, is shown in Figure 6.19.

The instrument background shown in Figure 6.19(a) was captured using a photo-editing tool, which enables the dial and markings to be drawn in high resolution on an instrument mounting showing a metallic texture, four mounting screws and shadow relief. The image is generated as a *png* file with a resolution 256 × 256 pixels, which is loaded at run-time. The use

Figure 6.18 Light Aircraft Instrument Panel (see Plate 2).

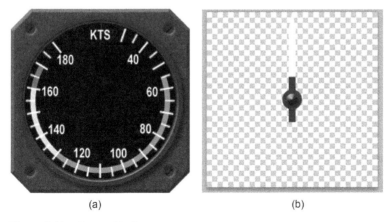

(a)   (b)

Figure 6.19 Airspeed Indicator.

of OpenGL textures enables the texture to be scaled to match the exact instrument size on a screen. The texture for the pointer, which is also created using a photo-editing tool, is shown in Figure 6.19(b). The implementation of the display requires computation of the angle of rotation of the pointer as a function of airspeed, rendering the background static texture and the rotated texture of the pointer. The code for the airspeed indicator for the Cessna 172 aircraft is as follows:

```
void Asi_Asi(int x0, int y0, float IAS, int dial, int needle)
{
    float Kts;
    int   MinSpeed = 30.0f;
    int   MaxSpeed = 180.0f;
    float needle_start = 20.0f; /* dial marking starting angle */
    float needle_rot = 0.0f;

    Kts = IAS * 1.944;
    if (Kts < 30.0)
    {
        Kts = 30.0;
    }
    if (Kts < MinSpeed)
    {
        Kts = MinSpeed;
    }
    else if(Kts > MaxSpeed)
    {
        Kts = MaxSpeed;
    }

    /* 1 Kt = 0.5 degrees, 90 Kt = 180 degrees */
    needle_rot = -(needle_start) - (Kts - MinSpeed) * 2.0f;

    Glib_SetTexture(dial);
    Glib_DrawTexture(x0 - 160, y0 - 160, 320, 320, 0.0, 0.0, 1.0, 1.0, 1.0);
    Glib_SetTexture(needle);
    Glib_DrawTextureRotated(x0, y0, 235, 235, 0.0, 0.0, 1.0, 1.0, needle_rot, 1.0);
}
```

The pointer angle **needle_rot** is computed from the indicated airspeed **IAS**. The specific background texture dial is selected and rendered by **Glib_DrawTexture**. The pointer texture needle is selected and the pointer is rendered as a rotated texture by **Glib_DrawTextureRotated**, which is rotated about the instrument centre (**x0, y0**). The two engine gauges at the bottom of the display are also drawn with rotating pointers, in this case, clipped to the visible part of the instrument.

Although most instruments have a static background, the heading indicator, is driven from a gyro (Helfrick, 1984), where the complete compass card rotates with the aircraft heading, with the aircraft heading (082°M) shown at the top of the compass card in Figure 6.18. In addition, the compass card can be aligned to the magnetic compass by rotating the HDG knob. The simplicity of rendering the compass is shown in the following code:

```
void Compass_Compass(int x0, int y0, float Hdg, int HdgBug, int texobj1, int texobj2, int texobj3)
{
    Glib_SetTexture(texobj1);
    Glib_DrawTexture(x0-160, y0-160, 320, 320, 0.0, 0.0, 1.0, 1.0, 1.0);
```

```
        Glib_SetTexture(texobj2);
        Glib_DrawTextureRotated(x0, y0, 320, 320, 0.0, 0.0, 1.0, 1.0, Maths_Degrees(Hdg), 1.0);
        Glib_SetTexture(texobj3);
        Glib_DrawTexture(x0-160, y0-160, 320, 320, 0.0, 0.0, 1.0, 1.0, 1.0);
    }
```

where `texobj1` is the background instrument bezel, `texobj2` is the compass card, which is rotated by the angle `hdg`, and `texobj3` is the large static aircraft symbol.

The AI, which rotates with pitch and roll, is possibly the most complicated instrument to render. In an aircraft, the AI is a stabilised platform. In the body frame (from the perspective of the pilot), the AI can be viewed as a sphere which rotates as the aircraft pitches and rolls. The AI is rendered as a three-dimensional sphere, located in the display panel and rotating about the pitch and roll axes. A sphere can be generated as several 3D quadrilaterals, so that the flatness of the surface of the quadrilaterals is not discernible (Song Ho, 2021). During rendering, a 2D texture can be applied to the surface of the sphere to 'paint' the pitch lines and blue and brown regions on the sphere. In addition, lighting can be applied to the sphere, to enhance the 3D appearance. Although the application of a 2D texture to a sphere could result in distortion at the top and bottom of the sphere, in the case of an AI, where the texture detail occurs towards the centre of the sphere, this concern can be ignored. Each quadrilateral is rendered as two triangles with strips of quadrilaterals from the top of the sphere to the bottom of the sphere, rendered as OpenGL triangle strips around the sphere. Figure 6.20 shows the sphere with a pitch angle of 20° and a roll angle of 30°. In this example, each square of the surface corresponds to 10° segments, rendered as two triangles, giving 1332 vertices. The sphere is represented by three arrays defining the position, surface normal and texture coordinates at each vertex. The arrays are constructed at the start of the simulation and then accessed during each frame, taking account of the pitch and roll angles of the sphere, rendering the sphere as a set of OpenGL triangle strips.

The code fragment to construct the sphere vertices is as follows:

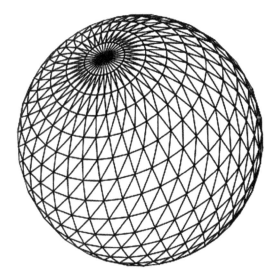

Figure 6.20 AI Sphere.

```
for (j=0; j<n/2; j+=1)   /* -90 -> +90 */
{
    float lat1 = (float) j * 2.0 * M_PI / (float) n - M_PI / 2.0;
    float lat2 = (float) (j + 1) * 2.0 * M_PI / (float) n - M_PI / 2.0;

    for (i=0; i<=n; i+=1)
    {
        float long1 = (float) i * 2.0 * M_PI / (float) n;
        GLfloat x, y, z;

        x = r * cos(lat2) * cos(long1);
        y = r * sin(lat2);
        z = r * cos(lat2) * sin(long1);
        sphere_pos_coords[k].x = x;
        sphere_pos_coords[k].y = y;
        sphere_pos_coords[k].z = z;
        sphere_nor_coords[k].x = x / r;
        sphere_nor_coords[k].y = y / r;
        sphere_nor_coords[k].z = z / r;
        sphere_tex_coords[k].u = (float) i / (float) n;
        sphere_tex_coords[k].v = 2.0 * (float) (j+1) / (float) n;
        k += 1;

        x = r * cos(lat1) * cos(long1);
        y = r * sin(lat1);
        z = r * cos(lat1) * sin(long1);
        sphere_pos_coords[k].x = x;
        sphere_pos_coords[k].y = y;
        sphere_pos_coords[k].z = z;
        sphere_nor_coords[k].x = x / r;
        sphere_nor_coords[k].y = y / r;
        sphere_nor_coords[k].z = z / r;
        sphere_tex_coords[k].u = (float) i / (float) n;
        sphere_tex_coords[k].v = 2.0 * (float) j / (float) n;
        k += 1;
    }
}
```

where n is the number of vertical and horizontal strips, and the position, normal and texture coordinates are stored in the arrays **sphere_pos_coords**, **sphere_nor_coords** and **sphere_tex_ccords**, respectively, and k is initially set to zero. The arrays contain $n*(n+1)$ vertices, which are cached in a similar way to 2D textures. The coordinates of the vertices are computed and stored at the start of the simulation. The fragment of code to render each strip is as follows:

```
glUseProgram(SphereProgram);
glBindVertexArray(SphereVAO);
glUniformMatrix4fv(uniformMatrixModelViewloc, 1, GL_FALSE, (float *) Transform);
glUniformMatrix3fv(uniformMatrixNormalloc, 1, GL_FALSE, (float *) Normal);
glUniformMatrix4fv(uniformMatrixProjectionloc, 1, GL_FALSE, (float *) ProjectionMatrix);
glUniformMatrix4fv(uniformMatrixMVPloc, 1, GL_FALSE, (float *) MVP);

glBindTexture(GL_TEXTURE_2D, CurrentTexture);
glUniform1i(uniformTex1loc, 0);

glBindBuffer(GL_ARRAY_BUFFER, SphereVBO);
glBufferData(GL_ARRAY_BUFFER, nsegments * 8 * sizeof(float), SegmentVertices, GL_STATIC_DRAW);
```

```
glDrawArrays(GL_TRIANGLE_STRIP, 0, nsegments);
nsegments = 0;
glBindVertexArray(0);
```

The vertex shader for a sphere is similar to the vertex shader used for textures except that the normal vector is an additional input and the three inputs are passed to the fragment shader, as shown in the following code:

```
#version 430

layout (location = 0) in vec3 VertexPosition;
layout (location = 1) in vec3 VertexNormal;
layout (location = 2) in vec2 VertexTexCoord;

out vec3 Position;
out vec3 Normal;
out vec2 TexCoord;

uniform mat4 ModelViewMatrix;
uniform mat3 NormalMatrix;
uniform mat4 ProjectionMatrix;
uniform mat4 MVP;

void main()
{
    TexCoord = VertexTexCoord;
    Normal   = normalize(NormalMatrix * VertexNormal);
    Position = vec3(ModelViewMatrix * vec4(VertexPosition,1.0));

    gl_Position = MVP * vec4(VertexPosition,1.0);
}
```

A Phong lighting algorithm (Watt, 1989) is implemented in the fragment shader to produce the appropriate ambient, diffuse and specular lighting conditions for a light source with a predefined position and intensity. The fragment shader is as follows:

```
#version 430

in vec3 Position;
in vec3 Normal;
in vec2 TexCoord;

layout (binding = 0) uniform sampler2D Tex1;

struct LightInfo
{
  vec3 Position;   // Light position in eye coords.
  vec3 Intensity;  // A,D,S intensity
};

uniform LightInfo Light;

struct MaterialInfo
{
    vec3 Ka;              // Ambient reflectivity
    vec3 Kd;              // Diffuse reflectivity
    vec3 Ks;              // Specular reflectivity
```

```
      float Shininess;    // Specular shininess factor
};


uniform MaterialInfo Material;

layout (location = 0) out vec4 FragColor;

void phongModel(vec3 pos, vec3 norm, out vec3 ambAndDiff, out vec3 spec)
{
    vec3 s        = normalize(vec3(Light.Position) - pos);
    vec3 v        = normalize(vec3(0.0,0.0,-1.0)); // view direction -> forwards from eyes is -ve z
    vec3 r        = reflect(-s, norm);
    vec3 ambient  = Light.Intensity * Material.Ka;
    float sDotN   = max(dot(s, norm), 0.0);
    vec3 diffuse  = Light.Intensity * Material.Kd * sDotN;
    spec          = Light.Intensity * Material.Ks * pow(max(dot(r,v), 0.0), Material.Shininess);
    ambAndDiff    = ambient + diffuse;
}


void main()
{
    vec3 ambAndDiff, spec;
    vec4 texColor = texture(Tex1, TexCoord);
    phongModel(Position, Normal, ambAndDiff, spec);
    FragColor = (vec4(ambAndDiff, 1.0) * texColor) + vec4(spec,1.0);
}
```

Figure 6.21 shows two versions of rendering the AI. In Figure 6.21(a), the sphere is rendered without lighting. In Figure 6.21(b), Phong shading is applied, where the specular component gives a clear impression of the 3D surface of the sphere. Note that as the lighting and the position of the sphere are constant, the lighting parameters are set at the start of the simulation.

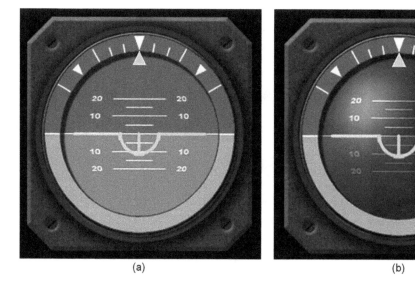

(a)   (b)

Figure 6.21 3D Rendering of an AI (see Plate 2).

6.7 Soft Panels

Many flight simulators are interfaced to aircraft sub-systems for radio, navigation, flight control system, fuel management, hydraulics and electrical sub-systems. In some training flight simulators, and in engineering flight simulators, the cost of installing actual aircraft parts or producing facsimile aircraft systems may be expensive and the use of software emulation provides a viable alternative. In such systems, which are often referred to as soft panels, the appearance of the displayed equipment closely resembles the actual aircraft system, albeit that the knobs, switches and lights are rendered using 2D graphics. The user interaction is by means of a mouse or touchscreen, which enables a knob to be turned, a switch to be selected, a button to be pressed or a selector to be turned to a specific position. In addition, specific panels can be displayed according to the context (e.g. displaying engine switches and fuel selectors during an engine-start sequence), which are removed at other times.

The choice depends on cost and the application. For example, teaching technicians to start a turbofan engine would be prohibitively expensive with actual jet engines and, moreover, any mistakes in operating the engine could result in costly repairs. In such applications, the training can

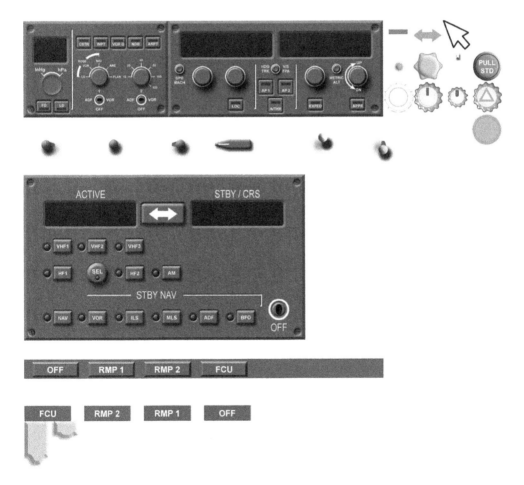

Figure 6.22 A Sheet of Textures for the FCU and Radio Panels.

be undertaken in a synthetic environment, enabling the trainee to understand and practise procedures, providing assessment of competency. This observation is also relevant to engineering flight simulators and research simulators, being able to switch between variants of aircraft panels without the cost of installing hardware equipment or taking the simulator out of service for an extended period.

Many of the concepts introduced in previous sections are also applicable to software panels:

- The panel can be captured as a texture from a photographic image or using photo-editing tools.
- Textures can be organised as an atlas of textures, selecting a specific texture by its coordinates.
- Textures can be rotated, for example, to display a rotating knob or selector.
- Equipment containing panels and lamps can be emulated using vectors and text strings.
- Static parts of a display can be rendered as textures.

To simulate an FCU and a radio panel, the textures were mapped onto a texture atlas and the coordinates of the panel elements were extracted using a photo-editing tool. The texture atlas, which contains all the elements of the FCU and the radio panel, is shown in Figure 6.22.

The radio panel and FCU panel are shown in Figure 6.23. The illuminated push switches are lit by overlaying small green filled rectangles. The LED lamps on the radio panel are illuminated by overlaying a lit LED texture over an unlit LED texture. The LCD panels of both units are implemented using specific LED fonts, with the font size matched to the panel dimensions. Note the use of shadows in the textures to simulate the effect of 3D relief.

The selection of individual components is similar to a texture atlas of characters, where the position and size of the component in the large texture enable the texture of an item to be extracted

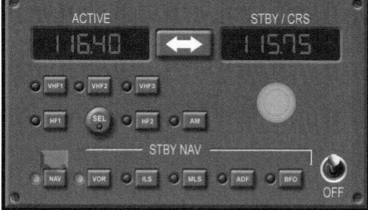

Figure 6.23 Computer-generated FCU and Radio Panels (see Plate 3).

and rendered. The code for the FCU follows, where the coordinates are relative to the bottom left-hand corner of the FCU panel:

```
void PanelLib_DisplayFCU(unsigned char FD, unsigned char LS, unsigned char CSTR,
                         unsigned char WPT, unsigned char VORD, unsigned char NDB,
                         unsigned char ARPT, unsigned char LOC, unsigned char AP1,
                         unsigned char AP2, unsigned char ATHR, unsigned char EXPED,
                         unsigned char APPR, int NAV1, int NAV2, int NAVSELECT,
                         int RANGE, float Baro, float Speed, float Hdg, float Alt,
                         float VS, unsigned char hPa)
{
    Glib_LoadIdentity();
    Glib_Translate((float) NFD_FCUX, (float) NFD_FCUY);

    /* FCU Panel */
    DrawSprite(0, 0, 768, 200, 1.0, 0);

    PushSwitch(28, -3, 32, 32, FD);
    PushSwitch(66, -3, 32, 32, LS);

    PushSwitch(114, 139, 32, 32, CSTR);
    PushSwitch(151, 139, 32, 32, WPT);
    PushSwitch(188, 139, 32, 32, VORD);
    PushSwitch(225, 139, 32, 32, NDB);
    PushSwitch(262, 139, 32, 32, ARPT);

    PushSwitch(438, 8, 32, 32, LOC);

    PushSwitch(502, 47, 32, 32, AP1);
    PushSwitch(545, 47, 32, 32, AP2);
    PushSwitch(523,  8, 32, 32, ATHR);

    PushSwitch(608, 8, 32, 32, EXPED);
    PushSwitch(698, 8, 32, 32, APPR);

    ThreeWaySwitch(129, 6, 64, 64, NAV1);
    ThreeWaySwitch(219, 6, 64, 64, NAV2);

    SelectorKnob(162, 85, 48, 48, NAVSELECT); /* rotated sprites need centre anchor */
    SelectorKnob(253, 85, 48, 48, RANGE);

    if (hPa)
    {
        BaroMark(62, 70, 64, 64, 45.0);  /* hPa or InHg? */
    }

    else
    {
        BaroMark(62, 70, 64, 64, -45.0);  /* hPa or InHg? */
    }

    Knob(62, 67, 38, 38, Baro);

    Knob(383, 83, 38, 38, Speed);
    KnobBlue(453, 83, 38, 38, Hdg);
    Knob(623, 83, 38, 38, Alt);
    Knob(714, 83, 38, 38, VS);
}
```

The code for **PushSwitch** and **Knob** is as follows:

```
void PushSwitch(int x, int y, int sx, int sy, bool state)
{
    if (state)
    {
        DrawSprite(x, y, sx, sy, 1.0, 1);
    }
}

void Knob(int x, int y, int sx, int sy, float angle)
{
    float rot = angle + 0.02;
    DrawSpriteRotated(x, y, sx, sy, rot, 1.0, 9);
}
```

The actual code to render the texture is given in DrawSprite and DrawRotatedSprite, which differ only in the rotation of the texture. The texture coordinates are given by (**tx1,ty1**) and (**tx2,ty2**), which are predefined from measurements of the texture maps.

```
void DrawSprite(int x0, int y0, int sx0, int sy0, float Alpha, int texnum)
{
    float tx1 = PanelTextures[texnum].tex_umin;
    float ty1 = PanelTextures[texnum].tex_vmin;
    float tx2 = PanelTextures[texnum].tex_umax;
    float ty2 = PanelTextures[texnum].tex_vmax;

    Glib_DrawTexture(x0, y0, sx0, sy0, tx1, ty1, tx2, ty2, Alpha);
}

void DrawSpriteRotated(int x0, int y0, int sx0, int sy0, float R, float Alpha, int texnum)

{
    float tx1 = PanelTextures[texnum].tex_umin;
    float ty1 = PanelTextures[texnum].tex_vmin;
    float tx2 = PanelTextures[texnum].tex_umax;
    float ty2 = PanelTextures[texnum].tex_vmax;

    Glib_DrawTextureRotated(x0, y0, sx0, sy0, tx1, ty1, tx2, ty2, R, Alpha);
}
```

For example, the details of the selector knob are given by

```
    PanelTextures[5].tex_vmax = 1.0 - 0.196;
    PanelTextures[5].tex_vmin = 1.0 - 0.304;
    PanelTextures[5].tex_umin = 0.384;
    PanelTextures[5].tex_umax = 0.491;
```

and similarly for the other textures. The point to appreciate is the simplicity of constructing these textures, which can be placed in a panel, where they can be illuminated or rotated, as appropriate.

A similar approach is used to emulate a radio stack, as shown in Figure 6.24. The texture in Figure 6.24(a) is overlaid with strings based on an LCD character font, with the knobs, selectors and switches shown as rotated textures in Figure 6.24(b). One small point is the presentation of concentric knobs, which are shown as circles. Textures of knobs showing the relief can be realistic, but at the expense of complicating the selection by mouse or touchscreen.

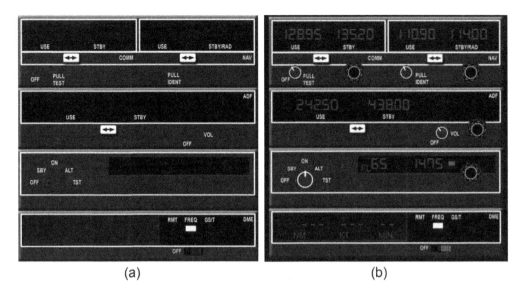

(a)   (b)

Figure 6.24   Radio Panel Textures (see Plate 3).

Very accurate and detailed facsimiles of actual aircraft panels can be produced and, with the use of modern photo-editing tools, for example, Blender, (https://www.blender.org), changes are relatively easy to implement. However, the use of soft panels is not without its problems and these mostly stem from the user interface. Modern touchscreens provide an interface which is similar to the conventional mouse, where touching the screen is equivalent to pressing a mouse key at a screen location. The event is detected by the user interface to derive the screen coordinates. While the user's finger is in contact with the screen, it is the equivalent of holding a mouse key down.

The problems faced by the simulator developer is the same as the human interface problems of the smartphone. How can the user interact with computer-generated buttons, sliders and knobs? The problem is only partially solved by ensuring compatibility of mouse and touchscreen inputs. For push switches, the cursor can be placed over the switch and a key press corresponds to pressing or releasing the switch. With a mouse, the valid region can be the bounds of the switch image, but for a touchscreen with finger motion, the bounds need to be extended slightly to allow for the relatively poor resolution of a finger press on a screen. With slider switches or up/down switches, the switch can be activated if a mouse key is pressed when the cursor is placed over the desired switch position. Alternatively, pressing the left mouse key could be detected as a left or up movement of the switch and a right or down movement for the right mouse key.

The situation with concentric knobs is more complicated, particularly for dual concentric knobs. The knob could be turned clockwise if the mouse key is pressed while the cursor hovers over the right side of the knob for clockwise motion and the left side of the knob for anti-clockwise motion. Alternatively, rotary movement of the finger or mouse cursor could be detected as clockwise or anti-clockwise for small turns of a knob. In addition, where knobs can be turned slowly or quickly, a software gain can be applied so that a knob turns slowly to start, speeding up while the selection is maintained. However, if there are two (or more) concentric knobs, careful detection of the finger or cursor position is needed to avoid incorrect selection of a knob. Strictly, there is no definitive solution and the developer needs to experiment with various algorithms, depending on the size, layout and conventions of specific panels.

The acceptance criterion in validating soft panels is to compare the effort or workload of an action in the aircraft with the same action in the simulator, for example, to set the FCU altitude to 25,000 ft or the barometric pressure to 1020 mb. It is certainly irritating if the interaction results in overshooting a desired value (requiring further selection) or the removal of finger pressure or releasing a mouse button introduces a spurious movement, unintentionally changing the set value. Having developed a soft panel user interface, the validation involves defining a set of tasks to be performed, which are closely monitored and timed during the user interaction. In such tests, user feedback can also highlight deficiencies or identify a convention that is unacceptable on a busy flight deck. It is important that the operation of a soft panel in a simulator is robust and consistent and, as far as reasonably possible, identical to operating the aircraft system.

References

Anon. (2013), *The Open-Source OpenGL Utility Toolkit (Freeglut 3.0.0) Application Programming Interface*, available from http://freeglut.sourceforge.net/docs/api.php (accessed 13 June 2022).

Anon. (2018), *OpenFlight® Scene Description Database Specification, Version 16.7, Revision A*, available from https://www.presagis.com/workspace/uploads/files/openflight16-7.pdf (accessed 13 June 2022).

Dawson, W. D. (1992), *C-130 Glass Cockpit Development Program*, AIAA Conf. Aerospace Design, Irvine, CA.

Helfrick, A. (1984), *Modern Aviation Electronics*, Prentice-Hall.

Kessenich, J., Sellers, G., and Shreiner, D. (2017), *OpenGL Programming Guide: The Official Guide to Learning OpenGL*, Version 4.5, ninth edition, Pearson Education.

Kilgard, M. J. (1996), *OpenGL Programming for the X Window System*, Addison Wesley.

Koniche, M. L. (1988), *747-400 Flight Displays Development*, AIAA Conference Aircraft Design, Systems and Operations, Atlanta, GA.

Pallett, E. H. J. and Coombs, L. F. E. (1992), *Aircraft Instruments and Integrated Systems*, Pearson Education Ltd, Longman.

Sellers, G. and Kessenich, J. (2016), *Vulkan Programming Guide: The Official Guide to Learning Vulkan (Opengl)*, first edition, Pearson Education.

Song Ho, A. (2021), *OpenGL Sphere*, http://www.songho.ca/opengl/gl_sphere.html (accessed 11 August 2021).

Spitzer, C. R. (1987), *Digital Avionics Systems Principles and Practices*, McGraw-Hill.

Watt, A. (1989), *Fundamentals of Three-dimensional Computer Graphics*, Addison Wesley.

Wolf, D. (2013), *OpenGL 4 Shading Language Cookbook*, Packet Publishing.

Woo, M., Neider, J. and Davis, T. (1996), *OpenGL Programming Guide: The Official Guide to Learning OpenGL*, version 1.1, second edition, Addison Wesley.

7

Image Generation Systems

The whole of physics, that is the whole of philosophy of nature is nothing but geometry.

Réné Descartes

The objective of designing an image generation (IG) system for a flight simulator is challenging. It is to create a visual illusion to convince a pilot that they are actually flying in an aircraft, where the view from the simulator flight deck is identical to the view from an aircraft flight deck.

7.1 IG Pipeline

In many flight simulators, the trend is to purchase a real-time rendering system off the shelf; for example, *OpenSceneGraph* (http://www.openscenegraph.org) is an open-source graphics toolkit capable of rendering complex scenes in real time. A number of commercial packages are also available including *Unity* (https://unity.com*)*, which provides a comprehensive library to integrate user software, and UnReal (https://www.unrealengine.com). In addition, simulation games including *X-Plane* and *Microsoft Flight Simulator* include a renderer which can be accessed independently. The alternative, of developing a renderer from scratch, although feasible, is generally impractical as the requirements of a real-time renderer for 3D graphics (Akenine-Moller et al., 2008) are particularly demanding:

- Rendering a detailed scene at least 50 times per second.
- Transforming millions of 3D textured triangles to 2D triangles per frame.
- Acquiring the position and orientation of the pilot eye-point from the simulator.
- Loading visual databases of various formats.
- Organising the visual database for a large area of operation.
- Applying level-of-detail to the scene content.
- Providing height above terrain information and detection of collisions.
- Providing dynamic entities in a scene.
- Illumination of the scene from sun, moon, clouds and individual lights.
- Providing a realistic sky including clouds and cloud base.
- Varying lighting conditions with time of day.
- Providing environmental effects, including fog, rain, snow and lightning.
- Ensuring real-time performance is maintained under all loading conditions.

Flight Simulation Software: Design, Development and Testing, First Edition. David Allerton.
© 2023 John Wiley & Sons Ltd. Published 2023 by John Wiley & Sons Ltd.
Companion Website: www.wiley.com/go/flightsimulationsoftware

Objects in the visual database are composed of polygons or triangles. In some legacy visual databases, where objects may be stored as polygons, they are converted to triangles (Lee and Schachter, 1980) during loading of the database. The overall process of extracting 3D triangles from a visual database and transforming each triangle to a 2D triangle, which is written as pixels to a framestore, is known as the IG pipeline. The term pipeline is apposite as the triangles are processed sequentially as they move from one graphic process to another. With the availability of the GPU since the 1990s, the majority of the pipeline operations are performed on the GPUs. The pipeline covers the following activities:

- geometry and transformations – converting coordinates (vertices) from the scene space to the viewing space;
- the camera and viewing – positioning the camera and setting the viewing angles;
- addition of light and shading effects;
- rasterisation – generation of pixels from vertices;
- implementation of textures as pixels are written.

The overall architecture of the pipeline for modern GPUs is shown in Figure 7.1.

An object or primitive is made up of several vertices, where each vertex is defined by its position (x, y, z), colour (RGBA), vertex normal (n_x, n_y, n_z) and texture. The rasteriser interpolates the vertices to form a set of fragments aligned on a grid. The vertex processor contains the vertex shader which transforms the vertices and performs per-vertex lighting. The fragment processor performs texturing and per-fragment lighting. The normal, defined at each vertex, is a normal to the plane of the triangle. The fragments are based on rasterised 2D coordinates but have other attributes, including texture coordinates, depth values and alpha (transparency) values. The vertex transformations are outlined in Figure 7.2.

Objects in the model space are transformed to the world space. Once the camera position and orientation are defined in each frame, the objects are transformed to the camera space. Knowing the field-of-view of the camera, the objects are projected to take account of perspective within the clipping volume. Finally, the rasteriser transforms the objects to pixels, to combine colour and texture.

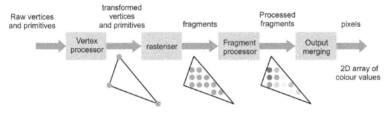

Figure 7.1   The IG Pipeline.

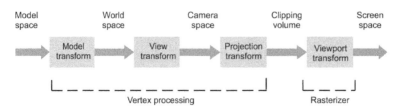

Figure 7.2   Vertex Transformations.

The model transform is also used in the construction of a database. Typically, a visual model may be built in a local frame, for example a rotor blade or an airport building. The rotor blade may be added to a rotor hub and then to a helicopter model or the airport building may be added to an airport model. These transformations are based on well-known matrix transformations between axes, using homogeneous coordinates.

The transformation from the world space to the camera space is a similar set of transformations, based on the camera position in the scene and the orientation (pitch, roll and yaw) of the camera (or pilot eye-point), and often these are combined with the transformation to the world space. In each frame, there is a viewing frustum from the pilot's eye to the four corners of the windscreen (assumed to be rectangular). Objects completely outside the frustum are discarded but objects which are only partially inside the frustum are clipped to the edges of the frustum. Finally, based on the field-of-view of the frustum and the distance of the pilot eye-point from the windscreen, the vertices of the 3D triangles are transformed to 2D screen coordinates, known as projection. In flight simulation, a perspective projection is used which takes account of distance. For example, looking along a railway track, the lines will appear to converge to a point in the distance.

Using homogeneous coordinates, these transformations of translation, scaling, rotation and projection can be combined to form a single 4×4 transformation matrix, which is applied to the vertices of every triangle during each frame. Nowadays, the operations are performed by the GPU cores of a graphics card, relieving the developer of the computations inherent in image generation. With the availability of real-time IGs, attention is focused on the fidelity (visual quality) of the visual database, the interface with the simulator software and ensuring that all rendering is completed within the frame time, under all conditions.

The final stage of visualisation is not included in this pipeline. Although images can be generated for a 2D screen located in front of the flight crew, more commonly, the image on a screen is projected optically to a curved mirror or set of screens. The warping or distortion of an image from a flat 2D screen to a curved 3D screen is provided, either by the projection system or by applying a further transformation to the frame-store, which may add a further frame to the latency.

From the user's perspective, these transformations are performed by the rendering package passing graphics operations to the GPU cores. The user provides a visual database, which is loaded by the renderer, updates the position of any dynamic entities in the scene, provides the current position and orientation of the view point (normally, the pilot eye-point), passes environmental information to the renderer, including visibility, time of day and cloud base, and monitors the rendering of each frame. The interaction with the graphics driver and the GPU cores of the graphics card is handled completely by the renderer.

7.2 Visual Databases

7.2.1 Constructing Visual Databases

The visual database in a modern flight simulator may contain several million surfaces, particularly if the database contains complex objects. Attempts have been made to define visual database standards with the specific benefits of interoperability and reuse of databases. The availability of a standard enables rendering packages to support databases complying with the standard and, moreover, allows databases supplied by different vendors to be imported into applications.

Users of flight simulators are also likely to need to modify a database to incorporate changes such as new roads or airport buildings. A number of packages are available to both create and modify visual databases. These packages are very similar to the tools used in computer-aided design. They enable a developer to create objects from lines, shapes and surfaces, to add colour and texture to surfaces, to position objects precisely, to apply graphic operations such as translation, rotation and scaling to objects and to replicate objects. In an editing mode, an object or set of objects can be selected and modified, for example, to lengthen or widen a runway and its edge markings and lights or to add a new building to an airport scene. The aim of these tools is two-fold, firstly to ensure the integrity of the database and, secondly, to minimise the effort needed to create or modify a visual database. For example, having created an individual runway light, the replication and placement of lighting can be achieved by selecting the spacing and location of lights in a single operation rather than placing several hundred lights individually.

The main activity of a rendering package is to extract objects from the database, apply the graphic transformations to write an image into a frame-store and generate the video signals for a display or projector. The emphasis has been to improve the processing rate of the graphical operations, particularly with multiple cores, to increase the number of objects rendered per frame. Although this requirement depends on the efficiency of the GPUs in a modern graphics card, there is a potential benefit in terms of the rendering quality if only the active objects, that is, objects which are directly visible, are rendered. In other words, objects which are not visible should be discarded at the earliest opportunity. The alternative, which is to process every object in the database and only discard objects after the graphics transformations, during the clipping process, is extremely wasteful of the potential GPU processing.

In a scene in a typical flight simulator, there are three classes of objects that can be discarded, if the visual database is structured to facilitate fast access to objects and contains information to guide the accessing of objects:

1) An object will contain surfaces which are not visible, for example, the back wall of a building seen from the front of the building or the far surfaces of an aircraft fuselage. This process is known as hidden-surface elimination. If a surface is constructed from triangles, then, if the angle between the line-of-sight from the viewer to the surface and the normal is greater than 180°, the surface is not visible (excluding transparent surfaces) and can be discarded. Normally, the surface normal is computed when the object is loaded and is stored with the object vertices.

2) Although the viewing frustum is used in the clipping process, the projection of the frustum onto the visual database is a good approximation to the space containing visible objects. Therefore, objects outside this region can be excluded from rendering during the current frame, taking into account the shape of the 3D footprint on sloping terrain.

3) If the objects in a visual database are defined with different resolutions depending on the viewing distance, they can be rendered with an appropriate level-of-detail (Luebke et al., 2002), where information is embedded in the visual database to indicate which set of surfaces should be rendered as a function of viewing distance. An obvious example is an airport building. At 5 nm, most of the detail would only occupy an area of a few pixels on a display and a coarse representation of the building can be rendered. However, taxiing back to the stand, the individual doors and windows of the building would be visible and a more detailed set of textures can be rendered, varying the scene content according to the position of the aircraft.

Consequently, the organisation and structure of a visual database have a major impact on the performance of the IG system. In many applications, it is possible that only one-fifth of the database is rendered in a frame. The time to exclude or discard the remaining triangles in the database must be minimal. Such methods are a combination of the algorithms used in rendering and the information provided in the visual database. For example, if the movement or orientation of the aircraft between consecutive frames is small, the group of visible surfaces will only change by a small amount and most of the surfaces from the previous frame are still active, further reducing the effort to search for surfaces to be excluded.

An IG system that is capable of excluding 80% of the objects in a database is the equivalent of an IG that is potentially five times faster than an IG which attempts to render every object in the database. However, the dynamic selection of objects in a scene must ensure that the rendering is smooth and continuous. For example, if a lamp post is excluded in one frame but reappears in a subsequent frame, the lamp post will suddenly seem to pop up, an event which can never occur in the real world. A similar effect is that a thin object can scintillate if it appears and disappears over a series of frames as a result of rounding errors in the rendering algorithms. The human eye picks up such aberrations, rather defeating the benefits of rendering highly detailed scenes. Because such events are dependent on specific conditions, they are difficult to replicate and may be difficult to detect (or repeat) in the validation of a visual system.

7.2.2 Visual Database Standards

Until around 2000, most simulator companies developed their own visual systems and, generally, all manufacturers adopted the standard known as the OpenFlight Scene Description Database Specification (Anon., 2018). The OpenFlight standard is over 180 pages and covers all aspects of visual database definitions. For manufacturers and developers, it defines how a database is organised and the content of objects, textures and surfaces, including the position of vertices of triangles and colour palettes used to produce colour. The advantage of OpenFlight is that it is an open standard and has been rigorously followed by the major manufacturers. The IG rendering packages used by most manufacturers include software to read OpenFlight-compatible databases and convert them to a local format used by the renderer. However, OpenFlight is not without its disadvantages. The standard has changed fairly frequently, causing problems with compatibility between versions. But the main limitation is the effort to produce a new converter for IG systems lacking software to read OpenFlight databases. A few organisations have produced translators between the various standards but often these are limited if specific features are not supported by the different IG systems. Note that *OpenSceneGraph* provides plug-ins to read 24 different file formats and write nine file formats (OpenSceneGraph, 2012).

One other consideration with visual databases is the classes of data provided. Clearly, the data will include latitude, longitude and elevation data to locate points and features in a database but should also define the type of terrain or surface. However, data derived from satellite or airborne capture of terrain may only include the colour of the terrain, at the resolution of the captured photographic data. With such data, it is possible to infer the type of vegetation, to distinguish fields, hedges, tracks and roads and, for generic terrain, accuracies of 5–20 m are generally acceptable. Nevertheless, for many simulator applications, other attributes of the terrain may be missing in such data. For example, multi-spectral sensor data used in radar, night vision or infra-red sensors cannot be derived from raw colour data. Moreover, the data will be captured at a specific time of day and also time of year. For example, the scene may contain shadows of tall trees that will not correlate with the simulated time of day or conditions (for example, strong shadows appearing in foggy scenes). Similarly, summer vegetation is different from winter vegetation, making it difficult to apply winter effects such as snow to scenes produced in summer conditions.

The availability of high-performance graphics cards in recent years has seen a growth in simulation, covering car driving and train driving games in addition to flight simulation. The manufacturers have mostly produced in-house definitions of the visual database structures and, because of the commercial interest in developing visual databases for established games markets, have been reluctant to publish details of the databases. The one exception is the FlightGear package (https://www.flightgear.org), which uses a proprietary format but releases details of its visual database.

The simulator developer is left with three choices:

1) Use an existing visual database which is a supported standard for a proprietary IG system; neither the database nor the IG system is likely to be open and the user will expect to procure both the database and the IG software.
2) Convert a visual database to a standard used by a proprietary IG system; the difficulty is locating tools which import an existing standard and export it in a standard supported by the target IG.
3) Develop a visual database from scratch using an editing tool, which exports the database using a standard supported by an IG system; the problem with this approach is the amount of effort needed to generate an acceptable visual database.

In recent years, this third option has become more practical with the availability of open software for digitised terrain data, satellite imagery and desktop 3D scene capture and editing tools. Digital elevation models (DEMs) are available to 30 m resolution via the OpenDEM website (https://www.opendem.info), with 1 arcsecond resolution data available for parts of Europe. Packages including SAGA, a System for Automated Geoscientific Analyses (https://saga-gis.sourceforge.io/en/index.html), provide terrain data for wide regions of the world in numerous formats. In addition, a number of organisations have released open map data applications which include extensive data, particularly for cities. It is left to the reader to investigate the fast-changing applications supporting the provision of publicly available terrain data. The important considerations are the resolution of the data, the classes supported (types of vegetation, roads, railway lines, rivers and lakes etc.) and the data formats supported, in order to minimise the work to import the data into database editing packages. The ideal solution is to acquire data from a DTED (Digital Terrain Elevation Data) or DTEM (Digital Terrain Elevation Map) package or mapping package and be able to apply this data to generate 3D terrain over an extensive region with sufficient content to be able to render the terrain with minimal additional work. In practice, these databases provide generic terrain, possibly extending to cities, leaving the developer to incorporate aviation-specific features, particularly airports and airfields.

7.2.3 Visual Database Editing Tools

A number of vendors provide tools to view and edit the databases used in a rendering system. A few organisations have also developed open tools, which enable a user to develop a complete terrain or region or individual items from scratch or to modify existing scenes and objects. It is not the purpose of this book to provide a user guide for such packages, but a number of the common features and capabilities are outlined to provide insight into the work involved in developing visual databases.

The Blender package (https://www.blender.org) is one example of an open tool used to construct detailed visual databases. Strictly, it is more suited to games animation than to the production of flight simulation visual databases, although this distinction is somewhat blurred as games applications demand increasingly realistic and detailed scenery. Before embarking on the development of a visual database or its objects, it is advisable to check that it is possible to export files from Blender in a format supported by the target IG system. Note that Blender supported the importing and exporting of OpenFlight files up to version 2.46 but this facility has since been withdrawn. Figure 7.3 shows a visual database of Manchester airport using Blender to produce an OpenFlight file.

Figure 7.3 A Visual Database Produced Using Blender (see Plate 4).

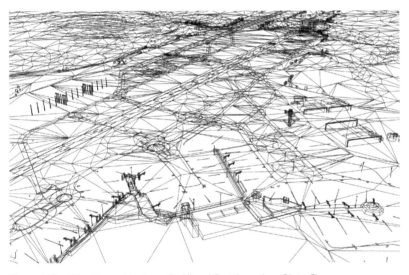

Figure 7.4 Wire-Frame Version of a Visual Database (see Plate 5).

Figure 7.4 shows the wire-frame triangles used to construct the terrain around the airfield, the main runways and taxiways and various runway markings, edited using the Blender package.

With the increased processing capabilities of tablets and mobile phones, there has been a growth in mapping tools and also an increase in open-source data used in mapping. Google maps use Graphical Information System (GIS) data derived from satellite data to provide detailed terrain mapping. The Shuttle Radar Topography Mission (SRTM) program captures elevation data for most of the world 1 arcsecond resolution (30 m). In Blender, it is possible to select an area of the world to produce a terrain and apply SRTM to the same region to import 3D terrain data in a few minutes which can be visualised directly. Blender includes tools to interpolate height data, modify colour values and saturations and apply a wide range of textures to a terrain, in order to produce realistic terrain.

Although 30 m resolution is acceptable for representative terrain, it is blurred and lacks sufficient detail close to the terrain. At this resolution, taxiway and runway markings and airport buildings could not be used in taxiing in a flight simulator. In addition, satellite data is derived as a plan view and cannot include the sides of buildings. Where detail close to an object is essential, the editing tools in Blender are used to add textured polygons to taxiways and runways and to replace the flat rectangles of buildings with 3D shapes, textured appropriately for each building. Similarly, runway markings and lighting can be added from either imported objects and textures or from objects constructed individually within Blender. In these cases, specific airport detail is added to generic terrain to reduce the effort to produce a visual database, while retaining an acceptable level of visual fidelity for the different phases of flight.

In the last few years, applications such as OpenStreetMap (https://www.openstreetmap.org) have been developed for satellite navigation for vehicles. In some cities, this coverage extends to 3D buildings, and Blender includes a facility to import OpenStreetMap data directly to add buildings to a terrain. Although the buildings are bland, it is possible to import building textures which can then be applied to selected buildings. The main advantage with Blender is that it is possible to create very realistic scenery of hills, valleys, roads, rivers, towns and cities in a matter of a few days using a handful of tools and importing data from open sources. Ten years ago, production of similar databases would have taken several months.

Although Blender has automated much of the capture of terrain imagery, the overall process still requires significant manual intervention. With variations in the geographic reference data used to capture terrain data, merging of data from different sources can result in geometric misalignment, where the accuracy of terrain elevation data may be acceptable for general terrain, but can result in a building placed 20 m above the pavement in a city. Similarly, applying texture to a building can result in an overlap of half a window between a wall and a roof. Nevertheless, these problems are well-known and Blender provides tools to realign textures and adjust the height and width of buildings.

7.2.4 Representative Visual Databases

For an airline, it is essential that their flight simulators contain databases of the airports where the airline operates. In this case, an airline will procure the visual databases for specific airports, modelled to a very high level-of-detail. The qualification of airline simulators includes assessment of the visual databases. For an engineering flight simulator, the situation is different. The purpose is not to train pilots to operate from specific airports and, consequently, the fidelity of the visual scenery and airport can be reduced and a general terrain and airport facility may be adequate for many applications. The main requirements are that the length and width of the runway and taxiways are appropriate for the aircraft used in the simulation and that the runway correlates with the corresponding data in a navigation database. In other words, it should be possible to have a single visual database containing one airport and to map this database to airports and runways defined in the navigation database. In such cases, a pilot will tune the ILS frequency for a runway and, on the approach, the runway will be visually aligned with the ILS, with the correct QDM and the correct altitude above sea level. With this simplification, the same visual database can be used for all the runways in a navigation database but adjusted according to the data for each runway.

In constructing a new visual database, the developer defines the units of measurement, the location of the origin or reference point and alignment of the terrain. Subsequent transformation of the aircraft position and orientation to the visual system coordinate frame is straightforward. The latitude, longitude, altitude and heading of the aircraft are provided by the flight model. The latitude,

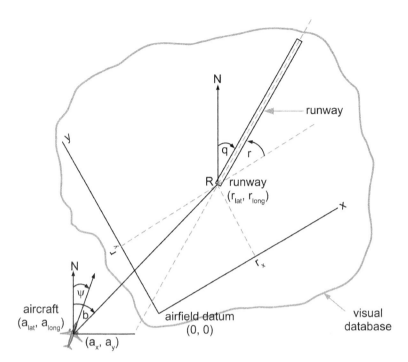

Figure 7.5 Transformation of a Generic Runway.

longitude and altitude of the runway reference point are available in the navigation database. The position (x, y, z) of the runway reference point, the alignment of the axes and the orientation of the runway are defined in the visual database coordinate frame.

A typical implementation is illustrated in Figure 7.5, where the aircraft position is defined in the world frame and the runway is defined in both the navigation database and the visual database frame.

The following information is computed in the flight model equations:

- the latitude, longitude and altitude, a_{lat}, a_{long} and a_z, respectively, of the aircraft C/G;
- the pitch, roll and yaw Euler angles, θ, ϕ and ψ, respectively, of the aircraft;
- the eye-position offsets from the C/G, E_x, E_y, E_z, respectively, transformed from the body frame to the Euler frame.

The navigation database contains the latitude, longitude and altitude, r_{lat}, r_{long} and r_z, respectively, of the runway threshold and the runway QDM q (degrees true). The visual database contains the position and height, r_x, r_y, r_z, respectively, of the runway threshold in database units, relative to the database origin and the rotation of the runway centreline r relative to the database axes.

The IG rendering algorithm is computed in the database frame using database units. The distance of the aircraft from the runway and its relative bearing are computed from the latitude and longitude of the aircraft and the latitude and longitude of the runway threshold. Rotating this vector by the relative bearing, the aircraft is effectively positioned in the Euler frame, relative to the runway threshold datum. In addition, the pilot eye-point offsets, which are defined in the body frame, are transformed to the Euler frame in the flight model and can be added directly to x, y and z, as these variables are also defined in the Euler frame. In Figure 7.5, a rotation by $2\pi - q - r$, and a translation by the runway coordinates (r_x, r_y), positions the aircraft in the database frame with the aircraft heading h set to $r - \psi + q$, where (in this example) the Euler frame is positive clockwise and the database frame is positive anti-clockwise. Furthermore, a heading of zero is north pointing in the Euler frame but aligned with the x-axis in the database frame. Assuming the visual database is level, pitch and roll angles are identical in both the world and the database frames. Similarly, the aircraft altitude is relative to the reference height of the origin of the database, which are both relative to sea level.

A simplified version of the code for the transformation from aircraft latitude, longitude and altitude to the Cartesian frame of the visual database follows. The pilot eye-point locations are defined as follows for the Boeing 747-100 flight model:

```
#define Aero_AcHeight      -8.76      /* all values in metres */
#define Aero_CGHeight      -5.1816
#define Aero_EyeXStation   25.19
#define Aero_EyeYStation   0.51
#define Aero_EyeZStation   (Aero_AcHeight - Aero_CGHeight)
```

The transformation of these components from the body frame to the Euler frame uses the DCM values described in Section 3.2.

```
void SetEyePosition()
{
    Model_Ex = Model_A11 * Aero_EyeXStation + Model_A12 * Aero_EyeYStation +
               Model_A13 * Aero_EyeZStation;
    Model_Ey = Model_A21 * Aero_EyeXStation + Model_A22 * Aero_EyeYStation +
               Model_A23 * Aero_EyeZStation;
    Model_Ez = Model_A31 * Aero_EyeXStation + Model_A32 * Aero_EyeYStation +
               Model_A33 * Aero_EyeZStation;
}
```

The database and world coordinates for runway 05L at Manchester airport are defined in the IG software as follows:

```
RunwayX         = 0.0;
RunwayY         = -1091.0;
RunwayZ         = -7.26;
RunwayRotation = 0.0;
RunwayAltitude = 72.0;

RLAT            = 53.347556;      /* runway threshold */
RLONG           = -2.287764;
RQDM            = 51.06;          /* 05L QDM true */
```

The transformation from the world frame to the runway frame is given by

```
q = Normalise(rads(RQDM));
d = Distance(AircraftLatitude, AircraftLongitude, RunwayLatitude, RunwayLongitude);
b = Bearing(AircraftLatitude, AircraftLongitude, RunwayLatitude, RunwayLongitude);
x = -d * sin(b) + (double) Ey;
y = -d * cos(b) + (double) Ey;
z = -AeroPkt.Pz + (double) Ez + GroundLevel;
```

where `GroundLevel` is the nominal height of the runway threshold above sea level and the aircraft altitude (-ve upwards) is computed in the flight model. The transformation to the visual database is given by

```
a  = normalise(TWOPI - q - RunwayRotation);
px = x * cos(a) + y * sin(a) + RunwayX;
py = y * cos(a) - x * sin(a) + RunwayY;
pz = z + RunwayZ;
p  = degrees(Pitch);
h  = degrees(RunwayRotation - Yaw + q);
r  = degrees(Roll);
```

where the values for `RunwayRotation`, `RunwayX`, `RunwayY` and `RunwayZ` are measured from the visual database. The position vector `px`, `py`, `pz` and the orientation vector `p`, `h`, `r` define the view port used by the IG.

7.2.5 Visual Database Organisation

A visual database is simply a data file defining the position, colour and texture of objects in the database. However, the database can be organised to reduce search times to locate objects in order to minimise delays in the rendering of database objects. Such structures are used widely in search algorithms. For example, a dictionary of two million words could be ordered as a linear list of words, where the average search would take one million accesses of the database. However, if the list is organised as an ordered tree, where each node contains each word as a string and left and right pointers to the sub-trees holding words alphabetically less than the word stored in the node and alphabetically greater than the word stored in the node, respectively, then the search time is proportional to $\log_2 10^6 \approx 21$, a speed improvement of nearly 50,000. The tree search has two requirements: the database must be organised as a tree structure and the search algorithm must support recursion. All nodes of the tree have an identical structure, with the data value and pointers to the left sub-tree and the right sub-tree and with the leaf nodes of the tree pointing to a NULL (or empty) node.

One method to discard objects in the visual database at the earliest opportunity is to project the viewing frustum onto the terrain. Objects outside these intersections should not be visible and, quite possibly, large areas of the terrain can be discarded without any significant processing, as illustrated in Figure 7.6.

The region ABCD covers the extremities of the frustum intersection and the search for active objects can be constrained to this area. To avoid possible discontinuities, the region may be expanded slightly to ensure that all visible objects are included. The main limitations with this method are, firstly, the results can be erroneous if the terrain is not flat, for example, a hill hiding objects that may be in view, and secondly, when the aircraft is very low, the frustum may cover most of the objects in the database, even though many are not actually visible. The actual intersection can change very quickly with changes to the aircraft Euler angles although the frustum intersection can be tracked from previous frames to estimate the directional trend of the frustum.

The database can also be organised as a tree structure to support level-of-detail (LOD) operations. For example, an airport contains an airfield, runways, taxiways, lighting, buildings and an airfield. Similarly, a runway includes a centreline, edge markings, skid mark textures and threshold and touchdown markings. A simple classification is illustrated in Figure 7.7.

At the leaf nodes, each object is defined in terms of its maximum detail, typically the maximum resolution of texture. But at higher levels, several objects can be reduced to a single object with significantly less detail. The trigger for changes of resolution is the distance from the pilot eye-point. For example, the airport can be represented by a coarse image of the airfield and runway. Nearer

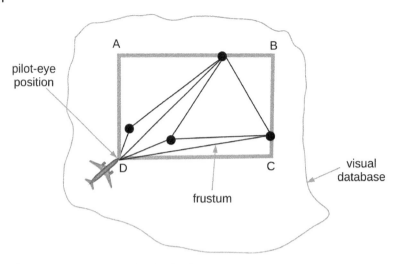

Figure 7.6 Projection of the Viewing Frustum.

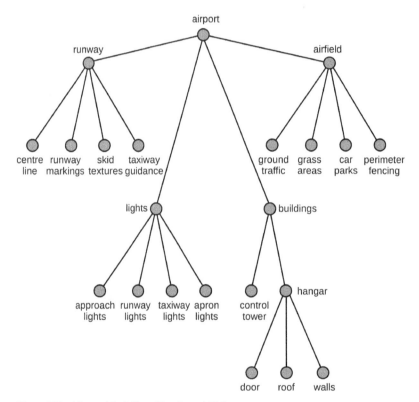

Figure 7.7 Hierarchical Classification of Objects.

to the airport, it is possible to distinguish buildings. Closer still, the control tower and a hangar are visible and, finally, taxiing the aircraft, the doors, walls and roof are rendered in detail. The editing tools enable objects to be defined at different resolutions and placed at the appropriate level of the tree. During traversal of the tree, the search stops when there would be no improvement in rendering quality by searching deeper. The advantage of using LOD methods is the considerable saving of time to render textures. The only drawbacks are that several versions of texture need to be created for objects and that care is needed to avoid abrupt and noticeable changes of the LOD. If these levels are chosen carefully, the viewer should be unaware of the transitions.

One particular area of visualisation where LOD methods are used is the representation of aircraft in scenes, as shown in Figure 7.8. The cylindrical shape of an aircraft fuselage may require several hundred textured triangles to appear smooth and rounded. However, with increasing distance, the number of triangles can be reduced without noticeable reduction in fidelity. There are also potential savings in increasing the mesh spacing used to render terrain where detailed resolution may only be needed close to the ground. However, the overhead of switching resolutions for meshes, triangles and textures may outweigh the potential gain in performance. Moreover, validation of such algorithms requires extensive testing to ensure no visible anomalies occur with changing levels of detail.

One method of reducing models to lower levels of detail is mesh compression, where a common characteristic of adjacent tetrahedrons enables their merging, based on an algorithm of edge collapsing (Cignoni et al., 2000), as shown in Figure 7.9.

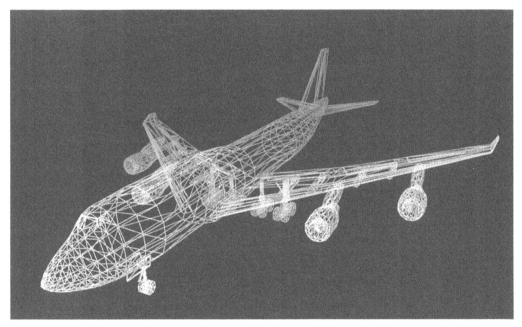

Figure 7.8   Mesh Form of an Aircraft Visual Model.

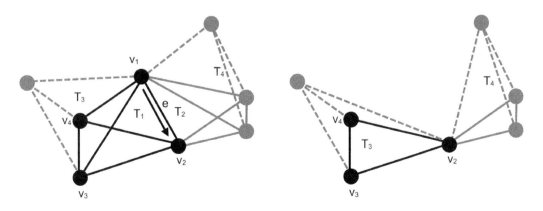

Figure 7.9   Mesh Collapsing.

Assume the vertices are a function of three variables (u, v, w). Removing the edge e and the vertex v_1 in Figure 7.9(a), the tetrahedrons T_1 and T_2 which share this edge are removed from the mesh, as shown in Figure 7.9(b), modifying the tetrahedrons T_3 and T_4. The errors in this mesh, as a result of the edge collapse, can be measured and compared with the original mesh on a node-by-node basis. Typically, for 3D models, this measure is the accuracy of the coordinates of the reduced mesh in comparison with the original mesh, given by

$$\varepsilon_f = \left| F(\sqrt{u^2 + v^2 + w^2}) - F'(\sqrt{u^2 + v^2 + w^2}) \right| \tag{7.1}$$

where ε_f is the local field error, F is the original field and F' is the reduced field. The simplification proceeds incrementally to locate an edge, where its removal causes the minimum disruption to the model. For example, an aircraft wing may be formed from 200 tetrahedrons, but as many of these are almost coplanar, if the wing is reduced to 20 tetrahedrons, the wing will still have a recognisable shape, but with a significant reduction in the number of triangles to be rendered.

7.2.6 Binary-spaced Partition Trees

One method to minimise the amount of processing to discard non-visible objects is to organise the scene as a tree structure, to satisfy the criterion that, if a node of the tree is not visible, then none of its sub-nodes are visible and the traversal of the tree can stop at the parent node. Consider a visual database containing the objects shown in Figure 7.10(a). Each dot represents an object such as a building or a light or a set of lights in close proximity. For the purpose of illustration, the tree contains 32 objects. The binary-spaced partition (BSP) tree in Figure 7.10(b) shows the partitioning of the tree with the cuts numbered in the right half of the tree to illustrate the process of sub-division. Cut 1 is chosen such that there are 16 objects in the left half of the tree *ABCHIJLKSTUVabcd* and 16 objects in the right half of the tree *DEFGNMOPQRWXYZef*. Cut 2 splits the sub-tree into two groups of eight,

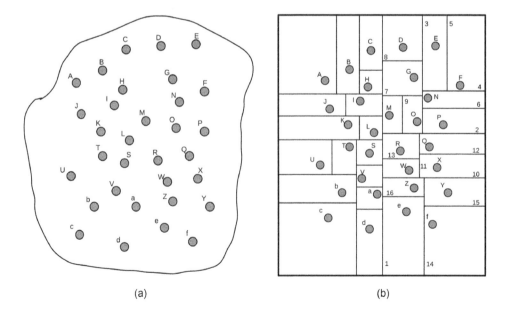

(a)   (b)

Figure 7.10   A BSP Tree.

the right node containing *DEFGNMOP* and the left node containing *QRXYYZef*. Cut 3 splits the sub-tree into two groups of four, the right node containing *EFNP* and the left node containing *DGOM*.

Cut 4 splits the sub-tree into two groups of 2, the right node containing *EF* and the left node containing *NP* and so on. The cuts to the left half of the tree are omitted in this illustration. The order of traversal of the tree in this case is the node, then the right sub-tree and then the left sub-tree. Depending on the criteria for partitioning the tree, the traversal stops when a specific condition is met. As the method is recursive, the search stops when all searches left and right are completed. The attraction of methods of this form is that the search time is $O(\log_2 N)$, where N is the number of nodes in the tree. Of course, in a worst case, every node may still be visited and the advantages of using a BSP tree are lost. However, in practice, database objects are unlikely to be ordered in a totally random way and tests to validate the algorithm are likely to identify such problems. Note also that the sub-division of the tree is arbitrary and different selection methods can be tested offline for a given database to find the optimal partitioning, in terms of the number of tree accesses, which will affect the run-time performance.

7.3 OpenSceneGraph

Although it is possible to write a complete rendering package in OpenGL (Kessenich et al., 2017) or DirectX (Gray, 2003), it is far more convenient to use a rendering package, which is also referred to as a graphics engine or games engine. Developers of these packages will optimise the rendering for modern graphics architectures and provide capabilities to meet the requirements of a wide range of applications. For users, the effort to develop 3D applications is significantly reduced and, in general terms, the user needs only to provide a visual database and an interface between the simulator and the rendering package. In many cases, these packages are proprietary and the source code is not accessible, for example *Unity*. Of course, for many users, this is not a restriction as the advantages of access to a cross-platform, real-time graphics engine outweigh the lack of open-source code.

One package which does provide completely open-source software for rendering and has gained wide acceptance for real-time graphics applications is *OpenSceneGraph* (OSG) which is written in C++ and is based on OpenGL (Wang and Xuelei, 2010). OSG is compatible with most visual database formats. The user can add cameras to a scene and provide sources of lighting. For the flight simulator developer, OSG offers a number of advantages:

- It can be integrated with OpenGL modules and exploits the capabilities of modern GPU architectures (Wang and Xuelei, 2012).
- It is available for a range of platforms including Windows, Linux and Android.
- It is written in C++ and supports linkage to modules written in C.
- It supports many visual database formats including OpenFlight.
- It provides LOD functions.
- It supports dynamic entities including aircraft, land vehicles and sea states.
- It includes multi-threading to exploit multiple core architectures.
- It manages the framebuffer and gives the user complete control to define the display resolution and field-of-view.
- It supports database paging extending to full-earth geospatial paged-terrain.

A particular advantage for flight simulation is that OSG can run on a standard PC with a modern graphics card, the visual database is managed completely by OSG and the user application is only required to provide the camera positions and environmental conditions such as lighting and

fogging levels. Although not strictly a real-time implementation, the performance of OSG is sufficient to render large and complex visual databases at frame rates in excess of 50 frames per second.
To illustrate the capability of OSG, the following C^{++} program loads a visual model of a Boeing-747
aircraft, which is displayed on the screen and allows the user to control the camera position and
orientation with mouse and keyboard inputs.

```cpp
#include <osgDB/ReadFile>
#include <osgViewer/Viewer>

int main(int argc, char **argv)
{
    osgViewer::Viewer viewer;
    viewer.setSceneData(osgDB::readNodeFile("b747sia.flt"));

    return viewer.run();
}
```

The `#include` statements declare the header files needed for the program. In the main function,
the structure `viewer` is declared, a specific OpenFlight model is loaded and the viewer function
is executed until the user terminates the session. In just a few lines, the user is provided with the
capability to visualise a potentially complex scene.

Renderers come in various forms. In some renderers, the user sets up conditions such as the sky
or weather conditions using a desktop GUI interface and is able to specify the database file to be
loaded. In other renderers, the user is provided with a toolkit and has complete control of the
resources and facilities provided by the package, typically in the form of calls to library functions.
One other potential requirement, particularly if software reuse is important, is the ability to embed
or integrate C code modules directly.

OSG is a renderer which provides a C^{++} toolkit (Wang and Xuelei, 2010). Although the learning
curve to produce the first scene is considerable, there are numerous examples of rendering applications
(Wang and Xuelei, 2012). In many ways, the toolkit is similar to GLFW; the graphics environment is
set up, the state of the scene and camera position of the camera are updated as a result of user input,
the frame is rendered and a main loop is provided, which is terminated by a user process when an
'end-of-session' condition is detected. Within OSG, there are numerous threads that are transparent
and managed by OSG to minimise any delays in rendering the frame. Although OSG is underpinned
by OpenGL, generally there is no need for the user to implement any OpenGL code. The interface
between a user program and the frame-store and the GPUs is managed completely by OSG.

OSG defines the header files according to the various modules contained in OSG. These include
header files relating to reading files, managing the aircraft position, setting visibility levels, defining lighting, computing geometry, defining culling values, defining the rendering of billboards
(e.g. trees) and so on. The general structures needed for the simulation are defined as follows:

```cpp
osgViewer::Viewer viewer;
osg::Vec3f vecPosAircraft;
osg::Vec3f vecAttAircraft;
osg::Vec3f vecPosTarget;
osg::Vec3f vecAttTarget;

osg::ref_ptr<osg::Fog> fog = new osg::Fog();
osg::ref_ptr<osg::LightSource> sunLight;
```

```
osg::ref_ptr<osg::Node> TargetNode;
osg::ref_ptr<osg::PositionAttitudeTransform> TargetXForm =
                                      new osg::PositionAttitudeTransform();
osg::ref_ptr<osg::Group> SceneRoot = new osg::Group();
```

where **viewer** will be used for the scene view, **vecPositionAircraft** and **vecAttitudeAircraft** define the aircraft position and attitude, respectively, and similarly for the 'target' aircraft (in this case, a single aircraft in the scene). Variables for fog and sunlight will be required. A node is added to the scene for the target aircraft and a transform is added to access the target. Finally, **SceneRoot** is a single pointer to the uppermost node of the scene, representing the whole scene.

A head-up display (HUD) is also added and its functions are covered in Section 7.5. A function to define the light source is given as follows:

```
osg::ref_ptr<osg::LightSource> createSunLight(void)
{
    osg::ref_ptr<osg::LightSource> sunLightSource = new osg::LightSource;
    osg::ref_ptr<osg::Light> sunLight = sunLightSource->getLight();
    sunLight->setPosition( osg::Vec4( 0.0f, 0.0f, 10000.0f, 1.0f ) );
    sunLight->setAmbient( osg::Vec4( 0.2f, 0.2f, 0.2f, 1.0f ) );
    sunLight->setDiffuse( osg::Vec4( 0.8f, 0.8f, 0.8f, 1.0f ) );
    sunLightSource->setLight( sunLight.get() );
    sunLightSource->setLocalStateSetModes( osg::StateAttribute::ON );

    sunLightSource->getOrCreateStateSet()->setMode(GL_LIGHTING,
        osg::StateAttribute::ON | osg::StateAttribute::OVERRIDE);

    osg::ref_ptr<osg::LightModel> lightModel = new osg::LightModel;
    lightModel->setAmbientIntensity(osg::Vec4(1.0f,1.0f,1.0f,1.0f));
    sunLightSource->getOrCreateStateSet()->setAttribute( lightModel.get() );

    return sunLightSource;
}
```

The source is allocated together with its position and the location of the ambient and diffuse light sources. This light source is turned on permanently and the lighting model is defined and set to full intensity and then included in the sunlight source. **sunLightSource** is a unique variable for the sunlight structure and functions.

In the main function, the terrain OpenFlight file is loaded in a single call as follows:

```
osg::ref_ptr<osg::Node> TerrainNode = osgDB::readNodeFile(filename);
if (!TerrainNode)
{
    std::cerr << "Failed to load terrain database!\n";
    exit(1);
}
```

The result of loading the file is checked and any failures result in terminating the visual system at this stage. The sky model is then loaded. Note that this is only one version of a sky dome.

```
osg::ref_ptr<osg::Node> SkyNode = osgDB::readNodeFile("sky.flt");
if (!SkyNode)
{
    std::cerr << "Failed to load sky database!\n";
    exit(1);
}

osg::ref_ptr<osg::Node> SkyBoxNode = osgDB::readNodeFile("models/skyb.ac");
if (!SkyBoxNode)
{
    std::cerr << "Failed to load skybox database!\n";
    exit(1);
}

osg::ref_ptr<osg::PositionAttitudeTransform> SkyBoxXForm =
                                    new osg::PositionAttitudeTransform();
SkyBoxXForm->addChild(SkyBoxNode.get());
osg::Vec3 SkyBoxPosit(0,0,-10000);
SkyBoxXForm->setPosition( SkyBoxPosit );

osg::ref_ptr<osg::ClearNode> backdrop = new osg::ClearNode;
backdrop->setClearColor(osg::Vec4(0.8f,0.8f,1.0f,1.0f));
```

Typically, a sky texture is applied to the surface of a hemispherical dome, constructed from a series of triangles. The dome is loaded as an OpenFlight file and defined as a node with a new transformation to add the sky node to the scene and position it in the centre of the scene. The dome should be sufficiently large (or distant) that the same view is seen at all positions in the scene. Finally, the sky node is given a background sky colour. A skybox used with an OpenFlight visual database is shown in Figure 7.11.

The terrain consists of a tile containing typical terrain with the skybox at the centre of the terrain. For simplification, a cube can be used for the skybox in preference to a hemisphere, with the

Figure 7.11 A Sky Dome.

proviso that the corners of the cube do not add any noticeable distortion. Note also that there is no need for any rotation or perspective transformation of the skybox; it always remains at the same distance from the viewer.

With the terrain file loaded, a HUD is added to the scene, the light sources are initialised and the sky dome is in place, the position and field-of-view of the camera are defined, the scene is set to point to the root node and the scene is enabled (or realised). The camera orientation is defined as follows:

```
osg::Matrixd myCameraMatrix;
osg::Matrixd cameraRotation;
osg::Matrixd cameraOffsetRotation;
osg::Matrixd cameraTrans;

cameraOffsetRotation.makeRotate(
    osg::DegreesToRadians(0.0), osg::Vec3(0,1,0),             // roll
    osg::DegreesToRadians(0.0), osg::Vec3(1,0,0),             // pitch
    osg::DegreesToRadians(ChannelOffset), osg::Vec3(0,0,1) ); // heading
```

setting initial roll, pitch and yaw angles of zero for a forward-looking (centre channel) camera and an offset viewing angle for side screens or an extended screen or mirror.

In the main loop, the camera position, camera orientation and state of the visual system are acquired from the flight simulator, for example, in a broadcast packet. For the most part, a visual system is passive unless it is required to transmit height above terrain, or collision with the terrain, or collision with other entities in the scene. The camera position is translated and rotated according to the aircraft position and OSG is invoked to render a frame, as shown in the main loop in the following code fragment:

```
while (!viewer.done())
{
    retval = socket_get_data();

    cameraTrans.makeTranslate(vecPosAircraft.x(), vecPosAircraft.y(), vecPosAircraft.z());
    cameraRotation.makeRotate(
    osg::DegreesToRadians(vecAttAircraft.z()), osg::Vec3(0,1,0),   // roll
    osg::DegreesToRadians(vecAttAircraft.y()), osg::Vec3(1,0,0),   // pitch
    osg::DegreesToRadians(vecAttAircraft.x()), osg::Vec3(0,0,1) ); // heading
    myCameraMatrix = (cameraOffsetRotation * cameraRotation) * cameraTrans;
    osg::Matrixd i = myCameraMatrix.inverse(myCameraMatrix);
    osg::Matrixd xxx = osg::Matrixd::rotate( -M_PI*0.5, osg::Vec3(1,0,0) );
    viewer.getCamera()->setViewMatrix(i * xxx);

    viewer.frame();
}
```

Additional information passed to the visual system includes the time of day to set the appropriate lighting conditions, visibility and cloud base to set the fogging level, the position and attitude of other aircraft and vehicles in the scene and possibly a different camera position, for example, if the scene is played back from another viewpoint. Note that a cloud base may also be set as partial transparency of clouds formed as 3D objects.

The function **socket_get_data** is the main interface between the simulator and the IG and performs two functions. Firstly, it captures UDP packets broadcast by the simulator which include the aircraft position (latitude, longitude and altitude) and the aircraft attitude (pitch, roll and yaw). Secondly, it transforms the aircraft position and attitude to the IG coordinate frame. The first part of the transformation is from world coordinates to database coordinates:

```
q = normalise(rads((double) (NavPkt.RunwayQDM)) + (double) (NavPkt.MagneticVariation)));
d = Distance(AeroPkt.Latitude, AeroPkt.Longitude,
             (double) NavPkt.RunwayLatitude, (double) NavPkt.RunwayLongitude);
b = Bearing(AeroPkt.Latitude, AeroPkt.Longitude,
             (double) NavPkt.RunwayLatitude, (double) NavPkt.RunwayLongitude);
x = -d * sin(b) + (double) AeroPkt.Ey;
y = -d * cos(b) + (double) AeroPkt.Ex;
z = -(AeroPkt.Pz + AeroPkt.Ez) + NavPkt.GroundLevel;
```

where **AeroPkt** is the UDP packet transmitted by the flight model, **q** is the runway QDM, **d** and **b** are the distance and bearing to the runway, respectively, and **x**, **y** and **z** are the aircraft coordinates in the visual database frame. **Ex**, **Ey** and **Ez** are the pilot eye offsets from the C/G in the body frame. The second transformation takes account of the runway orientation in the database frame, as follows:

```
a = normalise(TWOPI - q - RunwayRotation);
sina = sin(a);
cosa = cos(a);

px = x * cosa + y * sina + RunwayX;
py = y * cosa - x * sina + RunwayY;
pz = z + RunwayZ;
p = degrees(pitch);
h = degrees(RunwayRotation - yaw + q);
r = degrees(roll);

vecPosAircraft.set(px, py, pz);
vecAttAircraft.set(h, p, r);
```

where **RunwayRotation**, **RunwayX**, **RunwayY** and **RunwayZ** define the runway coordinates in the database frame. The OSG functions **vecPosAircraft.set** and **vecAttAircraft.set** set the eye-point position and attitude for OSG for the current frame, respectively. This method enables several OpenFlight databases to be used by the IG, where the runway details can be captured from a database viewer. For the Manchester airport OpenFlight database, the following details are included in the IG software:

```
RunwayX         = 0.0;
RunwayY         = -1091.0;
RunwayZ         = -7.26;
RunwayRotation  = 0.0;
RunwayAltitude  = 72.0;
```

Figure 7.12 shows the OpenFlight database for Manchester airport. A variant of the IG was developed to provide a view looking directly down on the database. As the cursor (not shown) is moved over the image, in this case at the runway threshold, the latitude and longitude are displayed together with the *x* and *y* coordinates of the database and the altitude. By panning and zooming, it is possible to obtain the database coordinates of any points in the database, which are needed for the transformations given earlier.

The interface between the simulator and the IG, *cgi.c*, is only 1400 lines, which includes the selection of databases, management of dynamic targets and responding to IOS commands. Although some of the details of the implementation seem complicated, they provide a template which is unlikely to vary significantly for most applications in flight simulation. More dynamic objects can easily be added to the scene and, with advances in GPU performance and memory

latitude, longitude

x, y, altitude

Figure 7.12   Manchester Airport Database Coordinates.

Figure 7.13   Approach to Bristol Lulsgate Airport in 5 km Visibility (See Plate 6).

capacity, the loading of the database, 3D rendering operations including lighting, shading, hidden-surface removal and LOD, is managed transparently by OSG. It also enables the camera settings to be matched to the screen or projector resolution and field-of-view. With the further advantages of high-performance rendering, compatibility with a wide range of visual databases and open software, OSG should not be dismissed as a potential candidate for a real-time IG in flight simulation. A typical image from the OSG IG is shown in Figure 7.13, which is based on an OpenFlight database for Bristol Lulsgate airport.

7.4   X-Plane 11

7.4.1   X-Plane 11 API

Although this textbook focuses on open software, there are examples of commercial software where the volume of sales is high, the unit cost is low and far more effort is invested in the software than any one individual could undertake. Two examples of this genre are covered in this book: the *Navigraph* package provides regular updates of worldwide navigation aids and airports, and X-Plane (https://www.x-plane.com) offers a software development kit (SDK) to enable a user to connect an interface to specific elements of the package. X-Plane is sold as a flight simulator and includes a relatively high-performance IG and the availability of several hundred visual databases of major airports and regions of the world and models of civil, military and general aviation aircraft. The flight model, navigation and displays can be disabled allowing a developer to use only the IG component as a separate real-time graphics engine. As a result of the popularity of X-Plane, a small but active community of developers has evolved, offering models of terrain, airports, cities, aircraft and ground vehicles which are compatible with X-Plane. Typically, these models are available for $20–50 and open tools are also provided to develop and modify these models. For that reason, while not intending to endorse X-Plane, it is worth investigating the degree to which it can be interfaced to an external flight simulator and its potential as a real-time IG.

Unlike OSG, the external environment is set via the desktop API provided with X-Plane, which includes the visibility, weather conditions and sky. In addition, the visual database and specific aircraft models are also loaded via the API. It is also necessary to disable the instrument displays, flight model and crosshairs. One further difference is that the (x, y, z) axes are orientated *north-east-down* in OSG, whereas they are orientated *east-up-south* in X-Plane. In OSG, the visual scene is rendered with respect to an origin defined in the visual database, which is tiled dynamically. With X-Plane, the user supplies latitude, longitude and altitude, which is converted to local (x, y, z) values, with respect to the current reference point, using functions provided in the SDK. Extensive notes are provided for developers in the SDK documentation (https://developer.x-plane.com/sdk). In particular, the following documents provide details to interface X-Plane to a flight simulator:

- **XPLMDisplay** provides an API to add an extra display and a call-back mechanism for mouse and keyboard interaction, including hot keys. A user is able to write directly to this display, for example to provide a moving-map display or a HUD.
- **XPLMGraphics** outlines the coordinate systems used by X-Plane and describes functions to translate between the user coordinates and X-Plane coordinates.
- **XPLMDataAccess** enables data to be read from and written to X-Plane variables, including the aircraft position and frame rate. X-Plane contains over 4000 'data references', which can only be identified by strings. A set of functions are provided to obtain and set data references covering all types of variables.
- **XPLMScenery** provides functions to interrogate the terrain data, including the height of the terrain directly below the aircraft.
- **XPLMCamera** allows the user to control the camera by means of a call-back to give the view from the flight deck or an external position.

The functions needed to interface to X-Plane are defined in six header files: **XPLMDisplay.h**, **XPLUtilities.h**, **XPLMCamera.h**, **XPLMDataAccess.h**, **XPLMPlugin.h** and **XPLMGraphics.h**.

7.4.2 An X-Plane 11 Plug-in

The user is able to exert surprisingly little control over the X-Plane API. As X-Plane is a compiled program, the user interacts with the run-time system by means of call-backs, which are described in the documentation. It also differs from OSG in that the API includes software for the flight model, navigation and the aircraft displays known as the 'physics engine', which controls the aircraft motion and must be disabled to allow the user code to update the aircraft position and attitude from information provided by the external simulator. This lack of control also includes the frame rate. Although the documentation provides guidance to minimise the time required for the graphics computations per frame, it is possible, particularly with highly detailed scenes, that the frame time may be exceeded.

The code to obtain a handle to an X-Plane variable is as follows:

```
XPLMDataRef GetRef(const char str[])
{
    XPLMDataRef r;
    r = XPLMFindDataRef(str);
    if (r == NULL)
    {
        printf("Cannot access %s\n", str);
        exit(-1);
    }
    return r;
}
```

where the handle returned is given by the X-Plane type **XPLMDataRef**. This check is necessary to ensure that the variable is found, otherwise the IG is terminated.

X-Plane calls **XpluginStart** for all active plug-ins, which initialises the user IG code as follows:

```
PLUGIN_API int XPluginStart(char *outName, char * outSig, char * outDesc)
{
    int t[1] = { 1 };

    strcpy(outName, " Flight simulator IG");
    strcpy(outSig,  "simcgi.camera");
    strcpy(outDesc, "A plugin providing an IG channel.");

    PhysicsEngine    = GetRef("sim/operation/override/override_planepath");
    JoystickOverride = GetRef("sim/operation/override/override_joystick");
    TargetXRef       = GetRef("sim/flightmodel/position/local_x");
    TargetYRef       = GetRef("sim/flightmodel/position/local_y");
    TargetZRef       = GetRef("sim/flightmodel/position/local_z");
    TargetPitchRef   = GetRef("sim/flightmodel/position/theta");
    TargetRollRef    = GetRef("sim/flightmodel/position/phi");
    TargetYawRef     = GetRef("sim/flightmodel/position/psi");

    gHotKey = XPLMRegisterHotKey(XPLM_VK_F8,
                                 xplm_DownFlag,
                                 "SimCGI External View",
                                 MyHotKeyCallback,
                                 NULL);

    XPLMSetDatavi(PhysicsEngine, t, 0, 1);  /* disable the physics engine */
    XPLMSetDatai(JoystickOverride, 1);

    UDP_Start();

    return 1;
}
```

GetRef provides access to the X-Plane variables **override_planepath**, **override_joystick**, **x**, **y**, **z**, **theta**, **phi** and **psi**. Note that the user-defined variables **PhysicsEngine**, **JoystickOverride**, **TargetXRef**, **TargetYRef**, **TargetZRef**, **TargetPitchRef**, **TargetRollRef**, **TargetYawRef** and **gHotKey** are not treated as variables, they are simply a handle to access the equivalent X-Plane variables. The function **XPLMRegisterHotKey** registers an F8 key press to invoke **MyHotKeyCallback**. The function **XPLMSetDatavi** disables the physics engine (by setting the fourth argument to 1) and the function **XPLMSetDatai** disables any spurious joystick input. Finally, in this version of the simulator, the aircraft data is passed in a UDP packet and the UDP library, used for the other simulator modules, is reused to manage the network transfers.

The user program can only start after the initial conditions have been set up in X-Plane; when the F8 key is pressed, the function **MyHotKeyCallback**, is called which contains the following initialisation code:

```
void MyHotKeyCallback(void * inRefcon)
{
    XPLMControlCamera(xplm_ControlCameraUntilViewChanges, SimCGIViewerFunc, NULL);

    BEGIN_UDPLib();
    UDPLib_Connect(1, &IOPkt1, sizeof(IOPkt1));
    UDPLib_Connect(3, &IGPkt, sizeof(IGPkt));
    etc for the other nodes
    UDPLib_Open(8); /* IG node = 8 */

    QueryPerformanceFrequency((void *) &freq);   /* initialise timers */
    QueryPerformanceCounter((void *) &Timer0);
    Timer1 = Timer0;
    Timer2 = Timer0;

    XPLMDebugString("Hotkey Callback: Defaults Initialised\n");
}
```

The **XPLMControlCamera** call passes control to the user function **SimCGIViewerFunc**. The UDP library acquires the packets broadcast by the other simulator nodes for the IG. Finally, the timer is initialised and the start time is copied to the timers **Timer1** and **Timer2**.

The incoming UDP packets are decoded in exactly the same way as the OSG version with the exception that a check is made for delayed packets. If the time since receiving the last frame exceeds 22 ms (assuming a 50 Hz frame rate), the packets are discarded and a new set of packets is acquired. The processing in X-Plane to render an image can exceed the frame time and, if the IG is synchronised to the incoming simulator packets, the delays will accumulate and the IG will lag increasingly behind the pilot inputs. However, if these late packets are discarded, this latency cannot build up as the aircraft position and attitude used in the current frame are guaranteed to have been computed in the previous frame. The loss of an occasional packet at 50 Hz is unlikely to be noticed and this delay is completely independent of the flight model computations. Although this situation is far from ideal, the number of discarded packets is logged and reported at the end of a session.

X-Plane will call **SimCGIViewerFunc** once every frame. It is the responsibility of the user code to detect any changes to the scene environment and to set the aircraft position and attitude in the X-Plane coordinate frame. Although the X-Plane database will be initialised to the loaded database coordinates, the flight simulator software will be unaware of these flight conditions, which will be reset following an IOS restore command (or explicit aircraft repositioning). Initially, the simulator sets the aircraft latitude and longitude to 0°, and if this condition is detected, the frame update is ignored to avoid the aircraft moving from the X-Plane referenced origin. If the aircraft position is updated, the following call is made to **XPLMWorldToLocal**:

```
XPLMWorldToLocal(IGPkt.Latitude * RAD2DEG,
                 IGPkt.Longitude * RAD2DEG,
                 -IGPkt.Pz,
                 &AircraftX,
                 &AircraftY,
                 &AircraftZ);
AircraftPitch = IGPkt.Pitch;
AircraftRoll  = IGPkt.Roll;
AircraftYaw   = IGPkt.Yaw;
```

where **RAD2DEG** ($180.0/\pi$) provides a conversion from radians to degrees. The aircraft latitude, longitude and altitude transmitted by the simulator are converted to local coordinates returned in **AircraftX**, **AircraftY** and **AircraftZ**. These values are relative to the X-Plane local origin, which is updated independently by X-Plane if the aircraft moves a significant distance from this origin. The X-Plane local coordinates are used to set the camera position and orientation, as shown in the following code fragment:

```
if (CameraPosition == 0)   /* place the camera in the aircraft cockpit */
{
    /* set the camera position to the pilot eye-point */
    outCameraPosition->x = (float) (AircraftX + IGPkt.Ey); /* east component */
    outCameraPosition->y = (float) (AircraftY - IGPkt.Ez); /* up component */
    outCameraPosition->z = (float) (AircraftZ - IGPkt.Ex); /* south component */
    outCameraPosition->pitch   = AircraftPitch * RAD2DEG;
    outCameraPosition->heading = AircraftYaw * RAD2DEG;
    outCameraPosition->roll    = AircraftRoll * RAD2DEG;
    outCameraPosition->zoom    = 1.0f;
}
```

The target position is similarly updated which may be used as a second aircraft in the scene or to position the camera at a specific location, for example as a side view of the aircraft or a view from the control tower. The code for these other locations is omitted for reasons of brevity but can be found in the companion code referenced in the Appendix.

The interface between the simulator and the IG, *cgi.c*, is only 550 lines, which includes reading incoming UDP packets and responding to IOS commands. Two typical images rendered by the IG interfaced to X-Plane are shown in Figures 7.14 and 7.15.

7.4.3 OSG versus X-Plane 11

There are two main differences between the OSG and X-Plane interfaces. Firstly, OSG maps the databases to the runways given in the navigation database, whereas with X-Plane, a specific visual database is loaded and the aircraft latitude and longitude correspond to that database. There is no requirement for a transformation to database coordinates. Secondly, OSG sustains a 50 Hz frame rate whereas the frame loading varies considerably with X-Plane, which can result in the frame times being exceeded. The following code is used in the packet handling to respond to frame overload:

Figure 7.14 Manchester International Approach 2 nm (see Plate 7).

Figure 7.15 London Heathrow International Dusk Approach 1 nm (see Plate 8).

```
p = UDPLib_GetPkt();
if (p == 1)  /* I/O RPi 1 */
{
    QueryPerformanceCounter((void *) &Timer2);
    if ((Timer2 - Timer1) > 220000)  /* frame > 22 ms? */
    {
        do
        {
            p = UDPLib_GetPkt();  /* wait for pkt3 */
            if (p == 3)
            {
                pkt3found = true;
            }
        } while (!pkt3found);
        pkt1found = false;  /* discard pkt1 */
```

```
        pkt3found = false;   /* discard pkt3 */
        Timer1 = Timer2;     /* remember time of last pkt1 */
        PktsLost += 1;
        continue;
    }
    else
    {
        pkt1found = true; /* now wait for pkt3 */
        Timer1 = Timer2;  /* remember time of last pkt1 */
    }
}
```

As X-Plane runs under Windows (msys2), the function `QueryPerformanceCounter` is used to measure time accurately. When a packet is received from node 3 (the flight model computer), a check is made that no more than 22 ms has elapsed since the previous packet and, if it has, the packet is effectively discarded as the loop will wait for the subsequent packet from node 3. Otherwise the packet is read and `pkt3found` is marked accordingly. Diagnostic information is recorded to provide some insight into the overall performance of X-Plane as an IG.

In many ways it is unfair to compare these two IG systems. OSG is designed explicitly as a real-time renderer for a wide range of visual database formats, whereas X-Plane is a flight simulation package (or arguably a game) with a proprietary database format. Both packages provide frame rates of 50 fps with extensive and detailed databases and both can be interfaced relatively easily to flight simulation software written in C. The choice is left to the developer and is likely to be influenced more by the availability of visual databases and models than features offered by the respective packages. There is no interoperability between these (and other) renderers and the primary advantage of X-Plane is the large number of visual databases, airports and aircraft available for relatively low cost in comparison with the man-hours of effort needed to develop equivalent models for OSG. Both packages have attracted enthusiastic contributors and both invest considerable effort in exploiting state-of-the-art graphics hardware.

It is likely that OSG, X-Plane and Unity will be used in real-time flight simulation packages for several years to come and will also exploit the performance gains of GPU hardware and graphics languages, including OpenGL, Vulkan and Metal. It is also likely that ray-tracing methods will supersede the traditional 3D rendering methods in the next five years, but it is probable that these changes will be transparent to the user, other than the increased scene content and detail, increased resolution per channel and higher frame update rates.

7.5 Head-up Displays

Head-up displays are commonplace in modern military aircraft (Hall, 1993) and are used increasingly in civil transport aircraft. In the aircraft, 2D graphics is projected onto glass plates directly in front of the pilot. The optical response of the HUD is optimised to display the graphics (typically green phosphor) while minimising the optical attenuation through the HUD. The advantage is that essential information is, in effect, overlaid on the outside scene in front of the pilot covering a lateral field-of-view of up to 20°. This information may include flight data, weapon guidance, terrain avoidance and flight plan and navigation data. Although the information is likely to contain a mixture of graphics and text, the geometry of the system of lenses used in a HUD provides a collimated image seen in the HUD. In other words, the HUD data appears to be displayed at an infinite distance and the pilot can see both the terrain and the HUD data without requiring any visual accommodation of focal length.

The installation of a HUD in a flight simulator is slightly different. There are four possible options:

1) Use an actual HUD – if the equipment is airworthy it will be expensive, and if the HUD is located in the same position as the aircraft, it may not be possible to align the collimation of the HUD with a projection mirror that is only a few metres in front of the pilot.
2) Fabricate a facsimile HUD – the cost is significant as the full combiner optics and a symbol generator are still required.
3) Install a flat-screen display in front of the pilot – although the HUD information is visible, it obscures the pilot's outside view and it is difficult to generate the hidden fragment of the 3D scene on the display. Only a small movement of the pilot's eye will cause the view in the display to differ from the outside view. In addition, the pilot's eyes are focused on the near display, whereas when viewing a collimated display, the pilot's view is focused at infinity. For most applications, this solution is unacceptable.
4) Render a 2D HUD as an overlay on the projected image – if the projected image is several metres in front of the pilot, both the scene and the HUD information will, effectively, be collimated. In addition, the transparent properties of the combiner can be simulated by providing a small optical attenuation of the HUD image overlaid on the scene, which is visually similar to an actual HUD and is achieved by modulating the alpha (transparency) component of HUD image. In addition, an empty frame can be placed in front of the pilot to replicate the HUD frame. However, if the pilot eye-position changes, there is no change in the parallax between the HUD and the external scene, which is contrary to the airborne situation, that is, the pilot may also see HUD information outside the HUD frame.

Invariably, the choice of implementation is a balance between the training benefits and the cost of the equipment. However, a 2D HUD overlay meets many of the requirements for a HUD in a flight simulator and only requires the mounting of a panel in the shape of the HUD frame and 2D rendering software to emulate the HUD. A typical HUD format, generated using 2D graphics, is shown in Figure 7.16.

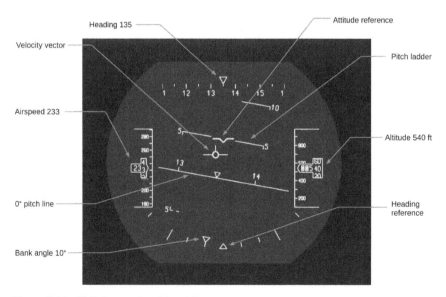

Figure 7.16 HUD Format (see Plate 10).

Figure 7.17 A HUD Overlay (see Plate 11).

The HUD is shown in a basic configuration, providing pitch, roll, altitude, airspeed and heading. The dark green background replicates the HUD shape in terms of the field-of-view and has a level of transparency consistent with an aircraft HUD, as shown in Figure 7.17.

A simple vector font is used to provide a representative font. The rolling digits for the airspeed and altitude are reused from the EFIS display software. The bright green colour selected for the vectors and the symbology is similar to the colour and intensity of an aircraft HUD, where the displayed information is significantly brighter than the background terrain seen through the HUD.

Development of the HUD graphics can be undertaken using a desktop computer before being integrated with the IG software. Care is also needed to turn off any lighting or perspective transformations in the IG code that do not apply to a 2D overlay. Generally, HUD formats are based on existing aircraft displays. However, the vector and character generators used for aircraft HUD displays are mostly implemented using calligraphic hardware, where the low drawing rate limits the HUD content, particularly character generation.

7.6 Digital Terrain Elevation Data

In recent years, with the availability of terrain data produced from satellite imagery and tools to construct models of terrain, airport, roads and buildings, visual databases contain both texture maps to replicate terrain features and airports and elevation data to enable areas to be correctly modelled in 3D. While the terrain texture is generally provided as photographic images, the height information, known as DTED, is provided as a grid of height posts, spaced at regular intervals. There are two considerations with height information provided by DTED. Firstly, the grid spacing can vary from 1 m for very detailed terrain to a 30–50 m grid for lower resolution terrain. Secondly, the height measurement may be taken from radar measurements and the vertical resolution can also vary from 1 m to 10–20 m, where measurements are nowadays acquired from satellite data and low-altitude aerial surveys.

A DTED is used in three ways in a flight simulator:

- to provide the elevation of the terrain when the aircraft is in contact with the ground;
- to provide the height above the terrain to simulate a radar altimeter;
- to detect possible collision with the terrain, for example, in the emulation of ground proximity warning systems (GPWS).

During taxiing (and touchdown), the aircraft C/G (and therefore the pilot eye-point) depends on the position of the undercarriage oleos. Very few airfields are completely flat and any slopes or humps of runways and taxiways will be visible to the flight crew during taxiing. The height and slope of the terrain are needed in modelling the undercarriage and this information is only available in the visual database. Although the IG could transmit this information to the flight model, this information will increase the network traffic, applies to the previous frame and may lack slope information. An alternative approach is to capture DTED information from the visual database offline and interpolate between grid points in the real-time computation of the flight model.

There are several potential sources of error in conversions and transformation from raw elevation data acquired by surveying or mapping from satellite data, including the various coordinate frames and the resolution of the measurements. By the time this data is used in a visual database, it may have been filtered, smoothed and aligned to specific grid coordinates. Consequently, there may be a difference between the elevation data in the navigation database and the elevation data used to define the terrain in the IG. Therefore, the elevation data referenced by the IG is used, avoiding discrepancies between the navigation database information and the terrain elevation seen in the simulator. In other words, the source of a DTED in a flight simulator is taken from the visual database rather than other sources of terrain elevation data.

For an engineering flight simulator, the primary use of a DTED is in taxiing firstly, to ensure the aircraft wheels never go below the terrain and, secondly, to determine the slope of terrain in order to include the components of acceleration resulting from any slope.

The first decision with using a DTED is the choice of coordinate frame. The visual database coordinates are used by the visual system but, as transformation from world coordinates (latitude, longitude and altitude) to database coordinates is straightforward, it is sensible to use the world coordinate frame used by the flight model. A variant of the IG software was written to be able to pan and zoom over the terrain, displaying both the latitude and longitude coordinates and database Cartesian coordinates. As an aircraft will only move on the ground in the vicinity of an airfield, it is reasonable to restrict the DTED to this region. A second decision is the spacing of the DTED grid points which will define the number of grid points needed for the DTED. As the DTED values correspond to latitude and longitude in radians, a scale of degrees $\times$ 100,000 was chosen for grid spacing in both the x and y directions. This scale is relatively close to metres. For example, at Manchester airport, the airfield covers an area from (53.327°, −2.230°) to (53.372°, −2.250°), or approximately 5010 m $\times$ 4652 m or 4500 $\times$ 7000 in scaled units. In other words, 1° of latitude, which is 110,946 m, corresponds to 100,000 units. This scaling is purely arbitrary and a coarser or finer value could be used, provided it is used consistently in both the capture and application of the DTED. The following data structure was developed for the capture of the DTED, which is also used to determine the terrain height in the flight model software:

```
typedef struct
{
    float bx1;
    float by1;
    float bx2;
```

```
    float by2;
    int    xPosts;
    int    yPosts;
} DTED_Record;
```

The extent of the DTED is given by the bottom left corner (**bx1**, **by1**) and the top right corner (**bx2**, **by2**) and these values are rounded to the nearest 10 units of measurement. The values **xPosts** and **yPosts** define the number of grid posts in the x and y directions, respectively. The DTED for the airfield is initialised as follows:

```
wx1 = (int) (DTEDlongmin * 100000.0);
wy1 = (int) (DTEDlatmin  * 100000.0);
wx2 = (int) (DTEDlongmax * 100000.0);
wy2 = (int) (DTEDlatmax  * 100000.0);

DTED.bx1 = (float) ((wx1 / 10) * 10);
DTED.by1 = (float) ((wy1 / 10) * 10);
DTED.bx2 = (float) ((wx2 / 10) * 10);
DTED.by2 = (float) ((wy2 / 10) * 10);

DTED.xPosts = abs(DTED.bx2 - DTED.bx1) / 10 + 1;
DTED.yPosts = abs(DTED.by2 - DTED.by1) / 10 + 1;
```

where the bottom left corner and the top right corner are defined by (**DTEDlatmin**, **DTEDlongmin**) and (**DTEDlatmax**, **DTEDlongmax**), respectively. Once the size of the DTED is established, the generation of the DTED is straightforward.

```
Posts = (double *) malloc(DTED.xPosts * DTED.yPosts * sizeof(double));
if (!Posts)
{
    printf("Insufficient memory for DTED (%d K bytes)\n",
           DTED.xPosts * DTED.yPosts * sizeof(double) / 1000);
    exit(1);
}

p = 0;

for (i=0; i<DTED.xPosts; i+=1)
{
    for (j=0; j<DTED.yPosts; j+=1)
    {
        float x = DTED.bx1 + (float) (i * 10);
        float y = DTED.by1 + (float) (j * 10);
        double longitude = rads(x / 100000.0);
        double latitude  = rads(y / 100000.0);

        Posts[p] = GetSpotHeight(latitude, longitude);
        p += 1;
    }
}
```

A linear array **Posts** is allocated to store the DTED and the visual database is scanned in the x and y directions, writing the elevation data sequentially to the array **Posts**. The OSG library function `osgSim::HeightAboveTerrain::computeHeightAboveTerrain` is used to obtain the database elevation as follows:

```
double GetSpotHeight(double latitude, double longitude)
{
    double hat;
    double x, y;

    GlobeToGrid(latitude, longitude, &x, &y);
    hat = osgSim::HeightAboveTerrain::computeHeightAboveTerrain(SceneRoot.get(),
                                                osg::Vec3(x, y, 5000.0));
    return 5000.0 - hat + RunwayAltitude;
}
```

where `GlobeToGrid` is the same transformation used by the visual system to convert (*latitude, longitude*) to (*x, y*) in the database frame. The altitude is returned (in database units) at the location (*x, y*). The variable `RunwayAltitude` defines any vertical offset between the runway reference in the visual database and the navigation database. Generation of the DTED file is as follows:

```
fwrite(&DTED, sizeof(DTED), 1, f);
fwrite(Posts, DTED.xPosts * DTED.yPosts * sizeof(double), 1, f);
```

where `f` is the file stream pointer to the DTED file being written. The elevation is written as a 64-bit floating-point value. Similarly, in the DTED code accessed by the flight model, the DTED is loaded by a single read operation.

```
dsize = DTED.xPosts * DTED.yPosts * sizeof(double);
fread(Posts, dsize, 1, f);
```

where `Posts` is the array allocated for the DTED posts and `f` is the input file stream pointer for the DTED file. The code for `PostHeight` in the flight model uses 2D interpolation of the DTED grid, as follows:

```
double PostHeight(double latitude, double longitude, bool *found)
{
    double res = 0.0;
    double x = Degrees(longitude) * 100000.0;
    double y = Degrees(latitude) * 100000.0;

    if (x < (double) DTED.bx1 || y < (double) DTED.by1 ||
        x > (double) DTED.bx2 || y > (double) DTED.by2)
    {
        *found = false;
    }
    else
    {
        int px = (int) (x - (double) DTED.bx1) / 10;
        int py = (int) (y - (double) DTED.by1) / 10;
        int p = px * DTED.yPosts + py;
```

```
        if (p < 0 || p >= (DTED.xPosts * DTED.yPosts))
        {
            *found = false;
        }
        else
        {
            double x1, y1, z1;
            double x2, z2;
            double z3;
            double z4;
            double z14;
            double z23;

            *found = true;

            x1   = (double) DTED.bx1 + (double) px * 10.0;
            y1   = (double) DTED.by1 + (double) py * 10.0;
            z1   = Posts[p];
            x2   = x1;
            z2   = (double) Posts[p+1];
            z3   = (double) Posts[p+1+DTED.yPosts];
            z4   = (double) Posts[p+DTED.yPosts];
            z14  = z1 + (x - x1) * (z4 - z1) / 10.0;
            z23  = z2 + (x - x2) * (z3 - z2) / 10.0;

            res = z14 + (y - y1) * (z23 - z14) / 10.0;
        }
    }

    return res;
}
```

A test is applied to ensure that the coordinate lies within the region covered by the DTED. Two values are returned, **res**, the elevation, and **found**, a Boolean variable denoting that the point lies within the DTED. Note that this DTED is not needed by the visual system and is only used by the flight model software and the undercarriage model.

In the flight model, the slope of the incline is computed in the DTED module if the aircraft is in contact with the ground (i.e. any one of the oleos in under compression) and has moved by more than 1 m since the previous measurement.

```
double dx = DTED_Distance(latitude, longitude, OldLatitude, OldLongitude);

if (dx > 1.0)   /* only update incline after 1m of movement */
{
    double dy = h - OldPostHeight;

    incline = atan2(dy, dx);
    OldLatitude = latitude;
    OldLongitude = longitude;
    OldPostHeight = h;
}
```

If the aircraft is in contact with the ground, the computation of the X force in the body frame is adjusted to include the slope component as follows:

```
Model_XForce -= Aero_Mass * Model_g * sin(DTED_Incline());
```

where the function **DTED_Incline** returns the current angle of the ground slope. In the under-carriage model, the height above ground is computed to take account of the terrain height above sea level as follows:

```
hat = DTED_PostHeight(Model_Latitude, Model_Longitude, GroundLevel);
h = Maths_Feet(-Model_Pz + (float) hat);
```

where **GroundLevel** is the reference altitude of the airfield.

The DTED library loads the DTED file at the start of the simulation and computes the terrain elevation and slope once per frame, as just shown. This information is used in computation of the longitudinal force in the body frame (the lateral slope force is negligible), where it affects the aircraft acceleration, and also in computation of the height above the terrain in the undercarriage model. Figure 7.18 shows the terrain elevation and runway slope computed from the DTED for runway 05L at Manchester airport, as the aircraft taxies along the runway, where the x-axis denotes time.

Although this application of a DTED is limited to aircraft taxiing on runways and taxiways around an airport, in battlefield simulation including land vehicles, a more extensive area of terrain would be used for manoeuvring. The DTED for Manchester airport covers an area of 24 km^2 comprising 700×451 grid points or approximately 2.5 MB of memory, based on 10 m spacing. One particularly important area in flight simulation is the synchronisation of motion and the visual system, where, for example, an aircraft taxies over bumps or joints in a concrete runway. If the terrain height measurement lacks resolution or the position in the database is too coarse, such cues are lost. While these effects may be secondary, airfields contain a surprising amount of sloping terrain, including runways and taxiways, and the use of a DTED can simplify the computation of

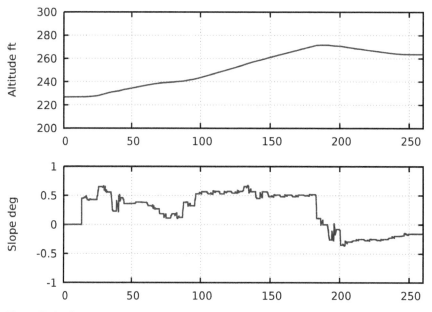

Figure 7.18 Runway Altitude and Slope.

aircraft altitude in taxiing. Some IGs may provide fast searching to obtain terrain height for the flight model and undercarriage computations or, alternatively, constrain the aircraft position in the IG, so that the aircraft never goes below the terrain, independent of the aircraft position computed in the flight model.

7.7 Visualisation

Although most IG systems are used in flight simulation to replicate highly detailed scenery, there are many applications where the emphasis is on visualisation rather than realism. In these applications, the content of real-time animation is more important than visual fidelity. For example, in aircraft accident investigation, the aircraft trajectory might be displayed in several orientations where fogging, detailed airports and varying time of day are irrelevant but visualisation of wind shear currents might aid the investigation. In such cases, there is a balance between developing a custom 3D renderer and adapting a general-purpose renderer.

Visualisation in engineering design can take two forms. Firstly, results of tests are plotted as graphs or strip charts. In some cases, the results are written to a data file and plotted offline. In other cases, the immediacy of flight test data justifies online analysis where the data is displayed while the simulator is running. In the latter case, the time to render plots must not interfere with the real-time performance of the simulator, implying that all data is plotted within the frame budget allocated to data recording and plotting. Secondly, aircraft displays can be emulated to replicate the flight deck or cockpit displays. In addition, displays can be constructed to display data that is not normally displayed but provides useful insight into the operation of the simulator, often for purposes of debugging.

The IG rendering time depends on the scene content and the performance of the graphics card. Although IG systems are generally used to display terrain and airfields appropriate to flight simulation, the use of visualisation allows other information to be displayed in 2D or 3D, where it can provide useful insight for the developer. For example, a view of the loading on an aircraft can be rendered in 3D, using colour coding to show the different load factors. Displays of this form are rarely used in training applications, but they can provide valuable information during engineering design. As an example, Figure 7.19 shows a real-time visualisation of an aircraft in a wake vortex encounter (Allerton and Spence, 2010). The orientation of the aircraft is shown, together with the direction and magnitude of the air flows and the resultant loading on the wings and tail. The loadings are colour-coded to indicate the severity in terms of possible structural damage.

One of the attractions of using OpenSceneGraph for visualisation is the ease with which C code and OpenGL can be integrated with the primary IG C^{++} code. In the case of 2D graphics, it enables the reuse of displays, possibly configured to display information specific to a set of flight tests. With 3D graphics, objects can be placed in the scene to augment the display. For example, the trajectory of the aircraft could be added to the scene or, in ATM applications, the threat of other aircraft could be shown in symbolic form, such as height posts or distance circles or traffic corridors.

7.8 Observations

Anyone who has sat in a train at a station and felt the train starting to depart only to realise it was the movement of a train on an adjacent platform that caused this sensation will appreciate the power of vision to create the illusion of motion. With a high frame rate, high-resolution IG and

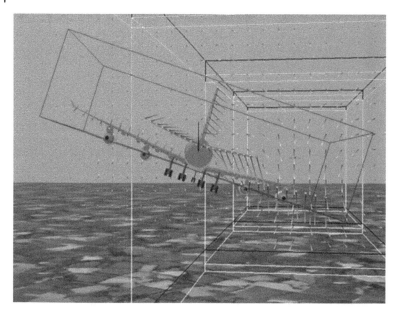

Figure 7.19 Visualisation of Air Flow and Forces during a Wake Vortex Encounter (see Plate 9).

detailed databases, even without any tactile sensing of motion or being aware of any applied motion, the scene is sufficiently convincing that the pilot response in the simulator is exactly the same as the airborne situation.

The relative value of motion versus visual cueing has been a long-running debate in flight simulation. But the reducing cost of IG systems and projectors, combined with increasing processing speeds and memory capacities, very impressive visual systems are now available at costs unimaginable ten or twenty years ago. In their contribution to fidelity, visual systems arguably offer far more cueing benefits than any other simulator component. Certainly, many military organisations have discarded motion platforms for wider field-of-view IGs with detailed scenery and dynamic agents. Surprisingly, there is very little information in the literature to analyse the benefits of visual systems versus motion systems.

Nowadays, many IGs are limited by the efficiency of the software rather than the processing power of modern IG systems. This constraint has seen the development of packages such as CUDA and Vulkan to exploit GPU architectures and a trend towards ray-tracing methods. The tools to develop visual databases have reduced development times by several orders of magnitude while, at the same, importing terrain data from public domain sources to increase the visual detail and fidelity of projected scenes. The performance of IG systems, which is based on advances in microelectronics, is likely to continue to follow Moore's Law (Moore, 1965) for the foreseeable future.

In claiming the importance of IG systems in simulation, it is essential that an IG meets its performance at all times, that the image is correctly rendered in terms of geometry, colour, texture, lighting, shading and fogging. Given all the variables influencing IG rendering, the importance of validation of IG systems cannot be overstated. Any lags in the visual system will appear to alter the perceived response of the simulated aircraft and any stalling or jittering of an image will appear unnatural and reduce pilot acceptance. In particular, any artefacts that could never occur in actual flight should never occur in the simulator. However, hidden-surface algorithms can contain errors, rounding errors can cause objects to change shape by small amounts from frame to frame and very small objects can disappear and reappear unnaturally, purely as a result of errors in graphics algorithms. Similarly, LOD changes can appear abrupt if not implemented carefully and, particularly with databases that are tiled, care is needed to ensure the smooth transition between tiles without introducing any delays or discontinuities.

It must also be remembered that flight simulation is an illusion. In other words, it is possible to 'fool' the brain and in such cases, there is a trade-off between software techniques and processing performance to increase the speed or content of rendering. One example is the LOD methods which reduce the number of triangles rendered as a function of distance. Other techniques extend to the use of flat billboards to represent 3D objects. For example, contrails or trees can be formed as billboards (Spence et al., 2007), which can be rotated so that the surface of the billboard is normal to the viewer. In the case of trees, each tree can be formed from a single texture, which rotates (indiscernibly) towards the direction of the viewer. Such methods significantly outweigh rendering trees as 3D objects and can improve fidelity significantly. Real-time rendering is one of the areas of computer science where the potential advancements result not only from improvement of an algorithm but also from elimination of unnecessary or redundant processing.

References

Akenine-Moller, T., Haines, E., and Hoffman, N. (2008), *Real-time Rendering*, A K Peters.

Allerton, D. J. and Spence, G. T. (2010), Wake Vortex Encounters in a Flight Simulator, *The Aeronautical Journal*, Vol. 114, No. 1159, pp. 579–588.

Anon. (2018), *OpenFlight Scene Description Database Specification*, Version 16.7, Revision A, available from https://www.presagis.com/workspace/uploads/files/openflight16-7.pdf.

Cignoni, P., Costanza, D., Montani, C., Rocchini, C. and Scopigno, R. (2000), Simplification of Tetrahedral Meshes with Accurate Error Evaluation, *Proc. IEEE Visualization*, IEEE, pp. 85–92.

Gray, K. (2003), *Microsoft DirectX 9 Programmable Graphics Pipeline*, Microsoft Press, U.S.

Hall, J. R. (1993), *The Design and Development of the New RAF HUD Format in Combat Automation for Airborne Weapon Systems: Man-machine Interface Trends and Techniques*, AGARD CP-520.

Kessenich, J., Sellers, G. and Shreiner, D. (2017), *OpenGL Programming Guide: The Official Guide to Learning OpenGL*, version 4.5, ninth edition, Addison-Wesley.

Lee, D. T. and Schachter, B. J. (1980), Two Algorithms for Constructing a Delaunay Triangulation, *Int. Journal of Computer and Information Sciences*, Vol. 9, No. 3, pp. 219–242.

Luebke, D. P., Reddy, M., Cohen, J. D., Varshney, A., Watson, B. and Huebner, R. (2002), *Level of Detail of 3D Graphics*, Elsevier Science & Technology.

Moore, G. E. (1965), Cramming More Components onto Integrated Circuits, *Electronics*, Vol. 38, No. 8, pp. 114–117.

OpenSceneGraph (2012), *Documentation osgPlugins – Invoking a Plugin to Load or Save Data*, Available at http://www.openscenegraph.org/index.php/documentation/user-guides/61-osgplugins (accessed 10 March 2022).

Spence, G. T., Le Moigne, A., Allerton, D. J. and Qin, N. (2007), Wake Vortex Model for Real-time Flight Simulation Based on Large Eddy Simulation, *AIAA Journal of Aircraft*, Vol. 44, No. 2, pp. 467–475.

Wang, R. and Xuelei, Q. (2010), *OpenSceneGraph 3.0 Beginner's Guide*, Packt Publishing Ltd.

Wang, R. and Xuelei, Q. (2012), *OpenSceneGraph 3 Cookbook*, Packt Publishing Ltd.

8

Sound Generation

Before I speak, I have something important to say.

Groucho Marx

8.1   Sound Waveforms

The cockpit or flight deck of an aircraft is a noisy environment. There are numerous sources of sounds including engines, background noise from the slipstream, electrical generation and air conditioning, occasional sounds from avionics systems and sounds of motor activation deploying flaps and raising and lowering the undercarriage. There is also a wide range of warnings and alarms. All these sounds need to be replicated accurately and combined together to give the sense of an aircraft cockpit or flight deck. Many of these sounds are also synchronised with events. For example, passing overhead a marker beacon activates a beeping sound when the aircraft is close to the beacon or the slipstream sound changes slightly when the undercarriage is lowered.

Strictly, there is very little training benefit from the provision of sound, with the exception of a few sounds that provide a cross-check and the warnings and alarms that enable emergency procedures to be practised. However, a completely silent flight deck would be very unnatural. In moving the engine levers, a pilot would expect to see a change in the relevant instruments and a corresponding change in the sound of the engines. In flight simulation, sound generation fits into the category of immersive or peripheral cues, which provide a sense of being airborne. Although impossible to quantify, the lack of aural cues or the provision of unrealistic sounds would break the illusion of flight and could prove to be counter-productive in a training environment.

The sounds of an orchestra, transmitted as radio waves, are amplified by a radio to drive the mechanical cones of a loud speaker. These vibrations of air enable the listener to hear sounds that resemble very closely the actual sounds heard in a concert hall. The output of each instrument in the orchestra is a waveform and all these waveforms are combined to form a single waveform that is transmitted and amplified. Ideally, the waveform listened to in a room is very close to the waveform produced in the concert hall in terms of blending the loudness of instruments and the pitch of notes produced. The human ear is remarkably sensitive. It can detect very quiet sounds and accommodate very loud sounds over an extremely large range. Similarly, it can detect very low frequencies, such as a bass drum, to very high frequencies such as a piccolo or a violin. In other words, we have an expectation of sound quality and are able to discriminate poor or unrealistic sounds and the same expectation applies to sound generated in a flight simulator.

Flight Simulation Software: Design, Development and Testing, First Edition. David Allerton.
© 2023 John Wiley & Sons Ltd. Published 2023 by John Wiley & Sons Ltd.
Companion Website: www.wiley.com/go/flightsimulationsoftware

Sound is a waveform in time defined by the amplitude and frequency of the sound. A simple waveform is shown in Figure 8.1.

The amplitude of this particular signal is ±1 (shown by the *y* axis). The signal is sinusoidal and there are 10 complete cycles over 1 s, giving a frequency *f* of 10 cycles/s or 10 Hz. Note that the period of each cycle is 0.1 s or 1/*f*. Although it is common to define frequency in hertz (Hz), frequency is also defined in rad/s; to convert from Hz to rad/s, the frequency is multiplied by 2π. The signal in Figure 8.1 was generated as a sine wave given by the mathematical function

$$V = \sin\left(2\pi\omega t\right) \tag{8.1}$$

where the output *V* is a function of frequency *ω* and time *t*. The amplitude of the signal corresponds to a voltage or to the power of the signal.

Although the signal appears to be continuous, in cases where the signal is generated digitally, it is formed from individual samples where the sampling rate is sufficiently high, that is, to the human ear, the gaps in the signal are indiscernible. Figure 8.2 shows the same sine wave over 0.1 s, which is generated at 0.001 s intervals or 1000 Hz. Note that the digitised signal has both a sampling frequency and the signal frequency.

The most important consideration is the sampling rate. At high sampling rates, the hardware to capture an audio signal is complex, a large amount of high-speed memory is needed to store the data and, arguably, there is no benefit if the sampled signal contains more information than can be perceived by the human ear. On the other hand, if the sampling rate is too low, the signal may no longer resemble a sine wave and some distortion is likely to be perceived by the human ear.

The human ear has a range of 120 dB, which is a remarkable range of one million. However, changes in loudness of less than 1 dB cannot be detected. In other words, there are approximately

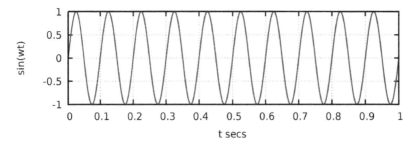

Figure 8.1 A Sound Waveform.

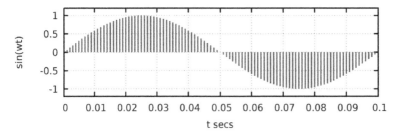

Figure 8.2 A Digitised Audio Signal.

120 levels of loudness ranging from a very quiet whisper to nearby thunder. Although the bandwidth of the human ear is generally accepted to extend from 20 Hz to 20 KHz, the ear is most sensitive between 1 KHz and 4 KHz, with the upper range reducing considerably with age. As sound in each ear is processed separately by the brain, it is possible to detect direction from stereo sound. In the case where identical sounds are heard in both ears, the brain suppresses any directional cues. Humans can discern variations in frequency of 0.3% at 3 KHz. These figures for discrimination of the amplitude and frequency of sound define the limits of digitisation of sound signals. For a frequency range 5–20 KHz, the sampling rate is 44.1 KHz with 16-bit values, giving a data rate of 705.6 Kbits/s for the highest quality of sound reproduction (Smith, 1997).

The other remarkable quality of human audiology is the ability to add or combine separate sound sources and to discern individual sounds from within a group of sounds. From the perspective of flight simulation, all the sounds in the cockpit or on the flight deck of an aircraft can be generated from individual sources, added together and then amplified as an audio signal heard by the flight crew. The requirement is to identify and generate sounds, which are closely matched to actual aircraft sounds in terms of amplitude and frequency. These sounds form two classes. Firstly, sounds are produced by the airborne environment resulting from the interaction between the aircraft and the atmosphere and mechanical sounds from the engines and electrical systems. Secondly, digital sounds are also produced by aircraft systems as indications and warnings. These latter sounds are likely to be generated synthetically by aircraft systems and their type and classification are important to flight crews, where the severity of warnings is defined for the different signals. The sounds found in a typical flight simulator are summarised in Table 8.1.

Table 8.1 Range of Aircraft Sounds.

Type	Source	Notes
Aerodynamic	Slipstream	*Varies with airspeed, flaps and undercarriage*
Engines	Engine start, propeller noise, rotor blades and turbofan whine	*Varies with engine RPM*
Ground movement	Runway rumble, tyre scuff and skid	*Rumble varies with ground speed, superimposed bumps and skid on touchdown*
Auxiliary power unit (APU)	Connection, disconnection and background motor	*Not applicable to all simulators*
Flight deck	Air conditioning and power generation systems	*Continuous background sound*
Indicators	Marker beacons and Morse ident	*Synchronised with navigation systems. Morse idents are selectable with adjustable volume*
Warnings	Fire, gear up, configuration and stalling	*Configuration warnings based on maximum speeds for flaps and undercarriage. Gear up warning below 500 ft. Stall warning linked to flight model*
Sub-systems	Flaps motor, gear motor and windscreen wipers	*Synchronised with selections. Event-triggered*
Emergencies	Engine fires, tyre burst and explosions	*Fire bell switched off by pulling extinguishers*
Text-to-speech	Air traffic and ATIS	*Computer-generated and recorded sounds*
Weapons	Missile launch, guns and threat warnings	*Applicable to fighter aircraft*

Although this table includes the major sources of sound generation in most flight simulators, it is not exhaustive and the requirements vary for different aircraft types and applications. The majority of sounds are continuous even though they vary with flight conditions. Other sounds are event-triggered, for example, selecting flaps, touching down on the runway or possibly a failure injected by the instructor. Event-triggered sounds either run for a predefined time or are stopped by complying with a flight condition; for example, a configuration warning cancels when the airspeed is less than the warning value. There is also a requirement to suppress all sounds, for example, in simulator 'flight-freeze' conditions. Particularly in training simulators, some sounds can be representative as the flight crew undergo training or respond to a critical event, for example, the sound of a tyre burst. In such cases the fidelity of the sound is secondary. In other cases, a warning may identify a specific problem and the flight crew need to feel confident they have correctly identified the warning and responded accordingly.

8.2   Sound Generation Methods

Up to the 1980s, most simulator manufacturers produced their own sound generation systems. Various aircraft sounds were analysed and electronic circuitry was developed to replicate these sounds. An oscillator can produce pure tones, and if the fundamental frequencies of a sound are known, the outputs from a set of oscillators can be merged, with the advantage that the frequencies of the components can be readily adjusted. Similarly, white noise generated with analogue circuitry and analogue filters can emphasise or supress certain frequency bands (Horowitz and Hill, 1989). Certainly, the ingenuity of these times produced realistic aircraft sounds. The main limitation was that they were analogue systems, often requiring hundreds of amplifiers with relatively limited programmability.

With advances in microelectronics in the late 1980s, several manufacturers produced integrated circuits (ICs) for sound generation. These devices were mainly used to simulate musical instruments, but the main advance was that they were programmable devices with 10–20 individual channels and could be adapted to generic sound generation. The waveform shape, the repetition frequency and the output amplitude could be defined for individual channels. The generality of such devices as the Yamaha OPL4 with 16-bit sine wave, half sine wave, saw-tooth signal generation and signal modulation enabled many aircraft sounds to be reproduced. Later, the Philips SAA-1099 sound generation IC provided envelope shaping, frequency generators, noise generators, mixers and a simple interface. These devices were adapted for a wide range of applications, including musical instrument emulation and arcade games. The programmability of these ICs encouraged their use in sound generation in simulation but with two limitations: they only contained 6–12 channels (depending on the configuration) and they were far from straightforward to program. It was left to the developer to set up the numerous registers to produce a specific sound.

By the 1990s, the SoundBlaster range of cards were introduced for PC applications and became popular for the playing of music stored on CDs. As memory capacities increased, these cards provided over 100 channels of digitised signals with amplitude control and the combining of channels. In particular, waveforms can be written to memory and then replayed with the significant advantage that the waveforms can be modified dynamically.

One specific feature of these cards was the recognition and adoption of standard waveform formats. The Waveform Audio File Format (*wav*) is recognised worldwide and used on most computers. Very little software is needed to play *wav* files, which are stored as binary files. The majority of the aural warnings on PCs and mobile phones are stored as *wav* files and numerous applications exist to capture, modify and store waveforms in this format. The use of this standard and the availability and relatively low cost of modern sound cards have largely changed their use from sound generation to sound replay, increasing their applicability to flight simulation. Having captured a sound, it can be replayed, adjusting the replay rate (the frequency) and the amplitude (the sound level) of individual channels of recordings. However, this approach introduces three problems. Firstly, it is necessary to obtain a recording of a specific set of sounds and acquiring such sounds in an airborne environment can be difficult. Secondly, if only a relatively short fragment of sound is available and the fragment is replayed repeatedly, it is necessary to join the end of a fragment to the start of the next fragment to avoid any discontinuity in the recording. Thirdly, a sound recording may include other unwanted sounds. For example, in recording engine sounds on the flight deck, an address by cabin crew may be superimposed on the recording or the flight crew may be given a clearance, neither of which would be appropriate if repeated continually.

8.2.1 WAV Format

Every *wav* file contains the following header block, followed by a raw waveform data block, which defines the content of the sound data:

```
struct Wav_Header
{
    char ChunkID[4];                    // "RIFF"
    unsigned int ChunkSize;             // overall file size (bytes)
    char Format[4];                     // "WAVE"
    char Subchunk1ID[4];                // "fmt "
    unsigned int Subchunk1Size;         // length of the format data
    unsigned short int AudioFormat;     // format type. 1=PCM
    unsigned short int NumChannels;     // number of channels
    unsigned int SampleRate;            // sampling rate (samples per second)
    unsigned int ByteRate;              // SampleRate * NumChannels * BitsPerSample/8
    unsigned short int BlockAlign;      // NumChannels * BitsPerSample/8
    unsigned short int BitsPerSample;   // bits per sample, 8=8bits, 16=16 bits etc
    char Subchunk2ID[4];                // "data"
    unsigned int Subchunk2Size;         // size of the data chunk (bytes)
};
```

This header is used for basic *wav* files and many of these fields are constant for a specific type of *wav* file. For example, PCM is the common form of modulation, one channel is used for mono sounds, the sampling rate is typically 44,100 bits/s and the audio is captured with a resolution of 16 bits. The sound data is then packed as 16-bit integers following the header block. Note that the 32-bit and 16-bit values in both the header block and the data block are stored in little-endian format.

One advantage of this simple structure is that it is straightforward to generate *wav* files, appending specific sound data to the file, including computer-generated waveforms. For example, the header can be written as follows (remembering to convert 16-bit and 32-bit values to little-endian format):

```
WriteString("RIFF");  // ChunkID
WriteInt(NumberOfSamples * NUMCHANNELS * BITSPERSAMPLE / 16 + 36);  // Chunksize
WriteString("WAVE");  // Format
WriteString("fmt ");  // SubChunk1ID
WriteInt(16);  // Subchunk1Size (PCM)
WriteShortInt(AUDIOFORMAT);  // AudioFormat = 1 (PCM)
WriteShortInt(NUMCHANNELS);  // NumChannels (Mono)
WriteInt(SAMPLERATE);  // SampleRate
```

```
WriteInt(SAMPLERATE *NUMCHANNELS * BITSPERSAMPLE / 8);   // ByteRate
WriteShortInt(NUMCHANNELS * BITSPERSAMPLE / 8);   // BlockAlign
WriteShortInt(BITSPERSAMPLE);   // BitsPerSample
WriteString("data");   // Subchunk2ID
WriteInt(NumberOfSamples * NUMCHANNELS * BITSPERSAMPLE / 8);   // Subchunk2Size
```

where **NUMCHANNELS**, **BITSPERSAMPLE** and **SAMPLERATE** are defined as constants and the variable **NumberOfSamples** is the number of samples in the sound block. The sound block is written directly after the header block, as follows:

```
for(i=1; i<=NumberOfSamples; i+=1)
{
    WriteShortInt(f(t));
}
```

where the function **f(t)** is the amplitude of the sound at time **t** and the functions **WriteInt** and **WriteShortInt** generate 32-bit and 16-bit integers in little-endian format, respectively. For example, to generate a 400 Hz sine wave, **f(t)** is given by **sin(2.0 * M_PI * w * t)**, where **M_PI** is the value of π in the system library *math.h*, **w** = 400.0 and **t** is time in seconds. For **n** samples

```
t = t_end * (float) i / (float) n;
```

where **t_end** is the duration of the recording.

To access the waveform data in a *wav* file, the **Wav_Header** header block is read, which includes the number of samples in the file; the waveform data can then be read as a stream of 16-bit integers. The following fragment of code reads the waveform data into an array **v** of 16-bit integers:

```
for(i=1; i<=header.NumChannels; i+=1)
{
    v[i] = ReadShortInt();
}
```

8.2.2   Fast Fourier Transform

Although sine waves can be used to produce pure tones, other waveforms are used in the production of aircraft sounds, including rectified sine waves, triangular and saw-tooth patterns, square waves and random noise. One way to determine the content of a specific sound is to perform a fast Fourier transform (FFT) on the data which transforms a signal from the time domain to the frequency domain. For example, consider a waveform given by $2\pi\omega_1 t + 2\pi\omega_2 t + 2\pi\omega_3 t$, where $\omega_1 = 400$ Hz, $\omega_2 = 600$ Hz and $\omega_3 = 800$ Hz. The following code writes the waveform data to the data block of a *wav* file:

```
for (i=1; i<=NumberOfSamples; i+=1)
{
    float t = (float) i / (float) NumberOfSamples;

    WriteShortInt((int) (10000.0 * (sin(2.0 * M_PI * 400.0 * t) +
                                    sin(2.0 * M_PI * 600.0 * t) +
                                    sin(2.0 * M_PI * 800.0 * t))));
}
```

The waveform is shown in Figure 8.3, where the amplitude of the three sine waves is multiplied by 10,000 to give nearly full-range sound values. The data is written as a signed 16-bit integer, in the range ±32,767, which can be extracted from a *wav* file or, alternatively, written directly to a file. The output from the FFT is shown in Figure 8.4. The three spikes occur at 400, 600 and 800 Hz of equal power, as expected for three pure sine waves of equal amplitude.

By contrast, the time series of one component of a turbofan engine sound is shown in Figure 8.5. The high noise content of the waveform over 1 s is clearly evident.

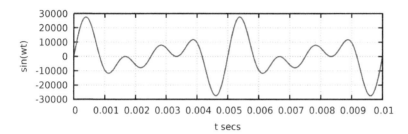

Figure 8.3 *Wav* Output of 400, 600 and 800 Hz Components.

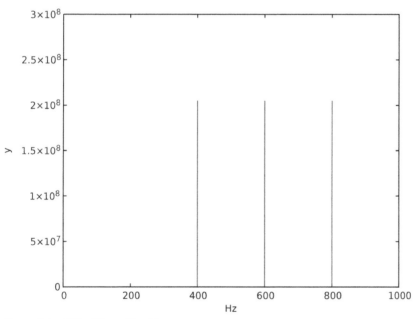

Figure 8.4 FFT of Three Sine Waves.

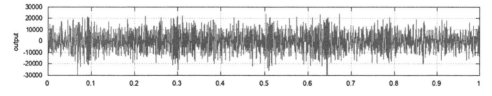

Figure 8.5 Engine Sound Waveform.

The same waveform is shown in Figure 8.6 over 0.12 s and the irregularity of the waveform is still shown clearly at this scale.

The FFT of this sound component is shown in Figure 8.7. The spread of relatively small components indicates the random noise content of the waveform, but with dominant components around 110 and 340 Hz. Generation of these sound components using saw-tooth, square or rectified waveforms at these fundamental frequencies may well produce sounds similar to the recorded sounds. The relative power of these components can be deduced by plotting the FFT output with a scale in dB.

8.2.3 FFTW

The examples of the FFT in this chapter were produced using the FFTW package developed at MIT (Frigo and Johnson, 2005). The open source library fftw3 (http://www.fftw.org) provides a suite of FFT algorithms written in C. The package includes tools for computing the discrete Fourier transform (DFT) including complex data, real data and the discrete sine and cosine transforms. The library is written to maximise the performance of the host computer with algorithm computation times $O(n \log n)$, where n is the number of samples. The three phases to produce an FFT using FFTW (version 3) comprise:

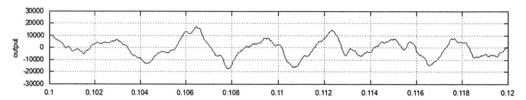

Figure 8.6 Engine Sound Waveform (Expanded).

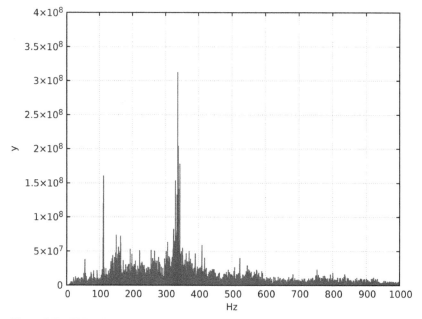

Figure 8.7 FFT of Turbofan Engine Sound.

1) Allocation of storage to contain the signal data.
2) Definition of details needed to execute the DFT, which is referred to as a *plan* in FFTW.
3) Execution of the DFT.

The main program to read the data file and generate an FFT is as follows:

```
int main()
{
    fftw_complex* signal = fftw_malloc(sizeof(fftw_complex)*N);
    fftw_complex* result = fftw_malloc(sizeof(fftw_complex)*N);

    fftw_plan plan = fftw_plan_dft_1d(N, signal, result, FFTW_FORWARD, FFTW_ESTIMATE);

    acquire_data(signal);
    fftw_execute(plan);
    printf_data(result);

    fftw_destroy_plan(plan);
    fftw_free(signal);
    fftw_free(result);

    return 0;
}
```

The input array **signal** and the output array **result**, both of type **fftw_complex,** are declared dynamically, which also ensures alignment of the arrays on different architectures. The *plan* for a one-dimensional FFT is set up, the input data is read from the *wav* data file, the FFT is computed, the results are written and the storage for the arrays is released. The fragment of code to read the data is as follows:

```
for(i=0; i<N; i+=1)
{
    char str[100];

    fgets(str, 1000, fs);
    signal[i][REAL] = (double) atoi(str);
    signal[i][IMAG] = 0.0;
}
fclose(fs);
```

where **signal** is the complex array used to store the values, which are read as 16-bit integers and converted to double values, and **fs** is the input file stream. Note that the imaginary part of the complex number is set to zero for real data. The data output is given by the magnitude of the elements in the complex array **result**:

```
for (i = 0; i < N/2; ++i)
{
    double mag = sqrt(result[i][REAL] * result[i][REAL] + result[i][IMAG] * result[i][IMAG]);

    printf("%f %f\n", i * (double) SAMPLERATE / (double) N, mag);
}
```

where **N** is the number of sampled values, the array **result** holds the transformed values and the constant **SAMPLERATE** is the sampling rate. Note that only the lower half of the array is used because, for real 1D inputs, the upper half contains no useful information (it is the complex conjugate of the lower half). The FFT output is organised as bins; the nth bin is given by $\dfrac{nS}{N}$, where S is the sampling rate and n is in the range $0..N-1$.

The attraction of using FFTW is the simplicity of setting up the data array, defining the plan and executing an appropriate DFT. The algorithms are optimised for speed of computation, although the library is only used for offline testing, and provides a library of functions for signal analysis. This package is particularly useful in the development of sound generation systems as it enables the constituent components of sound waveforms to be identified. In cases where the signal is generated synthetically, rather than from a sound recording, validation of the sound generation includes the capture of equivalent aircraft and simulator sounds and the comparison of their respective FFTs.

8.2.4 Filters

One method to replicate jet engine sounds is to combine different sources of noise. Although a white noise signal can be derived from the standard library *rand* function, in practice a 32-bit random number generator is not pure white noise, which would have equal power across the frequency spectrum of the signal. Nevertheless, computer-generated white noise is adequate for most sound reproduction applications. With mechanical systems such as engines, propellers and undercarriage assemblies, the underlying sounds tend to dominate at lower frequencies. Brown noise is a variant of white noise, with a low-pass filter applied to the signal. The gain reduces by 6 dB per octave so that higher frequencies are significantly attenuated above the cut-off frequency of the filter. The same filters can also be applied to recorded sound to remove unwanted high frequencies. Typically, three forms of filter are used: low-pass filters, high-pass filters and band-pass filters. Ideally, the gain of a low-pass filter is 0 dB up to the cut-off frequency and is then attenuated by 20 dB per decade. A simple passive low-pass filter is the RC network shown in Figure 8.8(a) with the perfect response shown in Figure 8.8(b).

The transfer function is given by

$$\frac{V_o}{V_i} = \frac{1}{1+sRC} \tag{8.2}$$

with the cut-off frequency ω_c at $\dfrac{1}{RC}$. Rearranging the transfer function

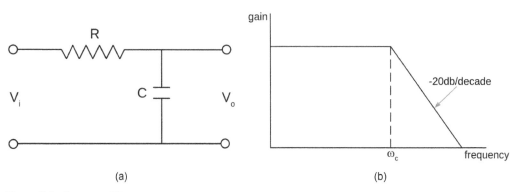

(a)   (b)

Figure 8.8 Low-pass Filter.

$$V_o = \frac{1}{s}\left(\frac{V_i - V_o}{RC}\right) \hspace{4cm} (8.3)$$

The code is readily implemented as follows:

```
Vout = integrate(Vout, (Vin-Vout) / (R * C));
```

Note that care is needed with the cut-off frequency, which is in radians, rather than Hz. The code should be modified as follows:

```
Vout = integrate(Vout, (Vin-Vout) * (Wc * 2.0 * M_PI));
```

where `Wc` is the cut-off frequency. As an example, the waveform shown in Figure 8.3 is passed through a first-order low-pass filter with a cut-off frequency at 600 Hz. The code fragment to write the filter waveform to the WAV data block is given by

```
for (i = 1; i <= NumberOfSamples; i+=1)
{
    double t = (float) i / (float) NumberOfSamples;
    double x = 10000.0 * (sin(2.0 * M_PI * 400.0 * t) +
                          sin(2.0 * M_PI * 600.0 * t) +
                          sin(2.0 * M_PI * 800.0 * t));
    Vout = Integrate(Vout, (x - Vout) * 600.0 * 2.0 * M_PI);
    WriteShortInt((int) Vout);
}
```

The FFT of the filtered waveform is shown in Figure 8.9.

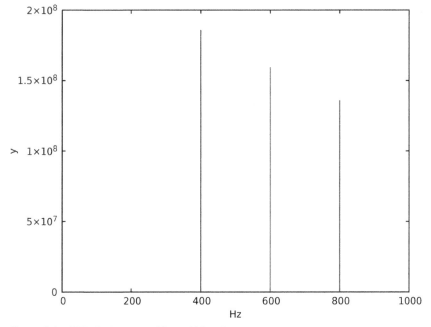

Figure 8.9 FFT of a Low-pass Filtered Waveform.

Note that `Vout` is defined as a static variable as it is updated at each iteration of the filter. Also note the attenuation of the 800 Hz component and that there is also some attenuation of the 600 Hz component, which is at the corner frequency of the 600 Hz cut-off frequency. The attenuation of 20 dB/decade with a first-order filter would be insufficient to suppress the 800 Hz component, and in practice, a higher-order filter with much better attenuation would be used. These examples are only presented to illustrate the ease of generating and filtering audio *wav* files. For more detailed analysis and design of audio filters, the reader is referred to the standard texts on the subject (Brigham, 1974; Rabiner and Gold, 1975).

A high-pass filter, allowing higher frequencies to pass while attenuating lower frequencies, is shown in Figure 8.10(a) with the ideal response shown in Figure 8.10(b).

The transfer function is given by

$$\frac{V_o}{V_i} = \frac{sRC}{1 + sRC} \tag{8.4}$$

also with the cut-off frequency ω_c at $\dfrac{1}{RC}$. Rearranging the transfer function,

$$V_o = V_i - \frac{1}{s}\left(\frac{V_o}{RC}\right) \tag{8.5}$$

which is implemented in code as follows:

```
t = integrate(t, Vout);
Vout = Vin - integrate(Vout) * 2.0 * M_PI * Wc;
```

A band-pass filter can be formed from the equivalent of a circuit consisting of an inductor, capacitor and resistor, as shown in Figure 8.11(a), which passes a signal which is close to the pass-band frequency ω_0. The gain near to this frequency is defined by the Q of the circuit, where a high Q value denotes a narrow pass-band and a low Q value denotes a broad pass-band. The ideal filter response is illustrated in Figure 8.11(b).

The transfer function is given by

$$\frac{V_o}{V_i} = \frac{R}{R + sL + \dfrac{1}{sC}} \tag{8.6}$$

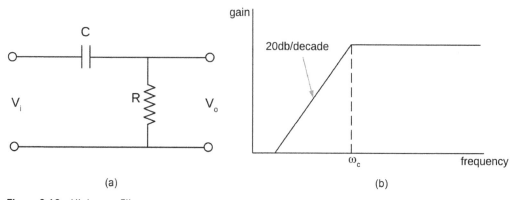

(a)   (b)

Figure 8.10 High-pass Filter.

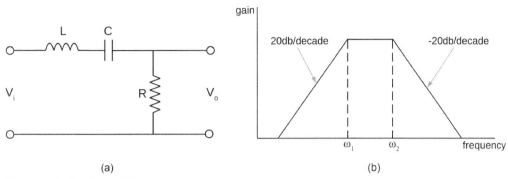

Figure 8.11 Band-pass Filter.

with the mid-band frequency $\omega_0 = \dfrac{1}{\sqrt{LC}}$ and $Q = \sqrt{\dfrac{1}{R^2C}}$. Rearranging the transfer function,

$$\frac{V_o}{V_i} = \frac{sRC}{sRC + s^2LC + 1} \tag{8.7}$$

giving

$$V_o = \frac{1}{s}\left(\frac{R}{L}(V_i - V_o) - \frac{1}{s}\frac{V_o}{LC}\right) \tag{8.8}$$

Choosing appropriate values of L, R and C for ω_0 and Q, the computation of the filter can be implemented by the following code:

```
t = integrate(t, Vout)
Vout = integrate(Vout, (R / L) * (Vin - Vout) - t / (L * C));
```

More complex filters are available in the literature, which are normally defined as transfer functions and can be implemented using the techniques described in Section 2.2. Given the various waveforms and a set of filter functions, it is possible to experiment with different waveforms to replicate aircraft sounds and generate an equivalent *wav* file. Alternatively, a wide variety of open tools are available in the public domain to edit sound waveforms, for example, the Audacity package (https://www.audacityteam.org).

8.3 OpenAL

8.3.1 OpenAL Application Programming Interface

In much the same way that OpenGL is an application programming interface (API) to enable users to render computer graphics, where the detail of the graphics hardware is hidden from the user, so OpenAL (Hiebert, 2006) is an API for rendering sounds but without the user needing to understand the low-level hardware of a sound card. Furthermore, as there are several manufacturers of sound cards and numerous versions of sound cards produced by the manufacturers, OpenAL provides a high degree of portability of software between different platforms and sound cards. OpenGL enables a scene of graphic entities to be defined and to visualise the scene from different positions.

Similarly, OpenAL enables the user to position sources of sound in 3D space and to hear these sounds from different positions, where the computations are performed with respect to the position of the listener. OpenAL provides a library of primitives to provide realistic sound generation (Peacock, 2007). Note that OpenAL outputs directly to a sound card and operates independently of any hardware codecs. OpenAL provides a library of function calls to define the listener, the sounds and the sound buffers. Each call to OpenAL primitives pushes an error message onto a stack (even if there is no error) which should be checked after each call.

OpenAL is invariably used in combination with ALUT, the OpenAL Utility Toolkit (Panne, 2006), which provides tools to load and create waveforms. Only three ALUT functions are used:

- **alutInit** initialises the toolkit, creates a context for OpenAL and makes this context the current thread;
- **alutExit** closes the toolkit, deletes the context and shuts down the audio device;
- **alutLoadWAVFile** loads a *wav* file and specifies how it will be used.

Once initialised, OpenAL runs continuously as a background thread and sounds are enabled by setting the sound output gain to the required sound level or disabled by setting the output gain to zero. The following code initialises an OpenAL environment:

```
intSoundInit(int *argc, char *argv[])
{
    alutInit(argc, argv);

    if (alGetError() != AL_NO_ERROR)
    {
        return 0;
    }

    alGenBuffers(MAX_SOUND_BUFFERS, soundBuffers);
    if (alGetError() != AL_NO_ERROR)
    {
        return 0;
    }

    return 1;
}
```

where **MAX_SOUND_BUFFERS** is the number of independent sounds to be used, which is dependent on the sound card hardware and available memory, and **soundBuffers** is an array of buffers. Typically, 50–100 individual sound channels are allocated for a flight simulator, depending on the fidelity of the sound modelling and the range of sounds specific to an aircraft. The sounds and their associated buffers are defined as an enumerated set, enabling individual sounds to be accessed via a relevant name. For example, in the library soundlib.c, the stall warning sound is defined as **SNDID_STALL_WARNING** and the stall warning sound buffer is **SNDSRC_STALL_WARNING**.

8.3.2 Loading Sound Files

Although OpenAL enables the position, velocity and orientation of the listener to be defined, which provides directional sound, Doppler effects and fading with distance, the default location is at the centre of the sounds, that is, an aircraft cockpit or flight deck in a flight simulator. The internal sound of, for example, an engine is very different from the external sound. Having initialised the sound context, the first step is to initialise the sound buffers, which fall into two categories: sound buffers loaded from a *wav* waveform defined in a file and sound buffers where the waveform is written directly into the buffer. For many sounds, the waveform is provided as a *wav* binary file. For example, the fire warning sound is loaded as follows:

```
int FireWarningInit()
{
    return ChannelInitWav("../wav/fire_warning.wav",
                          SNDSRC_FIRE_WARNING, SNDID_FIRE_WARNING, 0.8, true);
}
```

The first argument is the name of the *wav* file, **SNDSRC_FIRE_WARNING** specifies the buffer, **SNDID_FIRE_WARNING** identifies the source, the fourth argument is the default gain (amplitude) and the final argument is true if the sound is to be replayed repeatedly or false if the sound is only played once. The function **ChannelInitWav** in the library **soundlib.c** is called to initialise a sound channel for the majority of the generated sounds and is given as follows:

```
int ChannelInitWav(char wavfilename[], SoundSourceType src, SoundIdentType ident,
                   float defaultgain, bool looping)
{
    ALuint buf;
    buf = loadBufferFromFile(wavfilename);
    if(buf != AL_NONE)
    {
        soundBuffers[ident] = buf;
    }
    else
    {
        printf("ChannelInitWav: Unable to load wav file %s\n", wavfilename);
        return 0;
    }

    /* create a source */
    if (!createSourceWithProperties(src, looping))
    {
        printf("ChannelInitWav: CreateSourceWithProperties failure %d\n", src);
        return 0;
    }

    /* Attach a buffer to the source */
    sourceAttachSound(src, ident);
    if (alGetError() != AL_NO_ERROR)
    {
        printf("ChannelInitWav: sourceAttachSound failure %d %d\n", src, ident);
        return 0;
    }

    /* Modify gain to initial level */
    soundDefaultGains[src] = defaultgain;
    sourcePlay(src);
    if (alGetError() != AL_NO_ERROR)
    {
        printf("ChannelInitWav: sourcePlay failure %d\n", src);
        return 0;
    }
    alSourcef(soundSources[src], AL_GAIN, 0.0);  /* turn off for now */

    return 1;
}
```

LoadBufferFromFile calls **alutCreateBufferFromFile**, checking that the file is valid and that the buffer is created. **createSourceWithProperties** creates a source calling **alGenSources** and sets the sound as looping or single-shot. **sourceAttachSound** attaches a buffer to the source calling **alSourcei**, again checking that the call is valid and the buffer is set up correctly. The local array **SoundDefaultGains** retains a copy of the default sound level for each sound. Finally, the sound is activated by **sourcePlay**, which calls **alSourcePlay**. A check is made that the sound is playing, which is immediately turned off by setting its gain to zero by calling **alSourcef**, as it will be enabled subsequently in the simulator software.

Note the simplicity of setting up a *wav* file as a sound source which can be switched on and off by the simulator software. For example, Morse code signals are played at a specific frequency (750 Hz). The dots and dashes are implemented by turning on the Morse sound for defined periods; at other times, the Morse output is zero. Also, some background sounds (e.g. air conditioning) play repeatedly with no intervention. The only time they might be disabled is if the simulation is in a flight-freeze state.

An alternative method of generating a sound is to write the waveform to a buffer at run-time, which may occur if no *wav* file exists for a particular sound. Once the buffer is formed, the activation and enabling of the sound are exactly as already described. Typically, sounds formed in this way include sine waves, rectified sine waves and white noise. For example, pure tones can be generated as follows:

```
short int* create16bitSineTone(float toneFrequency, float sampleFrequency,
                               float peakAmplitude, unsigned int *pointsPerCycle)
{
    unsigned int i;
    short int*  buffer;
    unsigned int numberOfSamplePoints = (unsigned int) (sampleFrequency / toneFrequency + 0.5);

    if (numberOfSamplePoints & 1)
    {
        numberOfSamplePoints -= 1;
    }

    buffer = malloc(sizeof(short int) * numberOfSamplePoints);
    if (buffer == NULL)
    {
        return NULL;
    }

    for (i=0; i<numberOfSamplePoints; i+=1)
    {
        buffer[i] = (short int) (32767.0 * peakAmplitude *
                    (sin((2.0f * M_PI * toneFrequency) / sampleFrequency * (float) i)));
    }

    *pointsPerCycle = numberOfSamplePoints;
    return buffer;
}
```

ToneFrequency is the frequency of the generated sine wave (Hz), **sampleFrequency** is the waveform sampling frequency, **peakAmplitude** is the waveform amplitude in the range 0–1 and **pointsPerCycle** is the resolution of the waveform in terms of the number of samples per cycle. The buffer is dynamically allocated, the number of sample points is set to an even number to ensure symmetry of the cycles and the amplitude of the computed sound is scaled to the range ±32,767 (16 bits) and written to the buffer.

8.3.3 Dynamic Sounds

Some sounds used in flight simulation are constant sounds, which may be switched on or off but, once generated, are unchanging during a simulator session. However, other sounds may vary with airspeed or engine RPM, altering both the amplitude *and* the frequency of the waveform, which may be defined as plots or tabular values of the waveform. Particular sounds in this category include:

- runway rumble – the sound of the undercarriage along the runway varies with ground speed;
- slipstream – the sound of the wind interacting with the aircraft structure varies with airspeed;
- engines – the amplitude and frequency of the sound of a jet engine vary with the N1 and N2 rotational speeds of the engine;
- propellers – the amplitude and frequency of the sound of a propeller vary with engine RPM.

Figure 8.12 shows the variation of the frequency and amplitude of an engine intake sound component as a function of N1, which is given as a percentage of the maximum RPM. The actual intake

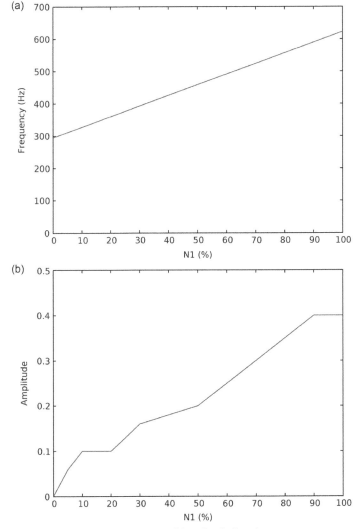

Figure 8.12 Sound Frequency and Amplitude Envelopes.

waveform is generated as notch-filtered white noise. However, as the waveform scales linearly from 300 Hz to 620 Hz, a nominal waveform centred on 460 Hz is used (with $Q = 1$). During each frame, the current N1 value is used to compress or expand the frequency. Similarly, the amplitude is also set as a function of N1, with the major increase occurring after N1 = 50%.

These modulating values are defined as two linear arrays:

```
float intake_gain_x[9] = { 0.0, 10.0, 20.0, 30.0, 50.0, 70.0, 80.0, 90.0, 100.0 };
float intake_gain_y[9] = { 0.0,  0.08, 0.08, 0.16, 0.2,  0.3,  0.35, 0.4,    0.4 };
```

The gain and frequency are interpolated from the tabulated values to enable the amplitude and frequency of the sound sources to be modified dynamically. Note that the frequency value is a ratio of the current frequency to the centred frequency:

```
Gain = interpolate(sizeof(intake_gain_x), N1, intake_gain_x, intake_gain_y);
PitchFactor = interpolate(sizeof(intake_frequency_x), N1, intake_frequency_x, intake_frequency_y);
alSourcef(soundSources[SNDSRC_ENGINE_INTAKE], AL_GAIN, Gain);
alSourcef(soundSources[SNDSRC_ENGINE_INTAKE], AL_PITCH, PitchFactor / 460.0);
```

This particular method simplifies the management of the sound generation system. Once the base waveforms are generated, the amplitude and frequency of the waveform are recomputed every frame, avoiding the need to re-compute any waveforms. However, care is needed as variation in the frequency range can result in distortion of the waveform perceived by the listener. In addition, several different sets of sound data may be acquired for different operating conditions and the joining of these sections must avoid any discontinuity. In other words, the amplitude of the end of one recorded track must be identical to the start of the next track, otherwise a distinct 'click' will be heard in the transition between recordings. Tools such as Audacity enable the waveforms to be trimmed to satisfy this criterion.

8.4 Tones

Tones are used in flight simulation to provide warning sounds and to replicate tones generated by avionics equipment. For example, passing over the middle marker on an approach, a 1300 Hz tone is generated by the ILS receiver. The flight crew are able to identify this tone as a middle marker indicator from its frequency and repetition rate, which is used as a procedural check during an approach. The actual tone is generated as a repeating sequence of two dashes per second. In addition, the sound is coincident with the flashing of the blue marker lamp on the instrument panel or EFIS display.

Although some warning sounds are continuous, the majority are modulated, in the sense that they are formed as pulse trains of constant sine waves. A repetitive sound clearly provides a better alerting or warning signal than a continuous sound. In addition, the repetition rate enables immediate identification of the source of the warning by the flight crew. In the case of pulse train generation, there are two options. Firstly, a set of pulses can be generated with the correct timing sequence, which is sufficiently long that the waveform repeats without any noticeable disruption. Alternatively, the waveform can be switched on or off using the frame rate as a timing reference. In practice, the latter method is simpler to implement for three reasons:

1) Several warnings take the form of time-varying pulse trains. A generalised function can provide the range of waveforms used in simulation.
2) The sounds are updated every frame and a frame time of 20 ms provides sufficient resolution to switch waveforms on and off, with the proviso that all waveforms can be synchronised to the simulator frame rate.
3) Morse waveforms are included in the class of pulse train waveforms, otherwise it would be necessary to generate separate waveforms for each Morse symbol.

The simplest method to inhibit or switch off a sound is to set its amplitude to zero. For historic reasons, many warnings can be traced back to the dot–dash sequences of Morse signalling. Strictly, as Morse signals were transmitted by human operators, there is no standard for Morse signal timing other than that a dash is three times the length of a dot and a gap between symbols is the same length as a dot. With a 50 Hz frame rate (20 ms) used in a real-time simulator, convenient values are 120 ms for a dot and 360 ms for a dash.

8.4.1 Outer Marker

The outer marker transmits a 400 Hz signal at two dashes per second. The duration of two dashes is 720 ms and the duration of two gaps is 240 ms, giving 960 ms. Although this value is not exactly 1 s the difference is indiscernible. The updating of all sounds occurs once per frame and the following data structure is provided for pulse code modulated sounds:

```
typedef struct
{
    bool         Enabled;
    bool         Sound;
    unsigned int Pulse;
    unsigned int Timeout;
    PulseProc    OnProc;
    PulseProc    OffProc;
    PulseProc    StopProc;
    WaveformType Waveform;
} PulseRecord;
```

The field **Enabled** is **true** if the sound is required and has been initialised. **Sound** is **true** if the sound is currently active. **Pulse** defines the current pulse in the pulse train. **Timeout** defines the number of frames before the signal is switched off. Three functions are used with each waveform: **OnProc** to turn the sound on, **OffProc** to turn the sound off and **StopProc** to suspend the sound. **Waveform** is an array of four integer values, defining the number of frames, where the waveform is switched on, then off, then on and finally off. A sound is started by calling the following function:

```
void StartPulseRecord(PulseRecord *p, bool On)
{
    if (On == p->Enabled)
    {
        return;
    }

    if (On)
    {
        p->Sound = false;
```

```
            p->Pulse = 4;
            p->Timeout = 0;
        }
        else
        {
            (p->StopProc)();
        }

        p->Enabled = On;
    }
```

A check is made that the sound is enabled. If the sound is to be turned on, it is initialised to a state where the field **Sound** is false (the sound is inactive), **Timeout** is zero (a new pulse will be activated) and **Pulse** is four, so that the pulse train will restart. During each frame, the various pulse trains are updated by calling the function **UpdatePulse**:

```
void UpdatePulse(PulseRecord *p)
{
    if (p->Enabled)
    {
        if (p->Timeout > 0)
        {
            p->Timeout -= 1;
        }
        else
        {
            p->Pulse += 1;
            if (p->Pulse > 4)
            {
                p->Pulse = 1;
            }
            p->Timeout = p->Waveform[p->Pulse];
            p->Sound = !p->Sound;
            if (p->Sound)
            {
                (p->OnProc)();
            }
            else
            {
                (p->OffProc)();
            }
        }
    }
}
```

If the pulse is enabled, and the current pulse has timed out, the next pulse is initiated, the pulse state is reversed and the sound is either turned on or off, depending on the current state. If the timeout is non-zero, it is simply decremented and the pulse state remains active for the current frame. For example, the outer marker update is called by

```
UpdatePulse(&OuterMarkerRecord);
```

where **OuterMarkRecord** is given by

```
const WaveformType OuterMarkerWaveform = {0, 18, 6, 18, 6};
```

In this case, the outer marker is enabled for 18 frames (360 ms), disabled for 6 frames (120 ms), enabled for 18 frames (360 ms) and disabled for 6 frames (120 ms) every 48 frames. The simplicity of the method is that it only requires the generation of a 400 Hz sine wave and a call to update the waveform every frame, with four values defined in a small table. One other point to note is that the display software can also access the state of the outer marker to synchronise the sound with the display of the outer marker panel lamp.

8.4.2 Middle Marker

The management of the middle marker is very similar to the outer marker. The waveform is a 1300 Hz sine wave modulated with dot–dash pairs. This sequence is achieved by setting the following values in the table **MiddlemarkerWaveform**:

```
const WaveformType MiddleMarkerWaveform = {0, 6, 6, 18, 6};
```

The middle marker is enabled for 6 frames (120 ms) for a dot, followed by a gap of 6 frames (120 ms), followed by a dash of 18 frames (360 ms) and a gap of 6 frames (120 ms). This sequence takes 720 ms, or 1.4 s for two dot–dash pairs.

8.4.3 Morse Code

For many years, pilots were required to be able to read Morse code as a means of identifying navigation transmitters. However, this requirement has been relaxed for two reasons. Firstly, with modern displays, the three- or four-digit 'ident' code is shown on EFIS displays, enabling flight crews to identify a transmitter. Secondly, many charts still include the small Morse sequence next to the transmitter to aid with identification. The concern with tracking, particularly for VOR, is that it is possible to incorrectly select a VOR frequency, which may be a valid VOR transmitter, and then track towards the wrong transmitter. Morse identification provides a further (and essential) cross-check that the correct station has been tuned. While a secondary requirement in most flight simulators (and Morse is often switched off in simulator sessions), Morse is simple to implement as it is a variant of the marker signals described previously. The table of the 26 alphabetic characters is as follows:

```
const unsigned int MorseWaveform [26][9] =
  {{DOT,  GAP,  DASH, GAP,  EOM,  EOM,  EOM,  EOM, EOM},  /* A .-    */
   {DASH, GAP,  DOT,  GAP,  DOT,  GAP,  DOT,  GAP, EOM},  /* B -...  */
   {DASH, GAP,  DOT,  GAP,  DASH, GAP,  DOT,  GAP, EOM},  /* C -.-. */
   {DASH, GAP,  DOT,  GAP,  DOT,  GAP,  EOM,  EOM, EOM},  /* D - .   */
   {DOT,  GAP,  EOM,  EOM,  EOM,  EOM,  EOM,  EOM, EOM},  /* E .     */
   {DOT,  GAP,  DOT,  GAP,  DASH, GAP,  DOT,  GAP, EOM},  /* F  .-. */
   {DASH, GAP,  DASH, GAP,  DOT,  GAP,  EOM,  EOM, EOM},  /* G --.  */
   {DOT,  GAP,  DOT,  GAP,  DOT,  GAP,  DOT,  GAP, EOM},  /* H .... */
   {DOT,  GAP,  DOT,  GAP,  EOM,  EOM,  EOM,  EOM, EOM},  /* I .     */
   {DOT,  GAP,  DASH, GAP,  DASH, GAP,  DASH, GAP, EOM},  /* J .--- */
   {DASH, GAP,  DOT,  GAP,  DASH, GAP,  EOM,  EOM, EOM},  /* K -.-  */
   {DOT,  GAP,  DASH, GAP,  DOT,  GAP,  DOT,  GAP, EOM},  /* L .-  .*/
   {DASH, GAP,  DASH, GAP,  EOM,  EOM,  EOM,  EOM, EOM},  /* M --   */
   {DASH, GAP,  DOT,  GAP,  EOM,  EOM,  EOM,  EOM, EOM},  /* N -.   */
   {DASH, GAP,  DASH, GAP,  DASH, GAP,  EOM,  EOM, EOM},  /* O ---  */
   {DOT,  GAP,  DASH, GAP,  DASH, GAP,  DOT,  GAP, EOM},  /* P .--. */
   {DASH, GAP,  DASH, GAP,  DOT,  GAP,  DASH, GAP, EOM},  /* Q --.- */
   {DOT,  GAP,  DASH, GAP,  DOT,  GAP,  EOM,  EOM, EOM},  /* R .-.  */
   {DOT,  GAP,  DOT,  GAP,  DOT,  GAP,  EOM,  EOM, EOM},  /* S ...  */
```

```
{DASH, GAP,  EOM,   EOM,   EOM,   EOM,   EOM,   EOM, EOM},  /* T -    */
{DOT,  GAP,  DOT,   GAP,   DASH,  GAP,   EOM,   EOM, EOM},  /* U  .-  */
{DOT,  GAP,  DOT,   GAP,   DOT,   GAP,   DASH,  GAP, EOM},  /* V ...- */
{DOT,  GAP,  DASH,  GAP,   DASH,  GAP,   EOM,   EOM, EOM},  /* W .-- */
{DASH, GAP,  DOT,   GAP,   DOT,   GAP,   DASH,  GAP, EOM},  /* X - .- */
{DASH, GAP,  DOT,   GAP,   DASH,  GAP,   DASH,  GAP, EOM},  /* Y -.-- */
{DASH, GAP,  DASH,  GAP,   DOT,   GAP,   DOT,   GAP, EOM}}; /* Z -- . */
```

Accessing the symbols is similar to the method used for the outer and middle markers. The particular sequence for a character is accessed. As there is a variable number of dots and dashes per symbol, each symbol is terminated with an **EOM** character, where the **DASH**, **DOT**, **GAP** and **EOM** characters are defined as follows:

```
#define DOT          6 /* 120 ms */
#define DASH        18 /* 360 ms */
#define GAP          6 /* 120 ms */
#define EOM         30 /* 600 ms */
```

where the values correspond to frames.

8.4.4 Warnings

Two other tones provide a gear warning if the undercarriage is raised and the aircraft is less than 500 ft above ground level and descending, and a configuration warning if the airspeed exceeds the maximum values for flap settings. The gear warning is a 540 Hz saw-tooth waveform, as shown in Figure 8.13, which is switched on and off two times per second.

The configuration warning is also a triangular waveform but amplitude-modulated, as shown in Figure 8.14. Both waveforms are switched at 2 Hz, as given by the following:

```
const WaveformType GearWaveform        = {0, 10, 10, 10, 10};
const WaveformType ConfigWaveform      = {0, 10, 10, 10, 10};
```

The stall warning is a 250 Hz saw-tooth waveform, as shown in Figure 8.15.

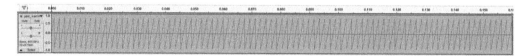

Figure 8.13   Gear Warning Waveform.

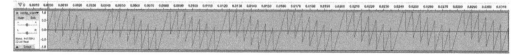

Figure 8.14   Configuration Warning Waveform.

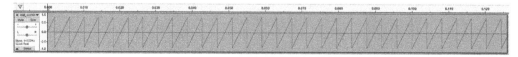

Figure 8.15   Stall Warning Waveform.

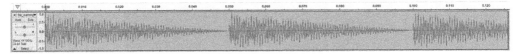

Figure 8.16 Fire Warning Waveform.

Figure 8.17 Electrical Background Noise.

The warning is generated as a constant waveform when the stalling angle of attack is exceeded. The saw-tooth provides a harsh warning signal. The fire warning is a ringing alarm, amplitude-modulated at 200 Hz, as shown in Figure 8.16.

The warning is generated as a constant waveform when an engine fire is detected (or activated) and is cancelled when the appropriate fire extinguisher switches are pulled by the flight crew.

8.4.5 Background Sounds

Three background sounds are generated, which are common to large transport aircraft: air conditioning, electrical systems and undercarriage lowering. Air conditioning is simply filtered white noise at low amplitude. The electrical sounds replicate the background noise from generators, converters and other ancillary aircraft electrical systems, as relatively quiet background noise. A typical waveform of electrical noise is shown in Figure 8.17.

These background sounds are played repeatedly without any need to adjust the frequency or amplitude. One other background sound is the raising and lowering of the undercarriage. This is a complex sound as it contains the opening and closing of the undercarriage doors, the motor driving system and the change in slipstream noise, and it is best implemented from a sound recording. This sound differs as it is a single-shot sound and is only activated when the undercarriage is raised or lowered.

8.4.6 Turbofan Sounds

The turbofan sounds comprise six tables defining the gain and frequency which are used for each engine:

```
float whn_0_frequency_x [2] = { 0.200000, 0.400000 };
float whn_0_frequency_y [2] = { 0.975000, 1.075000 };
float whn_0_gain_x [8] = { 0.000000, 0.100000,  0.157000,  0.221000,  0.305000,  0.370000,
                           0.439000, 0.606000 };
float whn_0_gain_y [8] = { 0.000000, 5.000000, 34.853001, 33.914001, 23.458000, 12.332000,
                           4.692000, 0.000000 };

float whn_1_frequency_x [2] = { 0.310000, 0.660000 };
float whn_1_frequency_y [2] = { 1.030000, 1.200000 };
float whn_1_gain_x [8] = { 0.282000,  0.341000,  0.379000,  0.423000,  0.497000, 0.517000,
                           0.560000, 0.626000 };
float whn_1_gain_y [8] = { 0.000000, 20.000000, 25.000000, 25.000000, 14.000000, 9.500000,
                           3.191000, 0.000000 };

float whn_2_frequency_x [2] = { 0.400000, 0.600000 };
float whn_2_frequency_y [2] = { 0.900000, 1.200000 };
```

```
float whn_2_gain_x [8] = { 0.333000,  0.414000,  0.483226,  0.556129, 0.581935, 0.618065,
                           0.669677, 0.731000 };
float whn_2_gain_y [8] = { 0.000000, 10.372000, 15.292553, 11.037234, 7.978724, 5.053192,
                           2.127660, 0.000000 };

float whn_3_frequency_x [2] = { 0.560000, 0.910000 };
float whn_3_frequency_y [2] = { 0.600000, 1.340000 };
float whn_3_gain_x [8] = { 0.530000,  0.603000,  0.641000,  0.672000,  0.732000,  0.808000,
                           0.881000,  0.990000 };
float whn_3_gain_y [8] = { 0.000000,  7.447000, 17.420000, 27.926001, 40.425999, 41.355999,
                          31.915001,  0.000000 };

float whn_4_frequency_x [2] = { 0.322000, 1.000000 };
float whn_4_frequency_y [2] = { 1.000000, 1.977000 };
float whn_4_gain_x [8] = { 0.400000, 0.600000, 0.700000, 0.765806,  0.800000,  0.850000,
                           0.900000,  0.950000 };
float whn_4_gain_y [8] = { 0.000000, 0.000000, 0.000000, 8.643617, 50.000000, 75.000000,
                         100.000000, 100.000000 };

float whn_5_frequency_x [2] = { 0.880000, 0.970000 };
float whn_5_frequency_y [2] = { 0.920000, 1.012000 };
float whn_5_gain_x[6] = { 0.875000,  0.890000,  0.905000,  0.920000,  0.950000,  0.970000 };
float whn_5_gain_y[6] = { 0.000000, 10.000000, 30.000000, 50.000000, 65.000000, 80.000000 };
```

The following fragment of code is called every frame to generate the sound for engine 0 (the port engine):

```
Gain = interpolate(sizeof(whn_0_gain_x), N1 / 100.0, whn_0_gain_x, whn_0_gain_y);
PitchFactor = interpolate(sizeof(whn_0_frequency_x), N1 / 100.0, whn_0_frequency_x,
                          whn_0_frequency_y);
alSourcef(soundSources[SNDSRC_WHN_0_0], AL_GAIN, Gain / 100.0);
alSourcef(soundSources[SNDSRC_WHN_0_0], AL_PITCH, PitchFactor);

Gain = interpolate(sizeof(whn_1_gain_x), N1 / 100.0, whn_1_gain_x, whn_1_gain_y);
PitchFactor = interpolate(sizeof(whn_1_frequency_x), N1 / 100.0, whn_1_frequency_x,
                          whn_1_frequency_y);
alSourcef(soundSources[SNDSRC_WHN_0_1], AL_GAIN, Gain / 100.0);
alSourcef(soundSources[SNDSRC_WHN_0_1], AL_PITCH, PitchFactor);

Gain = interpolate(sizeof(whn_2_gain_x), N1 / 100.0, whn_2_gain_x, whn_2_gain_y);
PitchFactor = interpolate(sizeof(whn_2_frequency_x), N1 / 100.0, whn_2_frequency_x,
                          whn_2_frequency_y);
alSourcef(soundSources[SNDSRC_WHN_0_2], AL_GAIN, Gain / 100.0);
alSourcef(soundSources[SNDSRC_WHN_0_2], AL_PITCH, PitchFactor);

Gain = interpolate(sizeof(whn_3_gain_x), N1 / 100.0, whn_3_gain_x, whn_3_gain_y);
PitchFactor = interpolate(sizeof(whn_3_frequency_x), N1 / 100.0, whn_3_frequency_x,
                          whn_3_frequency_y);
alSourcef(soundSources[SNDSRC_WHN_0_3], AL_GAIN, Gain / 100.0);
alSourcef(soundSources[SNDSRC_WHN_0_3], AL_PITCH, PitchFactor);

Gain = interpolate(sizeof(whn_4_gain_x), N1 / 100.0, whn_4_gain_x, whn_4_gain_y);
PitchFactor = interpolate(sizeof(whn_4_frequency_x), N1 / 100.0, whn_4_frequency_x,
                          whn_4_frequency_y);
alSourcef(soundSources[SNDSRC_WHN_0_4], AL_GAIN, Gain / 100.0);
alSourcef(soundSources[SNDSRC_WHN_0_4], AL_PITCH, PitchFactor);

Gain = interpolate(sizeof(whn_5_gain_x), N1 / 100.0, whn_5_gain_x, whn_5_gain_y);
PitchFactor = interpolate(sizeof(whn_5_frequency_x), N1 / 100.0, whn_5_frequency_x,
                          whn_5_frequency_y);
alSourcef(soundSources[SNDSRC_WHN_0_5], AL_GAIN, Gain / 100.0);
alSourcef(soundSources[SNDSRC_WHN_0_5], AL_PITCH, PitchFactor);

alSourcef(soundSources[SNDSRC_REVRS_0], AL_GAIN, 0.0);
```

where the gain and frequency of the sound components are derived by a linear interpolation of the engine RPM **N1** from the tabular values. Similar code is applied for the other three engines and a further set of tables is used for reverse thrust.

8.4.7 Real-time Sound Generation

The sound software is based on all the sound channels being updated every frame. The software consists of a library of the individual sounds (soundlib.c), which includes the specific OpenAL calls, and a module (sounds.c), which is independent of OpenAL, to activate specific sounds, depending on the state of the simulation. The function **SoundSystem**, which is called every frame, is as follows:

```
void SoundSystem(bool Hold)
{
    bool t;
    unsigned int e;

    OuterMarkerIdent(EngLink_NavPkt.OuterMarker);
    MiddleMarkerIdent(EngLink_NavPkt.MiddleMarker);

    t = Feet(EngLink_NavPkt.GroundLevel - (float) EngLink_AeroPkt.Pz) < 500.0 && /* below 500 ft */
        EngLink_AeroPkt.GearSelector == IODefn_GearUp &&                  /* gear up and */
        EngLink_AeroPkt.Vd > 1.0;                                         /* descending */
    GearWarning(t);

    if (EngLink_NavPkt.MorseChannel != 0)
    {
        Morse(EngLink_NavPkt.MorseIdent, EngLink_NavPkt.MorseChannel == 1);
    }

    ConfigurationWarning(EngLink_AeroPkt.ConfigWarning);
    StallWarning(EngLink_AeroPkt.Stalling);
    FireWarning(Systems_EngineFireSound);

    if (!Hold)
    {
        if (Engines_EngineType == EngDefn_Turbofan)
        {
            AirConditioning(true);
            ElectricalNoise(true);
        }

        if (OldGearSelector != EngLink_AeroPkt.GearSelector)
        {
            if ((OldGearSelector == IODefn_GearUp) || (OldGearSelector == IODefn_GearDown))
            {
                GearMotor(true);
            }
            OldGearSelector = EngLink_AeroPkt.GearSelector;
        }
    }

    Slipstream(Kts(EngLink_AeroPkt.Vc));
    GroundRumble(EngLink_AeroPkt.OnTheGround, Kts(EngLink_AeroPkt.Vc));
    GearBuffet(Kts(EngLink_AeroPkt.Vc), EngLink_AeroPkt.GearPosition);

    Sounds(!Hold, Engines_EngineType);

    for (e = 0; e <= 3; e += 1)
    {
        JetEngine(e, Engines_Engines[e].Rpm, Engines_ReverseLever[e] > 0.05);
    }
}
```

The arguments of the functions `OuterMarkerIdent`, `MiddleMarkerIdent`, `GearWarning`, `ConfigurationWarning`, `StallWarning` and `FireWarning` enable or disable specific sounds. For example, if the aircraft is below 500 ft, with the undercarriage not lowered and descending, the gear warning sound is activated. Unless the simulator is in a flight-freeze (held) state, the functions `AirConditioning` and `ElectricalNoise` are enabled. If the gear lever is selected to raise or lower the undercarriage, the gear sound is activated as a single non-repeating sound. The slipstream sound is activated as a function of the airspeed. If the aircraft is on the ground, the rumble sound is activated as a function of the ground speed. The sound of the undercarriage buffeting is activated as a function of the airspeed and the position of the undercarriage. The function `Sounds` controls all the sound buffers, enabling the sounds for normal operation and disabling the sound if the simulator is in a flight-freeze state. For each of the four engines, the frequency and amplitude in each channel are set as a function of the engine RPM (N1). Similar code is used to set turboprop and piston engine sounds.

8.5 Recordings

Ideally, the recording of sound in an aircraft would be made close to the pilot's ear position. The main difficulty with this approach is the separation of sound sources. For example, the recording may include a component from the slipstream and components from the aircraft engines. In the case of propeller-driven aircraft, the engine sound is combined with propeller noise, which is mostly a function of airspeed and engine RPM. Inadvertent or unintended background sound can be edited and removed from recordings, but otherwise a sufficient number of recordings are required to cover the operating range of the aircraft. Even with several hundred recordings of different engine RPM settings, there is still a requirement to select the appropriate recording and also to blend the transition between two recordings. This problem is compounded, as a recording will vary in both amplitude and frequency and, certainly, discontinuities in frequency will be noticed as spurious or unexpected noise.

An alternative approach is to take a recording at a centre frequency in the operating range and then modulate the frequency and amplitude to match, as best as possible, the actual recorded sounds. One attraction of this method is that OpenAL provides exactly this capability, allowing the frequency and amplitude of a recorded waveform to be adjusted every frame, where small variations between consecutive frames are imperceptible. The drawback is that scaling frequency over a relatively large range may give rise to unacceptable distortion.

8.5.1 Airspeed-related Sound

Several aircraft sounds vary with airspeed. The main contribution is from the slipstream, that is, the interaction between the fuselage, wings, engines and tail and the airflow. Generally, the hiss associated with the slipstream is modelled as white noise with the amplitude increasing up to approximately 400 Kt. A similar effect occurs with the interaction of the undercarriage and the slipstream, which is modelled as low-pass filtered white noise, with the amplitude varying with airspeed and with the proportion of the gear lowered into the slipstream. During taxiing, landing and take-off, the interaction between the runway surface and the undercarriage assemblies is a deep rumble, increasing with ground speed. Again, this is best achieved as filtered white noise with the amplitude increasing from zero at rest to maximum at take-off.

8.5.2 Turbofan Engines

The sounds produced by jet engines, including turbofans, turbojets and turboprops, are complex. In addition, the engine sounds vary significantly for different manufacturers and different engine ratings. Certainly, for the main simulator manufacturers, the aircraft company will provide recordings from flight tests covering the range of engine power settings over the full flight envelope. For turbofans, there are two primary sources of sound, emanating from the fan (N1) at higher power settings and the engine core (N2), which is associated mostly with engine start and shutdown. For a typical turbofan engine, the sound can be broken down into five components:

1) Engine whine, which is proportional to N1 and only increases from zero amplitude after 40%. It is commonly simulated by a rectified sine wave as a function of N1.
2) Secondary engine whine, which is proportional to N2 with an amplitude that only applies for N2 in the range 10–30% and is otherwise zero. It is also commonly simulated by a rectified sine wave as a function of N2. Its main contribution is during engine start and shutdown.
3) Engine intake, which relates to the disruption of airflow into the engine and varies with N1. It can be implemented as white noise with a notched filter ($Q = 1$), with frequency increasing with N1.
4) Engine efflux, the noise associated with the hot gases of the engine exhaust. This is often implemented as low-pass filtered white noise with the amplitude varying with N1. The cut-off frequency depends on the size and power of the engine.
5) Engine buzz, resulting from high frequency noise at high N1 speeds. One option is to generate a square wave with amplitude increasing nonlinearly from 60% N1.

In addition, some turbofan engines exhibit an engine squeal when accelerating from idle speed, which can be represented as a rectified sine wave as a function of the rate of change of N1.

8.6 Observations

It is possible to locate sources of aircraft sounds on the Internet; these are often developed for the games market but there are also vendors of sound files, which include aircraft, although primarily external sounds. Nevertheless, it is difficult to adapt aircraft sounds from individual sound sources for several reasons:

- The quality of the recording may not be known.
- The flight conditions may not be specified.
- It is difficult to separate actual aircraft sounds into individual components. For example, in cruise conditions, the slipstream is a likely to be rendered as white noise, and the efflux component and the intake components are filtered white noise. Separation of such sources can prove intractable.
- There is wide variation of sound over the complete flight envelope, where the instantaneous sound is a function of airspeed, engine settings and flight conditions. In order to cover taxiing, take-off, climb, cruise and descent, numerous sound recordings are required, with the additional requirement of blending sounds smoothly as the phases of flight change.
- It is possible to estimate the loudness and frequencies of sound from FFT analysis, but this is a very approximate method, particularly where the sound frequency changes dynamically.
- Subjective testing of sound is fraught with error. Even pilots experienced on specific aircraft types find it very difficult to compare trial recordings with aircraft sounds from memory.

The same problems apply to synthetically generated sound. The developer may have some insight into the major frequency components and variations in amplitude, but thereafter the requirement is to produce sine waves, rectified sine waves, square waves, triangular waves and saw-tooth patterns and apply low-pass, high-pass and notch filters to produce the desired sound.

Having pointed out the limitations of sound generation, it is worth considering the importance of the fidelity of sound cues in flight simulation. Although there is an expectation to synchronise events and sounds (e.g. stall warning), pilots do not use sound cues to control the aircraft. Consequently, sounds are background cues and their consistency is arguably more important than the detailed quality of sound replication. For example, in decreasing the throttle input, a pilot expects a reduction in engine RPM but the audible change of frequency is a secondary cue. Unless the sound cues are unusual or inappropriate, the pilot's brain will filter aural inputs, focusing on primary actions. Sound generation falls into the category of cues that sustain the illusion of flight rather than providing any benefit in pilot training or in the design and evaluation of aircraft and aircraft systems.

References

Brigham, E. O. (1974), *The Fast Fourier Transform*, McGraw-Hill.

Frigo, M. and Johnson, S. G. (2005), The Design and Implementation of FFTW3, *Proc. IEEE*, Vol. 93, No. 2, pp. 216–231.

Hiebert, G. (2006), *OpenAL Specification and Reference*, Version 2.8, available from https://www.openal.org/documentation/openal-1.1-specification.pdf (accessed 13 June 2022).

Horowitz, P. and Hill, W. (1989), *The Art of Electronics*, Cambridge University Press.

Panne, S. (2006), *The OpenAL Toolkit (ALUT)*, Version 1.1.0, available from http://distro.ibiblio.org/rootlinux/rootlinux-ports/more/freealut/freealut-1.1.0/doc/alut.html (accessed 13 June 2022).

Peacock, D. (2007), *OpenAL Programmer's Guide*, Revision 1.5, available from https://www.openal.org/documentation/OpenAL_Programmers_Guide.pdf (accessed 13 June 2022).

Rabiner, L. R. and Gold, B. (1975), *Theory and Application of Digital Signal Processing*, Prentice-Hall.

Smith, S. W. (1997), *The Scientist and Engineer's Guide to Digital Signal Processing*, California Technical Publications.

9

The Instructor Station

You can observe a lot just by watching.

Yogi Berra

The term *Instructor Operating Station* (IOS) is a generic term for the computer system and the software used to manage and supervise a simulator session. In a training situation, the supervision of a simulator session is primarily a training role. In airline simulators, the instructor is located on the flight deck directly behind the flight crew. In military simulators and in simulators where space is limited, the instructor is located externally. In both situations, the instructor manages and controls the session and monitors the flight crew actions, interacting with a computer screen, setting conditions, selecting options and monitoring information generated by the IOS (Ahn, 1997). In an engineering flight simulator, the purpose of the IOS is to set conditions, manage the session and collect and record data from tests. From a software perspective, in all these applications of flight simulation, the IOS is a graphical user interface (GUI).

9.1 Requirements

9.1.1 User Interfaces

Most users will have experienced GUIs in word-processing, spreadsheets, computer-aided design (CAD) applications and a wide range of desktop applications. The temptation is to assume that all GUIs are the same and that a general form of a GUI is appropriate as an IOS interface. Neither assumption is valid; the design and capability of a GUI are invariably dependent on the application. For example, in a word-processing package, the interface allows the user to enter and view text as it will appear on paper, cutting and pasting text, presenting data in diagrams and tables and applying actions such as sizing pages, checking spelling and outputting text in various formats. These actions are very different from an IOS.

One difficulty with the design of an IOS is that it lacks the objectivity of the other software components of a flight simulator. The design of a GUI is, by its very nature, subjective, with one user preferring the facilities offered by a particular GUI, while another user may find the same interface tedious, ambiguous or ineffective. Furthermore, a user accustomed to the conventions of one GUI

Flight Simulation Software: Design, Development and Testing, First Edition. David Allerton.
© 2023 John Wiley & Sons Ltd. Published 2023 by John Wiley & Sons Ltd.
Companion Website: www.wiley.com/go/flightsimulationsoftware

may be reluctant to change to an alternative GUI. Whereas a phugoid response can be used to validate aircraft dynamics, no such methods exist to assess the effectiveness of a GUI.

The design of a GUI (Galitz, 2007; Schneiderman, 1992) is determined by the requirements of the application. For example, a spreadsheet application provides numerical processing and tabulation of data, whereas a vending machine application focuses on the selection of items and entry of payment. In the case of applications where there is considerable user text input, it is obvious that a keyboard or keypad would be the most efficient form of input. In applications involving drawing operations, a mouse or pen provides precise movement and is clearly preferable to keyboard entry. In applications involving the selection of options, a touchscreen is probably more effective than a mouse or keyboard. In a flight simulator with a motion platform, where the instructor sits orthogonal to the main axis of rotation, a comment made by several experienced instructors is that, with the movement of the platform, a mouse or keyboard can readily slip off the instructor's table, making it difficult to select the buttons on a screen, leading to input errors and possible exasperation.

From a user's perspective, there are many desirable features of a user interface, some fundamental, some specific to the application and some preferable. The requirements of a GUI are very easy to overlook. A designer or developer may feel they can provide an ideal user interface, based on their experience of software design and knowledge of computer graphics, whereas the initial starting point of any design should be to understand and capture the user's requirements. Put another way, a GUI should reflect the user's needs rather than the developer's concept of what is best for the user. There are many examples of poor user interfaces where, invariably, the developer has overlooked the basic requirements of the user. For example, there may be functions where the user needs to input degrees magnetic or degrees true. Without consultation with the user, the developer may only provide entries in degrees magnetic.

The primary requirements of a user interface include:

- Structure – any hierarchy in the interface should be clearly evident and the user should know where they are at any level of the interface and be able to revert to a previous or base level. In simple terms, the package should be able to answer the questions: 'Where am I in this package?' and 'How do I get back to the start?'
- Effectiveness – the speed of performing operations should be acceptable to a user. For example, copy-and-paste functions allow an object to be replicated in a few selections rather than entering the same object multiple times. Most actions should appear to be instantaneous.
- Readability – operation of the user interface should be intuitive, implying a simplicity of selection and clear indications of the options available. A corollary of this requirement is that the user interface should be easy to learn.
- Navigation – the user should be able to reach a function quickly. There is invariably a hierarchy of operation in any user interface and this should be reflected in the structuring and presentation of commands and options.
- Context – the interface should provide means to enter text, numbers, settings and options and to access folders and files in an efficient set of operations.
- Consistency – there should be minimal variation to execute functions in an interface so that, for example, a set of selections for numeric input is similar to the selections for text input. Simple functions should be simple to use.
- Robustness – it should not be possible for the user to perform a function which leads to an error or exception. For example, out-of-range inputs should be flagged by the user interface and the user should be prompted to re-enter a valid value. Accidental deletion of an object should either be prompted for confirmation or it should be straightforward to reverse the action.

- Visibility – the user should be fully aware of the status of the application and the status of the user interface at all times. In simple terms, the user should never be left thinking: 'What is it doing now?' or 'Why did it do that?'
- Relevance – the user interface should relate to the application. Any symbols or designs should correspond to the relevant aspects of the application.

In their seminal textbook *Principles of Interactive Computer Graphics*, Newman and Sproull (1981) describe a conversation with a user trying to use a graphics package for the first time as an example of a badly designed user interface. It reiterates the pitfalls of assuming that the user has either the same level of knowledge or the same perspective on design as the developer. The user's apparently sensible actions result in chaos, as a consequence of the user having one perspective of the user interface and the developer having a completely different understanding. Nowadays, most projects invest considerable effort in capturing the user requirements before embarking on the design of the user interface. Even so, interface design is invariably an iterative process and involvement of the end-user at all stages of the development is strongly recommended. In particular, rapid prototyping tools enable trial versions of a GUI to be developed and emulated, without actual attachment to the application. These tools identify gaps or inconsistencies in the user interface and also provide some insight into its efficiency. For example, it may be possible to set 50 simple tasks and record the user's attempts to execute these tasks. The number of incorrect selections or failed attempts and actual user inputs will provide some insight into the effectiveness of the prototype interface, before committing to any coding.

9.1.2 Instructor Station Requirements

The role of the instructor (or supervisor) in a flight simulator is two-fold. Firstly, the instructor manages the session. For example, to place the aircraft at a specific location or to reduce the visibility. Secondly, the instructor monitors the session, for example, displaying the aircraft track or plotting flight data. There are several early design decisions based on the application and the user's requirements:

- What information is displayed?
- What form does the user input take – covering keyboard, mouse, tracker-ball, pad, touchscreen or voice?
- How are commands and options presented?
- How are commands grouped and prioritised?
- What is the interaction between the IOS software and the flight simulation software, bearing in mind that the simulator software is real-time?

By observing instructors in flight training simulators and engineering flight simulators, it is possible to gain insight into the scope of their actions and activities (Goode and Evans, 1992). Common activities include repositioning the aircraft, setting environmental values including the time of day and weather conditions, injecting failures and recording aircraft tracks and trajectories. In addition, in an engineering simulator, the management of data recording is a critical activity, ensuring that flight data is acquired and stored for subsequent analysis. Two primary observations were noted from operating an engineering flight simulator. Firstly, there is a natural grouping of commands, with several commands relating to a common activity. Secondly, a hierarchy of three levels is adequate for the majority of commands. A group of commands can be selected, then the relevant sub-command is selected from a list and, if appropriate, values or options associated with the sub-command can be entered or modified.

The classification of tasks in a flight simulator instructor station reduces to the following groups of tasks:

- Overall management – setting display modes, general options and terminating a session.
- Setting the state of the aircraft, for example, position and altitude.
- Management of the IOS display.
- Setting environmental conditions.
- Initiating failures.
- Management of air traffic.
- Provision of flight data recording.
- Reloading software, for example, to change the flight model or the visual database.

Although not an inclusive list, it summarises the main functions of an engineering flight simulator. It is worth noting that there are less than 10 primary groups of commands and within each group there should be no more than 20 sub-commands. For example, the environmental setting may include time of day, winds, turbulence, visibility and pressure settings. Similarly, the number of variables used in data recording is typically less than 30. There is one further facility provided in some simulators, which is known as scripting. Rather than initiating events by selecting commands interactively, a script of events and actions can be initiated from a script written as a text file. In military, civil and engineering flight simulators, scripts can be used to define complex environments and events, which would otherwise be difficult to implement (and repeat accurately) by selecting screen commands. Scripts are particularly useful in experiments and testing where they can specify events that occur in a session and define how the data is recorded. Importantly, they are repeatable, ensuring precise timing of events.

One advantage of a hierarchical classification is that unnecessary information is hidden from the user. It would be impractical to display all the possible commands and sub-commands without cluttering the display. Whereas a hierarchical display restricts the display to the current selection of options, command options are only required for the sub-group selected and irrelevant information is hidden. A similar approach is used in many commonplace packages. In the word-processing package shown in Figure 9.1, the *layout* option is selected from the top level menu. The second-level menu is activated and the user has selected the *columns* option. Finally, a third menu pops up, enabling the user to select the specific number of columns by clicking on the options offered. A further level is activated if the user selects 'More Columns'.

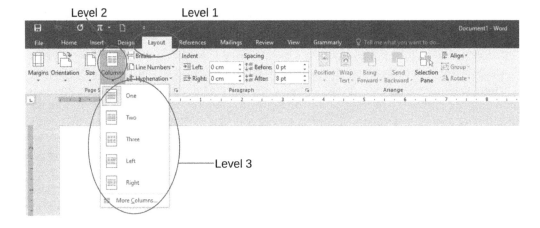

Figure 9.1   An Example of Hierarchy in a GUI.

One observation with this approach is that the top level is always active and obscures some of the scene space otherwise set aside for text presentation, which is adjusted as the menu options are selected. Also note that the selection of the main level commands is limited to a single row of 10 options, spaced equally, implying that accurate selection is required to activate a command and this structure of hierarchical menus is found in most modern applications. A variation of this theme is to place commonly used options on the left- and right-hand sides of the main screen, providing the selection of these commands in a single mouse click. The balance is between cluttering the display with many similar options, which are difficult to differentiate, and giving access to the most useful options so that the screen space is optimised for the application. GUIs of this form are common in CAD packages and graphics editing tools. The trade-off is between obscuring parts of the display and the need for extra mouse clicks to select commands. In some applications, certain commands are continually activated and there is a good case to provide a strip for commonly used commands. In other applications, maximising the displayed content is more important and, if commands are only activated occasionally, the overhead of an additional key press to activate the command prompts is insignificant.

A major difference between training flight simulators and engineering flight simulators is the importance of data recording. With modern computers, large quantities of flight data can be captured at the simulator frame rate and stored in main memory or on disk. What is important is to provide the user with access to as many variables as possible (both analogue and digital) and to be able to record and plot flight data in appropriate formats. The display of data as strip charts enables a user to confirm the result of a test during a simulator exercise and, ideally, to be able to modify the displayed variables and their scales and formats without needing to rerun the exercise. Moreover, if every item of flight data is captured and written to disk, the flight data can also be accessed and analysed offline and the data can also be played back though the simulator to replay simulator exercises or to debrief flight crews.

The overall organisation of an instructor station in a distributed system is shown in Figure 9.2. Notice that the interface with the simulator introduces the problem of reading packets from, and writing packets to, the simulator at the simulator frame rate, minimising the delay in responding to incoming packets. At the same time, user inputs can also modify the information displayed. The flight data is read and written to disk storage sporadically but should not introduce any delay into

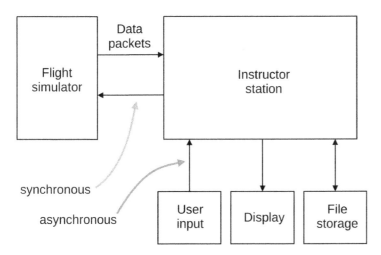

Figure 9.2 Organisation of the Instructor Station.

the simulation. The requirement to combine synchronous packet transfers and asynchronous user inputs poses a potential problem in the design of the user interface. From the user's perspective, the updating of the display should be smooth and continuous and the response to user inputs should be instantaneous. It is clearly unacceptable if the user interface introduces discontinuities into the real-time simulation or results in poor response to user interaction.

One further requirement of a modern flight simulator is to provide a realistic air space environment, in terms of the weather, communications, other aircraft and land and airborne assets. Generally, these items are not static and, in many cases, the instructor can introduce and control these elements to provide a more realistic operating environment. For example, in civil flight training simulators, air space constraints, air traffic and changing weather conditions can be introduced to provide a less benign environment. Similarly, in military simulation, ground and airborne threats and complex tactical activity of friendly and enemy forces can be added to replicate a realistic operational environment. In such cases, the instructor needs to introduce elements and initiate actions, possibly with automated or intelligent agents, appropriate to the training session. Although less important in an engineering flight simulator, provision of a representative environment needs to be included in the requirements of the IOS.

9.2 GUIs

9.2.1 User Inputs

Nowadays, a keyboard, mouse and touchscreen are commonplace with desktop computers, with touchscreens more commonly used on tablets and similar hand-held devices. These three forms of input are normally managed by the operating system device drivers so that there is a common response for all three forms of input, avoiding the need to modify the user software for different inputs. Nevertheless, these input devices have a significant impact on the effectiveness of user forms of input for different applications.

A keyboard is best suited to the entry of alphanumeric text. Although there are direction arrow keys and function keys, keyboard input to move a cursor or a graphical item on a screen is very restrictive. For mouse input, movement of a cursor over the full range of a screen is straightforward and the resolution of mouse motion is similar to the screen resolution. Typically, a mouse has two or three keys and a scroll wheel. In its simplest form, mouse motion can take one of two forms: movement of a cursor and pressing or releasing of a button. The main disadvantage of mouse input is the physical space needed to accommodate movement of the mouse and the disconnect between the cursor position on the screen and the hand movement. These limitations are overcome with the provision of a touchscreen and many touchscreen drivers provide inputs that imitate mouse functions. The main problem with a touchscreen is the relatively poor resolution in comparison with mouse input; touching the screen with a finger can cover a relatively large area of pixels. Consequently, touchscreens are ideal for option selection (e.g. a ticket-purchasing machine at a railway station) but restrictive in applications requiring accurate positioning. There is also no direct equivalent of a mouse button press, although this can be replicated by, for example, touching the screen twice or holding a finger on the screen for a short time. In addition, for both mouse and touchscreen input, the trajectory of user input, in terms of cursor movement, can be detected to differentiate between a single input and an implied linear or rotary action, inferring scaling or rotation. Some touchscreens also provide detection of more than one finger touch at a time, allowing inputs to imply expansion or contraction. However, most user interfaces provide only basic detection of mouse movement and key selection, and more advanced forms of input are left to the user application.

Taking these considerations into account, the user input device has a significant effect on the design of the GUI (Raskin, 2000). If not matched to the application, the wrong choice of device can reduce the effectiveness of the user interface and, consequently, its acceptance by users. The form of entry for most applications covers the following classes of input:

- entry of numeric data;
- entry of alphanumeric data;
- selection of options;
- confirmation or cancellation of actions;
- selection of folders;
- selection of files.

One important point to bear in mind is that the user interface is, for most purposes, programmable. For example, a filename could be magnified on a small screen, a numeric keypad could be displayed or, in selecting a file, the files could be arranged in order of the most recently or commonly selected files. In other words, the ingenuity of a GUI design can, to a limited degree, overcome the limitations of specific input devices.

9.2.2 Colour

The use of colour is, arguably, as important as the choice of the display content. Krebs et al. (1978) give an extensive list of considerations and guidance regarding the use of colour in displays, albeit for military displays, which are summarised as follows:

- Select the available colours and their contrast.
- Determine the ambient lighting conditions.
- Determine the relative contrast of colour to discard colours with poor contrast in the lighting conditions.
- Select the list of colours to be used, spacing them apart in terms of wavelength.
- Determine the operator's position relative to the screen: text should be at least 21 minutes of arc high and lines should be at least 4 minutes of arc wide; avoid blue for text or small symbols.
- Consider the choice of colours: to provide a contrast with the background; use red for danger, yellow for caution and green for safe; similar colours should never give contradictory information.
- Balance the display content with workload, reducing the amount of displayed information at times of high workload.
- Minimise the use of colour as far as possible, otherwise the benefit of using colour for attention-getting is, in effect, lost amongst similar colours.

The human factors of aircraft displays are covered extensively in the literature (Harris, 2004) and, clearly, the choice of colour for military aircraft displays in conditions ranging from bright sunlight to complete darkness is rather different from an office environment. Nevertheless, the use of colour, text and symbols is an important design consideration in developing user interfaces.

9.2.3 Prototyping

Once the functional requirements are captured and understood, a prototype user interface can be developed, usually as sketches. This is an important phase because it allows the user to imagine how the GUI will operate and can provide feedback to the developer. Note that no software is designed at this stage – the aim is purely to capture an acceptable user interface. The sketches in this section were generated using a drawing package and are used to show the functionality of the GUI rather than the visual representation.

One tool used to develop a GUI is Glade (https://glade.gnome.org), which affords the advantage of constructing prototype user interfaces that can be loaded at run-time. Earlier versions of Glade generated native C code but, for portability and language independence, Glade now generates descriptions of the user interface in XML, which can be translated to GTK code. The advantage of Glade is that it creates a GUI which, in visual terms, is very close to the working GUI, providing a realistic interface to highlight the advantages and problems associated with a specific design.

Figure 9.3 shows a preliminary sketch, where the grey dots and lines show the sequence of selections. The diagram shows a static top line for the main commands. The user has selected the *weather* option and a sub-menu has dropped down. The user has then selected *turbulence* and a sliding scale is presented. There are a number of observations:

- The diagram is only a preliminary sketch.
- The menus and selections are displayed in front of the background display.
- The top line menu is displayed at all times.
- The user has to select one of the sub-menu options.

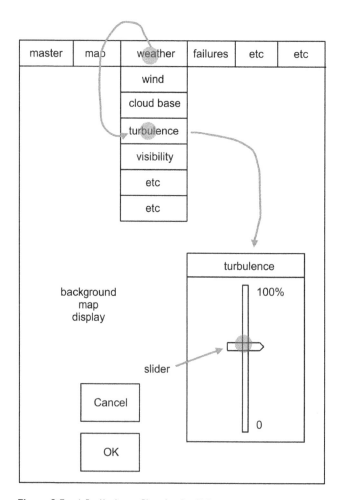

Figure 9.3 A Preliminary Sketch of a GUI.

- The turbulence is set from 0% to 100% by moving the scale pointer on a vertical slider.
- A *cancel* button and an *OK* button are included to abandon or complete the command.
- Two clicks are required to obtain the turbulence setting and then the slider is moved to set the turbulence value; a third click is required for the OK or cancel buttons.

This example also identifies a further problem. Small hand movements of a mouse may mean that the resolution of selection is accurate to 3–4 pixels. This may be acceptable to select the *weather* and *turbulence* options but it may prove difficult to select an exact value for turbulence. If the user input is a touchscreen, the situation is possibly worse as the linear resolution of a pressed finger may be 10–15 pixels. The selection of options depends on the height of the characters in the pop-up sub-menu and spacing of the options and movement of the slider may be relatively coarse. Alternatively, selection of turbulence may not be needed to the nearest percentage.

A further set of sketches is shown in Figure 9.4. Mouse key presses are shown as grey dots. Initially, no menus are displayed. In Figure 9.4(a), pressing a mouse key on the display brings up the main menu of commands. The user selects *weather* and the main menu is replaced by the sub-menu relating to the *weather* option, as shown in Figure 9.4(b). The user then selects *turbulence* and the sub-options are replaced with a keypad, as shown in Figure 9.4(c). The user enters a number via the keypad and has entered 25. Bearing in mind that these are only preliminary sketches, the following points should be considered:

- Three mouse presses are required to locate the turbulence-setting option.
- The buttons in Figures 9.4(a) and 9.4(b) are spaced to minimise incorrect selection.
- In Figure 9.4(c), the keypad enables an exact value of turbulence to be entered. For example, three mouse presses are required to input the keys 2, 5 and EXIT for 25%.
- As with the previous example, the menus and selections are displayed in front of the background display.
- A cancel and an enter option have been included.

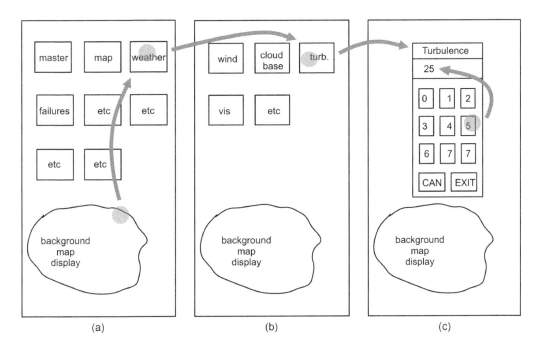

Figure 9.4 An Alternative GUI Design.

These examples are only rough sketches of possible GUI layouts and can be replaced with more acceptable graphics, but even so, they highlight both the advantages and deficiencies of the respective designs. In practice, sample pages would be sketched for all the possible variations of commands. However, these examples only apply to numeric input. It is likely that a user interface will also provide the selection of several options or accessing of files and folders, and that some commands (e.g. involving overwriting a file) may require an additional confirmation. This approach to design also identifies aspects of the GUI that might otherwise be overlooked, as prototype designs are assessed for every possible command.

There is one further consideration. It is far from clear how much of the structure and organisation of the user interface should be embedded in code or if this information would be better defined in a file. In the former case, changes to the GUI require recoding, whereas, in the latter case, the data file describing the commands and their format, can be edited and modified for a specific user's needs. Ideally, the user interface should be as generic as possible, so that changes are straightforward to implement and the GUI can be readily adapted or extended.

9.2.4 User Actions

The majority of the detail of a flight model, engine model and navigation software is embedded in the simulation software. The IOS provides a means of altering variables in these models without the need to edit or recompile the software, for example, to change the airspeed or altitude or to fail the flaps. The most common action is entering numeric values for specific variables. A second requirement is to select an option from a set of options offered, often to enable or disable an attribute. As with any user interface, some actions also require confirmation, particularly if the action could result in the loss of useful data. Finally, there is also a need to access information in files requiring an interface to inspect folders and select files.

One specific area that is unique to flight simulation is flight data recording. There is an analogy with physical data recording, which requires selection of input channels, setting of input gains, starting and stopping of the recording and annotating the recording with useful information. These operations are very similar in flight data recording, with the additional emphasis on saving acquired data to disk storage.

What is clear is that the number of different groups of actions for an IOS is far less than for commonly used desktop packages, which influences the design of the user interface and is covered in Section 9.3. The concept of a simpler user interface customised for the application of flight simulation is apparent from these background considerations.

9.2.5 Software Considerations

Before launching into software development, it is worth considering the software requirements of the user interface. Firstly, as the majority of the simulator software is written in C, it would be helpful to reuse as much of the simulator software as possible. Secondly, the interface should be applicable to all platforms, implying that platform-specific code should be avoided. Thirdly, the flight simulator software is real-time and the user interface should not introduce any delays into the overall frame rate, which is possibly the most important consideration. Some user interfaces are based on an internal iteration loop which is independent of the external computations. One option would be to run the user interface as a separate non-real-time process communicating with a dedicated real-time interface to the simulator software, but such methods could add considerable complexity to the IOS software. Fourthly, the instruction station provides the display of charts and flight data and it is necessary to be able to integrate OpenGL with the user interface.

Several open-source software user interfaces were considered in the design of the user interface: Qt was discarded owing to its size, complexity and dependence on C^{++}, Motif is somewhat dated and unsupported, wxWidgets is specific to the Windows platforms and Delphi is more of a programming language with the ability to create graphics for forms and screen displays. The Fox toolkit is supported by C^{++} rather than C. The main candidate GUI is GTK, which is an open source user interface written in C (Krause, 2007) and provides the basis for the Linux Gnome desktop. It can also be used with Glade, which affords the advantage of constructing prototype user interfaces using a graphical editing tool, which enables the design to be converted directly to GTK code, generated in C. Such methods use standard calls to GTK libraries and the automated code generation minimises the introduction of errors in the conversion from a graphic representation to C code. The main limitation with GTK is the difficulty of ensuring it meets the real-time requirement of interfacing with the simulator software, as the interface is intended for general-purpose applications. In other words, the time slot provided by GTK for user-specific interactions must not incur any delay, where the other flight simulator computers wait for a packet from the IOS. GTK has gone through several evolutions and GTK version 3, known as GTK+3, is covered in this book.

One other option for a user interface is to use OpenGL directly and generate the user interface as graphical 2D objects. The software developed for the generation of graphics and text for aircraft displays in Chapter 6 can be reused to produce a GUI for the IOS. The advantages of using GTK+3 and OpenGL to develop a user interface are summarised in Table 9.1.

An example of a very small GTK+3 program of a user interface is shown in Figure 9.5, in order to illustrate the simplicity of defining the widgets and setting call-backs. Only three menu items are shown; if the *Weather* button is pressed, a message is written to the terminal. The following listing illustrates initialisation of a GTK+3 window, creation of the three widgets and the two call-backs to respond to the *Weather* button and to close the window:

```c
#include <gtk/gtk.h>

void weather()
{
    printf("Weather option selected\n");
}

int main(int argc, char *argv[])
{
    GtkWidget *window;
    GtkWidget *fixed;

    GtkWidget *button1;
    GtkWidget *button2;
    GtkWidget *button3;

    gtk_init(&argc, &argv);

    window = gtk_window_new(GTK_WINDOW_TOPLEVEL);
    gtk_window_set_title(GTK_WINDOW(window), "Demo");
    gtk_window_set_default_size(GTK_WINDOW(window), 290, 200);
    gtk_window_set_position(GTK_WINDOW(window), GTK_WIN_POS_CENTER);

    fixed = gtk_fixed_new();
    gtk_container_add(GTK_CONTAINER(window), fixed);

    button1 = gtk_button_new_with_label("Weather");
```

```
        gtk_fixed_put(GTK_FIXED(fixed), button1, 10, 50);
        gtk_widget_set_size_request(button1, 80, 35);

        button2 = gtk_button_new_with_label("Restore");
        gtk_fixed_put(GTK_FIXED(fixed), button2, 100, 50);
        gtk_widget_set_size_request(button2, 80, 35);

        button3 = gtk_button_new_with_label("Failures");
        gtk_fixed_put(GTK_FIXED(fixed), button3, 190, 50);
        gtk_widget_set_size_request(button3, 80, 35);

        g_signal_connect_swapped(G_OBJECT(window), "destroy", G_CALLBACK(gtk_main_quit), NULL);
        g_signal_connect(button1, "clicked", G_CALLBACK(weather), button1);

        gtk_widget_show_all(window);

        gtk_main();

        return 0;
    }
```

This example illustrates the direct link between the physical presentation of the interface and the software, combining the geometry of the placement and appearance of the interface and the functionality, such as activating a widget to perform a function.

Clearly, there is no ideal GUI for use in flight simulation. The use of GTK+3 implies integration of the real-time simulation with the user interface. Alternatively, a custom user interface written

Table 9.1 Comparison of GTK+3 and OpenGL.

GUI	Pro	Con
GTK	*Rich set of widgets*	*Difficulty of combining menu displays with real-time OpenGL graphics*
	Widely used	*Synchronisation with the real-time simulator software may require separate processes*
	A prototyping tool is available (Glade)	*Map displays and flight data displays need to be adapted to the GTK environment*
	Call-backs are managed by GTK	*Lack of backward compatibility in different versions*
	Signals are used to detect the mouse state	
	The window can be resized during execution of a program, although a full-size window is likely to be used	
OpenGL	*Only a limited set of widgets is needed for an IOS*	*Widgets need to be developed from scratch*
	Real-time performance is assured	*Detection of the mouse or touchscreen inputs needs to be hard-coded*
	Reuse of graphics software (fonts and textures) from existing 2D graphics library	*The window size and positioning of widgets are static*

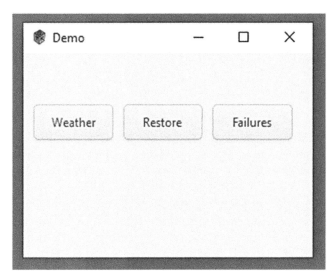

Figure 9.5 An Example of a GTK GUI.

in OpenGL provides total control over the sequencing of the interface and the interaction with the flight simulator. The OpenGL version of the GUI contains approximately 1900 lines of C code, whereas the GTK+3 version of the equivalent GUI, which generates its widgets internally, contains approximately 1200 lines of C code.

Examples of both approaches are covered in Section 9.3. The advantages and disadvantages of both approaches are illustrated for the range of functions appropriate to an engineering flight simulator. These examples of user interface software are not presented as a specific solution – they are intended to demonstrate the capability and limitation of both methods. The software provides a template for a user interface with widgets and call-backs and allows users to redevelop this software for other user interfaces, such as Qt. Bear in mind that the two versions of GUI covered in this book are merely an interpretation of a GUI for the specific application of an IOS in an engineering flight simulator.

9.3 Design of the User Interface

9.3.1 Classification of Operations

For flight simulation, the major functions of the IOS are setting conditions and monitoring the simulation. The group of commands to set conditions and events are based on the requirements set out in Section 9.1:

- Overall management of the session, loading flight models and terminating a session.
- Repositioning the aircraft.
- Management of the map display including approaches.
- Setting weather conditions.
- Setting the state of the simulator.
- Failure of systems and malfunctions.
- Management of air traffic.

The monitoring functions include:

- Management of the flight data recording including replay.
- Selection of variables for recording.
- Selection of display modes.
- Management of the flight data display, including the saving and printing of captured data.

In addition, extra commands include setting automatic flight modes, loading a flight plan, setting radio frequencies and changing the visual system viewpoint. These commands override user settings in the simulator and provide an alternative means to set specific simulator conditions.

The top-level commands are shown in Figure 9.6. This classification is arbitrary. The main objective is to choose a grouping that identifies functions for the user as clearly as possible and reduces the number of main commands to a manageable level. In addition, some commands replicate flight deck functions, such as setting autopilot modes and management of the flight plan and radios, and can be omitted. Similarly, the *Failures* option is equivalent to setting a state and could be grouped with *Settings*. Likewise, the selection of flight data variables could be included in *Data Recording*, and so on. In a training-needs analysis, observations and recording of user actions would indicate the frequency and grouping of operations, giving insight into the most appropriate structure. In effect, the menu structure is a tree and it is left to the developer to decide on the balance of height and breadth of the tree representing the hierarchy.

The *Master* sub-commands cover closing down the session, setting display modes, pausing and resuming the simulation, printing displayed information and selection of flight models, visual databases and links to external programs, as shown in Figure 9.7. The sub-commands are as follows:

- *Exit* – confirmation of termination of the session – closing down the simulator software and resetting all hardware to a standby state.
- *Display* – selecting the displayed data, including a map display, an approach plate, presentation of flight data or raw data.
- *Mode* – defining whether the simulator is running or suspended.
- *Print* – taking a snapshot of the current display.

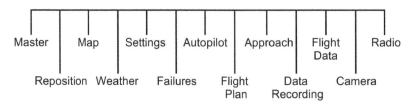

Figure 9.6 Main Commands.

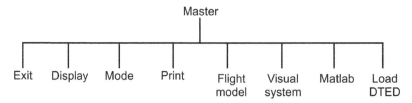

Figure 9.7 *Master* Sub-commands.

- *Flight model* – selecting a model from a library.
- *Visual system* – selecting a visual database from a library.
- *MATLAB* – enabling or disabling MATLAB (or Octave) linkage to the simulator.
- *Load DTED* – loading a digital terrain elevation database for applications needing the height above terrain.

The *Reposition* command alters the states of the aircraft, as shown in Figure 9.8:

- *Restore* – reload the aircraft state from a previously saved file. The user can select a file from a list of previously saved files.
- *Save* – save the current state of the aircraft, including engines and navigation settings, to a file in the user's current folder, where the user is prompted for a file name.
- *Position* – set the aircraft latitude and longitude, using the mouse pointer to position the aircraft on the displayed chart.
- *Altitude* – set the aircraft altitude.
- *Heading* – set the aircraft heading.
- *Airspeed* – set the aircraft indicated airspeed.

The *Map* command configures the map display as shown in Figure 9.9:

- *Centre* – the map is redrawn with the map centred at the current mouse pointer position.
- *Compass* – a compass rose is drawn on the map display at the mouse pointer position.
- *Find* – the map is redrawn with the map recentred at the current aircraft position.
- *Track* – a straight line track is entered between two points on the map.
- *Scale* – the scale of the displayed map is changed – this is the equivalent of zooming in or out (which can also be provided by a mouse scroll button).
- *Reset* – any items added to the map are removed. This command is likely to require confirmation.

The *Weather* command sets the environmental conditions as shown in Figure 9.10:

- *Turbulence* – set a turbulence value, from zero to maximum turbulence.
- *Wind speed* – set the wind speed.
- *Wind direction* – set the wind direction.

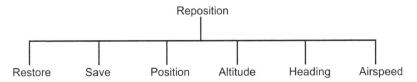

Figure 9.8   The *Reposition* Command.

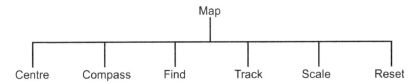

Figure 9.9   The *Map* Command.

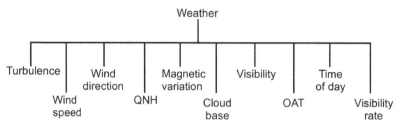

Figure 9.10 The *Weather* Command.

- *QNH* – set the regional QNH.
- *Magnetic variation* – set the magnetic variation.
- *Cloud base* – set the cloud base to enable the visual system to adjust the visibility as a function of altitude.
- *Visibility* – set the visibility range to enable the visual system to set the fogging level.
- *Visibility rate* – set the rate of change of visibility.
- *OAT* – set the outside air temperature, typically the temperature at ground level.
- *Time of day* – set the time of day to enable the visual system to adjust the lighting levels.

These commands have been presented to give a flavour of the commands and their grouping. The remainder of the sub-commands are not described for reasons of brevity. However, there are several points to note from the classifications outlined above. There is a consistency of two levels of command and a lower level of entering a numerical value or setting a condition. The classification is not always obvious and there is a clear advantage in encouraging the user to define the menu structure, classifying the commands as appropriate. Moreover, the commands and their structure may change with time or as the application is updated and the ease of being able to change the structure and commands of the menu is an important design consideration. In some commands (e.g. the weather command), rather than allocating the sub-commands to a second level, as the values are numeric, a menu could pop up, enabling any or all of the values to be changed. Finally, it is sensible to provide appropriate default values, for example, an OAT of 15°C (ISA value), to reduce the amount of unnecessary data entry.

9.3.2 Design and Implementation of Menus

Assuming a menu takes the form of a top level of commands, a lower level of sub-commands and the lowest level of a specific response to each sub-command, the menu can be structured as a file defining the commands, the type of command and the form of the expected response. Eight distinct formats of command response were identified for an engineering flight simulator:

- A question – confirming (or otherwise) a setting or condition.
- A numeric value – entering a specific value for a variable.
- An option – selecting one of a set of options to confirm a setting.
- Interaction with the map display.
- A file name – entering the name of a file to be read or written.
- A file list – a list of file names, enabling the user to select a particular file.
- A direct command – the selection of a command, which is acted upon immediately.
- Flight data settings – enabling flight data items to be plotted in specific units and scales.

Each command can be defined by five fields:

- The command name – to identify the command when it is displayed.
- The type of command, for example a numeric value or a choice of options.
- An identifier – a unique number used to access commands.
- A prompt – any information provided in a menu to assist the user.
- Default values or range limits.

The menu structure and format of the commands can then be organised as a text file, which can be modified to add or remove commands, to reclassify the grouping of commands, to change the options offered or to define the allowable range of an entered value. Although the name and the prompt may be similar, the prompt can provide additional clarification in the use of the command. The identifier is particularly important as it provides a unique value to identify each command. The actual number has no significance but must correspond to the value used to access the relevant software, typically given in a header file for the IOS. The following fragment of the menu file illustrates eight commands:

```
Exit\q               31 <OK to EXIT?>
Display\o            32 <Map Approach Flightdata Rawdata>
Print\f             34 <File>
Model\l             35
Compass\c           92
Find\x              93
Cloud Base\n       106 <Cloud Base (ft)> 0 30000 30000
Pitch rate\d       372 <deg/s rad/s> -20 20 10
```

Each command is given by its name, followed by a qualifier defining the type of command, its identification number, a prompt (if appropriate) and the arguments needed for the command. The **Exit** command is used to terminate a session; the type is defined by the \q qualifier, implying that it requires a question or confirmation and the user is prompted with **OK to EXIT?** The **Display** command specifies the type of display, with the \o qualifier implying that options are offered to the user; the user is able to select one of the four options **Map**, **Approach**, **Flightdata** or **RawData**. In this case, the prompt takes the form of the options offered. The **Print** Command has a \f qualifier, prompting the user for a file name to capture a copy of the screen. The \l qualifier in the **Model** command is used where the information is listed in a file. The file is accessed and the list of options is displayed. The \x and \c qualifiers are similar; both commands are executed directly but the \c qualifier implies that the map coordinates or compass are displayed before the command is actioned. The \n qualifier is the most common option and is used for the entry of numeric values. The list of numbers after the prompt provides the minimum value, the maximum value and the default value, respectively. Finally, the \d qualifier is used exclusively for flight data. The prompt defines the choice of units and minimum, maximum and default values, respectively.

The menu file is read for both the OpenGL and GTK+3 versions of the user interface and forms a data structure for the main menu, the sub-menus and the widgets. For each command, the name, qualifier, identifier and prompt are read. Depending on the qualifier, the options specific to each command are read and entered into the data structure. The four data structures **Buttons_t**, **Numeric_t**, **FileList_t** and **FlighData_t** are as follows:

```
typedef struct
{
    int               NumberOfButtons;
    GUI_ButtonsListType ButtonsList;
    int               ActiveButton;
} Buttons_t;
```

```
typedef struct
{
    float MinVal;
    float MaxVal;
    float Val;
} Numeric_t;
```

```
typedef struct
{
    int               NumberOfFiles;
    GUI_FileListString FileListName;
} FileList_t;
```

```
typedef struct
{
    int               NumberOfUnits;
    int               ActiveUnits;
    GUI_UnitsListType UnitsList;
    int               Plotting;
    float             ymin;
    float             ymax;
    float             yinc;
} FlightData_t;
```

A further union data structure, which is dependent on the qualifier, allows the dissimilar fields to be accessed in a single data structure **Data_t**, as follows:

```
typedef union
{
    Buttons_t         Buttons;
    Numeric_t         Numeric;
    FileList_t        FileList;
    GUI_FnameString   Fname;
    FlightData_t      FlightData;
} Data_t;
```

Each menu item (or command) is defined by the data structure **GUI_MenuItem** which contains the name of the item, its identifier, the prompt string and the qualifier in the **mName, Mval, Mprompt** and **Info** fields. The command-specific data is given in the **Data** field:

```
typedef struct
{
    GUI_MnameString   Mname;
    int               Mval;
    GUI_MpromptString Mprompt;
    GUI_MenuType      Info;
    Data_t            Data;
} GUI_MenuItem;
```

The complete menu reflects the hierarchy given in the data structure **GUI_MenuRecord**, which contains the array of menu items, the number of commands in the main menu and the name of each command in the fields **State, NumberOfItems** and **Title**, respectively. Pointers to the main menu widget and the commands widget are also provided, as follows:

```
typedef struct
{
    GUI_MenuItem          State[GUI_MenuSize];
    int                   NumberOfItems;
    GtkWidget             *Command;
    GtkWidget             *Menu;
    char                  Title[32];
} GUI_MenuRecord;
```

A text file **menu.dat** is loaded at run-time and contains the description of the menus and widgets and can be modified and adapted to a range of applications. It is worth noting that GTK+3 provides a similar mechanism to define the menu structure, using an XML notation. It is essential that the numbering of the commands defined in the file **menu.dat** is identical to the command numbering used by the user interface, given in the header file **iosdefn**.h. The complete list of commands is given in Table 9.2.

In contrast to the other computers used in the flight simulator, which transmit a constant packet size every frame, the content of the packet transmitted by the IOS varies dynamically, depending on the commands entered. As commands are activated by the GUI, the commands and their arguments are transmitted as an integer, floating-point number, Boolean or string of characters. In addition, numeric values are converted to SI units. At the receiving computers, the incoming commands are decoded according to the type of command and the number of arguments until a terminator is encountered. For the most part, the data transmitted by the IOS comprise variables to set states or conditions (e.g. visibility).

Table 9.2 Instructor Station Menu.

Main Group	Command	Type	Details
Master	Exit	confirm	Terminate the IOS session
	Display	option	Select the information displayed
	Mode	option	Control the simulator state (run, hold, freeze)
	Print	file	Print the screen to a file
	Model	file list	Load a flight model
	Visual	file list	Load a visual scene
	MATLAB	option	Activate MATLAB or Octave
	LoadDTED	file list	Load a digital terrain elevation data file
Reposition	Restore	file list	Restore a saved file
	Save	file	Save the simulator state to a file
	Position	cursor	Position the aircraft
	Altitude	value	Set the aircraft altitude
	Heading	value	Set the aircraft heading
	Airspeed	value	Set the aircraft airspeed

(Continued)

Table 9.2 (Continued)

Main Group	Command	Type	Details
Map	Centre	cursor	Centre the map display at the cursor position
	Compass	cursor	Draw a compass at the cursor position
	Find	command	Centre the display at the aircraft position
	Track	cursor	Draw a track line on the map using the cursor
	Scale	value	Set the map scale
	Reset	confirm	Reset the map display to the initial state
Weather	Turbulence	value	Set the turbulence conditions
	Wind speed	value	Set the wind speed
	Wind dir	value	Set the wind direction
	QNH	value	Set the regional QNH
	Magnetic var.	value	Set the magnetic variation
	Cloud base	value	Set the cloud base
	Visibility	value	Set the visibility
	OAT	value	Set the outside air temperature
	Time of day	value	Set the time of day
	Vis rate	value	Set the rate of change of visibility
Settings	RMI card	option	Set the RMI to a moving card or fixed card
	Morse	option	Enable or disable Morse idents
	Engines	option	Use a single engine lever for the engines
	CG position	value	Set the C/G position
	Flight controls	option	Select the centre-stick or side-stick
	HUD	option	Enable or disable the HUD
Failures	Aircraft systems	option	Fail or enable: Flaps, Gear, NAV1 LOC, NAV1 G/S, NAV2 LOC, NAV2 G/S, RMI 1, RMI 2, DME, Engines 1-4, ASI, AI, VSI, Altimeter, Turn, Compass, Flight Director, Engine Fire 1-4
Autopilot	A/P ALT	value	Set the autopilot altitude
	A/P HDG	value	Set the autopilot heading
	A/P SPD	value	Set the autopilot speed
	A/P VSPD	value	Set the autopilot vertical speed
	A/P ALT	option	Enable/disable autopilot altitude hold
	A/P HDG	option	Enable/disable autopilot heading hold
	A/P SPD	option	Enable/disable autopilot speed hold
	A/P VSPD	option	Enable/disable autopilot vertical speed hold
	Autoland	option	Enable/disable autopilot autoland
Flight Plan	Load flight plan	file	Load a flight plan
	Flight plan	option	Engage/disengage a flight plan
	Script	file	Activate a script

Table 9.2 (Continued)

Main Group	Command	Type	Details
Target	Target	file	Load a target
	Off	option	Enable/disable the target
	Script	file	Activate a target script
Approach	Range	value	Set the range for plotting
	Reset	option	Reset the approach plot
Data record	Record	option	Start, stop or continue recording
	Save	confirm	Set the recorded data to a file
	Next page	command	Move to the next page
	Previous page	command	Move to the previous page
	Next mark	command	Move to the next mark
	Previous mark	command	Move to the previous mark
	Mark	command	Mark the current position
	Goto	value	Start plotting from a specific time
	Time	value	Set the time axis
Flight Data	Data variables	data value	Plotting for recorded variables: Rudder, Aileron, Elevator, Airspeed, Altitude, Beta rate, Beta, Alpha rate, Alpha, Yaw rate, Roll rate, Pitch rate, Yaw, Roll, Pitch, g, LOC Error, G/S Error, Rate of climb, Flight Data 1-3
Camera	Record	option	Start, stop or continue recording
	Replay	option	Start, stop or continue replaying
	Position	option	Set the camera position
	Time	value	Set the recording time
	Load file	file	Load a data file to replay

9.3.3 Widgets

The term *widget* is used in most user interfaces and covers the graphical units needed to enable users to interact with the computer system and includes the entry of numerical data and text, the setting of options and access to files and folders. Typically, the user is able to organise widgets in terms of their structure, style and location and to define the linkage between selection of a widget and its associated response in software. In the OpenGL version, the widgets are hand-crafted as rectangles with space for numbers, text and selections. Similar sets of widgets are provided in GTK+3, but with the advantage that the widgets are pre-existing and the detection of interaction with a widget and its link to code is managed by GTK+3. In the OpenGL version, mouse interaction is detected for individual widgets and the response is hard-coded. By way of illustration, the main menus of the two GUIs are shown in Figure 9.11.

In the OpenGL version of the GUI, the main menu has been selected and the sub-menu of commands is shown. In the GTK+3 version, the main menu is displayed at the top of the screen at all times and the sub-menu for Weather has dropped down.

Typical widgets for numeric entry, option selection, selection of files and setting of flight data values are shown in Figures 9.12–9.15 for both OpenGL and GTK+3. Note that both GUIs are initialised from the same menu file and the commands are identical in each of the four cases. In

Figure 9.12, the OpenGL numeric widget is set by means of a keypad, which includes backspace (BS), clear (CLR), enter (ENT) and cancel (CANCEL), whereas the GTK+3 widget provides keyboard entry, where return or OK confirms the operation and CANCEL cancels the command.

In Figure 9.13, the function of the 'radio buttons' in setting the Record mode is the same for both GUIs. Only one option can be selected, giving the user the option to action the command or cancel it.

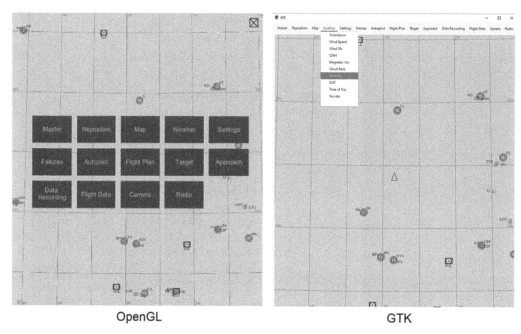

OpenGL   GTK

Figure 9.11   OpenGL and GTK+3 Widgets.

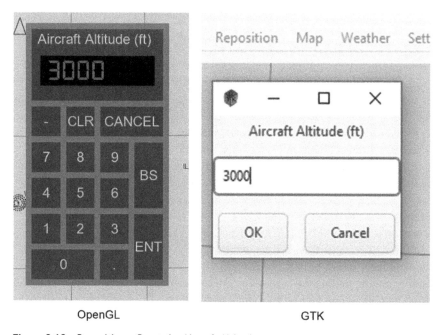

OpenGL   GTK

Figure 9.12   Reposition – Reset the Aircraft Altitude.

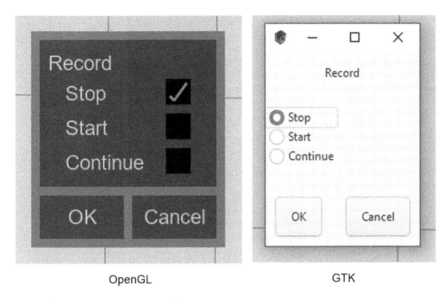

OpenGL   GTK

Figure 9.13 Record – Stop Data Recording.

In Figure 9.14, a list of saved files is presented to the user and one file can be selected to reset the simulator state to a previously saved state. In this case, the folder is searched for files in the current folder with an extension *.sav* and only these files are displayed. Selection is by pressing one of the file buttons. To abandon the command, the user can press a mouse key on any other part of the display for the OpenGL GUI or select Cancel on the GTK3 + GUI. Note that a confirmation could be added to avoid an incorrect selection.

In Figure 9.15, the display of flight data settings is also similar for both GUIs but contains three numeric values, a choice of units and a selection to enable plotting.

The main differences between the two GUIs are the appearance of the graphical formats and that the main menu is static (continually displayed) in the GTK+3 version but is activated by a

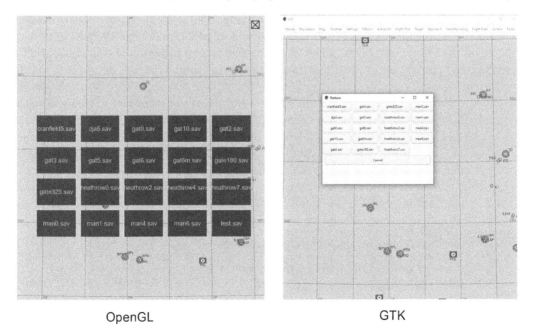

OpenGL   GTK

Figure 9.14 Restore.

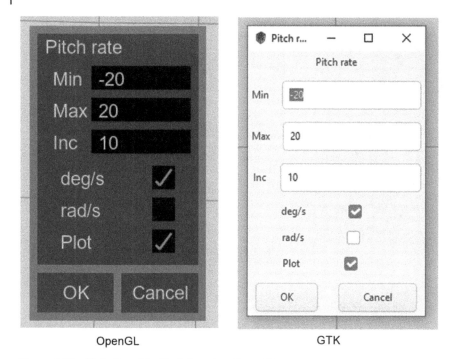

OpenGL   GTK

Figure 9.15   Flight Data Plot Pitch Rate (see Plate 12).

mouse click in the OpenGL version. One further requirement is the validation of entered values. Minimum and maximum are specified in the menu file and can be used to prohibit the entry of an invalid value. In Figure 9.16, the user attempts to set an airspeed of 550 Kt, which is detected as an out-of-range value and is offered the option of cancelling or modifying the entered value, until it is within the valid range.

The entry of numeric data poses a problem in GUI design. It can be argued that the entry of numeric values via a keypad or keyboard, as shown in Figure 9.12, is tedious and can be simplified by the provision of either a slider or a thumb-wheel widget, as shown in Figure 9.17.

In Figure 9.17(a), the spin button allows the value shown to be incremented or decremented by a defined amount. The advantage is that a specific value can be entered. The drawback is that, if the current value is 23, as shown, entering a value of 1223 could require 1200 presses of the mouse button. Alternatively, the display shown in Figure 9.17(b) allows values between a lower and upper range to be set by moving the slider. The advantage is that the range of admissible values is defined by the user and large changes are easy to enter. The drawback is that if, for example, the slider has a range of 3000, over a length of 127 screen pixels, the slider increments by 23.6, which is clearly inconvenient if the user needs to enter values in increments of 100.

The programming of a GUI can be summarised by the following pseudocode with the actual implementation varying with the application and platform:

```
Specify the widgets to be used
Place the widgets inside containers so that the visual layout is clear;
Specify the linkage between user interaction with the widget and the code to be activated;
Obtain the current values of the application and initialise the widget with these values;
Add the code to respond to the full range of widget events including termination of the widget.
```

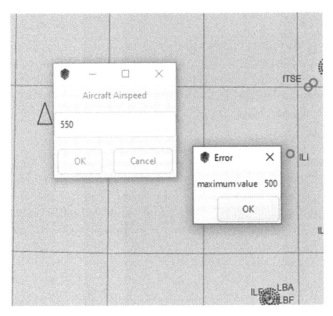

Figure 9.16 An Entry Error.

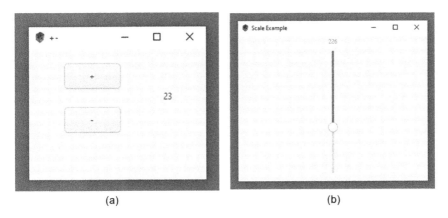

(a) (b)

Figure 9.17 Spin Buttons and Sliders.

The GUI code detects events, for example, pressing a cancel button, and invokes the code specified for that event. There is a clear division of labour with a GUI. The user provides the layout and the code to respond to events, while the GUI library maintains any changes to the GUI, for example, entering a numeric value or detecting user inputs such as keyboard entries or mouse button presses. Of course, in the case of an OpenGL version of a GUI, it is also the user's responsibility to detect user inputs, locate the GUI item corresponding to the input and activate the response in the relevant user code. The OpenGL version requires additional code to detect interaction with the mouse, whereas the attraction of using an existing GUI library is that it reduces considerably the amount of code required to implement a GUI.

9.3.4 Mouse Events

Generally, a GUI responds to user inputs from the keyboard, mouse or touchscreen to interact with the relevant menus and widgets. However, there are also many applications where the user interacts with a graphics display. For example, in a CAD package, the user may select widgets but also edit shapes and lines in the main display, where the user code responds to non-GUI inputs. In effect, once the GUI is displayed in the current frame, the user takes over control of the mouse, making changes to the displayed graphics.

In both GLFW and GTK+3, the mouse is accessed via call-backs. During initialisation, the user code informs the IDE of the functions to be called to respond to mouse or keyboard events. When the user moves the mouse or presses a mouse button, the operating system responds to the event, maintaining a list of the mouse states, particularly the position of the mouse cursor and the state of the buttons. This information is also acquired by the IDE and made available to the user via the call-back mechanism. In particular, the IDE defines the arguments of the call-back functions. Note that a call-back is never called directly by the user software, it is only invoked by the IDE and will execute the code provided by the user. For example, in GKT+3, the user can define a call-back for mouse movement as follows:

```
g_signal_connect(glarea, "motion-notify-event", G_CALLBACK (mouse_move), NULL);
```

The call-back only applies to the widget **glarea** and is called when the mouse is moved (a **motion-notify-event**). The function **mouse_move**, provided by the user, will be called if the mouse is moved (i.e. the mouse coordinates change). A typical response to mouse movement is given in the following user-defined function:

```
gboolean mouse_move (GtkWi.dget *widget, GdkEventMotion *event, gpointer data)
{
    Gui_MouseX = (int) event->x;
    Gui_MouseY = GUI_HEIGHT - (int) event->y;

    return TRUE;
}
```

In this example, the mouse coordinates are copied to the user program variables **Gui_MouseX** and **Gui_MouseY**, where the y coordinate is inverted so that $y = 0$ at the bottom edge. The function returns true if the code is successful, otherwise it returns false, to inform the GUI that a problem has occurred. Similar events can be defined for the actions of pressing or releasing mouse buttons, moving the scroll wheel or the entry of keyboard characters. As an example, it may be desirable to place a compass over a VOR station on the chart or to add a track to the chart, as shown in Figure 9.18. As the cursor is moved, the end of the track follows the cursor movement, indicating the length and bearing of the current line.

A 'rubber band' technique is used to display a track. The first point of the track is selected by a key press and, as the mouse is moved, the other end of the track follows the mouse cursor and the length and relative bearing (magnetic) of the track are displayed. When the mouse key is pressed for the second time, the track is fixed on the map display. In this example, two compasses have been placed on the display and the track from the VOR **GAM** to the VOR **OTR** is entered. The length and relative bearing of the track are displayed as the cursor is moved.

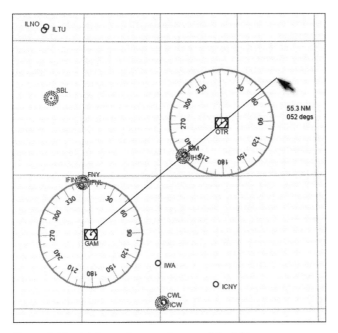

Figure 9.18   Adding a Compass Rose and Track Line.

There are two very useful functions with a map display in an IOS; panning moves the displayed map left or right or up or down and zooming changing the magnification of the map, in effect, zooming in or out. Panning is straightforward to implement in response to a mouse key press. In the following example, the map pans while the right mouse is held down, following the mouse movement:

```
gboolean mouse_move (GtkWidget *widget, GdkEventMotion *event, gpointer data)
{
    floatLat, Long;

    Gui_MouseX = (int) event->x;
    Gui_MouseY = GUI_HEIGHT - (int) event->y;

    if (Gui_MouseRightButton)
    {
        Map_ScreenToGlobe((float) (Map_CentreX - Gui_MouseX + PanningX),
                          (float) (Map_CentreY  - Gui_MouseY + PanningY) ,&Lat, &Long);
        PanningX = Gui_MouseX;
        PanningY = Gui_MouseY;
        Map_SetMapCentre(Lat, Long);
    }
    return TRUE;
}
```

The mouse coordinates are passed to the call-back **mouse_move**. The variables **Gui_ MouseRightButton, PanningX** and **PanningY** are set in the separate call-back for button presses. When the right mouse button is first pressed, **PanningX** and **PanningY** are set to the current mouse *x* and *y* positions, respectively. The map is recentred by deriving the offset of the mouse coordinates from the current map centre, relative to the panning coordinates to derive the latitude and longitude of the new map centre.

For zooming, it is possible to provide zoom-in and zoom-out commands or a separate command to set the map scale. Alternatively, the mouse scroll wheel can be used for zooming, as follows:

```
gboolean mouse_scroll (GtkWidget *widget, GdkEventScroll *event, gpointer data)
{
    if (event->direction == GDK_SCROLL_DOWN)
    {
        Gui_MouseScrollFactor *= 1.1;
    }
    else if (event->direction == GDK_SCROLL_UP)
    {
        Gui_MouseScrollFactor /= 1.1;
    }
    if (Gui_MouseScrollFactor < 1.0)
    {
        Gui_MouseScrollFactor = 1.0;
    }
    Map_SetMapScaleFactor(Gui_MouseScrollFactor);

    return TRUE;
}
```

The map scale is increased or decreased by 10% for each click of the mouse scroll wheel, which is limited to 1.0, the fully zoomed-in value, giving 10 discrete levels of zoom.

9.4   Real-time Operation

A GUI used in a flight simulator instructor station has to meet two real-time requirements. Firstly, the IOS should comply with the network protocol, so that no latency is introduced into the simulation by the IOS. Secondly, the IOS must respond to inputs from the simulator computers so that there is no noticeable delay in the response or discontinuity in the IOS displays. These requirements can be demanding as the provision of a GUI implies a considerable amount of graphics processing to maintain and update the user interface display. In addition, the user interface may also include map and chart displays, presentation of aircraft data and acquisition and recording of flight data, adding further graphics processing to each frame. During each frame, the IOS performs four functions:

1) Reading the incoming packets.
2) Transmitting the IOS packet from the previous frame.
3) Responding to user commands and forming a packet to be broadcast.
4) Updating the IOS display.

The response to user commands and updating of the IOS display must be completed within the current frame.

In many ways the requirement to sustain a real-time user interface is similar to the provision of flight displays, in the sense that each frame is a fixed time, data required by the graphics must be acquired and all the graphics operations must be completed to ensure that the frame time is never exceeded. Consequently, the developer is dependent on the efficiency of any GUI packages used in the application. The GUI software is responsible for the management and the layering of the graphics to ensure the occlusion (hidden surface elimination) of 2D objects in the display is correct at all times. For example, if a menu pops up over a display, or is resized or moved over the display, it must not interfere with the other elements of the display and the GUI must appear consistent at all times. To some degree, this separation can be achieved (or is simplified) by the ordering of rendering so that the top layers are rendered last in any frame. It is likely that the amount of processing will vary dynamically with the GUI content of the screen and tests should be undertaken to ensure that, under worst-case conditions, the computational delay imposed by a GUI never exceeds a predefined frame-time budget.

The main loop executed by an OpenGL GUI is called every frame and is summarised by the following actions:

```
Initialise the graphics settings
Clear the display to a background colour
Render
    the map, or
    the approach, or
    flight data, or
    raw data
Update the GUI for mouse inputs
Render the GUI widgets
Wait for the end-of-frame
```

This sequence is typified by the GLFW main loop. Mouse and keyboard interaction is outside this loop and is managed by GLFW call-backs. Having rendered the graphics, the GUI widgets (if any) are drawn on top of the display.

The main loop is slightly different with GTK+3, which is responsible for the sequencing of the loop. Having initialised the GUI and the OpenGL graphics interface, the GTK+3 function **gui_run** is called; this function never returns and control of the application is passed to the GUI. In addition to call-backs to respond to mouse inputs, a call-back is also provided by means of a signal, which is activated on completion of the previous frame.

```
g_signal_connect(glarea, "render", G_CALLBACK(on_render), NULL);
```

The call-back, in this case, is specific to the OpenGL region **glarea**, the argument **"render"** denotes that it is a rendering function and the user provided call-back is **on_render**. The fourth argument is **NULL** as no parameters are passed in this instance. The user-defined function **on_render** is as follows:

```
gboolean on_render (GtkGLArea *glarea, GdkGLContext *context)
{
    IOS_Update();

    glFlush();

    return TRUE;
}
```

Note that **on_render** is invoked once per frame and allows the user to supply any graphics (or indeed any computation) that is required during the frame. On return from **on_render**, GTK+3 completes the GUI operation, updating the widgets according to any mouse or keyboard inputs. For example, if a user is entering a number, the display and updating of the numeric value are managed by the GUI as characters are entered or deleted in the widget. **Ios_Update** contains the following code, which is very similar to the OpenGL version of the GUI:

```
void IOS_Update()
{
    IOS_FrameCount += 1;

    Update();

    glClearColor(0.83, 0.83, 0.83, 1.0);  /* off-white background */
    glClear(GL_COLOR_BUFFER_BIT | GL_DEPTH_BUFFER_BIT);

    Glib_LoadIdentity();
    Glib_SetTexture(2);  /* map symbols */

    Display();

    Glib_Flush();  /* process any pending objects */
}
```

The two main calls are **Update** and **Display**. **Update** involves no graphics; it reads incoming packets, broadcasts the packet formed in the previous frame, checks for a shutdown state, copies any data needed for flight data recording, connects or disconnects the interface to MATLAB or Octave (if requested), updates any aircraft tracks to be displayed, updates the flight plan and invokes any scripting (if active). The function **Display** contains the OpenGL code appropriate to the display modes.

```
void Display(void)
{
    Glib_ClipWindow(0, 0, GUI_WIDTH, GUI_HEIGHT);

    switch (IosLink_IOSMode)
    {
        case MapDisplay:
            Map_DrawMap();
            break;
        case ApproachDisplay:
            Approach_ShowApproach();
            break;
        case FlightDataDisplay:
            Plot_ShowPlot();
            break;
        case RawDataDisplay:
            DataView_Display();
            break;
    }
}
```

Note that, for the OpenGL version, having completed rendering the display graphics, the updated menus and widgets are also rendered, whereas with GTK+3, it is the responsibility of the GUI to update the menus and widgets. Also note that the OpenGL version renders both the display graphics and the GUI graphics using GLFW, whereas the graphics rendering for the GUI is provided by GTK+3 and any graphics rendered by the user software should only contain OpenGL (gl) calls, that is, GLFW is not used in a GTK+3 environment.

9.5   Charts and Maps

In recent years, the providers of navigation databases have produced digitised versions of charts for en-route navigation, approach plates and airport maps. Several vendors offer digitised maps in various forms, including open access, providing free downloading of maps. Generally, maps and charts of this form are too detailed to be useful in an IOS environment unless the training role involves VFR navigation exercises.

The main design consideration for map displays in an IOS is the choice between a digitised chart versus vector-based rendering of charts. The advantages of the digitised chart are that it can be rendered as a single texture (simplifying panning and zooming), contains all the relevant information and is likely to be updated on a regular basis. The disadvantages are that the chart may contain much unnecessary information that cannot be hidden and that charts cover a specific region, requiring some form of tiling to move smoothly across chart boundaries. A further major drawback is that, in order to overlay graphics on a textured map, the projection method and co-latitudes used to produce the chart must be known and consistent across a range of textured tiles. Alignment of a textured display with rendered graphics to support panning and zooming actions and to display an aircraft track is non-trivial. Alternatively, a vector-based display of map features is based on data derived from a navigation database. The advantage is that the display can be optimised for the application and that, mostly, the content of a map or chart for an IOS is far less detailed than standard navigation charts. The main drawback is that the chart is somewhat stylised and visually different from standard charts.

The important point to consider is that map displays should meet any requirements analysis undertaken in the design of an IOS. For most instructor stations, the main requirements are monitoring the session, setting states and conditions and recording data. A training-needs analysis will identify the visual information needed by an instructor in terms of the aircraft flight path for the different phases of flight. In the author's experience of designing engineering flight simulators, the major items to display are:

- an overhead view of the flight path, being able to centre the chart and set a scale of the displayed region, appropriate to the session;
- a view of the approach, showing the trajectory relative to the glide path and localiser and the airspeed during the approach.
- display of flight data in the form of strip charts, selecting specific flight variables and defining the units, scales and recording time;
- presentation of raw data, which is invariably used for debugging and fault-finding rather than direct observation of data.

A further proviso is that it should be straightforward to modify the instructor station displays, which is difficult if the charts are fixed textures.

Two navigation databases are used in this book: **navaids.txt** contains a worldwide list of navigation radio beacons, including the latitude, longitude, ident, frequency and range of each beacon; **airports.txt** contains a worldwide list of runways, including the latitude, longitude, length and QDM of each runway. In exercises in an engineering flight simulator, the maximum range of the navigation area is typically ±12° of latitude. As 1° of latitude is 60 nm, the extent of the region is approximately 700 nm. Any computations involving searching these databases can introduce significant delays if the complete database is accessed. For simulator operations involving distances greater than 700 nm, it would be necessary to devise a tiling scheme to continually update the local copies of the databases.

At run-time, a region is defined in terms of the minimum and maximum latitude and longitude and the databases are accessed to load radio beacons and runways within this area into arrays containing the data structures used for the navigation computations described in Chapter 5. At any time, the map can be recentred at a specific latitude and longitude. The two functions **GlobeToScreen** and **ScreenToGlobe**, described in Chapter 5, provide the mapping between world coordinates (*latitude, longitude*) and screen Cartesian coordinates (*x, y*). Each time the display is recentred, the co-latitudes are reset to compute the coefficients **Lambert_K** and **Lambert_N**, which are used in the Lambert conformal transformations, as given in the following function:

```
void Map_SetMapCentre(float Latitude, float Longitude)
{
    float Chi;
    float lat1;
    float lat2;
    float r;
    float r1;
    float r2;

    MapLatitude   = Latitude;
    MapLongitude  = Longitude;
    Chi           = PIBY2 - fabs(Latitude);
    Lambert_N     = cos(Chi);
    Lambert_K     = tan(Chi) * pow(tan(Chi / 2.0), -Lambert_N);
    lat1          = Chi - DEG6;
    lat2          = Chi + DEG6;
    r             = Lambert_K * pow(tan(Chi / 2.0), Lambert_N);
    r1            = Lambert_K * pow(tan(lat1 / 2.0), Lambert_N);
    r2            = Lambert_K * pow(tan(lat2 / 2.0), Lambert_N);
    msf           = D12 / fabs(r1 - r2);
    MapOffset     = r * msf;
}
```

where **DEG6** is 6° (radians) and **D12** is $\dfrac{2\pi R}{30}$, the distance on the earth's surface subtended by 12°, where R is the radius of the earth. Two additional global variables are computed: **msf** is the map scale factor and **MapOffset** is used in the functions **GlobeToScreen** and **ScreenToGlobe**. Notice that the two functions **GlobeToScreen** and **ScreenToGlobe** are the only places where it is necessary to take the hemisphere location into account to ensure that the map display is correct for both northern and southern latitudes.

The lines of longitude, which are straight lines on the chart, are drawn from the minimum longitude to the maximum longitude in half-degree intervals. The lines of latitude are drawn from the minimum latitude to the maximum latitude, between half-degree intervals of longitude as shown in Figure 9.19. Note the curvature of the lines of latitude.

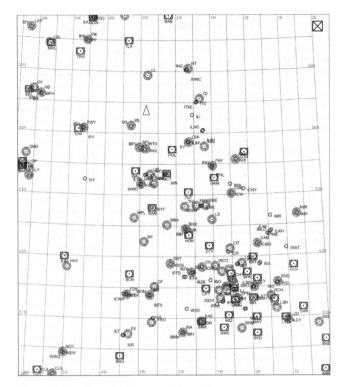

Figure 9.19 Chart Display (see Plate 13).

It is a reasonable assumption that a line of latitude subtending 0.5° can be drawn as a straight line. Note the orthogonal intersections of the lines of latitude and longitude. There is one further consideration. The lines are clipped to the extremities of the display region and can be discarded if a line is completely outside the display region, which includes:

- lines of longitude, where both x coordinates are to the left of the left edge or to the right of the right edge;
- lines of latitude, where both y coordinates are above the top edge or below the bottom edge.

Although these lines would be discarded during clipping by the GPU, the intersection of clipped lines of longitude and latitude is computed in the host computer, in order to annotate the lines of latitude and latitude on the chart at 1° intervals, as shown in Figure 9.19. The latitude and longitude values are placed adjacent to the intersection of the clipped lines with the edges of the display.

For each beacon and runway in the navigation databases, the appropriate symbol is checked to ensure that it is within the display region and rendered as a texture from a small texture atlas of map symbols, as shown for VOR, VOR/DME, DME, ADF, TACAN beacons and runways in Figure 9.20 (left to right).

Figure 9.20 Chart Symbol Textures.

The rendering of map symbols as textures rather than vectors is significantly faster, for example, the ADF symbol would otherwise be rendered as almost 60 dots. Waypoints from the flight plan, the aircraft track, other air traffic and compasses and tracks entered by the instructor are also rendered.

Again, the reader is reminded that these are rudimentary map displays, providing a template for further improvement, for example, integration with OpenStreetMap (https://www.openstreetmap.org) or other map displays. The main consideration is to ensure that such improvements are consistent with the requirements of map displays for an IOS.

9.6 Flight Data Recording

Flight data is generated by the main simulator computers at the simulator frame rate. The I/O computer broadcasts analogue data acquired by the A/D system (including inceptor positions) and digital inputs, the PFD computer provides aerodynamic data, the ENG computer produces engine data and the NFD computer broadcasts navigation data. Of course, not all flight data is relevant to an exercise and the choice is either to collect selective data, specific to the application, or to collect all the data produced during the simulation.

However, with the amount of main memory and disk storage available on modern computers, there is a persuasive case to store all the data at the simulator frame rate, ignoring the fact that much of this data may be redundant. If 1 Kb of flight data is acquired per frame, the storage requirement is 50 Kb/s, 3 Mb/min or 180 Mb/hour. With modern computers having memory capacities of over 8 Gb, albeit that not all of the memory is available for user applications, storing large amounts of flight data at the maximum frame rate is not a significant problem. The main advantage of this approach is that if all the flight data is acquired and stored, it avoids the situation where essential data items may be missed. Moreover, if the flight data is stored permanently, it is possible to view and analyse this data both online during a simulator session and offline on other computers.

One consideration with flight data recording is the potential overhead of copying data to memory or to disk, resulting in the introduction of undesirable delay into the simulation. In practice, the time to copy 1 Kb of data to memory is of the order of a few microseconds. Although writing data to disk could introduce delays of milliseconds, there are two options to avoid delays from the use of disk store: a background process could be used to access the main memory data storage and copy this data to disk; alternatively, the data could be written to disk when the simulator is not active. For example, on completion of a test, it is likely that the simulator is set to a 'flight-freeze' state and any delay at that time will not affect the real-time performance of the simulator.

What is paramount is the importance of flight data in simulator flight tests. Put simply, it should not be possible to lose recorded data. Possible causes of loss of data include software bugs in the simulator causing the simulator software to 'crash' or 'freeze', hardware faults such as an A/D card or network card so that data is corrupted and lost during transmission and power failures. Again, the design decisions depend on the value of the data. It may be possible to run an exercise in a trial and then write the data to disk store at the end of the exercise, reducing any loss of data to one exercise. If that option is unacceptable, it is necessary to write to non-volatile memory or to ensure that data is written to disk immediately, typically where a background process transfers the data to disk.

One other important aspect of flight data recording is the cost of flight trials, particularly in terms of time. Once data is acquired, it may need to be analysed to determine if further tests are required or to repeat earlier tests to confirm findings. The tools provided, particularly for online data analysis, can reduce the analysis time significantly, otherwise considerable time can be wasted searching through flight data to access particular events or data trends. The ability to skip pages (forwards and backwards) and to mark events, exercises or timing, so that the data can be fast-forwarded to a mark, can save considerable time in flight testing.

9.6.1 Data Recording

In the distributed simulator architecture described in Chapter 1, data packets are broadcast by the I/O computer (or computers), the PFD computer, the ENG computer and the NAV computer. The set of data in each packet is defined in the header files, **iodefn.h**, **aerodefn.h**, **engdefn.h** and **navdefn.h**, respectively. The data in these headers is combined in a single structure, defined as a flight recording packet in the instructor station header file **iosdefn.h**, which is written to memory and disk, given by the following data structure:

```
typedef struct
{
    AeroDefn_AeroDataPkt   AeroPkt;
    EngDefn_EngDataPkt     EngPkt;
    IODefn_IODataPkt       IOPkt1;
    IODefn_IODataPkt       IOPkt2;
    NavDefn_NavDataPkt     NavPkt;
} IosDefn_PlaybackDataPktRecord;
```

The typical storage (bytes) used by the individual packets is shown in Table 9.3. The gcc library contains two functions for reading and writing blocks of data from and to a disk file:

```
size_t fread(void *ptr, size_t size, size_t nobj, FILE *stream)
size_t fwrite(const void * ptr, size_t size, size_t nobj, FILE *stream)
```

where **ptr** is the address of the block of data to be transferred, **size** is the number of bytes to be transferred for each object, **nobj** is the number of objects to be transferred, **size** is the number of bytes in each object and **stream** is the file stream opened by **fopen**. Both functions return the number of objects transferred. Typical calls to read flight data from file to memory or to write flight data from memory to a file are as follows:

Table 9.3 Packet Storage.

Computer	Storage
I/O (1):	80
I/O (2):	80
PFD:	340
ENG:	196
NFD:	364
Flight data:	1060

```
rv = fread (&a, sizeof(IosDefn_PlaybackDataPktRecord), 1, f);
rv = fwrite (&a, sizeof(IosDefn_PlaybackDataPktRecord), 1, f);
```

where **rv=1** for a valid transfer, **a** is the array in memory holding the flight data and **f** is the file stream to the flight data file.

Similarly, to copy an individual block of flight data in an array containing blocks of flight data to or from a block of flight data, two functions **ReadBlk** and **WriteBlk** were used:

```
void ReadBlk(IosDefn_PlaybackDataPktRecord *dblk,
             IosDefn_PlaybackDataPktRecord *sblk,
             unsigned int slot)
{
    memcpy(dblk, sblk + slot, sizeof(IosDefn_PlaybackDataPktRecord));
}

void WriteBlk(IosDefn_PlaybackDataPktRecord *sblk,
              IosDefn_PlaybackDataPktRecord *dblk,
              unsigned int slot)
{
    memcpy(dblk + slot, sblk, sizeof(IosDefn_PlaybackDataPktRecord));
}
```

where **sblk** is the source block of flight data, **dblk** is the destination block of flight data and **slot** is the block of flight data to be read or written.

The attraction of using these four functions to manage captured blocks of flight data is that the currently accessed flight data can be written to any array containing sequential blocks of flight data, individual blocks can be accessed directly and the format of the flight data written to disk is identical to the flight data acquired from the incoming packets. There are three advantages from organising flight data on disk that is compatible with flight data stored in memory:

1) The tools that are used to view and analyse flight data online can be reused directly in the tools used for offline analysis.
2) Flight data stored on disk contains a copy of the original packets acquired by the IOS. Back-driving the simulator, by applying inputs from these packets, provides a replay facility requiring minimal additional software.
3) The structure of the data in a flight data block is defined by the header file for the flight data types and the respective header files for the packets, providing straightforward access to the items in the flight data packets for other applications.

Data acquisition based on acquiring blocks of flight data comprising raw simulator packets adds a negligible overhead to real-time simulation for the cost of a few gigabytes of memory. As the simulation software is a dedicated program, and a minimal set of background process are running, most of the main memory is available for user processes. Various examples of flight data recording are given throughout the book and the generation of strip charts is straightforward. The *x*-axis time scale, the *y*-axis units and the flight data parameters are selected via a widget in the flight data mode and linkage to the simulator variables is defined in the relevant header files.

The data is written to a data file with a filename **yymmddhhmm.sav**, which defines its date stamp, where **yy, mm, dd, hh** and **mm** denote the year, month, day, hour and minute when the data recording is written to the file, enabling flight data files to be indexed and accessed by reference to the date stamp.

9.6.2 Data Display

Flight data files can be accessed in several ways:

- Individual data plots can be viewed as strip charts during a simulator exercise.
- Data can also be viewed and analysed offline by retrieving the saved file.
- Displayed data can be captured as a screenshot and copied to a *png* file, which can be imported into a document.
- Various packages provide graphical displays of data, including MATLAB, Excel and gnuplot, which afford user-defined plotting methods and colour schemes.

It is straightforward to generate a script file and a data file in the specific format of a plotting package directly from the flight data and IOS settings. In Figure 9.21, a gnuplot script is generated for an undercarriage 'drop test', recording the oleo displacement of the left main gear, the right main gear, the nose gear and the height of the aircraft C/G. The generated script and a space-separated data file produce a *png* file. The first 10 frames of the data file are as follows:

```
0.02,    28.3753509521,    28.3753509521,    28.3753509521,    4.4608659744
0.04,    28.3551063538,    28.3717460632,    28.3717460632,    4.4609928131
0.06,    28.1730880737,    28.2169303894,    28.2169303894,    4.4649825096
0.08,    27.8455352783,    27.9183769226,    27.9183769226,    4.4726271629
0.10,    27.4108886719,    27.4953918457,    27.4953918457,    4.4833955765
0.12,    26.9142150879,    26.9733009338,    26.9733009338,    4.4966030121
0.14,    26.3925247192,    26.3775215149,    26.3775215149,    4.5115790367
0.16,    25.8794403076,    25.7446060181,    25.7446060181,    4.5274014473
0.18,    25.3899421692,    25.1010017395,    25.1010017395,    4.5434222221
0.20,    24.9287681580,    24.4654312134,    24.4654312134,    4.5591964722
```

The gnuplot script file, which is generated automatically by the IOS, specifies the general plotting options in terms of colours, file output type and plotting styles. For each of the four variables, the strip chart is positioned, the x and y axes are defined in terms of range and units and each variable is plotted from the corresponding field in the data file, as shown in Figure 9.21.

This example illustrates the ease of generating flight data plots in formats for commonly used plotting packages. Having specified the data to be plotted using the GUI flight test data options, the gnuplot files are generated with a single command selection. To adapt the output for other plotting formats is a relatively simple task.

9.7 Scripting

During a simulation exercise, the instructor can change settings, initiate events and record flight data. However, the features afforded by flight data recording may be excessive for certain flight tests or exercises and it is clearly difficult to repeat the operations exactly for repeated exercises. There is one further consideration in a flight simulator, the flight crew may be aware of an instructor's presence and detect that the instructor is about to initiate an event. In civil simulators, in addition to the normal roles, the instructor also injects events appropriate to the training session, including conflicting traffic, changing weather conditions and dialogue with an air-traffic controller. In military flight simulators, the instructor also introduces threats, tactical situations and may control both additional friendly and enemy ground and airborne forces. In both cases, the workload for an instructor to manage the session and provide realistic interaction with the flight crew can be very high.

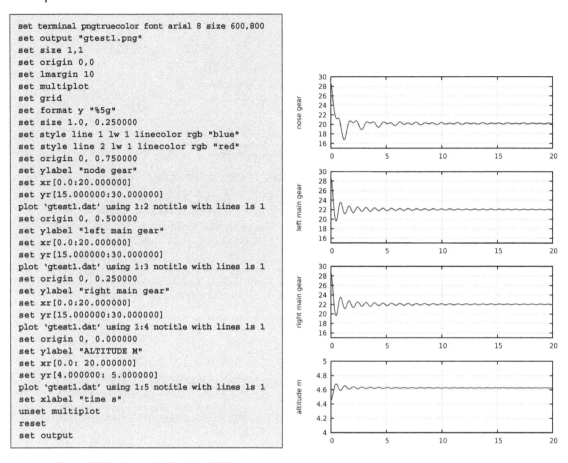

```
set terminal pngtruecolor font arial 8 size 600,800
set output "gtest1.png"
set size 1,1
set origin 0,0
set lmargin 10
set multiplot
set grid
set format y "%5g"
set size 1.0, 0.250000
set style line 1 lw 1 linecolor rgb "blue"
set style line 2 lw 1 linecolor rgb "red"
set origin 0, 0.750000
set ylabel "node gear"
set xr[0.0:20.000000]
set yr[15.000000:30.000000]
plot 'gtest1.dat' using 1:2 notitle with lines ls 1
set origin 0, 0.500000
set ylabel "left main gear"
set xr[0.0:20.000000]
set yr[15.000000:30.000000]
plot 'gtest1.dat' using 1:3 notitle with lines ls 1
set origin 0, 0.250000
set ylabel "right main gear"
set xr[0.0:20.000000]
set yr[15.000000:30.000000]
plot 'gtest1.dat' using 1:4 notitle with lines ls 1
set origin 0, 0.000000
set ylabel "ALTITUDE M"
set xr[0.0: 20.000000]
set yr[4.000000: 5.000000]
plot 'gtest1.dat' using 1:5 notitle with lines ls 1
set xlabel "time s"
unset multiplot
reset
set output
```

Figure 9.21 A gnuplot Script and Plot.

Often engineering flight simulators are used in flight trials to assess new displays or systems where the acquisition of flight data is critical and trials may be conducted with several pilots. These simulators may also be used in experiments to assess pilot workload or to devise responses to situations. In all these cases, the number of events and interactions needed to record flight data may be overwhelming and impractical for an instructor. An alternative approach is to write down all the events and actions and also the data and timing of data recordings in a form that can be activated by the simulator. This document, written in a formal (but not necessarily complicated) language, is known as a *script*. A script has a defined syntax and notation, which can be written and edited in much the same way as a computer program, and is loaded by the simulator. As the simulation progresses, any events or actions are correctly sequenced and executed to produce situations requiring a response from the flight crew. In addition, a script can also define the data to be recorded, when the recording is turned on or off, and the format of any output. Normally, the output from a script is analysed offline. A script can be thought of as a written list of commands and settings that could be initiated by the instructor, but provide a more convenient method of entering numerous commands.

9.7.1 A Simple Scripting Language

The basis of a scripting language is to define events, actions and monitoring. Events depend on external activity, for example, winds, visibility or time, or on the aircraft state or the engine state, for example, altitude, distance from a beacon or airfield or engine temperature. Actions cover settings such as failing systems or positioning the C/G or setting the fuel load. Monitoring involves the acquisition of variables used in the simulation and recording this information in an appropriate format.

On the one hand, a scripting language should be easy to write, requiring minimal computing skills, for example, allowing users to refer to common variables such as *altitude* or *airspeed* and use appropriate units, such as *Kt*, *m/s* or *mph*. On the other hand, the language should be exact, clear and unambiguous. For example, a statement defining the conditions for an event could include:

```
Altitude above 2000 ft and DME1 closer than 2 nm or airspeed less than 70 Kt
```

The ambiguity in this statement is clear and there is a requirement to be able to check the syntax and compile some form of code to be executed, while at the same time making the language as free and flexible as possible.

A script program takes the following form:

```
event:
The conditions to initiate an event
actions:
A set of actions to be executed when the event occurs
record:
A set of variables or states to be recorded and their units
end:
The conditions to terminate the actions and monitoring including time
```

Once loaded, and during each frame, the script checks a set of conditions. If they have not occurred or if the script is inactive, no action occurs. Otherwise the defined set of actions are invoked. The script then enters the second phase, where the requested data is recorded until the conditions to terminate the script have occurred. Once terminated, the script becomes inactive and is only reset by loading a new script. The organisation of a script is shown in Figure 9.22.

When the *script* command is activated by the instructor, the script program is loaded and compiled and the intermediate code is stored in memory. The script program is executed by the IOS software and has access to the packets broadcast by the other computers, which enables the script to check continuously for conditions. In much the same way that an instructor can invoke an action, the script can send similar actions in the packet broadcast every frame by the IOS, for example, to fail an engine or set visibility. During each frame, an intermediate code instruction is extracted and executed until a return instruction is encountered. An instruction can access flight data broadcast by the other computers or send an action in an outgoing packet, which will be responded to by the other computers. One particular point to note is that the script, which is running in parallel to the other computers, must not introduce any significant delay into the real-time frame transfers.

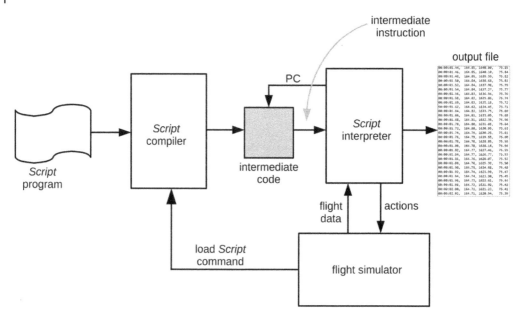

Figure 9.22 Organisation of a Script.

One option to implement a script is to write a program in C, which runs as a background process, sharing its inputs and outputs with the instructor station packet management software. However, this solution would need a library of functions to access flight data and to set simulator conditions. When this process has nothing to do, it waits for the IOS software, and is activated when a new packet arrives. Certainly, this is the best solution if the script needs to implement complex events that are dependent on numerous conditions. It has the advantage that the C code runs as a background task but with the complexity of interacting with the IOS software. The drawback is that expertise in C programming is needed to produce a script.

A simpler solution is to write a small expression compiler that processes the events and actions in the form of an interpretive language totally under the control of the IOS software. In other words, the IOS can execute instructions of the script. It has a list of instructions for each of the script phases; some of these instructions perform underlying arithmetic and other instructions access variables in the packets or activate an action or record data. An expression compiler is straightforward to implement, allowing the user to express events and actions in a relatively free format. Rather than generating machine instructions, the compiler can generate intermediate instructions that are interpreted. Although there is an obvious loss of speed in interpretation, it is unlikely that a script will contain more than a few hundred instructions and the delay associated with interpreting these instructions is negligible.

Note that in a conventional programming language the variables are defined by the user, whereas in the scripting language, the variables are restricted to the variables commonly used in flight modelling, flight control and navigation, and are predefined.

9.7.2 A Stack Machine

A common implementation of interpretive languages is a stack machine. The machine is very simple; it contains a program counter (PC), a stack and a stack pointer (SP). Pushing an item onto the stack, the item is placed at the memory location pointed to by the SP and the SP is incremented.

Removing an item from the stack, the SP is decremented and then the item at the memory location pointed to by the SP is copied to a variable. The machine also has a number of arithmetic and logical instructions, most of which operate on the two items at the top of the stack, popping the two items from the stack and pushing the result of the instruction onto the stack. For example, consider the expression $x = b^2 - 4ac$. The expression can be implemented with the following stack operations:

		SP
	initialise the stack	0
`Push b`	*push b onto the stack*	1
`Push b`	*push b onto the stack*	2
`Mult`	b^2 *now at the top of the stack*	1
`Push 4`	*push the number 4 on the stack*	2
`Push a`	*push a onto the stack*	3
`Mult`	*4a now at the top of the stack*	2
`Push c`	*push c onto the stack*	3
`Mult`	*4ac now at the top of the stack*	2
`Sub`	$b^2 - 4ac$ *now at the top of the stack*	1
`Pop x`	*store the result in x*	0

where the value shown for the SP is the value *after* the instruction is executed.

The intermediate code only requires a small set of instructions covering addition, subtraction, multiplication, division, negation, logical operations including 'and' and 'or' and arithmetic comparisons. The types and variables are limited strictly to the variables in the flight simulator. One final consideration is the provision of bracketing in expressions, to clarify the operations and to override any operator precedence. As an example, the following script program fails an engine when the aircraft is 200 ft above the ground and records the aircraft altitude, airspeed and heading for 30 s.

```
; engine failure after take-off at 200 ft
; runway altitude 210 ft ASL

event:
    altitude > 410 ft
action:
    fail engine1
    Log      TIME           SPD      ALT      HDG
    timestamp elapsedtime
record:
    airspeed kts
    altitude ft
    yaw degs
end:
    time > 30 secs
```

The first two lines are comments; the remainder of a line after a semicolon is ignored. The event to initiate the actions is an altitude exceeding 410 ft. The subsequent action is to fail the number one engine, log the message and specify that time stamping is included in the output based on the elapsed time (the time since the script was started). The record section defines the three flight variables to be recorded and their units. Finally, the end section defines the condition to terminate the script, which occurs after 30 s. An example of compiled intermediate code is shown in Figure 9.23, with a brief description of each instruction.

The return instruction implies completion of a phase. The script is a simple finite state machine; when an event is detected, an action is implied. Subsequently, it records flight data until the completion condition is detected. The record instruction pops the data from the stack and writes the data to memory. On completion of the script, the data in memory is written to a file in *csv* format, enabling the captured data to be used in various spreadsheet and plotting packages.

The attraction of this approach is the simplicity of the scripting language, which requires minimal programming expertise, while allowing a user to set up complex situations and record flight data. The flight data recording command simply records flight data, whereas a script allows the user to set up experimental conditions and record data without any intervention of the user. Consequently, scripting can be used as an alternative to flight data recording. In practice, scripting allows experiments to be repeated exactly and for data to be recorded, which depends on specific flight conditions. One final point is that the scripting covered in this section is only an example and it would be possible to extend this approach to more complex situations. For example, in military simulation, the script could initiate certain enemy action and react by changing the tactics of enemy forces in response to the actions of the pilot in the simulator. Similarly, in a civil training simulator, with voice detection and voice generation, it would be possible to produce realistic air-traffic dialogue based on the current flight conditions and artificially generated air traffic.

Figure 9.23   Script Intermediate Code.

The execution of a script program by the IOS is summarised by the following functions:

```
Read a script file
Compile the script source code to intermediate instructions
During each frame:
     If the script is active, execute intermediate instructions until a return is encountered
     Detect an event, then activate the action code and initialise the script
     Check for termination and write outputs to memory
Copy remaining outputs to memory
Copy the data written to memory to a file
```

The compilation is straightforward; the code is parsed to remove leading and trailing spaces to convert each text unit to a symbol with a type and a name in the form of a text string. These symbols cover keywords, the names of variables, the names of units, numeric constants and arithmetic operators. The main part of the compiler is to read statements until an error is detected or an end-of-file symbol is encountered, implying that the compilation is completed without detecting any errors.

```
do
{
    ReadStatement();
} while (!(File_EOF || Script_Error));
```

In `ReadStatement`, the code is compiled for each of the four phases. If an *EVENT* declaration is detected (`k_EVENT`), the code is compiled as follows:

```
case k_EVENT:
    Phase = Phase_Event;
    Script.EventStart = PC;
    Script.Status = Inactive;

    Op = NextSymbol(Str);
    if (Op > k_END && Op != k_EOF)
    {
        ReadbooleanExpression();
    }
    GenCode1(i_RETURN);
    return k_EVENT;
```

The function `NextSymbol`, which is the same as `ReadSymbol`, looks ahead into the input stream to find the next token, repositioning the input stream so that the text can be reread by `ReadSymbol`. If the keywords `ACTION`, `RECORD` or `END` are found or if an end-file-symbol is detected, the compilation of the *EVENT* phase is completed. Otherwise a single Boolean expression is compiled and a return instruction is added to the compiled code. The initial `PC` value is set in `Script.EventStart` and the script status is set to `Inactive`. The code for the *ACTION* phase, where an `ACTION` keyword is detected, is similar:

```
case k_ACTION:
    Script.ActionStart = PC;
    Phase = Phase_Action;

    while (1)
```

```
{
    Op = NextSymbol(Str);
    if (Op <= k_END || Op == k_EOF)
    {
        break;
    }
    Op = ReadSymbol(Str);
    ReadAction(Op);

}
GenCode1(i_RETURN);

return k_ACTION;
```

In this case, a set of actions are compiled by **ReadAction**. Mostly, actions take the form of an action keyword and a variable name; for example, **fail flaps** or **set visibility**. Once activated, the script *RECORD* phase is implemented until the terminating conditions are detected. The code for the *RECORD* phase is as follows:

```
case k_RECORD:
    Phase = Phase_Record;
    Script.RecordStart = PC;

    while (1)
    {
        Op = NextSymbol(Str);
        if (Op <= k_END || Op == k_EOF)
        {
            break;
        }
        ReadExpression();
        p = CheckUnits(Str);
        if (p > 0)
        {
            if (ConversionFactor[p] != 1.0)
            {
                GenCode3(i_PUSHC, 0, ConversionFactor[p]);
                GenCode1(i_DIV);
            }
        }
        GenCode1(i_RECORD);

    }
    GenCode1(i_RETURN);

    return k_RECORD;
```

In the repeat loop, a check is made to detect if the end statement or an end-of-file condition is encountered. The function **ReadExpression** performs the compilation of arithmetic and logical expressions, which may be as simple as a variable name or a constant but may contain bracketing and arithmetic and logical operators. The resultant value is pushed onto the stack. The function **CheckUnits** looks ahead to see if specific units are defined and, if so, pushes a conversion factor onto the stack and adds a division instruction to convert from user-specified units to SI units. The instruction will pop the value from the stack and add it to the outgoing list of recorded data.

The final phase, terminating the script, is similar to the event phase. A single Boolean expression is compiled and if this expression is true, the script terminates, as the following shows:

```
case k_END:
    Phase = Phase_End;
    Script.EndStart = PC;

    Op = NextSymbol(Str);
    if (Op > k_END && Op != k_EOF)
    {
        ReadbooleanExpression();
    }
    GenCode1(i_RETURN);

    return k_END;
```

A Boolean expression takes the form <expression> <comparison operator> <expression>, for example, **airspeed < 70 Kt**, where an expression can contain a variable or a constant and brackets. In the function **ReadExpression**, the brackets are handled by a recursive call, as follows:

```
if (NextSymbol(Str) == k_LBRACKET)
{
    Op = ReadSymbol(Str);
    ReadbooleanExpression();
    Op = ReadSymbol(Str);
    if (Op != k_RBRACKET)
    {
        Complain(") missing", Null);
    }
}
```

Any nesting of brackets is resolved by the recursion, checking that a matching right bracket is found. Otherwise the first expression is compiled, a left or right bracket is read for the comparison operator and then the second expression is compiled, as shown in the following code fragment from the compiler:

```
ReadExpression();
Op = ReadSymbol(Str);
if (Op == k_LESS || Op == k_GREATER)
{
    SwapArgs = Op == k_LESS;
    ReadExpression();
    if (SwapArgs)
    {
        GenCode1(i_SWAP);
    }
    GenCode1(i_COMP);
}
else
{
    Complain("Relation operator expected in a bool expression", Str);
}
```

There are two points to note with the comparison. Firstly, a single compare instruction (i_COMP) is used and the arguments are swapped by the compiler to ensure the correct order of the comparison. Secondly, only 'greater than' and 'less than' operators are supported as the variables are floating-point.

The code for the **ReadExpression** function is similar to **ReadBooleanExpression**. A statement takes the form <variable> <operator> <variable> and, as these are simple expressions, any bracketing and operator precedence is resolved by the recursive expansion of the statement. Negation of variables and constants with a preceding minus sign is again provided by recursively calling **ReadExpression**, where the result of the expression will be at the top of the stack and a negate instruction (i_NEG) is implemented in the interpreted code, as follows:

```
if (Op == k_MINUS)
{
        ReadExpression();
        GenCode1(i_NEG);
        return;
}
```

With a simple scripting language, an expression will contain a bracket, a variable name or a constant. Bracketing is resolved as already outlined. Valid variable names are defined in a table of strings and easily searched. Constants are strings of numeric characters and readily converted. For variable names, the expression is compiled as follows:

```
else if (Op == k_VAR)
{
    p = FindVariable(Str);
    if (p > 0)
    {
        GenCode2(i_PUSHV, p);
    }
    else
    {
        Complain("Unknown variable in Expression", Str);
    }
}
```

where the function **FindVariable** searches for the variable name. If found, a push-variable instruction (i_PUSHV) is generated with the argument **p**, the number of the variable in the table of variables. Similarly, the code to read a constant generates a push instruction, where the string is converted to a value. If the constant value is appended with a units qualifier, the constant is converted prior to generation of the push-constant instruction (i_PUSHC).

```
else if (Op == k_NUMBER)
{
    x = ConvertNumber(Str);
    p = CheckUnits(Str);
    if (p > 0)
    {
        x = x * ConversionFactor[p];
    }
    GenCode3(i_PUSHC, 0, x);
}
```

Note that the simulator variables used in a script are in SI units and the conversion of constants allows users to specify constants in other units, such that the conversion to SI units is transparent. The only arithmetic operations are addition, subtraction, multiplication and division. For example, the compilation of an addition is as follows:

```
Op = NextSymbol(Str);
if (Op == k_PLUS)
{
    Op = ReadSymbol(Str);
    ReadExpression();
    GenCode1(i_ADD);
}
```

In this case, where a plus symbol is detected, the second argument of the expression is compiled and the code for the expression is generated followed by generation of the addition instruction (i_ADD).

Although the compilation of a script may take several seconds, the normal practice is to compile a set of scripts needed for a simulator session or a series of tests, which subsequently can be activated by the instructor in a single command.

The script compiler illustrated here describes a basic scripting language. It provides the basis of a compiler which can be extended to include more complex constructions, user-defined variables, other operators and additional types to extend the capability of the scripting language. The purpose of showing these examples of a scripting language and a compiler is to show the power of scripting in real-time simulation and the extensibility of the compiler. However, the developer must not lose sight of the instructor station design requirements. There are simulators where only minimal flight data recording is required and events are generated within the simulation package rather than in an instructor station. Nevertheless, with the growing interconnection and integration of simulators in synthetic environments, access to a compiler of complex scripts is a particularly useful tool in the development of a flight simulator.

References

Ahn Y. H. (1997), Advances in Instructor Operating Stations, Royal Aeronautical Society Conference. *Flight Simulation – Expanding the Boundaries*, London.

Galitz W. O. (2007), *The Essential Guide to User Interface Design: An Introduction to GUI Design Principles and Techniques*, John Wiley & Sons.

Goode M. and Evans D. (1992), An Instructor Station for the Instructor, Royal Aeronautical Society Conference. *European Forum – Matching Technology to Training Requirements*, London.

Harris D. (2004), *Human Factors for Civil Flight Deck Design*, Ashgate, Farnham.

Krause A. (2007), *Foundations of GTK+ Development*, Apress.

Krebs M. J., Wolf J. D., and Sandvig J. H. (1978), *Color Display Design Guide, ONR Report No. ONR-CR213-136-2F*, Office of Naval Research, Arlington, V. A.

Newman W. M. and Sproull R. F. (1981), *Principles of Interactive Graphics*, McGraw-Hill.

Raskin J. (2000), *The Human Interface: New Directions for Designing Interactive Systems*, Addison Wesley.

Schneiderman B. (1992), *Designing the User Interface*, Pearson Education.

10

Validation

It is necessary to look at the results of observation objectively, because you, the experimenter, might like one result better than another.

Richard P. Feynman

Many years ago, the author developed a real-time flight model for a 16-bit microprocessor using integer arithmetic, with many of the variables computed using look-up tables. During a demonstration, as the aircraft climbed through 20,000 ft, it suddenly stalled. Subsequent inspection of the weather module showed that the entry in the table for air density around 20,000 ft was 0.0080 rather than 0.0880, reducing the aircraft lift by a factor of 10. The error should have been detected during testing of the weather module, but the testing coverage had failed to spot this anomaly.

Recalling points made in Section 2.2 that testing can take up to 50% of project development time, online debugging, that is, on the simulator, can take up valuable simulator time. Moreover, only one set of tests can be run at any time and may require pilot input to undertake specific tests. An alternative approach is to transfer much of the development and testing to offline facilities, such as the SimPlot tool described in Section 4.2, which is used to obtain the results of many of the examples in this chapter.

10.1 Software Verification

In developing software for any project, verification is a measure of the rigour of software development, whereas validation ensures that the software is fit for its intended purpose. As Boehm observed, verification answers the question 'Are we building the product right?' and validation implies 'Are we building the right product?' (Boehm, 1979). In other words, verification addresses the methods used to develop the software whereas validation is intended to confirm that the design meets the requirements.

In software development, it is assumed that software bugs are inevitable and emphasis is given to their detection. However, it is important to note Dijkstra's famous comment 'Program testing can be used to show the presence of bugs but never to show their absence' (Backhouse, 1986). Unexpected behaviour or erroneous outputs may be reported by developers and users, but with the implication that the further into the development that a bug is encountered, the more difficult it is

Flight Simulation Software: Design, Development and Testing, First Edition. David Allerton.
© 2023 John Wiley & Sons Ltd. Published 2023 by John Wiley & Sons Ltd.
Companion Website: www.wiley.com/go/flightsimulationsoftware

to locate and the more expensive it is to rectify. Bugs can be detected by inserting additional code to monitor the sequencing of the software and the state of variables, typically by adding print statements to the code to provide a 'commentary' of the running code or by using debugging tools, for example *gdb*, *CodeBlocks* or *Eclipse*.

Debugging tools can ascertain if specific points are reached in the code and indicate the state of the program during execution of the code and offer the capability to:

- set breakpoints on specific lines, where the code runs until a breakpoint is reached;
- detect that a variable is accessed or modified during execution of the code;
- inspect and modify the value of variables at specific points in a program.

In addition to debugging tools, syntax checking tools, such as *Splint* (https://splint.org), are used to statically check coding errors and vulnerabilities. Profiling tools including *gprof* (https://www.gnu.org) are used to assess software performance; the number of times a line has been reached or a function has been called is monitored, to identify the dynamic behaviour of the software. Once a program has run, it is possible to identify regions of code where the most time is spent and where optimisation of the code will give the greatest benefit.

A significant issue with software testing, particularly for real-time systems, is the *probe effect*, where additional code, added to provide debugging, may alter the behaviour of the software. Debugging points are detected in the software and replaced with in-line calls to the debugging API. In these instances, one or two machine instructions are replaced, possibly by several tens or hundreds of instructions, providing a powerful debugging aid but introducing an overhead, which has the effect of slowing the software and potentially altering its behaviour. While debugging is a very powerful method of detecting bugs, it is essential to ensure that the addition of debugging does not alter the behaviour of the software, defeating the benefit of the debugging tool.

In debugging a system, important debugging information may be written to the screen or to disk. However, there are three considerations:

- Writing to a screen is a relatively slow process, adding significant delay to the running software.
- Writing to disk can have an unpredictable effect on timing, as data may be buffered with negligible delay, but when a block or blocks are written to disk, there is a significant delay.
- The time maintained by local clocks in a distributed system can drift considerably, making it difficult to ascertain the order of events on different computers.

A structured approach to both the coding and testing of software and adherence to coding and testing rules is the recommended method of ensuring the quality and reliability of code (Pomberger, 1984). Software engineering methods focus on minimising the introduction of bugs rather than their subsequent detection by debugging techniques. The remainder of this chapter focuses on validation rather than verification, which is covered by numerous textbooks on software engineering.

10.2 Static Validation

Although a major part of validation of a flight simulator covers the dynamic response of the flight model, the major components of the simulator require different forms of validation to ensure that their contribution to the simulation does not introduce any anomalies or errors. These tests for conformance include ensuring the correctness and robustness of each component and also check that the computations do not affect the real-time constraints of the simulation and cover validation of the following components:

- the I/O system;
- the control loading system;
- the weather module;
- the navigation systems;
- the display modules;
- the IG system;
- the instructor station;
- network transfers.

Although, as far as is possible, these tests should be objective, it is inevitable that some tests are subjective. For example, validation of the instructor station requires a manual check that all options are activated, that any errors are detected or displayed and that the response to any action is immediate. Similarly, the sound generation system is likely to be assessed by comparing recorded sounds with the equivalent computer-generated sounds. The observations on testing conventional software (Sommerville, 1995) also apply to simulator software. The cost of fixing a fault detected at the operational stage is many times more expensive to rectify than during the development phase. Errors in the non-dynamic components of a simulator can affect the dynamic response perceived by a pilot and may be difficult to isolate; consequently, it is essential to validate these components prior to undertaking validation of the dynamic components of the simulator. In the following sub-sections, the validation processes are outlined for the non-dynamic components. However, covering these activities in detail is beyond the scope of this book.

10.2.1 I/O Systems

An I/O system is responsible for capturing analogue and digital inputs and for the generation of analogue and digital outputs. The tests of analogue inputs should ensure that:

- the input channels are connected correctly – it may be possible to configure connectors incorrectly;
- the polarities are correct – for example, it is possible to rewire a potentiometer the wrong way round;
- the analogue inputs cover the full range – for example, if an input is 12 bits, a range covering only 8 or 9 bits is a significant reduction in resolution;
- the time for acquisition meets any real-time requirements and is guaranteed – the amount of frame time occupied by data acquisition should be negligible;
- the noise levels are acceptable – poor grounding can lead to spurious noise and the introduction of 'mains hum'.

The basis of these tests is repeatability. If any changes are made to the hardware, these tests can confirm that the inputs are consistent or alternatively, during any maintenance schedule, it should be possible to confirm that there are no changes to the inputs. A reduced set of these requirements applies to digital inputs, particularly channel connection, polarity checks and noise immunity. A typical I/O system test is shown in Figure 10.1.

In Figure 10.1(a), the 32 channels are identified by the channel number. In this example, each channel is 12 bits and is displayed together with a noise estimate based on variation of previous samples. This display also ensures the correct polarity of each input. The small circle rotates to show that the sampling is active. The row of red and blue indicators shows the state of each digital input, again ensuring that each input is connected to the specific channel with the correct polarity. Figure 10.1(b) shows a display capturing the noise levels of the analogue inputs. In this example,

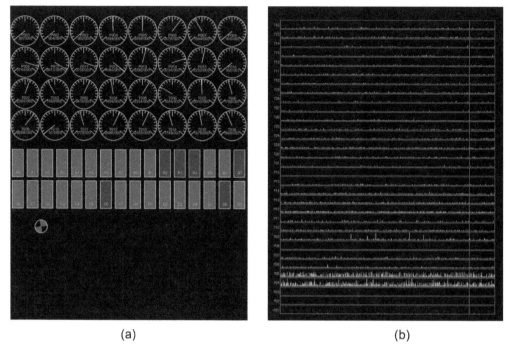

(a)

(b)

Figure 10.1 Diagnostic Tests.

channels P04 and P05 exhibit unacceptable levels of noise and there are discernible spikes in channel P09, which turned out to be a faulty component.

10.2.2 Control Loading Systems

Most control loading systems are computer-controlled and tests are needed to ensure both the response and accuracy of actuation of the centre-stick (or control column), rudder pedals, trim wheel, engine levers and so on. There is a further overriding requirement to validate the safety of these systems that can cause major injuries under conditions of malfunction. As the position of these inceptors can be acquired, it is possible to measure and display the step response to determine any overshoot or resonance. Typically, the simulator provides analogue outputs to set the position of each actuator and the actuator controller is responsible for movement of an actuator.

Test patterns are used to move the controls over predetermined trajectories which can be recorded to detect variability over time or for comparison with previous installations or updates. Most simulations include graphs of force versus displacement for the primary controls. It may be possible to attach a strain gauge to an inceptor and measure both applied force and displacement as the inceptor is moved. Alternatively, static measurements can be recorded using a spring balance, where the displacement is measured by the I/O system, the force is read from the balance and the results are plotted to confirm the relationship. One particular measurement is to confirm that any backlash in the system is below an acceptable limit and this can be confirmed from plots of force versus displacement.

The safety of hydraulic and electrical actuation systems is based on measuring the velocity of an inceptor to detect if it exceeds a predetermined limit. In such cases, the power to the actuator is switched off until the safe conditions are restored. The velocity is computed from successive position measurements and should be set to a level slightly below the maximum human input. In other words, it should be possible to manually trip the system to test that the overload detection system is functioning correctly.

10.2.3 Weather Module

The equations to determine temperature, pressure, air density, Mach number and so on should correspond to the tables published in textbooks on aerodynamics (McCormick, 1979). The importance is to check the consistency and continuity of values. As these terms are used in the aerodynamic equations and in the engine models, any errors will affect the validity of these models. It should also be noted that, particularly in the modelling of military aircraft and missiles, the flight envelope can cover altitudes up to 100,000 ft and Mach numbers in excess of 3. Validation of the weather model equations is straightforward but is also easy to overlook. The following code computes the International Standard Atmosphere (Anon., 1976) values for temperature, pressure, air density and speed of sound as a function of altitude, from ground level to 15,000 m, calling the function **Weather_Model** shown in Section 3.1.1. The output is given in Table 10.1.

```
#include <stdio.h>
#include <stdlib.h>
#include <sim/weather.h>

int main(int argc, char *argv[])
{
    double Z;     /* geometric height m */

    printf("  h     T         P        rho     a\n");

    for (Z=0.0; Z <= 15000.0; Z+=1000.0)
    {
        Weather_Model(false, -Z, 0.0);
        printf("%5.0f %6.2f %11.4f %6.4f %6.2f\n",
            Z, Weather_Temperature, Weather_Pressure, Weather_Rho, Weather_SpeedOfSound);
    }
}
```

It is far harder to validate wind shear models, turbulence models and wind models and, invariably, these are assessed subjectively with an experienced pilot flying a cross-wind approach or setting specific levels of turbulence. If actual flight data is available, the magnitude and underlying frequency of the gusts can be compared with data acquired during simulation.

Table 10.1 Computation of Temperature, Pressure, Air Density and Speed of Sound.

| h | T | P | rho | a |
m	K	Pascals	kg/m^3	m/s
0	288.15	101325.0000	1.2250	340.29
1000	281.65	89876.2891	1.1117	336.43
2000	275.15	79501.4219	1.0066	332.53
3000	268.66	70121.1641	0.9093	328.58
4000	262.17	61660.4453	0.8193	324.59
5000	255.68	54048.2852	0.7364	320.55
6000	249.19	47217.6406	0.6601	316.45
7000	242.70	41105.2734	0.5900	312.31

(Continued)

Table 10.1 (Continued)

h	T	P	rho	a
m	K	Pascals	kg/m^3	m/s
8000	236.22	35651.6289	0.5258	308.11
9000	229.73	30800.6973	0.4671	303.85
10000	223.25	26499.9004	0.4135	299.53
11000	216.77	22699.9609	0.3648	295.15
12000	216.65	19399.6211	0.3119	295.07
13000	216.65	16579.7734	0.2666	295.07
14000	216.65	14170.5059	0.2279	295.07
15000	216.65	12111.9346	0.1948	295.07

10.2.4 Navigation Systems

The validation of navigation modules is also largely subjective. It is possible to select frequencies for ADF, VOR, DME and ILS receivers and position the aircraft at specific locations to check the information displayed for each navigation aid. Specifically, such tests should include the effects of magnetic variation, range limits (transmitter power), line-of-sight effects and also the failure modes which apply to the equipment. For VOR receivers, the movement of the track pointer, heading bug and to/from indications should also be checked. Any device-specific characteristics, such as needle motion passing overhead a beacon or ADF dip during turning should also be checked. Most instruments have a dynamic response, particularly the needle movement when a navigation aid is tuned, which should be confirmed together with the angular displacement of VOR and ILS instruments as a function of aircraft position relative to a beacon. It is very easy for the incorrect units to be used, or the order of information in data packets or a database format to be changed, which can introduce an error into the navigation equations. For these reasons, the navigation data should be checked thoroughly before the simulator becomes operational.

For GPS and INS modelling, the characteristics are normally validated during development and testing of these modules. In cases where GPS equipment is used in a simulator, tests similar to the VOR and ILS tests above are applied to ensure the correct interfacing and connection of the equipment, particularly where navigation systems are connected via a databus. In practice, the validation of instruments involves flying known procedures and checking the functionality of the navigation instruments and displays. These systems can also be tested by generating aircraft tracks, recording both the aircraft position and the display indications.

10.2.5 Display Modules

In cases where aircraft displays are emulated using 2D graphics, the displays can, for the most part, be tested offline, where the tests include applying a comprehensive range of values to be displayed. As most of the EFIS components and individual instruments are independent, these items can be tested separately. The major validation is to confirm that each display is correct throughout the range of possible values. Also note that some components include the display of digits, including rolling digits in the case of an EFIS display. Generally, anomalies are easy to spot by visual inspection.

Nevertheless, displays should be tested for worst-case conditions, for example, with an attitude indicator if the pitch angle exceeds ±90°, or the roll angle exceeds ±180° or the yaw angle is outside the range 0–360°. In particular, with singularities occurring around pitch angles of 90° and –90°, it is important to ensure that the display transitions correctly near these values. The provision of an offline test program to exercise the displays over the full range of all displayed parameters should ensure that all display errors are detected. The tests of all input variables should be methodical and cover the full variations and be documented in a form that enables tests to be repeated exactly.

The other consideration with computer-generated displays is the real-time performance. The rendering of a display may occupy only part of the overall computations per frame but must not exceed the budget allocated to a display, in terms of frame time. However, the performance depends on the specific data to be displayed, the display resolution, the performance of the graphics software library and the graphics hardware. The display time increases with display resolution and the display content can vary according to the display modes and display parameters, for example, the amount of clipping needed. The delay resulting from the software library and the graphics hardware will also vary with display content and the developer has minimal control over this constraint. What is important is to base any frame time usage on worst-case display timing. Based on the overall computations per frame, a budget is allocated for the display rendering which must not be exceeded. Measurements can be taken using a system clock with sufficient resolution (typically 1 μs) from the start to completion of the rendering functions, during every frame. Care is needed to ensure that these measurements include all the graphics functions, including, for example, management of the frame buffer. In addition, the timing measurements should be written to an array rather than a file, to avoid any significant delays during the acquisition of the frame times. Figure 10.2 shows frame time measurements for the Boeing 747-400 PFD and NFD displays and the associated instructor station.

All the measurements were taken using a standard Raspberry Pi model 4B. As can be seen, the displays were rendered well within the 20 ms frame limit and mostly in under 7 ms. Note the presence of spikes immediately after the program is started, which are caused by the loading of various database files during initialisation. The variations in frame loadings are caused by exercising the different display modes, some of which are more intensive than others, in terms of graphics operations. It is important to appreciate these variations, which might not be noticed during normal

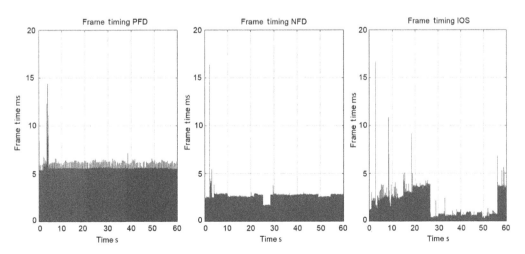

Figure 10.2 Frame Timings.

operations of the displays and are only highlighted by the per-frame timing measurements. The testing should include worst-case examples of frame overload and also check that occasional slow frames can be detected and accommodated.

10.2.6 Visual Systems

Many of the observations in the previous section are also relevant to IG systems. The rendering time must leave a small margin in each frame and the frame time should never be exceeded. However, the rendering time can vary with the number of visible objects, the lighting conditions and visibility. It will also vary with the detail of different visual databases. Again, the best testing strategy is to exercise worst-case conditions, for example, varying the angular rates (roll, pitch and yaw) and linear speeds (airspeed and vertical speed) to check for any inconsistencies in the IG rendering, typically by applying flight profiles of aircraft position and attitude. One worst-case condition is to position the aircraft at 50,000 ft, pitched down 90°, where all the objects in a database are probably in view. In addition, checks on level-of-detail should be made, slowly varying the distance to complex objects to observe the changes to the displayed textures. During such tests, the rendering time should be recorded to ensure the IG meets its frame time constraints under all conditions.

Both X-Plane11 and OSG can display frame timing statistics. Figure 10.3 shows the output from X-Plane11 for the visual database for London Heathrow in poor visibility conditions, as shown in Figure 7.15.

The overall frame rate is 49.878 fps, with a frame time of 0.02 s, of which the CPU and the GPU occupy 0.0197 s and 0.0166 s of the frame, respectively. The frame performance for the images in Figures 7.14 and 7.15 is shown in Figures 10.4(a) and 10.4(b), respectively. Notice the variation in the IG performance for the two images. While the rendering time for both examples is close to the 20 ms limit, there are significant excursions above 20 ms in Figure 10.4(b) and a couple of missed frames in Figure 10.4(a). This particular IG is close to overloading with such a detailed visual database.

Figure 10.3   X-Plane 11 Frame Timing Statistics.

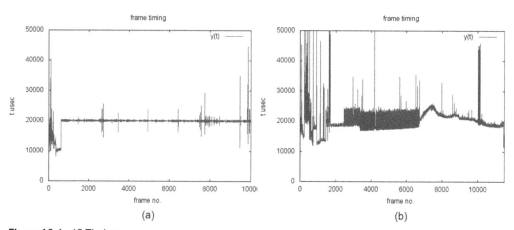

(a)   (b)

Figure 10.4   IG Timing.

There is a further consideration with IG systems and their projection systems, which is easily overlooked. The projection system will have fixed horizontal and vertical fields-of-view. The resolution of the projectors is likely to be matched to the graphics hardware resolution. The IG software will therefore need to generate an image matched to the field-of-view of each projector. The problem is further compounded because the aspect ratios of the projector screen and the graphics hardware may differ. Furthermore, if more than one projector is used, it is normal practice to overlap the vertical edges of the projectors to avoid any gaps appearing between adjacent projector channels.

As an example, Figure 10.5 shows the geometric layout for an IG system with three projectors and a screen with a field-of-view $150° \times 40°$. Figure 10.5(a) shows the geometric layout of the screen and the projectors in plan view. Notice that the projector lens is offset from the projector vertical mounting point. However, the important consideration is to meet the required field-of-view for each channel which includes an overlap with the adjacent channel, as shown in Figure 10.5(b). Although each channel is $50°$, the actual projected horizontal field-of-view for each channel is $60.42°$. The centre channel covers $-30.21°$ to $+30.21°$, the right channel projector covers $+14.58°$ to $+75°$ and the left channel projector covers $-75°$ to $-14.58°$.

In OpenSceneGraph, it is not possible to set both fields-of-view directly; rather an OpenGL transformation **setProjectionMatrixAsPerspective(40.0,1.599070,1.0,200000.0)** is used for the geometry just given, where the first argument is the vertical field-of-view in degrees, the second argument is the aspect ratio, and the third and fourth arguments are the near and far fields, respectively, in metres. The aspect ratio is given by

$$ar = \frac{\tan\left(\frac{fov_x}{2}\right)}{\tan\left(\frac{fov_y}{2}\right)} \tag{10.1}$$

where fov_x and fox_y are the horizontal and vertical fields-of-view, respectively.

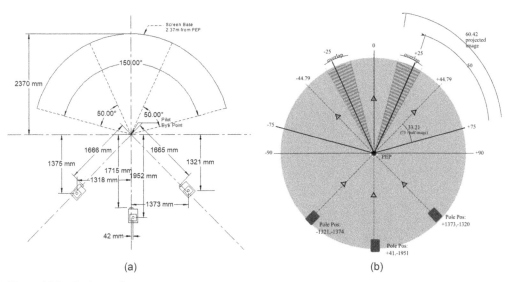

$$\text{(a)} \qquad\qquad\qquad\qquad \text{(b)}$$

Figure 10.5 Projector Geometry.

There are several sources of potential errors in the installation and calculation of the various geometric ratios. An axiom of 'real world' perspective is that orthogonal angles are preserved. For example, a white square with a red cross, positioned directly in front of the pilot eye-point should be a square with the diagonals intersecting at 90° when seen in the simulator. Any errors in setting the fields-of-view, the projector overlaps or the pilot eye-point are likely to cause distortion to the 3D views, even for simple shapes. Errors in the geometry of the visual system can introduce significant effects, which may be sensed as errors in aircraft handling by a pilot, incorrectly implying an error in the flight model.

In the case of the spherical screen described previously, one further form of validation is to generate sections of a sphere for each channel. Figure 10.6 shows a 2D test pattern which is projected for the centre channel. Projected from the pilot eye-point, the squares should appear orthogonal with equal sides. The white dotted line indicates the start of the overlap region. The importance of setting up the IG parameters, locating the screen origin relative to the pilot eye-point and matching the projector warping to the screen geometry cannot be overemphasised. Any errors caused at this stage of development will be particularly difficult to detect from inspection of the visual system in flight tests. Nowadays, the calibration of an IG is performed automatically, using a camera to detect projected patterns, dynamically modifying the warping to match the desired projection.

10.2.7 The Instructor Station

Validation of the instructor operating station (IOS) mainly involves checking the functionality of each command, which should be undertaken in a logical way, testing each command from the set of commands defined for the IOS, for example, testing that the failure modes of the flaps are fully implemented or that the simulator state is correctly reset after restoring from a saved position. There are two other requirements of the IOS that should be tested. Firstly, the graphics to display maps, flight data and aircraft information should not exceed the frame time. Clearly, the content of the display depends on the application and worst-case situations; for example, zooming out on a map display, containing several hundred runways and radio beacons, should be tested to ensure that the display rendering is within the frame time under all conditions. Secondly, the execution of

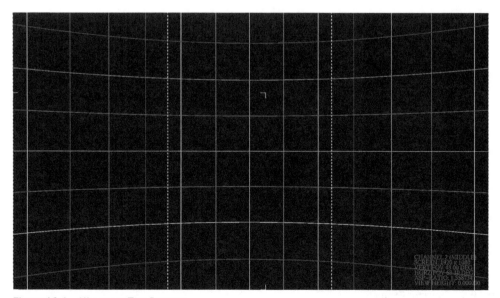

Figure 10.6 Alignment Test Patterns.

instructor commands may involve disk transfers or acquisition of flight data and, in a distributed system, it is important to test the response of the IOS to incoming packets, so that no delay is introduced into the schedule of packet transfers.

The most important aspect of validation by inspection is the provision of a schedule containing the complete lists of tests and testing values. For example, to test the reposition altitude command, a series of altitude values would be entered, including worst-case and invalid values, for example, 0, 50, 100, 30,000, 60,000, 200,000 and –500. After each test, the aircraft altitude computed in the flight model is measured. If there are any minimum or maximum values, test values should be given to check it is not possible to set an altitude outside the valid range. Knowing the altitude of a specific runway, small changes to the altitude setting should be made to check that the aircraft altitude is correctly shown by the visual system. Note that setting altitude could be referenced to the wheel height, the aircraft C/G or the pilot eye height. Checks should be made that the pilot eye height is correct, with the aircraft positioned on a runway, for example, by temporarily adding height posts to the runway. Similarly, if the altitude can be set in different units, these values should also be checked for consistency. The checklist for these tests provides a degree of repeatability if changes are subsequently made to the instructor station.

One potential source of problems with an instructor station is the possible introduction of delay caused by file transfers initiated by a command, for example, to save flight test data. Opening a file and then transferring a large amount of data is likely to exceed the frame time, causing the simulator to stall. However, commands of this form can be invoked when the simulator is in a flight-freeze mode, where any delay will not affect the real-time processing. Alternatively, a separate background process can be used to acquire data from the main simulator processes, transferring data to disk in the background, with negligible disruption to the real-time processing. The simulator update rate should be monitored throughout these tests to detect any frame overrun introduced by the IOS software.

10.2.8 Networking

Many modern simulator configurations are based on a distributed network of computers. There are two requirements of the network transfers, firstly, to ensure the integrity of transmitted data and, secondly, to minimise the overhead of the data transfers. With a token-passing protocol, problems can arise if a node is slow to respond to an incoming packet, which delays the onward transmission of its packet (or packets). It is straightforward to test the protocol, independent of the simulator, where an equivalent set of packets is transmitted over the network. In testing of this form, the content of packets can be checked for corruption, for example, using a checksum, and logging the arrival and transmission times for packets. A distributed network is vulnerable to the introduction and potential accumulation of delays, and corruption of data can result in undefined errors. If such errors are intermittent, it may be particularly difficult to locate the source of the problem when the simulator is operational.

One test of network reliability is to emulate the network packet transfers of the simulator, in terms of the number of network nodes, packet sizes and transfer rates. Each packet is then filled with random numbers and a 32-bit checksum is computed and added to the packet. When each node receives a packet, the checksum of the packet is computed and compared with the checksum in the packet. With a 32-bit checksum, the probability of corrupted data giving a falsely correct checksum is of the order of 1 in 2^{32} or 4.3×10^9 or, at 50 fps, approximately one undetected error per 2.7 years. Table 10.2 shows the timing of captured packets on a network with five nodes, numbered

Table 10.2 Frame Timing.

4561	17.	719746	192.168.1.1	192.168.1.255	UDP	122	34349	→	54321	Len=80
4562	17.	719904	192.168.1.3	192.168.1.255	UDP	382	38782	→	54321	Len=340
4563	17.	720118	192.168.1.4	192.168.1.255	UDP	238	51765	→	54321	Len=196
4564	17.	720316	192.168.1.5	192.168.1.255	UDP	406	48956	→	54321	Len=364
4565	17.	720583	192.168.1.6	192.168.1.255	UDP	1450	35701	→	54321	Len=1408
4566	17.	739699	192.168.1.1	192.168.1.255	UDP	122	34349→		54321	Len=80
4567	17.	739915	192.168.1.3	192.168.1.255	UDP	382	38782	→	54321	Len=340
4568	17.	740164	192.168.1.4	192.168.1.255	UDP	238	51765	→	54321	Len=196
4569	17.	740383	192.168.1.5	192.168.1.255	UDP	406	48956	→	54321	Len=364
4570	17.	740656	192.168.1.6	192.168.1.255	UDP	1450	35701	→	54321	Len=1408
4571	17.	759782	192.168.1.1	192.168.1.255	UDP	122	34349	→	54321	Len=80
4572	17.	759899	192.168.1.3	192.168.1.255	UDP	382	38782	→	54321	Len=340
4573	17.	760129	192.168.1.4	192.168.1.255	UDP	238	51765	→	54321	Len=196
4574	17.	760314	192.168.1.5	192.168.1.255	UDP	406	48956	→	54321	Len=364
4575	17.	760629	192.168.1.6	192.168.1.255	UDP	1450	35701	→	54321	Len=1408
4576	17.	779785	192.168.1.1	192.168.1.255	UDP	122	34349	→	54321	Len=80
4577	17.	779917	192.168.1.3	192.168.1.255	UDP	382	38782	→	54321	Len=340
4578	17.	780104	192.168.1.4	192.168.1.255	UDP	238	51765	→	54321	Len=196
4579	17.	780304	192.168.1.5	192.168.1.255	UDP	406	48956	→	54321	Len=364
4580	17.	780573	192.168.1.6	192.168.1.255	UDP	1450	35701	→	54321	Len=1408
4581	17.	799745	192.168.1.1	192.168.1.255	UDP	122	34349	→	54321	Len=80
4582	17.	800001	192.168.1.3	192.168.1.255	UDP	382	38782	→	54321	Len=340
4583	17.	800058	192.168.1.4	192.168.1.255	UDP	238	51765	→	54321	Len=196
4584	17.	800360	192.168.1.5	192.168.1.255	UDP	406	48956	→	54321	Len=364
4585	17.	800670	192.168.1.6	192.168.1.255	UDP	1450	35701	→	54321	Len=1408

with IP addresses 192.168.1.1 to 192.168.1.6 (node 2 is omitted). The sequence of broadcast transfers is 1, 3, 4, 5, 6 at the start of each frame and the respective packet sizes are shown. The timing, indicated in the outlined section for one frame, shows the first packet was transmitted at 17.759 782 s with the final packet transmitted at 17.760 629 s. In other words, the broadcasting of five packets of 80, 340, 196, 364 and 1408 bytes (2.4 Kb) occupied 847 μs, slightly less than 1 ms of the 20 ms frame. Running the test for 8 hours or 1,440,000 frames produced no data errors. In fact, Ethernet includes a built-in hardware checksum logic to detect data errors. For a dedicated local network, the probability of data errors resulting from data transfers is negligible and the data rates are sufficiently fast to capture and record relatively large amounts of data. Note that the packet protocol in these tests is UDP, which does not provide any retry mechanisms. The assumption is that the data broadcast and read by the other network nodes is guaranteed to be error-free. As an aside, AFDX, a variant of Ethernet, is used in some Airbus aircraft in safety critical applications, albeit with multiple channels for fault tolerance.

10.3   Aircraft Performance

Pilots of modern civil aircraft refer to 'flying by numbers'. What is meant by this aphorism is that, for example, after take-off, the pilot will set a specific power setting and climb at a particular pitch attitude. If these settings are maintained, the airspeed and climb rate will be constant and will match the recommended performance for the aircraft in this flight segment. These 'numbers' are defined in the operational notes for the aircraft and will vary for versions of the aircraft and the engines. If the aircraft is configured correctly, in terms of flap setting, speed brake and undercarriage, and the correct power and attitude are applied, the aircraft performance will match the values in the operational notes. This use of power setting and airspeeds covers all aircraft from light aircraft to military aircraft and commercial jet transport aircraft. For example, a Cessna-150 will have figures in the aircraft handbook for take-off speed, initial climb speed, cruise speed, stall speed, best rate-of-climb speed, descent speed, approach speed and so on, with power settings given in RPM.

Several performance figures are referred to as reference speeds (V_{ref}) such as an approach and touchdown airspeed. Many of these figures apply to safe operation of the aircraft; for example, following an engine failure, there will be a minimum safe speed to climb, together with a power setting for the working engine (or engines) and the pitch attitude. The flight crew will ensure that the airspeed does not drop below this minimum speed. Of course, flight crews practise for such events in a flight simulator and an instructor will check that the crew respond correctly to such events. Similarly, maximum operating speeds are also defined; these are airspeed values where manoeuvring may possibly cause structural damage to the aircraft and must never be exceeded.

The simulator developer must ensure that the simulator meets the performance data published for the aircraft. This information is far less sensitive than the data for the flight model or engine model and, generally, is in the public domain. However, these values depend on many variables, and if the contribution of any one variable is incorrect, then the aircraft performance will be in error in a particular part of the flight envelope. Moreover, if, having produced a simulation, the resultant performance is incorrect, it may prove difficult to isolate terms causing the problem. For example, if the aircraft climb rate is too high, the problem may be caused by the lift equation terms, the drag equation terms, the contribution of the flaps or the modelling of the engines. The validation of performance, which covers all the flight envelope, including ground handling, is to ensure that the simulator meets the performance figures published for the aircraft, and covers the following segments of flight operations:

- Taxiing, including acceleration, braking and turning tests.
- Take-off, particularly the distance from brake release to rotation and V_1 and V_2 speeds.
- Climb – the rate of climb, pitch attitude, airspeed and power settings for different flap configurations.
- Cruise – the airspeed and Mach number in the cruise as a function of altitude and aircraft mass.
- Descent and approach – the rate of descent, pitch attitude, airspeed and power settings for different flap configurations, including lowering of the undercarriage.
- Touchdown – the reference speed and pitch attitude with the engine power at idle.
- The landing roll including braking and reverse thrust.
- Turning, confirmation that turn rates correspond to bank angles and steady-state sideslip angles in response to aileron and rudder.

In addition, the airspeed, Mach number and rates of climb and descent can also be controlled via the FCU and tests should be made to check the performance values provided by the FCU. Note that the checks covered in the *Evaluation Handbook* (RAeS, 2005) occupy some 170 pages and the limitation of space in this chapter restricts the number of examples covered.

10.3.1 Taxiing

The taxiing performance is influenced by the engine power, ambient air temperature, the aerodynamic drag of the aircraft, the coefficient of friction and slope of the taxiway or runway and the coefficient of friction of the brakes and the tyres. The spool-up time of a turbofan engine from idle to full power, which may be 5–10 s, is also important as it affects the take-off distance. In recent years, the use of reverse thrust has reduced with advances in carbon braking systems. Note that, for airline simulators, the temperature rise of brakes as a function of braking force and speed should be modelled to simulate brake overheat and possible reduction in braking efficiency.

Taxiing tests ensure that a pilot can maintain the centreline of a taxiway or runway at speeds below V_1. The tiller, rather than rudder, is used at taxiing speeds typically below 20 Kt and Figure 10.7 shows the application of full rudder at 20 Kt. The maximum yaw rate is 4°/s in response to an applied doublet of full rudder for 15 s. The SimPlot test code overleaf was used for the tests.

Two tests were made, applying both positive and negative rudder to check the symmetry of the response. An autotrim function used for taxiing selects the engine lever position to maintain 20 Kt. The tests confirm the linearity of the nose-wheel steering and the response of the nose-wheel to a step input. In practice, both low-speed and high-speed steering checks are made by attempting to reposition the aircraft on the centreline after deliberately steering to an offset from the centreline. Such tests are mostly to check the pilot workload and effort in comparison with similar manoeuvres in the aircraft. The dynamic response of the oleos was covered in Section 3.7. Further tests would be made to confirm turning circles for different settings of ground speed and rudder input, particularly to check limits of scuffing. A test of thrust and braking is shown in Figure 10.8.

The aircraft speed increases from rest to 150 Kt in 35 s. Note the slight reduction in thrust with increasing ground speed. The aircraft covers approximately 1000 m before maximum braking is applied to stop the aircraft in 25 s covering a total runway distance of 2100 ft. The application of braking simulates an aborted take-off just below V_1.

10.3.2 Take-off

A third taxiing test is shown in Figure 10.9, to check the rotation speed and the elevator effectiveness to rotate the aircraft to 12.5° nose up at V_{ref} (155 Kt).

The flaps are set at 20° for take-off and 6° of back elevator is applied at 35 s at 150 Kt. In the initial climb, the aircraft should settle out at 155 Kt at a pitch attitude of 12.5°, with the undercarriage raised and flaps at 20°. The aircraft becomes airborne 42 s from brake release. Although the offline values used in SimPlot correspond to pilot inputs during take-off, extensive manual flight tests would also be undertaken.

10.3.3 Level Flight

Cruising covers level flight, mostly at altitudes above 20,000 ft for commercial aircraft, but can include lower altitudes where the aircraft is required to fly a holding pattern or altitudes where the initial fuel load reduces the maximum cruising altitude. In the cruise, altitudes are selected for several reasons:

- engine efficiency – the efficiency of a turbofan varies with altitude and Mach number; typically, commercial aircraft fly at the highest possible altitude, which is limited by the engine power rating, the aircraft mass (which reduces significantly as fuel is burnt) and the flight envelope of the aircraft;
- to avoid adverse weather conditions, particularly storms or turbulence;
- to comply with air-traffic requirements.

```
set altitude 0 ft
set TAS 20 kts
set flaps 20
set gear 1.0
set engine_lever 0.3
set heading 0

plot TAS kts 0 30
plot yaw degs -60 60
plot yaw_rate deg/s -10 10
plot rudder degs -30 30

time 60 secs
input rudder doublet 5 30 -25.0
autotrim
```

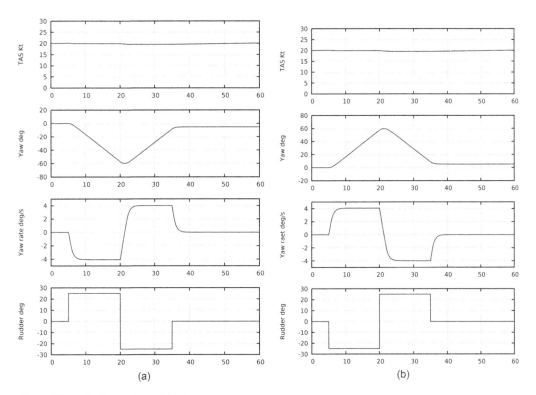

Figure 10.7   Taxiing Tests Rudder Input.

For all aircraft in straight-and-level flight at a specific airspeed, there is a unique angle of attack, which increases as the airspeed reduces towards the stalling speed. The complication with validation of aircraft performance in level flight is the number of contributing terms, which include mass, airspeed, altitude, flap position, undercarriage, C/G, engine thrust and ambient temperature. If validation is performed over the complete flight envelope from Mach 0.2 to Mach 0.9 and from sea level to 45,000 ft, the number of data points to check, including configuration settings below Mach 0.3 (200 Kt), is overwhelming.

```
set altitude 0 ft
set TAS 0 kts
set flaps 20
set gear 1.0
set engine_lever 0.2
plot TAS kts 0 200
plot thrust lbf 0 130000
plot distance 0 4000
plot engine_lever 0 1.2
plot left_brake 0 1.2

time 120 secs
input engine_lever pulse 10 70 0.8
input left_brake pulse 80 30 1.0
input right_brake pulse 80 30 1.0
```

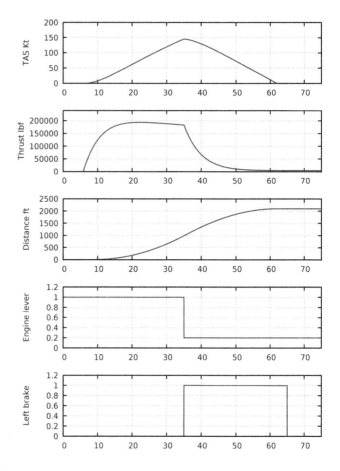

Figure 10.8 Thrust and Braking Tests.

In practice, the flight crew will use values from the aircraft operating manual for pitch attitude and engine EPR to obtain a specific airspeed (or Mach number) for a given altitude. Alternatively, if the autopilot and altitude hold modes are engaged by the flight control system, the flight crew can select the airspeed and altitude for level flight. Figure 10.10 shows the values of pitch attitude and engine EPR for level flight for a Boeing 747-100 over the flight envelope as a function of Mach number from 0.2 to 1.0 and altitude from sea level to 40,000 ft.

Notice that the plots reflect the complete flight envelope, where the values shown as zero are conditions where the aircraft cannot fly at high speed at low altitude or at low speed at high altitude. The plots show the variation in pitch angle and EPR, but also show that there are no inconsistencies in the interpolation of the data over the flight envelope. To obtain these values in a flight simulator would be extremely time-consuming, as the aircraft would need to be configured and trimmed for specific altitudes and airspeeds. However, the autotrim function in SimPlot can be used to derive the steady-state conditions for level flight at a given altitude and airspeed and these values can be compared with published data for the aircraft. Table 10.3 gives a range of values for aircraft mass, configuration, altitude and Mach number.

An important point to stress is the need to ensure that the settings of all the variables correspond to the values used in aircraft flight tests. A major difficulty for the developer is identifying (or isolating) errors in the flight model if the data match with the aircraft data is poor. For example, a high Mach value may result from errors with the lift or drag coefficients or the engine model. In these

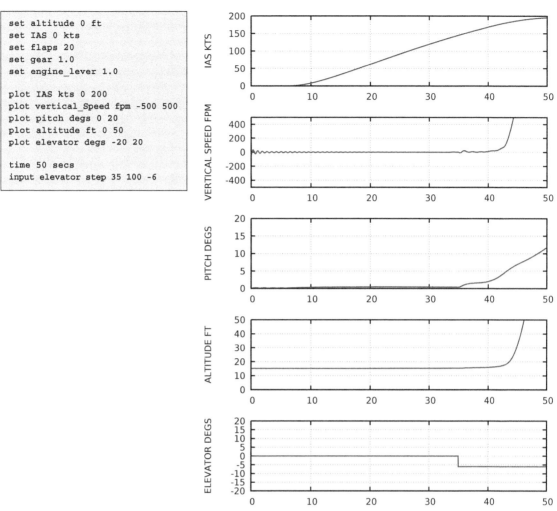

```
set altitude 0 ft
set IAS 0 kts
set flaps 20
set gear 1.0
set engine_lever 1.0

plot IAS kts 0 200
plot vertical_Speed fpm -500 500
plot pitch degs 0 20
plot altitude ft 0 50
plot elevator degs -20 20

time 50 secs
input elevator step 35 100 -6
```

Figure 10.9   Take-off Test.

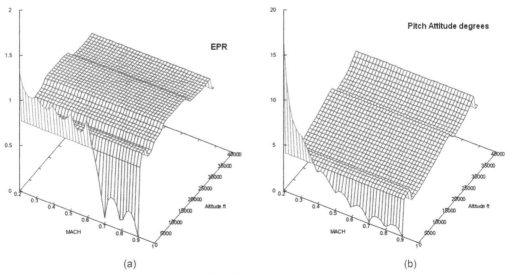

Figure 10.10   Engine Performance versus Mach Number and Altitude.

Table 10.3 Steady-state Level Flight.

Mass (lbs)	Gear	Flap (deg)	Altitude (ft)	M	θ (deg)	EPR
500,000	down	20	2000	0.28	2.63	1.10
500,000	down	30	50	0.23	1.43	1.07
600,000	up	0	5000	0.42	4.26	1.11
600,000	up	0	20,000	0.78	1.47	1.03
564,000	up	0	30,000	0.83	2.29	0.96
564,000	up	0	40,000	0.95	3.16	0.90

circumstances, offline nonlinear simulation offers considerable advantages. With faster than real-time simulation, it is possible to automate the scheduling of hundreds of tests per hour, attempting to match the performance to the data provided. Even so, one change may improve the accuracy in one part of the flight envelope but reduce the accuracy in other parts of the flight envelope. Multivariate optimisation of four or five variables in a nonlinear simulation is far from straightforward.

10.3.4 Climbing Flight

The validation of climb performance is similar to the tests used in the previous section. However, these are compounded by the variation in air density with altitude, such that to maintain a constant rate of climb or descent, the flight path will need to change, implying a change of pitch angle and elevator. Three parameters are used to validate the climb performance:

- the maximum rate of climb available – the maximum climb rate at which the airspeed can be maintained;
- the variation in rate of climb with aircraft attitude – a pilot will select a power setting and a pitch attitude, with an expectation that the actual climb rate corresponds to values in the aircraft handbook;
- the variation in rate of climb with altitude, which reduces with increasing altitude and is zero at the maximum altitude for the aircraft.

Note that the maximum rate of climb is not the same as the maximum angle of climb, where the flight path angle depends on both the airspeed and the rate of climb. In practice, climbing and descending above 5000 ft are normally performed by the aircraft flight control system, which is able to maintain a constant rate of climb or descent. Figure 10.11 shows the climb performance of a Boeing 747-100, including the EPR and pitch values, with an initial climb at 250 ft below 5000 ft shown in Figure 10.11(a) and a climb from 5000 ft (FL 50) to 40,000 ft (FL 400) at 300 Kt (IAS). Note the increasing Mach value and reducing EPR in Figure 10.11(b).

10.3.5 Approach and Touchdown

In a manually flown approach, the power and attitude settings are critical. For a large transport aircraft, the approach must be stabilised in terms of airspeed and rate of descent by 4–5 miles from touchdown because large corrections are generally avoided in the final few minutes of an approach, owing to the relatively slow response of the aircraft. Figure 10.12 shows an approach using an autothrottle with the autoland engaged.

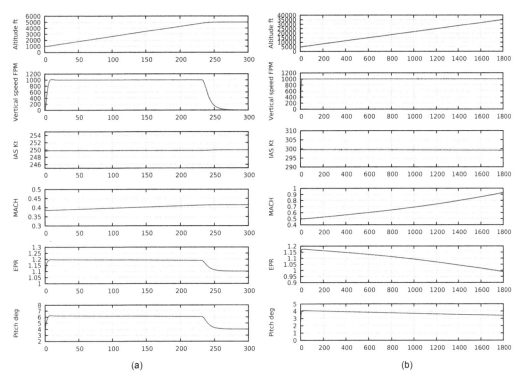

Figure 10.11 Climb Performance.

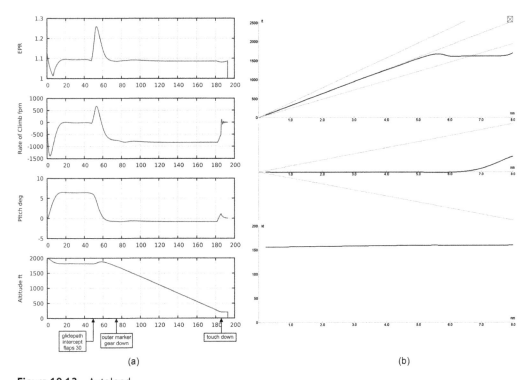

Figure 10.12 Autoland.

Figure 10.12(a) shows the change in engine EPR, the rate of climb, the pitch attitude and the altitude during the autoland. Figure 10.12(b) shows the glide slope, the localiser and the airspeed, which is maintained by the autothrottle. The approach is shown 8 nm from touchdown at 2000 ft and is flown at 156 Kt with an initial flap setting of 20°. Initially, the aircraft is not trimmed for the approach and flies level at 1800 ft until the glideslope is intercepted at approximately 5 nm. The flaps are lowered to the landing flap position (30°) at 6 nm from touchdown. Note the slight ballooning as the flaps are lowered, causing a change in power to maintain 156 Kt and level flight. The aircraft commences its descent at 820 ft/min at the glide path, the undercarriage is lowered and the EPR stabilises to 1.09. Note the stability and accuracy of tracking the glide path and localiser, once established on the glide path. At 50 ft above the runway (270 ft ASL), the power is reduced to idle power and the pitch increases from −1° to +1° to reduce the rate of descent at touchdown to approximately 50 ft/min. Autoland examples would be undertaken for different approaches, including the effect of wind for various approach speeds and configurations to provide reference data for future comparisons.

10.3.6 Turning Flight

Aircraft are turned by applying the ailerons to bank the aircraft. The difference between the aircraft heading (yaw) and the flight path is the angle of sideslip. In nil wind, when these two angles are identical, the sideslip is zero. In manual flight, particularly for gliders, light aircraft and aerobatic aircraft, the rudder is applied to reduce the sideslip which occurs in turning flight. For larger aircraft and fighter aircraft, the turn is coordinated by a flight control system to balance the turn. The effectiveness of the ailerons enables an aircraft to turn away quickly from a threat and to make precise adjustments to heading. The rate of turn depends on the angle of bank and the roll rate depends on the aileron angle and it is important to ensure that the roll performance matches the aircraft data over the flight envelope.

One complication with testing the roll performance of transport aircraft is the suppression of the yaw damper and the turn coordinator. In other words, it is important to assess the effectiveness of the ailerons and rudder of an airframe, excluding any flight control inputs. Of course, for many military aircraft, the basic response may be unstable and the required stability and agility are provided by the flight control systems. The response to aileron input is shown in Figure 10.13, for a Boeing 747-100.

In both cases, a step input of 20° is applied to the ailerons, with the aircraft trimmed for level flight. Figure 10.13(a) is at an entry airspeed of 180 Kt at 3000 ft with flaps 20, and Figure 10.13(b) is at an entry airspeed of 250 Kt at 5000 ft with the flaps up. Notice the onset of sideslip in both examples, with the roll rate approximately proportional to aileron input and increasing with airspeed. This example also illustrates the difficulty of taking static measurements in validation tests. Both the sideslip and roll rate show the presence of the sinusoidal Dutch roll response.

In order to assess the aircraft response in turning flight, the aircraft can be flown manually, with the pilot balancing the turn with the rudder, while maintaining a precise angle of bank, allowing any transients to decay. Alternatively, offline tests can be used with the yaw damper and turn coordinator engaged to measure the roll rate and yaw rate. The situation is illustrated in Figure 10.14 for a heading change from 30° to 80° at 3000 ft, at 180 Kt, with flaps 20 and undercarriage fully down.

In Figure 10.14(a), the yaw damper and turn coordinator are disengaged and the transients in the roll rate and aileron input are far more pronounced. However, both plots show that at a sustained bank angle of 25°, the aircraft has a turn rate of 2.4°/s and an initial roll rate of 10°/s.

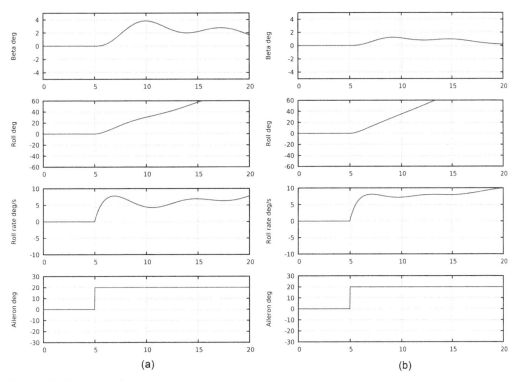

Figure 10.13 Aileron Response.

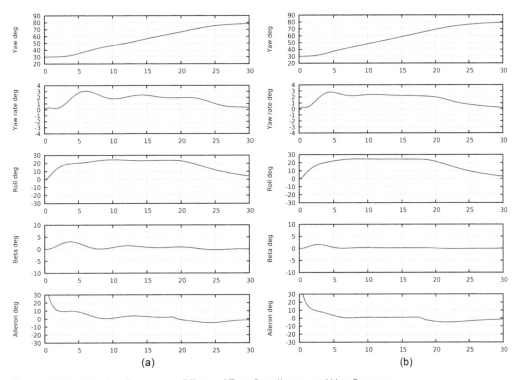

Figure 10.14 Heading Response – Effects of Turn Coordinator and Yaw Damper.

10.3.7 Sideslip

The comments on the effects of transients on steady-state measurements in Section 10.3.6 also apply to measurements of sideslip. In fact, the situation is exacerbated because, in addition to a yawing moment, the application of rudder also introduces a significant rolling moment, which produces a noticeable rate of turn, making it difficult to assess the rudder versus sideslip relationship. Arguably, these tests are simpler flown manually, with the angle of sideslip controlled by rudder and the resultant roll rate corrected by aileron. The effect of sideslip is particularly important following an engine failure in a multi-engine aircraft. In response to the failure, the pilot applies rudder to counter the yaw rate, aileron to counter the roll rate and selects an attitude to maintain an airspeed safely about the stall speed (blue line). Figure 10.15 shows the response to sideslip generated by rudder input for a Boeing 747-100. Full right rudder is applied, producing an initial sideslip of 9°, with the aircraft climbing at 160 Kt with flaps 20. The constant sideslip is countered by a bank angle of approximately 4°, which is stabilised after 15 s. The purpose of the test is to confirm the bank angle response to a given rudder input and also to confirm that the pilot workload corresponds to an equivalent exercise in the aircraft.

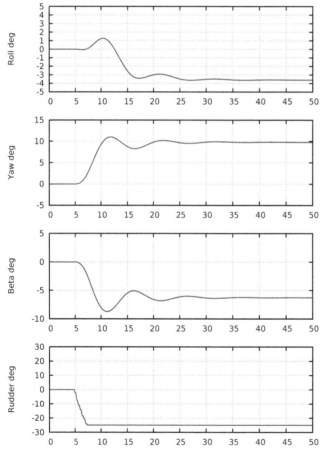

Figure 10.15 Response to Rudder Input.

10.3.8 Observations

One of the most important aspects of validation of aircraft performance is repeatability. The use of offline simulation with precise inputs is essential in cases where either the hardware or software is changed. The alternative, using pilot input, can introduce an element of doubt, if sets of data are compared with slightly different pilot inputs. A similar problem can occur in comparing simulator test data with actual flight data. It is particularly important that the altitude, aircraft mass, the C/G, the flight conditions and the flap and undercarriage positions are identical in both sets of tests. Validation of aircraft performance invariably comprises two stages, firstly a general check that the model is approximately correct and secondly, matching all the data points of all the tests to the reference data. The balance between these two phases depends very much on the application. For an engineering simulator used to develop control laws, for example, the tests can be less stringent. However, for an airline full-flight simulator, the tests are extensive and the aircraft performance must be within a small tolerance (typically one per cent) in order to meet the qualification criteria of the regulator. Note that this section has only contained a few examples of validation, simply because there is not sufficient space to include the results of several thousand tests, which are required for simulator qualification.

10.4 Dynamic Response

A main concern of any flight simulator developer is *Does the simulator fly like the aeroplane*? More precisely, the handling qualities of the simulator, particularly its responses to pilot input and disturbance, should be the same as the simulated aircraft. Hundreds of equations, coded in thousands of lines of software are used to generate the illusion of flight. One solution is to take airborne flight test data, execute the same tests in the simulator and then compare the results from the simulator exercises with the airborne flight test data. There is also an intermediate form of validation which is to treat the aircraft, its undercarriage and engines as mechanical systems. By observing the relationship between the system inputs and outputs, the dynamic response of an aircraft can be compared with the equivalent relationships of the simulation, developed as mathematical models and coded in software.

10.4.1 Longitudinal Dynamics

Aerodynamicists were aware of the dynamic response of an aircraft from the early days of flight. If the pilot pulled the centre-stick back by a small amount, the aircraft would start to climb, but as it did so, it lost speed, reducing lift to an extent that the aircraft then started to descend, but with increasing airspeed, which in turn increased the lift causing the aircraft to climb. This porpoise-like motion, known as a phugoid, meant that, for some aircraft, the motion might take a long time to damp out, or be difficult for the pilot to cancel, adding to the pilot workload.

  This dynamic response is, in many ways, a fingerprint of the behaviour of an aircraft and will vary significantly for different aircraft types. The pilot of an aerobatic aircraft requires a very responsive aircraft whereas a large transport aircraft has a lot of inertia, making it far less responsive but, at the same time, providing a more stable platform. These aircraft dynamic characteristics affect the handling qualities of an aircraft and, for an aircraft designer, it is important to match the handling qualities to the aircraft requirements. Certainly, it is undesirable to develop an aircraft that is difficult to fly for its intended purposes. The analysis of the aircraft modes of motion is critical to aircraft design, but it is also valuable to the simulator developer to ensure that the aircraft dynamics have been captured correctly. As all aircraft exhibit pronounced short-term (over a few seconds) and long-term (over 50 s) responses, the equations of motion developed for a simulator should produce an identical response.

In Section 10.5.1 the characteristic equation of the transfer function between the pitch output and the elevator input can be formed as a fourth-order differential equation, which comprises two second-order differential equations, where one equation describes the long-period phugoid and the other describes the short-period response. In simulation, both responses can be excited by an elevator pulse (or doublet) and are clearly evident from changes in the aircraft altitude, airspeed and angle of attack, as shown in Figure 10.16, which illustrates the phugoid response with the test initiated at 3000 ft at 180 Kt, with flaps 20 and the undercarriage up. A (backward) elevator pulse of 10° is applied for 5 s after 5 s, with the aircraft trimmed on entry to the test.

In Figure 10.16, assuming this data is generated by a second-order system with a characteristic polynomial of the form $s^2 + 2\zeta\omega_n + \omega_n^2$, then, as the plot contains an oscillation, the system is underdamped and the damping ratio ζ and the natural frequency ω_n can be determined by measuring the maxima (peaks) of the waveform as follows:

$$\omega_d = \frac{2\pi}{t_2 - t_1} \tag{10.2}$$

where ω_d is the damped natural frequency, t_1 is the time of the first local maximum and t_2 is the time of the second maximum. ζ and ω_n can be computed as follows:

$$\zeta = \frac{\Delta / 2\pi}{\sqrt{1 + \left(\Delta / 2\pi\right)^2}} \tag{10.3}$$

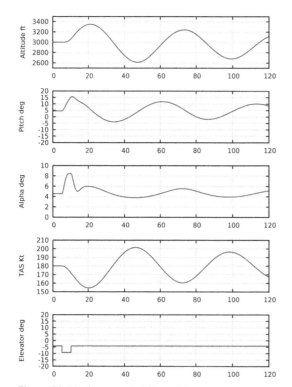

Figure 10.16 Long-period Phugoid.

$$\omega_n = \frac{\omega_d}{\sqrt{1-\zeta^2}} \tag{10.4}$$

where

$$\Delta = \ln\left(\frac{x_1}{x_2}\right) \tag{10.5}$$

and x_1 and x_2 are the values at t_1 and t_2, respectively, relative to the initial (trimmed) airspeed at the start of the test. For the phugoid shown in Figure 10.16, based on the airspeed plot, the following values were obtained: the first maximum occurs at (49.079 998, 0.363 427) and the second maximum occurs at (106.560 005, 0.254 393), giving $\zeta = 0.056\,679$ and $\omega_n = 0.109\,487$ rad/s. In other words, the phugoid is very lightly damped with a period of 57.4 s. This specific motion is initiated by a relatively large step input and, generally, this longitudinal motion poses no problems for pilots or flight control systems. However, the characteristics of the phugoid are critical to simulator validation. If the aircraft phugoid and the simulator phugoid differ significantly, then the handling qualities of the simulator will also differ from the handling qualities of the aircraft.

One final point to note with analysis of the phugoid is the effect of thrust from propellers or engines during the phugoid, which can dampen the phugoid significantly. The phugoid should be based on the airframe dynamics, with power effects excluded as far as possible. This is a further illustration of the benefits of autotrimming the simulator prior to a test. The aircraft should enter the phugoid in the trimmed state, with the measurements of excursions of pitch, airspeed or altitude referenced to the trimmed datum.

The response for the short period of the Boeing 747-100 shown in Figure 10.17 is very different, with the pitch transients mostly decaying within 4–6 s. Approximation of the transfer function directly from data is also far from straightforward. Although the pitch rate is cancelled within 8 s, the rise time and decay of the pitch angle are asymmetrical and the nonlinearities in lift and drag close to the stall angle affect the response. The validation of the short period is mostly based on timing measurements and displacements. The tests were initiated at 3000 ft at 180 Kt, with flaps 20 and the undercarriage up.

Figure 10.17(a) shows the response to a pulse input of 10° for 2 s. From the steady-state trim value, the angle of attack reaches 10° after 2.5 s, the pitch rate increases to 6° and then decays to −2° in 2 s. The pitch attitude rises to 15° in 3 s and decays to 10° in 5 s. A similar response is observed in Figure 10.17(b) where the elevator input is a doublet for 2 s. Note the nonlinearity which is evident from the pitch rate response. Linearisation of the pitch response will omit such nonlinearities.

Validation mainly covers checks on the rise and fall times and displacements for different values of altitude, airspeed, mass and configuration. In addition to measuring the control input (elevator), the control force may also be sensed by means of a strain gauge, in order to check the consistency and repeatability of these tests. Also notice that the input in this example is a perfect pulse, whereas in practice, in an airborne test, the control inputs are applied manually. However, it is also possible to capture data from airborne tests, which are replayed through the offline simulator to ensure that tests are conducted with identical inputs and flight conditions.

10.4.2 Lateral Dynamics

Although aileron is applied to bank an aircraft and rudder is applied to yaw an aircraft, the roll and yaw motion is cross-coupled so that aileron can induce yawing motion in response to a rolling moment and rudder can induce rolling motion in response to a yawing moment. This situation is

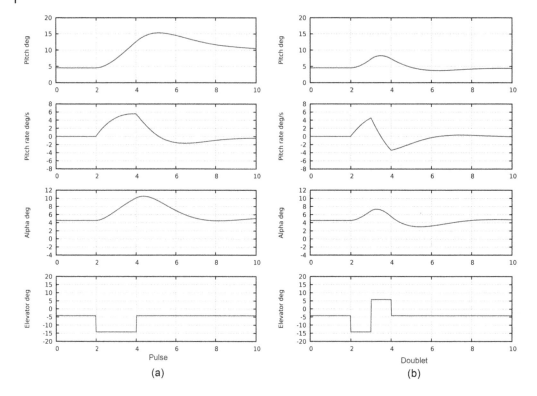

Figure 10.17 Short-period Mode.

further compounded because roll rate and yaw rate are generally feedback terms in the lateral equations of motion. In addition, asymmetric thrust from engines or propellers can also introduce rolling and yawing motions.

Generally, additional yawing motion is undesirable and both piloting techniques and flight control systems minimise these effects. However, if inputs are applied that cause an unexpected change of heading, then the pilot workload is increased and these effects can affect the aircraft handling qualities, making accurate manoeuvring difficult. In much the same way that the phugoid characterises longitudinal motion, the Dutch roll response is specific to an aircraft and depends on the aircraft aerodynamics, the mass, airspeed, altitude and configuration of the aircraft, as well as the inceptor inputs. In flight simulator validation, the Dutch roll observed in the simulator should be identical to the motion in the aircraft. Tests similar to the phugoid testing are used to confirm the relationship between the rudder and aileron inputs and the lateral motion, particularly changes in heading and bank angle, roll and yaw rates and the angle of sideslip. As with the phugoid tests, the lateral responses vary throughout the flight envelope.

Figure 10.18 shows a typical response to rudder and aileron input. Both tests were initiated at 5000 ft at 180 Kt, with flaps 20 and the undercarriage up, with an initial heading of 360°. Also, the yaw damper and turn coordinator were both disengaged for these tests. In Figure 10.18(a), the rudder input generates an initial sideslip of ±4° with a significant rate of yaw that takes over 60 s to damp out. During this time, the nose of the aircraft will oscillate by several degrees with a period of approximately 9 s. Although the aileron input in Figure 10.18(b) produces less sideslip than the rudder input, there is still a significant oscillation in the roll rate and bank angle. These

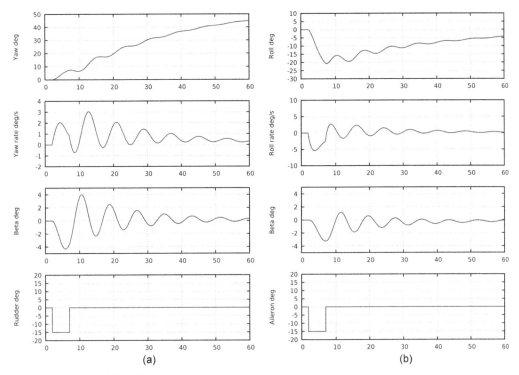

Figure 10.18 Dutch Roll Response.

effects are reduced by coordinated aileron and rudder inputs in manual flight to reduce the resultant sideslip and by the turn coordinator and yaw damper systems used in civil aircraft.

For the response shown in Figure 10.18(a), based on the sideslip plot, the following values were obtained: the first maximum occurs at (10.62, 3.989 422) and the second maximum occurs at (18.82, 2.494 583), giving $\zeta = 0.074\,519$ and $\omega_n = 0.768\,379$ rad/s. In other words, the Dutch roll mode is lightly damped with a period of 8.18 s. Similar to the phugoid mode, the lateral dynamics can be formed as a fourth-order transfer function where two of the poles define the Dutch roll response and the other two poles define the spiral mode and roll subsidence. In practice, the spiral mode and roll subsidence are more relevant to the design of lateral flight control systems and have less significance in validation, particularly for large transport aircraft, where the spiral mode is negligible.

10.4.3 Engine Failure

A particularly important test is the aircraft response to an engine failure, typically just after take-off, which is a critical failure in an aircraft. Multi-engine aircraft have sufficient power to climb with one failed engine, and the rudder and aileron have sufficient authority to cancel the initial roll and yaw, in order to maintain a safe airspeed and climb rate and to be able to control the aircraft heading. However, the correct pilot response must be applied within a second or so, to avoid the bank angle increasing to a dangerous (and potentially irrecoverable) value at low altitude. A response to an engine failure is shown in Figure 10.19.

The plots show the take-off roll followed by the initial climb, with the undercarriage raised once a positive rate of climb is established. The outer port engine of the Boeing 747-100 is failed at an altitude 200 ft above ground level, 52 s into the test, with the aircraft climbing with full

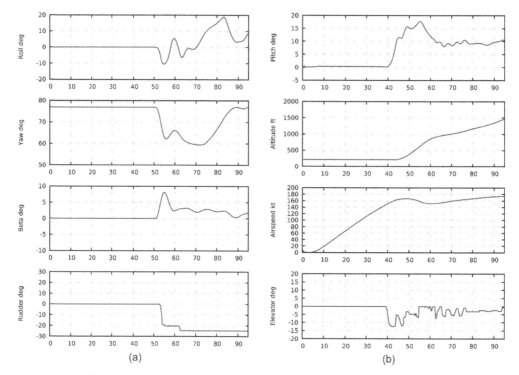

Figure 10.19 Pilot Response to an Engine Failure.

power at 156 Kt with flaps 20. In Figure 10.19(a), the pilot responds with full right rudder, to reduce the angle of sideslip and with the application of aileron to reduce the bank angle. The pilot is able to regain a heading close to 78°T. In Figure 10.19(b), the pilot maintains a positive rate of climb during the recovery and also ensures that the airspeed is safely above 150 Kt. The pilot inputs would be checked against the data for the aircraft, and the workload to execute the manoeuvre should also correspond to a similar operation in the aircraft. Even with the yaw damper and turn coordinator engaged, the effect of the Dutch roll response is evident in Figure 10.19. One very important point is that the rudder force provided by the control loading system should be very close to the aircraft values. Otherwise the pilot response will be very different from the response in the aircraft, even though the aircraft dynamics are correctly modelled.

This particular exercise is very dependent on the response of the pilot to the engine failure and requires correct input of the rudder, aileron and elevator. Clearly, there will be a wide variation of responses, which needs to be taken into account in this specific test.

10.4.4 Observations

The phugoid, short-period mode and Dutch roll response play a very important part in simulator validation. It is easy to introduce errors into a flight model or engine model, particularly incorrect sign conventions, or the mixing of units or poor estimates of aerodynamic derivatives. In such cases, the response of the simulated aircraft will be very different from the actual aircraft and this should be evident from the modes of motion. Moreover, a phugoid or short-period response can be excited by a step input of the elevator control or rudder to produce a pitch or roll response in an aircraft, in order to acquire flight data. The main consideration is to ensure that the aircraft mass, flight conditions, aircraft C/G, configuration, entry speed and altitude are carefully recorded, that the test is initiated in the trimmed state and that the inceptor forces, provided by the control loading system, match the aircraft. If these responses in the simulator match the equivalent aircraft response over the flight envelope, the handling of the simulator should correspond to the handling of the aircraft. In addition to matching the dynamics, the aircraft performance must also be captured accurately in the simulator. One other aspect of validation of aircraft dynamics is to check that the nonlinear dynamics have been modelled correctly and particularly that any linearisation has not introduced discrepancies into the modelling.

10.5   Octave and MATLAB

The importance of offline simulation was stressed in Section 4.2. The SimPlot package enables software modules to be interchanged between the offline simulation and the simulator and enables the simulation parameters to be recorded and plotted both offline and during simulation. However, there are numerous packages available for the simulation of system dynamics, including Dymola, Mathematica, Simio, MATLAB, Simulink, Scilab and Octave. These packages, and others, offer differing and wide-ranging capabilities. However, there are two classifications that are relevant to flight simulation: firstly, the availability of open-source software and, secondly, the ability to provide state-space modelling. Scilab and Octave (Eaton, 1997) provide an open-source alternative to Simulink and MATLAB, respectively. Both Simulink and Scilab enable systems to be captured in a block diagram notation and use MATLAB and Octave, respectively, as the underlying modelling packages.

MATLAB and Octave support state-space notations, enabling systems to be described using matrix and vector operations. They also support transfer function notation, algebraic formulation and provide a library of common mathematical functions, including numerical integration methods. To a considerable degree, Octave provides compatibility with MATLAB although it lacks the wide range of toolboxes that support the variety of applications covered by MATLAB (Quarteroni and Saleri, 2006). Both packages compile source code to an intermediate form that is interpreted and this is a significant limitation: their run-time performance is considerably slower than code compiled to native machine instructions. Although additional tools are available to create real-time code from MATLAB scripts, in general, MATLAB and Octave are limited to offline analysis.

The major advantage afforded by both MATLAB and Octave is the analysis tools they provide to determine the stability or response of complex systems. Methods include root locus techniques and frequency response methods. They also enable state-space descriptions to be converted to transfer function form and provide various numerical methods for time domain simulation. Control systems can be added to models, allowing the combined systems to be analysed to design a specific controller. The corollary is that, for the majority of control design applications, a model must be formulated in a state-space notation, requiring the system to be linearised about a specific operating point. These design tools are very powerful tools to design control systems but with the proviso that the controller may not necessarily function correctly at other operating points. This limitation is particularly significant in the implementation of flight models and flight control laws, owing to the wide range of speed and altitude of an aircraft flight envelope and the relatively fast time-constants in many flight control laws. Applying control laws developed for linear models to a nonlinear system is a potential recipe for disaster.

In this section, examples are limited to Octave, firstly because of its availability as an open-source package for most platforms and, secondly, because the examples use standard Octave functions and notation, which are very similar to MATLAB. In the aerodynamic modelling covered in Chapter 3, non-dimensional aerodynamic derivatives have been used exclusively. However, state-space forms of the equations of motion use dimensional stability derivatives and care is needed with the conversion from non-dimensional derivatives to dimensional derivatives. Babister (1980) refers to the two main notations as the American and British units, clarifying their usage. Dimensional derivatives can also be defined in SI units, which is the practice adopted in this book. The reader should also be aware that some aerodynamic notations are based on the wing span and others are based are the wing semi-span, introducing a potential scaling error of a factor of 2.

10.5.1 Longitudinal Model

The conversion from non-dimensional aerodynamic derivatives to the equivalent dimensional derivatives for longitudinal motion is given in Table 10.4.

One important consideration is that the non-dimensional derivatives should be applied in the trimmed state. The SimPlot package, which computes the non-dimensional aerodynamic derivatives for the trimmed state, also generates the dimensional aerodynamic derivatives. SimPlot is particularly useful as it enables the aircraft mass, C/G, configuration and flight conditions to be defined for specific values of airspeed and altitude. For example, a SimPlot script generates the following output containing both dimensional and non-dimensional aerodynamic derivatives:

```
set altitude 3000 ft      Autotrim: converged after 4656 steps
set TAS 200 kts           Thrust=215082 Drag=214860 Lift=2540835 mg=2550600
set flaps 20.0            trim: alpha= 2.60 deg, pitch= 2.60 deg, de=-3.98 deg, dp= 0.54 EPR= 1.09
set gear 0.0             TAS=103.0 m/s (200 Kt) Mach=0.31 Alt= 914 m ( 3000 ft)
set cg_position 0.2       Cxu=-0.141469 Cxw=0.715119
autotrim                  Czu=-1.735258 Czw=-6.258735
                          Cmu=0.067278 Cmw=-1.528000 CmwDot=-3.400000 Cmq=-20.280441
                          Xu=-0.016048 Xw=0.081124
                          Zu=-0.196850 Zw=-0.710001
                          Mu=0.000376 Mw=-0.008530 MwDot=-0.078968 Mq=-0.471030
                          Xde=0.000000 Mde=-0.790511 Zde=-4.719900
                          Yb=-0.110492 Lb=-2.010807 Lp=-1.314288 Lr=0.207274
                          Nb=0.501470 Np=-0.044896 Nr=-0.208585
                          Ydr=0.911268 Lda=0.653952 Ldr=0.072812 Nda=0.006997 Ndr=-0.131198
```

The aircraft is trimmed for level flight at 3000 ft, 200 Kt (TAS), with flaps 20, the undercarriage up and the C/G = 0.2, a typical configuration for a holding pattern.

Assuming an aircraft is symmetrical about the x-z plane in the body axes and all the engines produce identical thrust, then the longitudinal and lateral responses can be considered to be independent and can be solved separately. The general form of the system of equations of a dynamic system is given by

$$\dot{x} = Ax + Bu \tag{10.6}$$

where the vector x represents the state variables, u is the vector of inputs and, for aircraft dynamics, the matrices A and B contain aerodynamic derivatives. For longitudinal dynamics,

$$x = \begin{bmatrix} u \\ w \\ q \\ \theta \end{bmatrix} \quad \text{and} \quad u = \begin{bmatrix} \delta_e \\ \delta_t \end{bmatrix} \tag{10.7}$$

Table 10.4 Non-dimensional Longitudinal Derivatives.

Non-dimensional derivative	Multiplier	Dimensional derivative
C_{x_u}	$\rho VS/2m$	X_u
C_{x_w}	$\rho VS/2m$	X_w
$C_{x_{\dot{w}}}$	$\rho S\bar{c}/4m$	$X_{\dot{w}}$
C_{z_u}	$\rho VS/2m$	Z_u
C_{z_w}	$\rho VS/2m$	Z_w
$C_{z_{\dot{w}}}$	$\rho S\bar{c}/4m$	$Z_{\dot{w}}$
C_{z_q}	$\rho VS\bar{c}/4m$	Z_q
$C_{z_{\delta e}}$	$\rho V^2 S/2m$	$Z_{\delta e}$
C_{m_u}	$\rho VS\bar{c}/2I_y$	M_u
C_{m_w}	$\rho VS\bar{c}/2I_y$	M_w
$C_{m_{\dot{w}}}$	$\rho VS\bar{c}^2/4I_y$	$M_{\dot{w}}$
C_{m_q}	$\rho VS\bar{c}^2/4I_y$	M_q
$C_{m_{\delta e}}$	$\rho V^2 S\bar{c}/2I_y$	$M_{\delta e}$

where δ_e and δ_t are the elevator and throttle inputs, respectively, and the A and B matrices are defined as follows:

$$A = \begin{bmatrix} X_u & X_w & 0 & -g \\ Z_u & Z_w & V & 0 \\ M_u + M_{\dot{w}}Z_u & M_w + M_{\dot{w}}Z_w & M_q + M_{\dot{w}}V & 0 \\ 0 & 0 & 1 & 0 \end{bmatrix} \tag{10.8}$$

$$B = \begin{bmatrix} X_{\delta_e} & X_{\delta_t} \\ Z_{\delta_e} & Z_{\delta_t} \\ M_{\delta_e} + M_{\dot{w}}Z_{\delta_e} & M_{\delta_t} + M_{\dot{w}}V \\ 0 & 0 \end{bmatrix} \tag{10.9}$$

where V is the true airspeed and g is gravitational acceleration. As the aircraft is trimmed, $M_{\dot{w}} = 0$, the engine thrust can be assumed to be constant for a jet engine, and $X_{\delta_e} = 0$ as the elevator produces negligible drag in the X-axis of the body frame. The simplified equations are given by

$$\begin{bmatrix} \dot{u} \\ \dot{w} \\ \dot{q} \\ \dot{\theta} \end{bmatrix} = \begin{bmatrix} X_u & X_w & 0 & -g \\ Z_u & Z_w & V & 0 \\ M_u & M_w & M_q & 0 \\ 0 & 0 & 1 & 0 \end{bmatrix} \begin{bmatrix} u \\ w \\ q \\ \theta \end{bmatrix} + \begin{bmatrix} X_{\delta_e} & X_{\delta_t} \\ Z_{\delta_e} & Z_{\delta_t} \\ M_{\delta_e} & M_{\delta_t} \\ 0 & 0 \end{bmatrix} \begin{bmatrix} \delta_e \\ \delta_t \end{bmatrix} \tag{10.9}$$

Although the non-dimensional derivatives C_{xu}, C_{xw}, C_{zu} and C_{zw} are not usually provided in a simulator data package, the following approximations can be applied for a jet engine aircraft:

$$C_{xu} = -2C_d \tag{10.10}$$

$$C_{xw} = C_{ld} \tag{10.11}$$

$$C_{zu} = -2C_l \tag{10.12}$$

$$C_{zw} = -C_d \tag{10.13}$$

The output from SimPlot in the previous example, with the aircraft trimmed at 3000 ft, 200 Kt and flaps 20, with an elevator step input of $-1°$, produced the following values:

U_0	103.0
X_u	-0.015879
X_w	0.081072
Z_u	-0.193912
Z_w	-0.709851
M_u	0.000376
M_w	-0.008529
M_q	-0.470992
X_{de}	0.0
M_{de}	-0.790365
Z_{de}	-4.719028

The Octave program (omitting the axis labelling and legend) for a step input is as follows:

```
A = [-0.015789    0.081072      0.0       -9.81;
     -0.193912   -0.709851    103.0        0.0;
      0.000376   -0.008529     -0.470992   0.0;
      0.0         0.0           1.0        0.0];

B = [0.0; -4.719028; -0.790365; 0.0];

C = [0.0 0.0 0.0 -1.0];

D = 0;

Lambda = eig(A);

[num, den] = ss2tf(A, B, C, D);
dtf = tf(num, den);
[wn, zeta] = damp(dtf);

[y, t] = step(dtf, 100);
plot(t, y);
```

The corresponding plots generated by Octave and SimPlot are shown in Figure 10.20.

Note that the Octave plot is relative to an initial pitch attitude of $0°$ whereas the SimPlot plot is relative to the trimmed pitch attitude of $2.48°$. However, the amplitude and period are within 2%. SimPlot gives values of $\zeta = 0.022316$ and $\omega_n = 0.114853$ rad/s. In the Octave example, the

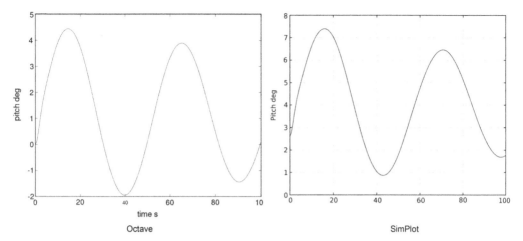

Figure 10.20 Longitudinal Response.

eigenvalues are at $-0.59473 \pm 0.03249j$ and $-0.00359 \pm 0.12406j$, giving $\zeta = 0.028955$ and $\omega_n = 0.12411$ rad/s, with the following transfer function given by the Octave function `ss2tf`:

$$\frac{\theta(s)}{\delta_e(s)} = \frac{0.790365s^2 + 0.533272s + 0.020792}{s^4 + 1.196632s^3 + 1.247186s^2 + 0.027102s + 0.018843} \tag{10.14}$$

If the contributions of w and q are omitted, equation 10.9 reduces to

$$\begin{bmatrix} \dot{u} \\ \dot{\theta} \end{bmatrix} = \begin{bmatrix} X_u & -g \\ \dfrac{-Z_u}{V} & 0 \end{bmatrix} \begin{bmatrix} u \\ \theta \end{bmatrix} \tag{10.15}$$

where the eigenvalues of the determinant of the matrix A are given by

$$\lambda^2 + X_u\lambda - \frac{Z_u}{g} = 0 \tag{10.16}$$

The natural frequency and damping ratio are then given by

$$\omega_n \approx \sqrt{2}\frac{g}{V} \tag{10.17}$$

$$\zeta \approx \frac{1}{\sqrt{2\dfrac{L}{D}}} \tag{10.18}$$

In other words, the natural frequency depends on the entry speed to the phugoid and the damping ratio depends on the lift–drag relationship, which for civil aircraft is often chosen to maximise range. A similar simplification can be used for the short-period mode, where the terms for u and θ are omitted, giving

$$\begin{bmatrix} \dot{w} \\ \dot{q} \end{bmatrix} = \begin{bmatrix} Z_w & V \\ M_w + M_{\dot{w}}Z_w & M_q + M_{\dot{w}}V \end{bmatrix} \begin{bmatrix} w \\ q \end{bmatrix} \tag{10.19}$$

Substituting

$$Z_\alpha = VZ_w \tag{10.20}$$

$$M_{\dot\alpha} = VM_{\dot w} \tag{10.21}$$

$$\begin{bmatrix} \dot\alpha \\ \dot q \end{bmatrix} = \begin{bmatrix} \dfrac{Z_w}{V} & V \\ M_\alpha + M_{\dot\alpha}\dfrac{Z_\alpha}{V} & M_q + M_{\dot\alpha} \end{bmatrix} \begin{bmatrix} \alpha \\ q \end{bmatrix} \tag{10.22}$$

Solving for the eigenvalues

$$\lambda^2 - \left(M_q + M_{\dot\alpha} + \frac{Z_\alpha}{V}\right) + M_q\frac{Z_\alpha}{V} - M_\alpha = 0 \tag{10.23}$$

The natural frequency and damping of the short-period mode can be approximated by

$$\omega_n \approx \sqrt{M_q\frac{Z_\alpha}{V} - M_\alpha} \tag{10.24}$$

$$\zeta \approx \frac{-\left(M_q + M_{\dot\alpha} + \dfrac{Z_\alpha}{V}\right)}{2\omega_n} \tag{10.25}$$

The main usefulness of these approximations is that they provide a first-order approximation for the phugoid and short-period mode in the early stages of development of a flight model. Examples of simulating and analysing aircraft modes of motion in MATLAB and Simulink are covered in Klee (2007).

10.5.2 Lateral Model

A similar approach can be used to analyse the lateral dynamics. Unlike the longitudinal dynamics, exciting the lateral dynamics can result in a change of airspeed and altitude, inducing a change in the longitudinal dynamics, and care is needed in lateral tests to minimise such effects. SimPlot also generates the dimensional and non-dimensional aerodynamic derivatives needed for a linearised state-space model of the lateral dynamics in Octave, which provides a comparison with the nonlinear modelling and insight into the damping ratios and natural frequencies of the lateral dynamics, particularly the Dutch roll response. The conversion from non-dimensional aerodynamic derivatives to the equivalent dimensional derivatives for lateral motion is given in Table 10.5.

For the state-space model, the longitudinal motion is omitted. Although this simplification reduces the order of the model, care is needed as it is likely that the aircraft pitch, airspeed and altitude will diverge from the trimmed state in turning flight. The state-space form of the lateral equations of motion is given by

$$\begin{bmatrix} \dot\beta \\ \dot p \\ \dot r \\ \dot\phi \end{bmatrix} = \begin{bmatrix} Y_\beta & 0 & -1 & \dfrac{g}{V} \\ L_\beta & L_p & L_r & 0 \\ N_\beta & N_p & N_r & 0 \\ 0 & 1 & 0 & 0 \end{bmatrix} \begin{bmatrix} \beta \\ p \\ r \\ \phi \end{bmatrix} + \begin{bmatrix} 0 & Y_{\delta_r} \\ L_{\delta_a} & L_{\delta_r} \\ N_{\delta_a} & N_{\delta_r} \\ 0 & 0 \end{bmatrix} \begin{bmatrix} \delta_a \\ \delta_r \end{bmatrix} \tag{10.26}$$

Table 10.5 Non-dimensional Lateral Derivatives.

Non-dimensional derivative	Multiplier	Dimensional derivative
C_{y_β}	$\rho V^2 S / 2m$	Y_β
C_{L_β}	$\rho V^2 S / 2I_x$	L_β
C_{L_p}	$\rho V S b^2 / 4I_x$	L_p
C_{L_r}	$\rho V S b^2 / 4I_x$	L_r
C_{N_β}	$\rho V^2 S b / 2I_z$	N_β
C_{N_p}	$\rho V S b^2 / 4I_z$	N_p
C_{N_r}	$\rho V S b^2 / 4I_z$	N_r
$C_{y_{\delta r}}$	$\rho V^2 S / 2m$	$Y_{\delta r}$
$C_{L_{\delta a}}$	$\rho V^2 S b / 2I_x$	$L_{\delta a}$
$C_{L_{\delta r}}$	$\rho V^2 S b / 2I_x$	$L_{\delta r}$
$C_{N_{\delta a}}$	$\rho V^2 S b / 2I_z$	$N_{\delta a}$
$C_{N_{\delta r}}$	$\rho V^2 S b / 2I_z$	$N_{\delta r}$

where δ_a and δ_r are the aileron and rudder inputs, respectively, g is gravitational acceleration and V is the true airspeed. The characteristic equation is a fourth-order polynomial of the form

$$\left(s + \frac{1}{T_s}\right)\left(s + \frac{1}{T_r}\right)(s^2 + 2\zeta\omega_n + \omega_n^2) \tag{10.27}$$

where T_s is the time constant of the spiral mode, T_r is the time constant of the roll subsidence mode and ζ and ω_n are the damping ratio and natural frequency of the Dutch roll mode, respectively. Note that the roots at $1/T_s$ and $1/T_r$ are real and, therefore, there is no oscillatory motion associated with these terms. For some aircraft, the spiral mode damping term can be positive, in which case the spiral mode diverges, causing an aircraft to enter an increasingly steep (and possibly insidious) turn if not noticed and corrected by the pilot.

In the following SimPlot script, the aircraft was trimmed at 5000 ft, at 180 Kt (TAS) with flaps 20 and gear down, and a rudder step input of 15° was applied:

```
set altitude 5000 ft
set TAS 180 kts
set flaps 20
plot yaw degs 0 50
plot yaw_rate deg/s -2 4
plot beta degs -5 5
plot rudder degs -20 20
time 60 secs
input rudder step 0 5 -15
autotrim
```

```
Autotrim: converged after 4544 steps
Thrust= 212345 Drag= 211548 Lift=2532212 mg=2550600
trim: alpha= 4.97 deg, pitch= 4.97 deg, de=-4.55 deg, dp= 0.51 EPR= 1.09
TAS= 92.9 m/s (181 Kt) Mach=0.28 Alt= 1524 m (5000 ft)
Cxu=-0.181628 Cxw=0.722880
Czu=-2.246196 Czw=-6.278814
Cmu=0.061146 Cmw=-1.528000 CmwDot=-3.400000 Cmq=-20.605795
Xu=-0.017509 Xw=0.069685
Zu=-0.216533 Zw=-0.605276
Mu=0.000290 Mw=-0.007249 MwDot=-0.067105 Mq=-0.406691
Xde=0.000000 Mde=-0.606234 Zde=-3.619638
Yv=-0.093893 Lb=-1.643476 Lp=-1.116848 Lr=0.231940
Nb=0.384571 Np=-0.061943 Nr=-0.177250
Ydr=0.698841 Lda=0.501508 Ldr=0.041689 Nda=0.005366 Ndr=-0.100615
```

The following values were used for the lateral non-dimensional derivatives:

V	92.9
Y_v	−0.093 893
L_β	−1.643 476
L_p	−1.116 848
L_r	0.231 940
N_β	0.384 571
N_p	−0.061 943
N_r	−0.177 250
$Y_{\delta r}$	0.698 841
$L_{\delta a}$	0.501 508
$L_{\delta r}$	0.041 689
$N_{\delta a}$	0.005 366
$N_{\delta r}$	−0.100 615

The Octave program (omitting the axes labelling and legend) for a step rudder input is given as follows:

```
A = [-0.093893    0.0        -1.0       9.81/92.9;
     -1.643476   -1.116848    0.231940  0.0;
      0.384571   -0.061943   -0.177250  0.0;
      0.0         1.0         0.0       0.0];

B = [ 0.698841;
      0.041689;
     -0.100615;
      0.0];

C = [1.0   0.0   0.0   0.0 ];

D = 0;

Lambda = eig(A);

[num, den] = ss2tf(A, B, C, D);

dtf = tf(num, den);
[wn,zeta] = damp(dtf);

[y, t] = step(dtf, 60);
plot(t, y);
```

The corresponding plots generated by Octave and SimPlot are shown in Figure 10.21.

From the second and third peaks of the SimPlot plot, $\zeta = 0.079\,912$ and $\omega_n = 0.764\,969$ rad/s. From the Octave example, the eigenvalues of the Dutch roll terms are at $-0.047\,36 \pm 0.745\,24j$, giving $\zeta = 0.063\,424$ and $\omega_n = 0.746\,747$ rad/s, with the following transfer function given by the Octave function **ss2tf**:

$$\frac{\beta(s)}{\delta_r(s)} = \frac{0.6988410s^3 + 1.0049837s^2 + 0.2677400s - 0.001840}{s^4 + 1.387991s^3 + 0.718406s^2 + 0.724792s + 0.021342} \qquad (10.28)$$

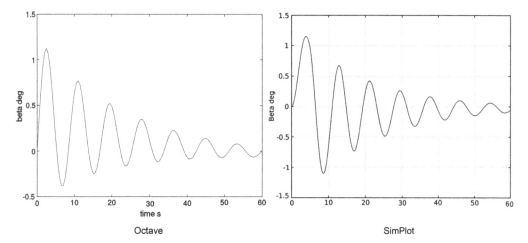

Figure 10.21 Lateral Response.

Note that the contribution of the aileron is omitted from the Octave simulation. The period of the spiral mode, which is positive and therefore not divergent, is 207.3 s. For the SimPlot simulation, the yaw damper and turn coordinator were disengaged. For the two simulations, the damping values are within 20% and the natural frequencies are within 3%. The variation in damping is possibly caused for two reasons: firstly, SimPlot will include the variation in airspeed during the 60 s manoeuvre, whereas the Octave equations omit the effect of airspeed in the lateral dynamics; and secondly, the Octave example is not initially in the trimmed state.

10.6 Simulator Qualification

In the preceding sections, emphasis has been given to measuring the response of the simulator dynamics and comparison of simulator data with data obtained from aircraft flight tests. During the manufacture of a flight simulator, the simulator company will have undertaken extensive tests prior to delivery of the simulator to an airline. After the initial installation, both the airline and the regulator need to assess the accuracy and fidelity of the simulator. For the airline, their flight crews will be operating aircraft based on training and skills learnt in the simulator and need to be confident there are minimal differences between the handling of the aircraft and the simulator. The regulator has a responsibility to the public to ensure that training is conducted to the highest possible standards, which includes the fidelity of the flight simulator.

Consequently, both the airline and the regulator will use pilot assessments to ensure that the simulator fully meets its requirements. They will have access to data produced during development of the simulator and also to the data packs provided by the airframe, engine and avionics manufacturers. The main problem they face is that the piloted validation should be as analytical and objective as possible but the very nature of manual testing is subjective. One concern is that one pilot might feel that the roll rate during an approach is too slow which may contradict data provided by the simulator manufacturer. In such cases, the airline may be tempted to request the manufacturer to increase the simulator roll rate. To do this, the roll effectiveness could be increased or the roll damping could be reduced. In fact, neither action may be correct and, worse still, in subsequent flight tests, other pilots may advise that the roll rate is too high. The question posed by the simulation industry was whether it is possible to undertake piloted assessments using methods that are as objective as reasonably possible, while reducing the variation in errors introduced by this process.

As recently as the 1980s, having procured a new flight simulator, an airline faced the problem of having the simulator training recognised by the multiple regulatory authorities of the countries where the aircraft and flight crews would be operating. From 1989 to 1992, the Royal Aeronautical Society (RAeS) convened a series of meetings with working groups formed to try to establish international regulations for the operation of flight simulators. This work culminated in the production of a handbook entitled *International Standards for the Qualification of Airplane Flight Simulators* in 1992. These standards were adopted by ICAO in the first edition of the *ICAO Manual of Criteria for the Qualification of Flight Simulation Training Devices*. A subsequent companion document was then produced under the auspices of the RAeS, known as *The Handbook*. The ICAO document and *The Handbook* were intended for use on Level-C and Level-D simulators and were restricted to commercial fixed wing aircraft, omitting general aviation aircraft, military aircraft and rotary wing aircraft. It is important to stress that these documents were intended to provide guidance to developers and assessors of these standards.

In 2006, the FAA asked the Flight Simulation Group of the RAeS to establish an International Working Group (IWG) to review and expand the technical criteria. This document became the third edition of the *ICAO Manual of Criteria for the Qualification of Flight Simulation Training Devices, Volume 1 – Aeroplane* and was published in 2009. The fourth edition was published in 2015 (ICAO, 2015). It is not possible to cover these documents in depth and readers needing to understand the practices in simulator qualification are strongly advised to review the methodologies described in these manuals.

It is important to appreciate the context of validation of airline simulators. An airline will have invested in excess of $10 million in procuring a simulator and expect to operate the simulator 23 hours/day for 365 days/year for recurrent checking and type conversions, possibly including zero flight time (ZFT) training. The procurement includes data packs for the airframe, engines and systems, containing very detailed data for every possible event throughout the flight envelope. The checking of simulator data against the aircraft data must be undertaken in a structured and methodological manner and, furthermore, must be repeatable. For example, if there is a software update to the engine FADEC system or the control loading system is replaced, it is essential that these changes do not introduce variations from the aircraft data or previously approved simulator baseline data. Moreover, for commercial aeroplane simulators, such modification will also need to be approved in advance by the regulatory authority.

The management of validation procedures, compliance with published standards and the rigorous maintenance of detailed documentation underpin validation. With the publication and acceptance of the ICAO manual, it was appreciated that guidelines were needed to implement the evaluation and qualification of airline simulators, in order to interpret the intentions of the ICAO document and assist in its implementation. Although the ICAO manual sets out the validation tests that should be conducted, it was felt that a handbook explaining how these tests should be arranged and conducted would provide a useful companion to the ICAO manual. For an airline, using flight crews (with experience of the specific aircraft type) to undertake the evaluation or qualification, the handbook does not need to explain the mathematics or software of the simulator but provides procedures and protocols to ensure that the simulator data matches the aircraft data and to resolve situations where there is a variation in the data.

The task of production of a handbook was taken on by the RAeS, which formed a working group to review the methodology of the process of simulator qualification. This group produced the *Aeroplane Flight Simulator Evaluation Handbook*, which was published in two volumes by the RAeS. Volume I advises on the use of 'engineering judgement' in the use of data, providing guidance for constructing a Qualification Test Guide (QTG) and conducting simulator evaluation tests.

Volume II provides guidance for subjective testing, where simulators are assessed by experienced line pilots and regulators during the acceptance and qualification tests. One concern addressed by Volume II was the reliance of some organisations on subjective evaluation, based on the skills and expertise of specific pilots participating in an acceptance programme, in particular the 'tweaking' of the software to match the opinions of individual pilots. This volume outlines a structured method using a standardised approach, albeit making use of the skills and experience of pilots in an acceptance team. In broad terms, Volume I covers objective testing and Volume II covers subjective testing.

10.6.1   Aeroplane Flight Simulator Evaluation Handbook – Volume I

Volume I of the handbook currently relates to the tests defined in Appendix B of the ICAO manual, second edition. Each test defines the variables to be measured and the required tolerance. The handbook makes an important clarification: 'it provides guidance for the application of judgement in the evaluation as well as guidance in the conduct of the evaluation'. The handbook introduces the concept of the baseline simulator performance, which is used in the initial and recurrent evaluations; the baseline should match the aircraft data and provide a reference for subsequent testing. The handbook also discusses the requirement for a QTG which is a reference document containing all the tests to be undertaken, how they are to be undertaken and the expected outcomes. The evaluation is driven by the QTG; any apparent or perceived inconsistencies or deviations are referred back to the QTG. The parameters recorded must include data with defined tolerances, initial conditions, flight conditions, inputs and other parameters which are required for a test.

The handbook identifies three forms of test:

1) Automatic tests – mostly performance and handling qualities testing, but can also include systems, for example, motion and sound.
2) Manual tests – these may be performance or handling qualities tests, based or relying on correct system functionality, for example, the time to climb in an autopilot mode.
3) Subjective tests – performed by a qualified pilot where exact data may not be available, or where expert knowledge of the aeroplane or geographic location is needed.

The intent is to test the simulator as a complete device rather than individual systems and all three types of testing should be performed in an integrated manner, as far as reasonably practical.

In automatic testing, specific inputs may be back-driven through the simulator to check the performance or response against the baseline (for the same inputs). Normally, these tests are to demonstrate tolerance matching. There is scope for the evaluator to fine-tune a model if, in consultation with the manufacturer or regulator, it is felt that the change, which should be limited to relatively short durations in the flight data, is justified. The advantage of automatic tests is that a large number of tests can be undertaken in a short time and the tests are repeatable.

Manual testing is performed by a pilot as an 'end-to-end' test without any automatic testing. Ideally, the tests should match the automatic testing but in practice there will be a variation of pilot input and the handbook provides guidelines to account for variations in the data which conform to the baseline test. Tests can be rerun to provide a closer match to the data. Given typical time constraints during evaluations, far fewer tests are possible than with automatic testing and there is some scope for interpretation of the results.

Integrated testing implies that all the simulator systems are functioning during the test, including flight control systems, automatic modes and avionics. While sub-systems can be tested separately, it is necessary to include these systems in integrated testing.

Although automatic testing is widely used, manual testing is still recognised as an important aspect of validation and is also covered in the handbook. It is understood that it is not possible to replicate every test and some leeway (or interpretation) is given to data tolerances, particularly where it may be difficult for a pilot to replicate exact manoeuvres in terms of inputs and timing. Tests should focus on one single parameter and allowance can be made to replicate as closely as possible the flight conditions to establish the initial conditions needed for the test.

The handbook also suggests levels of resolution for recorded data, noting that data from a simulator is likely to have much more resolution than data captured from sensors in airborne flight tests. Advice is also given for the presentation of data, particularly in plots, to avoid any ambiguity in data matching.

The *Evaluation Handbook* contains over 200 pages of validation tests, which are set out in a corresponding manner to the *ICAO Manual*. The handbook describes the test procedure, the flight conditions, the recorded parameters, the evaluation notes, the tolerances, manual testing methods, the flight and manoeuvre envelope protection functions and example plots for each test. For example, to assess the Dutch roll characteristics, a rudder doublet is applied, to produce at least six cycles of lateral oscillation, for two flight conditions. A list of recorded variables is given and the tolerance of the period and damping ratio of the response are ±0.5 s and ±0.02, respectively. In this case, the manual testing does not require matching of the rudder input; although the period, damping ratio and relationship between the bank angle and sideslip are required to meet the tolerances, this requirement does not apply to the roll rate and bank angle.

10.6.2 Aeroplane Flight Simulator Evaluation Handbook – Volume II

The intention of the *Evaluation Handbook*, with regard to subjective testing, is to provide a framework for standardised assessment in order to ensure consistency. Although the evaluators will be experienced line pilots, it cannot be assumed that they have an engineering background or are qualified test pilots. The handbook therefore provides schedules and processes to be followed and gives insight into the data capture methods needed for an evaluation and, to some extent, the potential pitfalls. With objective testing, it is possible to base the evaluation on physical measurements and tolerances, whereas with subjective testing, care is needed to minimise any biases in the assessment, which arise from recommendations based on perception.

The *ICAO Manual* refers to function tests and subjective tests as follows: 'Function tests are objective tests of systems using airplane documentation. Subjective tests provide a basis for evaluating FSTD capability to perform over a typical training period and to verify correct operation and handling characteristic of the FSTD.' An aim of the subjective testing is that a pilot, undertaking a future evaluation, should be able to repeat the same actions and acquire the same parameters to confirm that the system is compliant. The evaluation is essentially to test that all the systems function correctly in the training environment where they will be used.

For functions and subjective testing, each evaluation focuses on a specific training task and therefore reflects the segments of an end-to-end flight plan. One very important point is for the evaluator to avoid compensating for deficiencies in the simulator. The role of the evaluator is to detect such deficiencies. Bearing in mind that line pilots are used to flying safely, complying with performance figures for the aircraft, in an evaluation they are being asked to assess manoeuvres that rarely occur in normal operations.

In situations where a problem is detected, the evaluator needs to provide the essential information to the engineer charged with solving the problem, which includes the initial conditions, a description of the inputs, the expected behaviour and the actual behaviour. The information provided must be sufficient to enable an engineer to identify the cause of the problem.

The main distinction between objective testing and subjective testing is that objective testing is used to demonstrate that the aircraft performance and dynamics have been captured correctly, whereas subjective testing is undertaken to confirm that the simulator can be used for the training tasks required for pilot training. Inevitably, there is some overlap between these requirements, but for functions and subjective testing, the *Evaluation Handbook* provides guidelines to devise the test plan and test profiles, firstly, to enable an evaluator to devise a test programme that covers every aspect of simulator operations, secondly, to provide guidance in the method to run the test and, thirdly, to advise evaluators of the relevant issues of each test.

Volumes I and II of the *Evaluation Handbook* are widely used by manufacturers, operators and regulators in all aspects of flight simulator qualification. Volume II is a very detailed guide which is provided to enable an airline to conduct an evaluation by line pilots. It is not possible to cover this level-of-detail in a few pages and readers who are likely to undertake a simulator evaluation would be well advised to review these documents carefully.

References

Anon. (1976), *U.S. Standard Atmosphere*, NASA-TM-X-74335.

Babister, A. W. (1980), *Aircraft Dynamic Stability and Response*, Pergamon Press.

Backhouse R. C. (1986), *Program Construction and Verification*, Prentice-Hall.

Boehm B. W. (1979), Software Engineering: R&D Trends and Defense Needs, in *Research Directions in Software Technology* (Wegner P., ed.), MIT Press.

Eaton J. W. (1997), *GNU Octave Manual*, Network Theory Ltd.

ICAO. (2015), *Manual of Criteria for the Qualification of Flight Simulation Training Devices*, Doc. 9625, Vol. 1, Aeroplanes, ICAO.

Klee H. (2007), *Simulation of Dynamic Systems with MATLAB and Simulink*, CRC Press.

McCormick B. W. (1979), *Aerodynamics, Astronautics and Flight Mechanics*, John Wiley.

Pomberger G. (1984), *Software Engineering and Modula-2*, Prentice-Hall.

Quarteroni A. and Saleri F. (2006), *Scientific Computing with MATLAB and Octave*, Springer-Verlag.

RAeS (2005), *Flight Simulator Evaluation Handbook*, Vol. 1, third edition, Royal Aeronautical Society, London.

Sommerville I. (1995), *Software Engineering*, fifth edition, Addison Wesley.

Appendix

The following software is available from http://www.wiley.com/go/flightsimulationsoftware for free download.

A1   System-wide #include files

`c:/msys64/mingw64/include/SIM`

aerodefn.h	*Aerodynamic model*
clocks.h	*Clock management*
dted.h	*DTED functions*
engdefn.h	*Engine model*
fileio.h	*File I/O functions*
glib.h	*Graphics*
igdefn.h	*IG functions*
inslib.h	*INS functions*
iodefn.h	*I/O functions*
iosdefn.h	*Instructor operating station*
maths.h	*Maths functions*
navdefn.h	*Navigation functions*
navlib.h	*Navigation database functions*
pnglib.h	*.png file generation*
protodefn.h	*MATLAB/Octave interface*
shader.h	*GPU shader loader*
soundlib.h	*OpenAL sound functions*
splinelib.h	*Spline fitting*
target.h	*Target management*
udplib.h	*UDP transfers*
weather.h	*Weather functions*

Flight Simulation Software: Design, Development and Testing, First Edition. David Allerton.
© 2023 John Wiley & Sons Ltd. Published 2023 by John Wiley & Sons Ltd.
Companion Website: www.wiley.com/go/flightsimulationsoftware

A2   Libraries

`/~/sim/libs/`

Makefile	*Build file*
clocks.c	*Clock management*
dted.c	*DTED functions*
fileio.c	*File I/O functions*
glib.c	*Graphics library*
inslib.c	*INS functions*
maths.c	*Maths functions*
navlib.c	*Navigation database functions*
pnglib.c	*.png file generation*
shader.c	*GPU shader loader*
soundlib.c	*OpenAL sound functions*
splinelib.c	*Spline functions*
target.c	*Target management*
udplib.c	*UDP transfers*
weather.c	*Weather functions*

A3   Boeing 747-100

`/~/sim/pfd/b747`

Makefile		*Build file*
aero.c	aero.h	*Aerodynamic model*
aerolink.c	aerolink.h	*Network interface*
ai.c	ai.h	*Attitude indicator*
alt.c	alt.h	*Altimeter*
asi.c	asi.h	*Airspeed indicator*
b747.c		*Boeing 747-100 module*
compass.c	compass.h	*Compass card*
demo.c		*Display test*
diagnostics.c	diagnostics.h	*Performance monitor*
eicas.c	eicas.h	*Engine displays*
fcs.c	fcs.h	*Flight control system*
fma.c	fma.h	*Flight Mode Annunciator*
gear.c	gear.h	*Undercarriage model*
iolib.c	iolib.h	*Analogue and digital input*
model.c	model.h	*Equations of motion*
pfd.c	pfd.h	*Primary flight display management*
systems.c	systems.h	*System functions (flaps, undercarriage)*
vsi.c	vsi.h	*Vertical speed indicator*

/~/sim/eng/b747

Makefile		*Build file*
b747.c		*Boeing 747-100 module*
engines.c	engines.h	*Engine model*
englink.c	englink.h	*Network interface*
sounds.c	sounds.h	*Sound generation*
systems.c	systems.h	*System functions*

/~/sim/nfd/b747

Makefile		*Build file*
b747.c		*Boeing 747-100 module*
demo.c		*Display test*
diagnostics.c	diagnostics.h	*Performance monitor*
fcu.c	fcu.h	*Flight control unit interface*
iolib.c	iolib.h	*I/O functions*
nav.c	nav.h	*Navigation functions*
navinfo.c	navinfo.h	*Digital data display*
navlink.c	navlink.h	*Network interface*
nfd-compass.c	nfd-compass.h	*Compass*
nfd.c	nfd.h	*Navigation flight display management*
panel.c	panel.h	*Soft panel management*
panellib.c	panellib.h	*Soft panel functions*
radio.c	radio.h	*Radio functions*
systems.c	systems.h	*System functions*

/~/sim/ios/ios

Makefile		*Build file*
approach.c	approach.h	*Approach display*
dataview.c	dataview.h	*Data recording display*
gui.c	gui.h	*Graphical user interface*
iolib.c	iolib.h	*I/O functions*
ios.c		*Instructor operating station module*
ioslink.c	ioslink.h	*Network interface*
map.c	map.h	*Map displays*
plot.c	plot.h	*Plotting functions*
scan.c	scan.h	*Folder access*
script.c	script.h	*Scripting functions*

A4   Cessna-172

/~/sim/pfd/c172

Makefile		*Build file*
aero.c	aero.h	*Aerodynamic model*
aerolink.c	aerolink.h	*Network interface*
ai.c	ai.h	*Attitude indicator*
asi.c	asi.h	*Airspeed indicator*
c172.c		*Cessna-172 module*
blank.c	blank.h	*Blank (spare) instrument*
clock.c	clock.h	*Clock instrument*
compass.c	compass.h	*Compass*
demo.c		*Display test*
diagnostics.c	diagnostics.h	*Performance monitor*
egt.c	egt.h	*Exhaust gas temperature gauge*
eicas.c	eicas.h	*Instrument support functions*
enginegauges.c	enginegauges.h	*Engine gauges*
fcs.c	fcs.h	*Flight control system (not used)*
gear.c	gear.h	*Undercarriage model*
iolib.c	iolib.h	*Analogue and digital input*
model.c	model.h	*Equations of motion*
pfd.c	pfd.h	*Primary flight display management*
rpm.c	rpm.h	*RPM gauge*
systems.c	systems.h	*System functions (flaps, undercarriage)*
turnslip.c	turnslip.h	*Turn and slip indicator*
vsi.c	vsi.h	*Vertical speed indicator*

/~/sim/eng/c172

Makefile		*Build file*
c172.c		*Cessna-172 module*
engines.c	engines.h	*Engine model*
englink.c	englink.h	*Network interface*
sounds.c	sounds.h	*Sound generation*
systems.c	systems.h	*System functions*

`/~/sim/nfd/c172`

Makefile		*Build file*
adf.c	adf.h	*ADF instrument*
alt.c	alt.h	*Altimeter*
c172.c		*Cessna-172 module*
demo.c		*Display test*
diagnostics.c	diagnostics.h	*Performance monitor*
enginegauges.c	enginegauges.h	*Engine gauges*
fcu.c	fcu.h	*Flight Control Unit (not used)*
knobs.c	knobs.h	*Knobs (mouse tracking)*
magcompass.c	magcompass.h	*Magnetic compass*
mp.c	mp.h	*Manifold pressure gauge*
nav.c	nav.h	*Navigation functions*
navlink.c	navlink.h	*Network interface*
nfd.c	nfd.h	*Navigation flight display management*
panel.c	panel.h	*Soft panel management*
panellib.c	panellib.h	*Soft panel functions*
radio.c	radio.h	*Radio functions*
rpm.c	rpm.h	*RPM gauge*
systems.c	systems.h	*System functions*
vor.c	vor.h	*VOR instrument*
vsi.c	vsi.h	*System functions*

A5 Supporting Files

`/c/~/sim/files/`

Airports.txt[1]	list of worldwide airports
Manchester.dtd	DTED of Manchester Airport
Navaids.txt[1]	list of worldwide navigation aids
menu.dat	IOS menu definitions
models.lst	list of flight models
targets.lst	list of visual targets
visual.lst	list of visual databases

---

1 The files Airports.txt and Navaids.txt are provided by Navigraph and their generosity is acknowledged. The data must only to be used for flight simulation by a private individual for the purpose of game-based learning and are subject to Navigraph's licensing conditions (https://navigraph.com/legal/terms-of-service).

A6   SimPlot

`/~/sim/Simplot`

Makefile		*Build file*
aero.c	aero.h	*Aerodynamic model*[1]
aerolink.c	aerolink.h	*Network interface*[1]
clocks.c		*Clock management*[1]
dted.c		*DTED functions*[1]
engines.c	engines.h	*Engine model – minor difference from simulator*
englink.c	englink.h	*Network interface – emulation*
fcs.c	fcs.h	*Flight control system*[1]
gear.c	gear.h	*Undercarriage model*[1]
iolib.c	iolib.h	*Analogue and digital input - emulation*
maths.c	maths.h	*Maths functions*[1]
model.c	model.h	*Equations of motion*[1]
simplot.c		*SimPlot module*
simulate.c	simulate.h	*Simulation management functions*
splinelib.c		*Spline functions*[1]
stab.c	stab.h	*Stability derivatives functions*
systems.c	systems.h	*System functions - emulation*
weather.c		*Weather functions*[1]

A7   Raspberry Pi

`/~/sim/rpi/iosystem`

Makefile		*Build file*
	igdefn.h	*IG header file – needed for XPlane IG*
iolink.c	iolink.h	*Network interface*
iosystem.c		*IO system management*

`/~/sim/rpi/libs`

Makefile	*Build file*
iolib.c	*IO functions*
udplib.c	*UDP transfers – Linux version*

---

1 The module is identical to the simulator module.

A8   Diagnostics

`/~/sim/rpi/calibrate`

Makefile		*Build file*
iolink.c	iolink.h	*Network interface*
iotest.c		*Analogue and digital capture*

`/~/sim/calibrate`

Makefile		*Build file*
display.c	display.h	*Analogue and digital displays*
iotest.c		*IO test module*

A9   MSYS2

MSYS2 is an IDE that runs under Windows to provide a Linux-like terminal interface, and can be downloaded from https://www.msys2.org, to install a set of Linux commands to support software development. The Arch Linux software management tool *pacman* can be used to install the libraries used in this book. After downloading and installing MSYS2, enter:

```
pacman -Syu
pacman -Su
pacman -S --needed base-devel
```

The base development tools include the *gcc* compiler and software development tools. The following libraries for the *mingw-w64-x86_64* 64-bit toolchain can then be installed as follows:

```
pacman -S mingw-w64-x86_64-glfw
pacman -S mingw-w64-x86_64-glew
pacman -S mingw-w64-x86_64-freetype
pacman -S mingw-w64-x86_64-freeimage
pacman -S mingw-w64-x86_64-openal
pacman -S mingw-w64-x86_64-freealut
pacman -S mingw-w64-x86_64-gtk3
pacman -S mingw-w64-x86_64-cglm
pacman -S mingw-w64-x86_64-putty
```

`putty` provides an Ethernet SSH link to enable a user to connect a PC to a Raspberry Pi (RPi). In addition, the FTP client *WinSCP* provides direct access to the RPi (also via SSH) to access files and folders of an RPi. MSYS2 provides limited editing tools (e.g. Nano), and editors such as Programmer's Notepad (https://www.pnotepad.org) or Geany (https://www.geany.org) are recommended for software development.

A10   Miscellaneous

`/~/sim/misc`

Tustin	*conversion of transfer functions to Z-transform format*
Smetana	*translation of Smetana's FORTRAN software to C*
SVG	*generation of png files from SVG descriptions*
GTK	*version of the instructor station developed using GTK*
ThrustMaster	*test software for ThrustMaster controls with a USB interface*
Splines	*generation of spline codes from csv files produced by PlotDigitizer*
Misc	*miscellaneous software developed to support the book*

Index